General, Comparative and Clinical Endocrinology of the Adrenal Cortex

VOLUME 3

General, Comparative and Clinical Endocrinology of the Adrenal Cortex

Edited by

I. CHESTER JONES and I. W. HENDERSON

Department of Zoology
University of Sheffield
Sheffield

VOLUME 3

1980

Academic Press

London   New York   San Francisco

A Subsidiary of Harcourt Brace Jovanovich, Publishers

ACADEMIC PRESS INC. (LONDON) LTD.
24/28 Oval Road,
London NW1 7DX

United States Edition published by
ACADEMIC PRESS INC.
111 Fifth Avenue
New York, New York 10003

General comparative and clinical endocrinology of the adrenal cortex.
Vol. 3
1. Adrenal cortex
I. Chester Jones, Ian   II. Henderson, Ian William
596'.01'42   QP188.A28   75-19625

ISBN 0-12-171503-5

Text set in 11/12 pt Linotron 202 Baskerville, printed and bound
in Great Britain at The Pitman Press, Bath

Contributors

R. J. Balment, *Department of Zoology, University of Manchester, Manchester, M13 9PL, England.*

M. Beato, *Institut für Physiologische Chemie I, Institutsgruppe Lahnberge der Medizinischen Fakultät, Philipps–Universität, D–355, Marburg/Lahn, Germany.*

B. Bohus, *Rudolf Magnus Institute for Pharmacology, Medical Faculty, University of Utrecht, Vondellaan 6, Utrecht, Holland.*

I. Chester Jones, *Department of Zoology, University of Sheffield, Sheffield S10 2TN, England.*

D. R. Cullen, *The Royal Hallamshire Hospital, Sheffield S10 2JF, England.*

de Wied, D. *Rudolf Magnus Institute for Pharmacology, Medical Faculty, University of Utrecht, Vondellaan 6, Utrecht, Holland.*

D. Doenecke, *Institut für Physiologische Chemie I, Institutsgruppe Lahnberge der Medizinischen Fakultät, Philipps–Universität, D–355, Marburg/Lahn, Germany.*

H. O. Garland, *Department of Physiology, University of Manchester, Manchester M13 9PL, England.*

I.W. Henderson, *Department of Zoology, University of Sheffield, Sheffield S10 2TN England.*

R. Kilpatrick, *School of Medicine, University of Leicester, Medical Science Building, University Road, Leicester LE1 7RH, England.*

D. E. Kime, *Department of Zoology, University of Sheffield, Sheffield S10 2TN, England.*

E. H. McLaren, *Stobhill General Hospital, Glasgow, Scotland.*

P. W. Major, *Department of Pharmacology and Therapeutics, University of Sheffield, Sheffield S10 2TN, England.*

W. Mosley, *Department of Zoology, University of Sheffield, Sheffield S10 2TN, England.*

N. W. Nowell, *Department of Zoology, The University, Hull HU6 7RX, England.*

J. P. D. Reckless, *The Royal Hallamshire Hospital, Sheffield S10 2JF, England.*

P. G. Smelik, *Laboratorium voor Farmacologie, Der Vrije Universiteit, Van der Boechorststraat 7, Amsterdam, Holland.*

v

I. Vermes, *Laboratorium voor Farmacologie, Der Vrije Universiteit, Van der Boechorststraat 7, Amsterdam, Holland.*

G. P. Vinson, *Department of Biochemistry and Chemistry, The Medical College of St. Bartholomew's Hospital, Charterhouse Square, London, EC1M 6BQ, England.*

Preface

This is the third and final volume designed to bring together basic knowledge about the adrenal cortex and its homologues. Not all aspects are covered but the volumes serve as a guide to an enormous literature. The arrangement of chapters is not the one we had in mind when we asked for contributions. However it was imposed by the disparate times in receipt of manuscripts. Now that the three volumes have appeared, the reader may make his own choice of sequential examination.

We are very grateful to Mrs Nansi Chester Jones for compilation of the Subject Index and to Mrs Ruth C. Memmott, Mrs Jenny Noon and Miss Lesley Mayson for secretarial assistance. We were greatly helped by Mr Warwick Mosley in the preparation of many figures and by Mr D. Hollingworth for photography. Many scientists provided both published and new material and whilst these are acknowledged in the legends to Tables and Figures we should like to record our grateful thanks to: B. Anderson, J. N. Ball, J. M. Dodd, P. H. Greenwood, T. Gustafsson, W. Hanke, M. D. Lagios, L. O. Larsen, N. A. Locket, P. Meurling, A. M. Neville, M. Ogawa, M. Oguri, H. Onozato, M. J. Roscoe, K. Seiler, R. Seiler, G. Sterba, M. Weisbart, K. Yamamoto, J. H. Youson.

We record our thanks to the following for the use of previously published material: Pergamon Press Ltd., The New York Academy of Sciences, S. Karger AG, Swets and Zeitlinger N.V., Springer-Verlag, Harper and Row, Journal of Endocrinology, Wistar Institute of Anatomy and Biology, Philadelphia, Kimpton, London, Academic Press, Masson et Cie, J. B. Lippincott, Hungarian Academy of Sciences, Cambridge University Press, Federation of American Societies for Experimental Biology, *Acta Endocrinologica, Adriatico Editrice*, Bari, *Anatomische und entwicklungsgeschichtliche Monographien*, Leipzig, *Anatomical Record, American Journal of Anatomy, Beiträge zur pathologischen Anatomie und zur allgemeinen Pathologie, Bulletin of the Johns Hopkins Hospital, Cell Tissue Research, Endokrinologie, FEBS Letters, General and Comparative Endocrinology, Journal of Cell Biology, Journal of Experimental Molecular Pathology, Zeitschrift für Zellforschung und mikroskopische Anatomie*, The Carnegie Institute Washington Publications, Biochemical Endocrinology Series, Appleton-Century-Crofts, Plenum Publishing Corporation, New York, Academy of Sciences, Budapest, *Endocrinology, British Medical Journal, Medicine* (2nd series), *Clinics in Endocrinology and Metabolism*, W. B. Saunders and Co Ltd.,

Quarterly Journal of Medicine, Radiology, Handbuch der experimentallen Pharmakologie, Research in Reproduction, Journal of Clinical Endocrinology and Metabolism Marine Biology, Gunma Symposium on Endocrinology, *Annales de l'Institut océanographique, Comparative Biochemistry and Physiology, Journal of Steroid Biochemistry,* MTP Press, Lancaster, *Nature,* Macmillan Journals Ltd., *Bulletin of the American Museum of Natural History.*

June, 1980   *I. Chester Jones*
                                                                                 I. W. Henderson

Contents

Chapter 2. Clinical Disorders involving Adrenocortical Insufficiency and Overactivity. D. R. CULLEN, J. P. D. RECKLESS and E. H. McLAREN 57

PART 1. ADRENOCORTICAL INSUFFICIENCY

PART 2. ADRENOCORTICAL OVERACTIVITY

Chapter 5. Pituitary–Adrenal System Hormones and Adaptive Behaviour. BELA BOHUS and DAVID DE WIED 265

Chapter 6. Adrenocortical Function in Relation to Mammalian Population Densities and Hierarchies. N. W. NOWELL 349

1. The Regulation of the Pituitary–Adrenal System in Mammals

P. G. Smelik and I. Vermes

*Department of Pharmacology, Free University, Faculty of Medicine
Amsterdam, Holland*

1. Introduction

The regulation of the hypothalamus–pituitary–adrenocortical axis has been studied most extensively in mammals, particularly in the rat. Although the literature gives us no reason to believe that the control of this system in man is fundamentally different, nevertheless there are a number of endocrine mechanisms which do differ from those in the rat, and this accounts for at least quantitative and temporal differences. A few examples are: the reversed daily rhythm in the nocturnal rat; the oestrous cycle of 4 to 5 days; and the strong mineralocorticoid action of the predominant adrenal hormone in the rat, i.e. corticosterone. However, it is generally accepted that, essentially, the control of the system is the same in all eutherian mammals, and that the rat can stand as a paradigm for the whole group.

The existence of regulation of a system means that its output is controlled, implicating a central mechanism. Feedback is the simplest type of control whereby the output is fed back as the decisive input of the controller. The controller is activated when the output is too low, and shut off when too high. This thermostat principle has been applied to the pituitary–adrenal system (Sayers and Sayers, 1948), but it could not explain the experimental data. More recently, a feed-forward signal was assumed, which is supposed to be a hypothalamic neurohumour (corticotrophin releasing factor or CRF) released by certain stimuli. This notion introduced the modern view of the system, in which the controlling centre is thought to be situated in the hypothalamus. The controller should then integrate feed-forward and feedback signals and, after summation of these inputs, precisely tune the system in a quantitative manner by producing a releasing signal of determined intensity. This neurohumoral signal can be considered as a transducer, since it converts a digital (neural) signal into an analogue (humoral) one. In this view, the neural elements producing

CRF are considered as the controlling device or centre. Hence, the framework of this chapter is as follows (Fig. 1): (2) the feed-forward

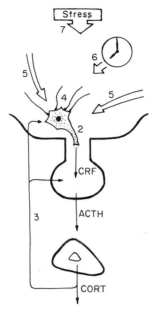

Fig. 1. Schematic representation of the regulation of the pituitary–adrenal axis. Numbers refer to respective sub-chapters (see text).

control of the system by the hypothalamic neurohumoral factor; (3) the feedback control by the output of the system, the corticosteroids; (4) the neural inputs to the hypothalamic controller; (5) the extra-hypothalamic brain structures modulating the activity of the system; (6) rhythmicity of the system and its regulation; and (7) the activation of the system as a whole by environmental stimuli ("stress").

2. Hypothalamic neurohumoral control of ACTH secretion

A. THE CRF THEORY

De Groot and Harris (1950) and, independently, Hume and Wittenstein (1950) were the first to obtain evidence for a significant role of the hypothalamus in the control of the pituitary–adrenal axis. They showed that electrolytic destruction of the hypothalamus inhibited the stress-induced lymphopenia, which was induced by electrical stimulation of the hypothalamus.

One of the classical observations arguing against such a role was the fact that pituitary stalk transection did not impair adrenal function (Uotila, 1939). However, Harris and his co-workers showed that after transection complete revascularization of the anterior lobe occurred within 6–15 days. If regeneration of the portal vessels was prevented by inserting a thin plate between pituitary gland and median eminence, adrenal atrophy developed (Harris, 1955). These findings not only were convincing evidence for the controlling function of the hypothalamus, but also provided evidence that the vascular supply of the anterior lobe conveyed the controlling signal to the hormone-producing cells.

In this way, the concept of neurohumoral control of the pituitary gland became generally accepted. The first indication of the existence of a hypothalamic substance involved in the regulation of the pituitary–adrenal axis was provided by Guillemin and Hearn (1955), and of a substance of neurohypophysial origin stimulating ACTH release by Saffran and Schally (1955). The latter workers proposed the term "corticotrophin releasing factor" (CRF). At present the identity of a hypothalamic and/or neurohypophysial CRF is still obscure. Many controversial observations have been made during the last two decades, without reaching a general consensus as to the nature of the CRF, although its existence has been generally accepted.

B.   HYPOTHALAMIC CRF

The classical approach in endocrinology in order to detect, localize and isolate a new hormone has been applied extensively to the hypothalamus as if it were an endocrine gland: demonstration of a deficiency after its removal or destruction, and of its presence in tissue extracts. Hypothalamic lesions were shown in a number of animal species to result in adrenal atrophy, reduction of compensatory adrenal hypertrophy and inhibition of the adrenal stress response. Conversely, electrical stimulation of the hypothalamus induced adrenal activation in many species. The essential role of the portal vessel system was also demonstrated in that pituitary grafts, reimplanted under the median eminence of hypophysectomized rats, restored adrenal function (Nikitovitch–Winer and Everett, 1958, 1959; Matsuda et al., 1964a; Halász et al., 1965).

Hypothalamic extracts were shown by a number of groups to possess ACTH-releasing properties, either in vivo or in vitro, on pituitary tissue. Further purification has also been attempted by many authors (Guillemin *et al.*, 1956; McCann and Haberland, 1959; Schally *et al.* 1958, 1967; Royce and Sayers, 1960; Leeman *et al.*, 1962) but the general experience appeared to be that at some step in the fractionation of the extract CRF-activity gets lost. Several groups observed that CRF-activity is ultimately found in two separate fractions, suggesting the existence of two factors or co-factors. An important criterion for the demonstration of a hypothalamic releasing hormone would be its detection in portal vessel blood. Porter, as early as 1956, gave evidence of the existence of a CRF in portal blood collected in hypophysectomized dogs (Porter and Jones, 1956; Porter and Rumsfeld, 1959). More recently, CRF activity has been found in portal vessel blood of the rat by Fink *et al.* (1971).

C. VASOPRESSIN AND CRF

As early as 1933 it was suggested by Hinsey and Markee that "pathways from the hypothalamus must activate the posterior lobe of the hypophysis, which in turn must exert an influence on the anterior lobe by humoral transmission". Much later, the observation by Nagareda and Gaunt (1951) that vasopressin induced adrenocortical activation, prompted numerous further studies, all essentially confirming this ACTH-releasing action of vasopressin. The posterior lobe hormone is capable of inducing ACTH release in rats bearing hypothalamic lesions (McCann and Brobeck, 1954), in corticosteroid-blocked animals (De Wied *et al.*, 1958), after pharmacological blockade rendering the system unresponsive to stress (Briggs and Munson, 1955; De Wied *et al.*, 1958), and from pituitaries incubated *in vitro* (Guillemin *et al.*, 1957; Saffran, 1959).

Since vasopressin is also released by stressful stimuli (Verney, 1947; Mirsky *et al.*, 1954), it was believed by a number of investigators for some time that vasopressin could represent the physiological CRF. However, this concept was not corroborated by a series of other findings. The amount of vasopressin needed to induce ACTH release was found to be very high, in the order of 100 to 1000 times higher than needed for an antidiuretic effect (Dekanski, 1952; Guillemin, 1957). Moreover, ACTH secretion could be elicited when vasopressin release was blocked and *vice versa* (Daily and Ganong, 1958; Nichols and Guillemin, 1959).

Rats with diabetes insipidus showed normal or even hypertrophied adrenals (McCann and Brobeck, 1954; Hume and Nelson, 1957; Smelik, 1960) and studies with the Brattleboro strains of rats, which have a hereditary deficiency to synthetize vasopressin, demonstrated that the pituitary–adrenal system of such animals still responds to many noxious stimuli, albeit often somewhat diminished (Arimura *et al.*, 1965; McCann *et al.*, 1966; Yates *et al.*, 1971; Wiley *et al.*, 1974).

Several attempts have been made to clarify the role of vasopressin in the pituitary-adrenal response to stress.

(a) Vasopressin may stimulate the release of CRF from its nerve endings. Hedge *et al.* (1966) found that micro-injection of 2 µg of vasopressin into the median eminence region induced ACTH release, whereas this amount injected directly into the anterior pituitary had no effect. This result was corroborated by Dhariwal *et al.* (1969). Hedge and Smelik (1969) showed that in dexamethasone-blocked rats, vasopressin was the only drug capable of releasing ACTH. Repeated administration of vasopressin had no effect, suggesting that CRF synthesis was blocked by dexamethasone and that a single dose depleted the releasable pool of CRF completely. Recently, the problem was re-studied in our laboratory using a cascade flow-system (Fig. 2)

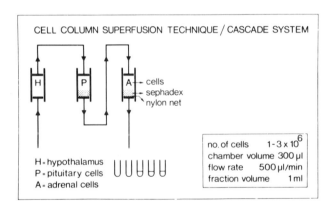

Fig. 2. The hypothalamus–pituitary–adrenal system of the rat *in vitro*. CRF, released from a hypothalamic block, stimulates ACTH release from isolated pituitary cells, which in turn provokes corticosterone production in adrenocortical cells. Corticosterone is determined in the outflow samples.

consisting of superfusion columns connected in series, the columns containing hypothalamic tissue, dispersed pituitary cells and dispersed adrenal cells respectively (Mulder *et al.*, 1976). Lysine vasopressin at a concentration of 0.27 or 2.7 mU/ml was capable of releasing CRF from the hypothalamic block, whereas a concentration of 27 mU or higher increased ACTH release from pituitary cells. It may be, therefore, that vasopressin displaces CRF from its storage granules in the nerve endings, possibly because it resembles CRF structurally (as amphetamines release noradrenaline).

(b) Vasopressin may potentiate the action of CRF on the anterior lobe of the pituitary. Yates *et al.* (1971) reported that the pituitary response to CRF was increased by subthreshold doses of vasopressin, though not by oxytocin. Lutz *et al.* (1975) confirmed this view by showing that preincubation of anterior lobes *in vitro* with vasopressin potentiated the response to CRF. Vasopressin alone stimulated ACTH release only after 2–3 h incubation, whereas CRF had an immediate effect. The authors suppose that vasopressin and CRF may have different binding sites, which may have a cooperative effect.

(c) Vasopressin may act as a CRF only in the mechanism responding to emotional stimuli. Mialhe–Voloss (1958) and Rochefort *et al.* (1959) reported evidence that the posterior lobe of the hypophysis contained ACTH, which is released after neurogenic stress (noise), whereas anterior lobe ACTH was released after systemic stress (histamine). Smelik (1960) found that in posterior lobectomized rats, the adrenal response to emotional stimuli was indeed impaired. De Wied (1961a) confirmed these results, but found that substitution with a vasopressin preparation restored this response. Subsequent work questioned the relevance of the small amount of ACTH present in the posterior lobe and suggested a permissive action of vasopressin in the pituitary–adrenal response to emotional stress (Smelik *et al.*, 1962). More recently, attention was drawn by De Wied to a possible role of vasopressin in the memory processing of emotional stimuli, suggesting that the "permissive" action may take place at a higher brain level (De Wied *et al.*, 1976).

The data reviewed strongly suggest that vasopressin is not a physiological CRF, but may participate in some still unknown way in the regulation of pituitary–adrenal function. A structural resemblance to the still unidentified CRF, or a close relationship of the vasopressin-producing neurons with the CRF neurones may perhaps in part account for this role.

D.   POSTERIOR LOBE CRF

In 1955, Saffran reported the partial purification of a factor of neurohypophysial origin, differing from vasopressin, that would stimulate ACTH release *in vitro* (corticotrophin releasing factor, CRF) (Saffran and Schally, 1955; Saffran *et al.*, 1955). Similar results were obtained by Guillemin *et al.* (1959). Schally *et al.* (1958) reported that a polypeptide from the neurohypophysis was effective in releasing ACTH at a dose of 1 μg, having a pressor activity of only 14–40 U/mg.

However, further studies questioned the existence of a posterior lobe CRF, distinct from vasopressin, and general agreement was reached within a few years that the CRF is of hypothalamic origin (Rumsfeld and Porter, 1962; Guillemin, 1964; Schally *et al.*, 1967). This would be in accordance with the reports that in posterior lobectomized rats the pituitary–adrenal response to a number of noxious stimuli is still present (Smelik, 1960; De Wied, 1961a; Itoh, 1964; Miller, 1973).

E.   EXTRA-HYPOTHALAMIC OR "TISSUE" CRF

Although it is generally assumed that the control of ACTH secretion is mediated by a hypothalamic factor, a number of reports have suggested that peripheral factors also may play a role. Fortier (1951) found that ACTH release can be induced by cold stress or bioactive substances like adrenaline and histamine from anterior lobes transplanted into the eye. He suggested that neurogenic stimuli would induce ACTH secretion via central neural pathways and that systemic ones may activate the pituitary via tissue factors, reaching the pituitary through the systemic circulation. This hypothesis has been corroborated by Brodish and Long (1956); Rochefort *et al.* (1959) and Smelik *et al.* (1962). Egdahl (1960, 1961) made the unexpected observation that in dogs with pituitary islands, having most of the brain removed, the adrenocortical response to stress still exists. He suggested that the brain may exert an inhibitory influence on the hypothalamo-pituitary system, and that tissue factors may release ACTH from the pituitary directly (Egdahl, 1962). Also, Brodish and his co-workers found, in the rat, that after complete destruction of the central hypothalamus or forebrain removal, the pituitary–adrenal axis still responds to stress, albeit in a slow and delayed fashion (Brodish, 1963, 1964). Lymangrover and Brodish (1973, 1974) found a humoral factor in the blood of such animals which induced a massive and prolonged secretion of corticosterone. They used the term "tissue CRF" for this factor.

These observations show that peripheral factors may, in certain conditions, be capable of stimulating the pituitary directly, and may perhaps arise from tissue damage, acting because of their vasoactive properties. They could for instance be kinin-like peptides, or amines like histamine and adrenaline, or perhaps prostaglandins.

F. ANATOMICAL LOCALIZATION OF CRF-PRODUCING ELEMENTS

Attempts to localize the production sites of CRF have been numerous, but these have not yet been identified. Most studies have used ablation or stimulation techniques, and the most relevant approaches will be reviewed.

1. *Lesion and stimulation experiments*

The most effective lesion in blocking pituitary–adrenal activity has to include the entire median eminence in rats (McCann, 1953; De Wied, 1961b) and especially the anterior part of the infundibular region in cats (Laqueur *et al.*, 1955) dogs (Ganong and Hume, 1954) and monkeys (Porter, 1954). Also electrical stimulation is most effective at the level of the median eminence (Porter, 1953; Anand and Dua, 1955; Endröczi *et al.*, 1956). However, it could be argued that the median eminence region is the final common pathway for all CRF-producing nerve fibres but their cell bodies may extend to much greater and rather diffuse areas. It has not been considered likely that the cell bodies of CRF-producing neurons are situated within the median eminence itself, since lesions outside have little inhibitory effect (Smelik, 1959; Mess *et al.*, 1966, 1970; Witorsch and Brodish, 1972). A rather diffuse localization of the CRF neurons has been assumed by most workers.

2. *Intra-hypothalamic pituitary grafts*

The implantation of anterior-lobe fragments into the hypothalamus delineated a "hypophysiotropic area" (Halász *et al.*, 1962), comprising a half-moon shaped part of the ventral hypothalamus between the optic chiasma, anterior commissure, mammillary bodies and pituitary stalk. Such grafts maintain the basal function of the adrenal cortex and the stress response in hypophysectomized rats (Halász *et al.*, 1965), provided the graft remains in direct contact with the median eminence (Csernus *et al.*, 1975).

3. *Corticosteroid implants*

Implants of crystalline adrenal steroids into the hypothalamus block ACTH secretion, presumably by a feedback action on the CRF-

producing elements. The most effective site is the ventral hypothalamus (Endröczi *et al.*, 1961; Smelik and Sawyer, 1962), especially the
anterior part of the median eminence and the area between the optic
chiasma and paraventricular nuclei (Smelik, 1965; Dallman and
Yates, 1968).

Although it has been argued that the steroid material can be
transported to other areas and even to the pituitary, a precise localized
effect can be obtained if sufficient precautions are taken. In a recent
more extensive study on the distribution of effective implantation
sites, we found that the area coincides with the so-called hypophysiotropic area (Fig. 3). Should the steroid effect be directly on CRF cells,
it would suggest their distribution over a considerable area.

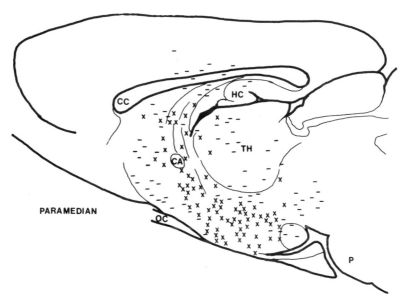

Fig. 3. Localization of implants of crystalline cortisol (20 μg). ×: sites where implants
prevented the stress-induced ACTH secretion. −: implants which did not inhibit
ACTH secretion. CC: Corpus callosum; HC: Hippocampus; TH-Thalamus;
OC: Optic chiasma; P: Pons.

4. *Histological observations*

Several groups have studied nuclear volume changes in the hypothalamic cell bodies after manipulating pituitary–adrenal function. Adrenalectomy, administration of corticoids or stress resulted in changes of
nuclear size in the ventromedial and arcuate nucleus (Halász and

Szentágothai, 1959; Palkovits and Mitro, 1968; Palkovits and Stark, 1972).

The amount of Gomori-positive granules in the external zone of the median eminence has been studied by Rinne (1960, 1972), Bock (1966) and Wittkowski and Bock (1972). These granules may represent the elementary granules of CRF-producing nerve terminals. Evidence for the presence of CRF in such granular fractions, using differential gradient centrifugation, was obtained by Ishii *et al.* (1969) in horse material, by Mulder *et al.* (1970) in rat tissue, by Fink *et al.* (1972) in bovine median eminence and by Edwardson and Bennett (1974) in sheep hypothalamus. Small neurons lying in the subependymal (internal) layer of the rat median eminence have been described and suggested to be responsible for the production of CRF (Rethelyi, 1975).

5. *Microdetermination*

The amount of CRF activity present in small areas, cut out from hypothalamic slices, has been determined by Lang *et al.* (1976). They found the highest concentration of CRF in the median eminence region, but also appreciable amounts in a large area which ranged from the posterior hypothalamus to the septal region. Similar findings were published by Krieger *et al.* (1977).

6. *Conclusion*

Only circumstantial and suggestive evidence for the localization of CRF-producing cells has been collected. The studies leave us with the impression that these cells are scattered over quite a vast area of the hypothalamus, but that in the rat they may be concentrated mainly in the anterior and ventromedial part of the hypothalamus. Direct proof must await the elucidation of the chemical identity of the CRF.

G. BIOCHEMICAL STUDIES TOWARDS THE PURIFICATION AND ISOLATION OF CRF

All studies performed so far have started from the idea that CRF should be a peptide, related in some way to the neurohypophysial hormones, both chemically and functionally.

Although CRF has been the first hypothalamic factor to be sought for, its identity appeared to be quite elusive. Early work with neurohypophysial material yielded a few promising leads. Schally *et al.* (1960), Schally and Guillemin (1963) and Guillemin (1964) claimed that two active peptides had been found, one related to MSH (called α-CRF) and one related to vasopressin (β-CRF). Definite isolation, however, met with insurmountable difficulties and this work has never been substantiated.

Work starting from hypothalamic tissue indicated that the CRF is a small, labile peptide which can be destroyed by pepsin and thioglycollate (Schally *et al.* 1962; Dhariwal *et al.* 1966; Chan *et al.* 1969). Again, two peaks of CRF activity were found by several groups after fractionation. Attempts to identify CRF starting from synthetic peptide material have also been unsuccessful (Arimura *et al.*, 1969; Saffran *et al.* 1972; Pearlmutter *et al.* 1974), although a tentative structure for CRF has been proposed by Saffran (1974), based on this type of work. Apparently, the CRF molecule is highly unstable and present in very low concentrations which renders it very difficult to collect enough material. Moreover, specific and sensitive tests for CRF activity have always been a matter of much concern (De Wied *et al.*, 1969).

H.   PHYSIOLOGICAL SIGNIFICANCE OF CRF

In spite of our ignorance about the chemical nature of CRF, the existence of such a factor is widely accepted. CRF activity can easily be demonstrated in crude acid hypothalamic extracts and a number of studies indicate that it can be produced and released by the hypothalamus. One of the most direct and earliest proofs of its role in the regulation of the pituitary–adrenal axis was presented by Porter, who could measure CRF activity in portal blood collected *in situ* from hypophysectomized dogs. It was demonstrated that surgical stress increased the CRF content of portal blood (Porter and Jones, 1956; and Porter *et al.* 1969). Fink *et al.* (1971) were able to demonstrate CRF activity in portal blood collected in the rat.

Changes in hypothalamic CRF content have been found by many workers. A circadian rhythm, preceding that of ACTH, was reported by Ungar (1964), David-Nelson and Brodish (1969), Hiroshige and Sakakura (1971) and Takabe *et al.* (1971). The Japanese workers made extensive studies of the CRF content and found that adrenalectomy or hypophysectomy did not abolish this rhythm.

Stress stimuli were reported to cause an increase in hypothalamic CRF content within 1–2 minutes (Vernikos–Danellis, 1964; 1965; Hiroshige and Sato 1971; Sato *et al.*, 1975), which could be prevented by administration of corticosteroids (Vernikos–Danellis, 1965; Sato *et al.*, 1975). Data on the effects of adrenalectomy or hypophysectomy on

CRF content are more conflicting. A permanent increase after adrenalectomy was found by Vernikos–Danellis (1965), Motta et al. (1968) and Takabe et al. (1971). No increase was found by Hiroshige and Sakakura (1971), Seiden and Brodish (1971) and Vermes et al. (1977). Since this parameter seems to depend very much on the time schedule, it is at present difficult to draw any conclusion from these studies. Moreover, the determination of the content does not give much information on synthesis and release mechanisms and therefore could be very misleading.

I. MECHANISM OF ACTION OF CRF

Earlier work indicated that activation of the pituitary–adrenal system by stress primarily resulted in an increased synthesis of ACTH. Vernikos–Danellis (1963, 1965) found that within a few minutes after surgery or ether anaesthesia pituitary ACTH content was increased, concomitant with a sharp rise in plasma ACTH levels. Both increases could be prevented by pretreatment with cortisol. Other indications for an early increase in synthetic activity came from the observation that the incorporation of ^{32}P *in vitro* was stimulated by CRF preparations (Hokin et al., 1958) and that pituitary glycogen content rapidly decreased after stress (Jacobowitz and Marks, 1964). These data supported the theory that release of ACTH is secondary to synthesis ("spill-over theory").

However, this theory is at variance with results obtained with protein synthesis blockers. Estep et al. (1967) reported that puromycin treatment did not alter stress-induced ACTH release and Arimura et al. (1969) found that actinomycin D did not affect the release of ACTH by pituitaries *in vitro*.

Our group studied the dynamics of ACTH by isolated pituitary cells in a superfusion system. It appeared that the lag time between the arrival of a CRF preparation at the cells and the rise in ACTH release was about 12 seconds.

Addition of the protein synthesis blocker cycloheximide to the superfusion medium did not prevent CRF-induced ACTH release, unless the blocker was present for more than 20 minutes prior to stimulation with CRF (Fig. 4). These data suggest again that CRF acts primarily on a releasible pool of ACTH and that the size of this pool is dependent on repletion by synthesis.

It is still unknown whether the primary effect of CRF is on the cellular membrane, although a few studies implicated cAMP as a mediator (Vernikos–Danellis and Harris, 1968; Fleischer et al. 1969;

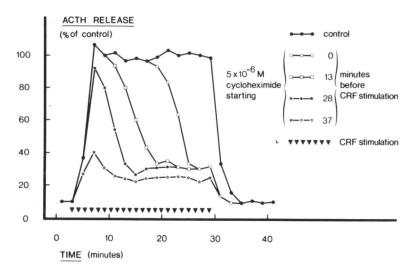

Fig. 4. Release of ACTH from isolated pituitary cells superfused continuously with a CRF preparation, during addition of the protein synthesis blocker cycloheximide, introduced at different times before CRF stimulation.

Hedge, 1971). Since it has been proposed that hormone secretion is dependent of Ca^{2+} influx (Douglas, 1968), the role of calcium has been studied by several groups. Stimulation of ACTH release *in vitro* is prevented by omission of Ca^{2+} from the medium (Kraicer *et al.*, 1969; Zimmerman and Fleischer, 1970; Portanova and Sayers, 1973). However, an increase in intracellular Ca^{2+} levels could not be demonstrated after addition of hypothalamic extract, vasopressin or cAMP (Milligan and Kraicer, 1971; Eto *et al.*, 1974). Milligan and Kraicer (1974) made the interesting observation that incubation of pituitary cells in Krebs–Ringer solution increases the intracellular Ca level three-fold. Mild washing in a Ca^{2+}-free medium abolishes ACTH release in response to elevated K concentration or by vaso-pressin, but not to hypothalamic extract or theophylline. They propose therefore that an intracellular redistribution of Ca^{2+} may be relevant to the process of physiological activation of ACTH secretion.

3.   Feedback control of the pituitary–adrenal system

It is generally accepted that adrenocortical hormones exert a negative feedback action on the secretion of ACTH from the anterior pituitary.

However, the role of such action in the regulation of the pituitary-adrenal system is not clear. At present it is difficult to conceive a model of the system.

There have been several theories concerning the control of the system which proposed a dominant role for the corticoid feedback, but it appeared that they were not entirely correct. What is then the evidence for the existence of a feedback action? A true negative feedback would exist if two conditions are fulfilled: (a) a decrease in corticosteroid blood levels which should provoke an increase in ACTH secretion; (b) an increase in corticosteroid levels which should inhibit ACTH release.

In the early days it seemed that both suppositions were valid, for it had been demonstrated both experimentally and clinically that removal of the adrenal glands resulted in an increased production of ACTH, and conversely administration of corticoid hormones blocked ACTH secretion. Consequently, a feedback theory on the control of ACTH secretion was proposed (Sayers and Sayers, 1948). It implied a balance system between the pituitary and the adrenal cortex: a stressful situation would increase the need for corticoids and due to the increased utilization of corticoids in the peripheral tissues the blood levels would fall; this would present a stimulus for the pituitary to secrete more ACTH and consequently corticoids would be produced in greater amounts and blood levels would increase, thereby gradually inhibiting ACTH release. This inverse relationship between ACTH and corticoid levels was supposed to be the only controlling system; the only decisive factor was the feedback by the peripheral corticoid levels on the ACTH-producing cells in the pituitary. This simple theory failed, however, on one incorrect assumption, i.e. that stress would cause an initial fall in corticoid levels. As soon as direct measurements of plasma corticoid titres could be performed, it appeared that such an initial fall does not take place.

During the nineteen-fifties it became increasingly clear that the central nervous system dominated the pituitary–target organ systems and that the controlling centre was to be sought in the hypothalamus. In 1961 Yates introduced a modern version of Sayers' feedback control theory (Yates et al., 1961). They argued that a hypothalamic controlling device would have a certain set-point, and that the corticoid levels in the blood would feed back information to this controller as to the degree of activity of the system. In terms of a central heating system: the variations in room temperature are recorded by a thermostat which compares the actual temperature with the desired temperature (the set-point), and which switches on and off the heater system. In such a system the heater can be activated by two means: first, a decrease in room temperature, and secondly a higher set-point of the thermostat. This theory proposed that stressful stimulation of the system would in fact cause a reset of the set-point at a higher level, thus inducing not an actual but a virtual decrease in corticoid levels.

This attractive theory was not substantiated by more recent experimental data. It implied that the system would not be activated if the corticoid levels were raised artificially to meet the new required set-point. However, we could show that an infusion of corticosterone in the rat, raising the blood levels to those reached by a stress stimulus, does not block the ACTH release due to that stimulus. Only when much higher levels were induced was a gradual inhibition of the stress response observed (Smelik, 1963a). Moreover, such an inhibition by pharmacological doses of corticoids is only established at the time when the induced blood levels have already returned to normal (Smelik, 1963b). This means that inhibition does not correlate with the blood levels, and a considerable delay occurs. This period of delay is not constant, but is dependent on unknown parameters. If one compares inhibition after subcutaneous injection with corticosterone with that after intraperitoneal injection, it appears that there is more reduction of the time lag than could be accounted for by the difference in rate of appearance in the blood. It also seems that the degree of inhibition may perhaps not be proportional to the absolute level of corticoids in the blood, but to the time integral of concentration.

Apart from a level-sensitive feedback, a rate sensitivity may also be present. As for the level-sensitive feedback, one may conclude from the earlier studies that this feedback has a rather slow action and low capacity. It seems likely that, in resting conditions, the adjustment to a certain basal level of activity depends on the blood levels of corticoids. There are indications that the system in steady state is more sensitive for feedback control. For instance, the circadian rhythm is easily suppressed by small amounts of corticosteroids. However, in an activated state, the produced increase in corticoid levels is insufficient to suppress or terminate the sytem's activity. One might say that a stressful stimulus activates the controlling centre irrespective of the existing corticoid levels and overrides the stabilizing influence of corticoid feedback.

This would imply that feedback inhibition can be achieved by extremely high corticoid levels during a short period, and also by moderate levels over a long period. This might explain why the circadian peak in resting levels is readily suppressed during treatment with relatively small amounts of corticosteroids.

Thus we may conclude that variations in corticoid blood levels influence to a degree the activity of the system but that there is a considerable delay in the stabilizing feedback action, with limited capacity. Only extreme changes in corticoid levels (e.g. by adminis-tration of pharmacological amounts or by adrenalectomy) have a clear-cut effect on the system's responsiveness. The time lag, which appears to be at least 30 min even after intravenous administration of corticoids, is difficult to explain. It may indicate that some cellular events are intercalated between the arrival of the steroid and its suppressive action, that conversions have to take place, or that the feedback action is on synthesis rather than on release of CRF or/and ACTH.

We have observed that it takes about 20 to 30 minutes before the response to CRF is inhibited after the addition of corticosterone to a medium containing isolated pituitary cells (Mulder and Smelik, 1977). In this study, the cellular content of ACTH increased consider-ably when corticosterone and CRF were added together, suggesting that at the pituitary level not synthesis but release of ACTH is impaired by corticoids.

A.   LEVEL-SENSITIVITY AND RATE-SENSITIVITY

The idea has been brought forward that a rapid feedback inhibition may be induced by the rate of increase in corticoid levels, and these are well within the physiological range (Dallman and Yates 1969). Jones et al. (1972) have shown that during the rise in corticoid levels, after administration of corticosterone, the system is unresponsive to stress. Moreover, the short peak in plasma ACTH levels after laparotomy is very much prolonged after bilateral adrenalectomy, indicating that ACTH secretion is shut off rapidly during the rise of endogenous corticoid levels. Administration of physiological amounts of corticosterone at the time of adrenalectomy prevents the persistence of elevated plasma ACTH levels (Dallman et al., 1972). Rate-sensitive feedback seems to act primarily on the release of CRF (Sato et al., 1975; Sakakura et al., 1976). This feedback mechanism may serve the turning-off of the reflex ACTH secretion during stress.

Taken together, feedback inhibition may occur during two periods: fast rate-sensitive inhibition during the rise of corticoid levels, and slow level-sensitive inhibition when there is a persistent high level of plasma corticoids. These two periods are separated by a "silent period" (Yates and Maran, 1975). It has been assumed that in the acute adrenalectomy experiments described above, the fast rise in ACTH release was caused by the surgical stress and not by the rapid fall in corticoid levels *per se*. We have tried to separate these effects, avoiding the stress of operation by working with rats in which one adrenal vein was ligated loosely. A sudden drop in corticoid levels was induced by tightening the ligature. Plasma corticosterone levels were monitored by serial samples from the carotid artery through an indwelling T-shaped cannula. It appeared that after shutting off the output of one adrenal, plasma corticosterone levels started to oscillate with a period of about 12 min. The oscillations died out after about one hour, and were prevented by hypophysectomy or hypothalamic implants of corticosterone (Papaikonomou and Smelik, 1974). These experiments also suggest that negative rate sensitivity may exist: the initial fall in corticoid levels induces CRF release, until corticoid levels are back to normal as a result of stimulation of the remaining adrenal by ACTH (after 5–8 min). CRF release then subsides, and the corticosterone levels fall off again within 3–4 min, causing the next CRF pulse. An indication that the fall in corticoid levels in itself may provoke ACTH release has been found by Bohus and Endröczi (1964) in dogs, in which a short period of shunting between the adrenal vein and the liver resulted in increased pituitary–adrenal activity.

Thus, it cannot be excluded that the rate of change in corticoid levels, both in positive and in negative direction, may exert a fast effect on CRF release. This mechanism may prevent overshoot of the system's activity when rapid changes in circulating corticoid levels might occur.

B.   SITE OF ACTION OF CORTICOID FEEDBACK

There has been much controversy expressed in the literature as to whether the feedback action is on the pituitary or on the hypothalamic level. Initially, the pituitary was considered as the target for feedback, but after the discovery of the hypothalamus as the controlling structure, it was generally accepted that the neural elements (and presumably the CRF-producing neurons proper) were sensitive to negative corticoid feedback (Porter and Jones, 1956; McCann *et al.*, 1958).

Studies with hypothalamic implants of crystalline steroids showed convincingly that hypothalamic sites for feedback could be traced (Smelik and Sawyer, 1962; Davidson and Feldman, 1963; Endröczi *et al.*, 1963). Although the method has been criticized (Bogdanove,

1963), later studies have confirmed the suggestion that the hypothalamus contains feedback receptors. After placing dexamethasone implants in the hypothalamic region, the pituitary response to CRF preparations remained unimpaired at the time that the stress response was blocked (Smelik, 1969). A distribution pattern of hypothalamic corticoid-sensitive sites has already been illustrated in Fig 3.

On the other hand, evidence for a pituitary feedback site has been collected over many years (De Wied, 1964; Kendall and Allen, 1968; Russell et al., 1969), and a general consensus developed that the central nervous system as well as the pituitary must be considered as a target for corticoid feedback. Since almost all studies were performed with doses or local concentrations far exceeding the physiological range, it remained an open question whether these findings could disclose a participation of either level in feedback inhibition of the system's response to stress stimuli.

C. *In vitro* STUDIES

A more detailed analysis of the sensitivity for time sequence and mechanism of action of corticoid feedback could be started with the improvement of *in vitro* techniques. The inhibitory effect of dexamethasone (Arimura et al., 1969) and corticosterone (Kraicer et al., 1969) was first shown on incubated pituitaries, and later on incubated pituitary cells, which were dispersed by treatment with trypsin (Portanova and Sayers, 1973). The most remarkable finding was that inhibitory action of corticosterone was prevented by actinomycin D in pituitary cells from adrenalectomized but not from intact rats. This would imply that corticosteroids promote the synthesis of a protein involved in the inhibition of ACTH release (Portanova and Sayers, 1974).

The development of a superfusion system for pituitary cells (Lowry, 1974) opened up possibilities for more detailed dynamic studies, because in such a system the outflow can be monitored continuously. Our group adapted this system for the study of the dynamics of the response to several positive or negative signals. It was found that addition of $0.2\,\mu g/ml$ corticosterone to the superfusion medium, which also contained a CRF-preparation, rendered the cells unresponsive for CRF only when corticosterone has been present for at least 30 min. After withdrawal of corticosterone from the medium the blockade lasted for another hour. This strongly suggested that the feedback action is not an immediate effect on ACTH release (Mulder, 1975; Mulder and Smelik, 1977).

In a cascade system in which a piece of hypothalamic tissue was superfused and its outflow was connected to the pituitary cell column, the amount of CRF released *in vitro* could be monitored. When corticosterone was added to the hypothalamic tissue during a 15 min pre-incubation period, the release of CRF was completely prevented (Smelik *et al.*, 1976; Vermes *et al.*, 1977a). Similar results were obtained when pituitary or hypothalamic tissues were taken from rats which had been treated with corticosterone.

The most extensive studies with hypothalamic tissue *in vitro* were done by the Jones group. They found that corticosterone immediately abolished the release of CRF induced by acetylcholine, possibly by an action on the neuronal membrane. Of a number of steroids tested, only corticosterone, cortisol and dexamethasone had this feedback effect at the hypothalamic level. Delayed feedback action, presumably inhibiting both CRF synthesis and release, was obtained by incubation of the hypothalamus with a greater variety of steroids, including 11-deoxycorticosterone and 11-OH-progesterone. These structure–activity studies suggest that different receptors exist for fast and for delayed feedback (Jones *et al.*, 1976a, 1977); they are confirmed *in vivo* by Jones and Tiptaft (1977).

D.   CORTICOSTEROID EFFECTS ON OTHER BRAIN STRUCTURES

Steroid-sensitive areas outside the hypothalamus have been found in the rostral midbrain reticular formation, medial thalamus, hippocampus, amygdala and septum (Bohus *et al.*, 1968). Corticosteroid implants in these regions may influence ACTH secretion, but it is likely that this effect manifests itself on facilitatory or inhibitory limbic-midbrain tracts involved in emotional behaviour. This implies that corticoids may modulate brain mechanisms which are not primarily involved in the control of pituitary–adrenal activity, but which will affect in general the neuro-endocrine and neurovegetative outflow. Therefore, we do not consider such corticoid effects as a part of the feedback action.

Since the hippocampus has been shown to contain specific corticosteroid receptors (McEwen *et al.*, 1969), and as this structure exerts a clear inhibiting influence on ACTH secretion, it has been speculated that the hippocampus is specifically involved in feedback mechanisms (McEwen *et al.*, 1972). Nevertheless, it has not been demonstrated either that corticoids block ACTH secretion through activation of hippocampal inhibitory pathways or that the hippocampus is indispensable for feedback inhibition. In conclusion, we think that midbrain-limbic structures are not intrinsic parts of the pituitary–adrenal control system, and that the terminology of feedback should not be applied to extra-hypothalamic brain areas.

4. Synaptic inputs to CRF-producing neurons

A. INTRODUCTION

Although the hypothalamic nerve elements elaborating the CRF have not yet been identified, the evidence for their existence appeared to be rather convincing. If these neurons can be considered as the "common final pathway" for the central regulation of ACTH secretion, it must be accepted that they are susceptible to inputs from other regions of the brain. Since the hypothalamic peptidergic neurons are located in a terminal area of neural pathways converging at the diencephalic level (Raisman, 1970; Kobayashi and Matsui, 1969), it can be anticipated that synaptic contacts with these neurons may be made by several neuronal systems.

During the last decade it became clear that a number of neurotransmitter substances are present in different nerve endings within the hypothalamus. Among them are acetylcholine, noradrenaline, dopamine and serotonin, all of which could be localized within nerve fibres either biochemically (e.g. Wurtman, 1970; Kuhar, 1973) or histochemically (e.g. Fuxe and Hökfelt, 1970; Björklund et al., 1973). Other neurotransmitters may be present but their localization and role are as yet uncertain.

Neural signals may be conveyed to the CRF neurons via such transmitters, and thus modify their activity. Very little is known as yet about their synaptic contacts, which may be at the level of the dendrites, cell bodies, axons or terminals of the neurosecretory elements. A general assumption is that the neurosecretory neurons can be either stimulated or inhibited by different neuronal systems through their respective neurotransmitters; in this way their activity can be modulated by a number of afferents. If this hypothesis is correct, it can be expected that specific postsynaptic receptors are present on the cell membrane of the neurosecretory elements. However, it should be pointed out that the demonstration of a response to a certain neurotransmitter does not necessarily involve a role of that neurotransmission system in the control of a neurosecretory function. It may well be that these neurons are sensitive to a great number of endogenous substances and drugs, especially when the local concentration is made high enough. In general, the demonstration that blockade or elimination of a particular neurotransmitter system interferes with the normal physiological control of a pituitary hormone is, in our opinion, a much better indication of the involvement of that neural system in regulatory processes.

With the pituitary–adrenal system, a complicating factor is that many experimental interventions such as handling, surgery or drug administration will activate the system by their stressful character. It can be expected that any drug produces toxic effects if given in high doses, so that the demonstration of pituitary–adrenal activation in itself does not produce evidence for involvement of a particular neurotransmitter system. In a survey of the literature, one is impressed by the overwhelming amount of contradictory results and conclusions, and by the virtual absence of well-controlled pharmacological studies which take into account unspecific and indirect effects, dose–response relationships and localization of the site of action. These general shortcomings of the experimental work in this area will prevent us from reviewing the literature in great detail, and we shall mainly focus on those studies which produced the most relevant results.

Before considering the role of several neurotransmitter systems in the production of hypothalamic CRF, it should be pointed out that actions on higher brain structures and on the pituitary itself are not included in this section. Extra-hypothalamic mechanisms are treated later on, and pituitary effects have not been encountered, unless clearly unphysiological doses have been used. A CRF-like activity of several neurotransmitters has been screened by several authors (Guillemin, 1955; Schally and Guillemin, 1963; Dhariwal *et al.*, 1969, Hiroshige and Abe, 1973), without results.

B.   ACETYLCHOLINE (ACh)

ACh, its synthetizing enzyme acetyltransferase and its metabolizing enzyme acetylcholinesterase have been demonstrated within the hypothalamus (Shute and Lewis, 1966, 1967). Histochemical localization of the enzymes showed that cholinergic neurons are found in the preoptic area, the paraventricular nucleus, the dorso-posterior part of the hypothalamus and the median eminence (Shute, 1970; Jacobowitz and Palkovits, 1974). However, the exact localization and course of these neurons is not known. Early reports did not favour the idea that a cholinergic link would be involved in the control of ACTH release, since even massive doses of atropine did not affect ACTH secretion (Dordoni and Fortier, 1950; Suzuki *et al.*, 1964; Otsuka, 1966).

However, intra-hypothalamic injection of ACh or carbachol elicited pituitary–adrenal activity in rats (Endröczi *et al.*, 1963) and in cats (Krieger and Krieger, 1964). Since such powerful drugs may stimulate CRF release in an unspecific manner, the demonstration that blockade of hypothalamic cholinergic receptors by local implants of atropine prevented stress-induced rise in plasma corticosterone levels (Hedge and Smelik, 1968) gave more convincing evidence for an intrinsic role of cholinergic transmission. In a further study, it was shown that strictly localized bilateral implants in the anterior hypothalamus blocked the effect of a number of stresses, but not of a CRF preparation. The onset of inhibition was very fast (10 min) and lasted for about 90 min. Implants of local anaesthetics and other drugs were ineffective (Kaplanski and Smelik, 1973a, but see Makara and Stark, 1976).

Other studies have confirmed the involvement of a cholinergic link (Porter *et al.*, 1969; Naumenko, 1969; Krieger and Krieger, 1970). *In vitro* studies re-established the role of ACh. Edwardson and Bennett (1974) found that 10^{-9}–10^{-11} M ACh is capable of releasing CRF from synaptosomal fractions obtained from the sheep hypothalamus. Hillhouse *et al.* (1975) showed a dose-dependent release of CRF from incubated rat hypothalamus in the 10^{-14}–10^{-11} M dose range. This effect could be prevented by atropine and hexamethonium. Our group, using a superfusion cascade system of hypothalamus, pituitary cells and adrenal cells, found that 10^{-9} M atropine inhibited basal CRF release (Smelik *et al.*, 1976). Although atropine in high concentrations may well exert a local anaesthetic effect, the recent indications of effectiveness in very low concentrations favour the idea of a specific cholinergic blockade. At present, there is a general consensus that a cholinergic link is involved in the release of CRF. Since its blockade prevents the effects of all other drugs or noxious stimuli, it is thought that this cholinergic link is the common final pathway to the CRF neurons. It is not clear, however, whether the cholinergic receptors are located on the CRF cell bodies or the axon terminals. The former possibility is favoured by the localization of the atropine effects in the anterior hypothalamus, the latter by the ACh effect on synaptosomes.

C. NORADRENALINE (NA)

Participation of catecholamines in the control of the pituitary–adrenal axis has been suspected ever since Cannon's concept of the "emergency reaction" was incorporated in the stress concept of Selye. The idea that adrenal medullary hormone release, as part of the alarm reaction, induced in its turn adrenocortical hormone production was formulated by Long and his associates (Long and Fry 1945; McDermott *et al.* 1950). Systemic administration of the catecholamines clearly induced ACTH release.

Catecholamines were found in the brain by Von Euler (1946) and in the hypothalamus by Vogt (1954). Decisive progress was brought about by the histochemical fluorescence method, and the exact localization of NA neurons was mapped out by Fuxe (1964, 1965) and Fuxe and Hökfelt (1969). The groups of NA cell bodies are located outside the hypothalamus and their axons enter the hypothalamus mainly laterally (Weiner *et al.*, 1972). NA terminals are found over the entire hypothalamus, including the median eminence (Fuxe, 1965; Palkovits *et al.*, 1974a).

The possible role of brain catecholamines in the control of ACTH secretion has been reviewed extensively by Van Loon (1973). Intracerebral administration of NA generally results in adrenocortical activation (Endröczi *et al.*, 1963; Krieger and Krieger, 1970; Bhargava *et al.*, 1972; Hiroshige and Abe, 1973). However, there seems to be a dose dependency, since lower doses have been reported to block the stress response. (Schiaffini *et al.*, 1970; Ganong, 1970; Motta *et al.*, 1971; Van Loon *et al.*, 1971a). Attempts to study the effect of interference with central NA release have been numerous, but the results from these studies appeared to be quite contradictory. Most frequently reserpine has been studied as a NA depletor, and α-methyl-paratyrosine (α-MT) as a synthesis blocker. Both drugs lack specificity for NA, and produce many side effects. When given systemically in high doses they activate the pituitary–adrenal system, presumably because of their noxious effects. Reserpine implants into the basal hypothalamus, not producing the generalized syndrome, did not prevent the stress-induced rise in plasma corticoid levels, including the rise induced by systemic reserpine administration (Smelik, 1967). This suggests that its effect was not due to depletion of hypothalamic monoamines. The systemic injection of α-MT causes NA depletion together with adrenal activation (Scapagnini *et al.*, 1970; Van Loon *et al.*, 1971b). However, adrenal activation can be prevented by repeated administration of small doses or by pentobarbitone sodium anaesthesia, without preventing severe NA depletion (Kaplanski *et al.*, 1972). This dissociation suggests again that the toxic effects of α-MT caused ACTH release. Chronic degeneration of central NA neurons can be induced by intracerebral administration of 6-OH-dopamine. This procedure has no effect on the resting levels or adrenal stress response (Kaplanski and Smelik 1973b; Kaplanski *et al.*, 1974; Abe and Hiroshige 1974). However, the combination of α-MT and 6-OHDA treatment appeared to be effective in activating the pituitary–adrenal axis, suggesting that a small residual pool of NA may be sufficient for tonic inhibition (Scapagnini and Preziosi, 1973).

In certain conditions the central administration of sympathico-mimetic drugs like l-dopa and MAO-inhibitors has been reported as preventing stress-induced ACTH secretion (Van Loon *et al.*, 1971a), which could be antagonized by the α-blocker phenoxy-benzamine (Ganong, 1972). A number of these studies suggested that a central NA system may exert an inhibitory influence on ACTH secretion, although the evidence is not very convincing and often dubious. *In vitro* studies corroborate that hypothesis, since NA in low concentrations (10^{-9} M) reduces CRF release from the incubated (Hillhouse *et al.*, 1975) or superfused (Smelik *et al.*, 1976) hypothalamus. It remains to be seen, however, how such an inhibitory function would operate during stress-induced adrenocortical activation. After application of stressful stimuli, NA concentration in the hypothalamus decreases (Bliss *et al.*, 1968; Corrodi *et al.*, 1968; Palkovits *et al.*, 1975), and NA turnover increases (Thierry *et al.*, 1968; Corrodi *et al.*, 1968), and these are not modified by corticosteroids (Fuxe *et al.*, 1970). This suggests that the central NA system is activated during stress, independent of the status of adrenocortical activity.

If the NA system did exert a tonic inhibition on CRF release, it might be expected that, during stress, either its inhibitory action is removed or its inhibiting activity shuts off the stress response. The former possibility is unlikely, since its turnover is increased during stress. The second assumption lacks the conclusive evidence that elimination of the NA system prolongs or intensifies the stress response.

D. DOPAMINE (DA)

DA neurons are found in the tuberal region of the hypothalamus, the cell bodies lying in the arcuate nucleus and their terminals in the outer zone of the median eminence and the intermediate lobe of the pituitary (Fuxe, 1964; Fuxe and Hökfelt, 1966; Björklund *et al.*, 1970; Palkovits *et al.*, 1974a). The involvement of this tubero-infundibular DA system in gonadotropic regulation and in prolactin secretion has been advocated by many groups, but so far no clear indications have been found for a role in the control of ACTH release. DA terminals within the basal hypothalamus have been described, which originate from DA cells outside the hypothalamus (Versteeg *et al.*, 1976; Björklund *et al.*, 1975). Their function is still unknown.

Experiments with l-dopa or DA suggested a role of NA rather than DA, because they can act as precursors of NA. Local depletion of the DA system by implants of reserpine, which remove all fluorescence due to catecholamines in the histochemical procedure, leaves basal and stimulated ACTH secretion unchanged (Smelik, 1967). DA added *in vitro* to the hypothalamus did not induce CRF release (Hillhouse *et al.*, 1975; Smelik *et al.*, 1976). An exceptional observation came from Upton and Corbin (1975) in that the DA-receptor blocker, pimozide, appeared to decrease pituitary–adrenal activity in rats and man. Edwardson and Bennett (1974) found that DA blocked the acetylcholine-induced CRF release from hypothalamic synaptosomes. The general impression remains that the hypothalamic DA system is not intrinsically involved in pituitary–adrenal regulation.

E. SEROTONIN (5-HT)

With the aid of the histochemical fluorescence method the 5-HT system in the brainstem has been located. Its cell bodies are situated in the area of the raphe nuclei, and terminals are found within the hypothalamus, especially in the suprachiasmatic nucleus (Dahlström and Fuxe, 1965; Carlsson and Lindquist, 1972). Other hypothalamic regions, including the median eminence, also contain appreciable amounts of 5-HT (Palkovits *et al.*, 1974b; Saavedra *et al.*, 1974a, b). Although systemic administration of 5-HT causes an increase in adrenocortical activity, its central administration yielded conflicting results. Stimulation of the pituitary–adrenal action was reported by Naumenko (1968) and Krieger and Krieger (1970), but inhibition was found by Van Loon *et al.* (1971a), Schiaffini *et al.* (1970), and Vermes and Telegdy (1972, 1973a). Systemic administration of the 5-HT precursors tryptophan or 5-OH-tryptophan also exerted an inhibitory effect (Vernikos–Danellis *et al.*, 1973, Berger *et al.*, 1974). Studies with the synthesis inhibitor p-chlorphenylalanine (pCPA) appeared to be contradictory. Basal plasma corticosterone levels were reported to be essentially normal (Preziosi *et al.*, 1968; Barchas and Vernikos–Danellis, 1970; McKinney *et al.*, 1971) or increased (Scapagnini *et al.*, 1971; Van Delft *et al.*, 1973), concomitant with a disturbance of the circadian rhythm. The adrenal stress response was found to be increased (Rosecrans, 1968; Berger *et al.*, 1974; Vermes and Telegdy, 1973b), unchanged (Bhattacharya and Marks, 1970; Dixit and Buckley, 1969) or decreased (Van Delft *et al.*, 1973). Since pCPA treatment exerted strong side effects (e.g. sleep disturbance, low food intake), a chronic stress effect may have interfered with normal pituitary–adrenal function (Van Delft et al., 1973).

Degeneration of 5-HT terminals by 5,6-dihydroxy-tryptamine injection into the raphe nuclei resulted in increased pituitary–adrenal activity (Fuxe et al., 1973, 1974). Electrolytic lesion of the same area had similar effects (Scapagnini and Preziosi, 1972; Vermes et al., 1974).

The effect of stressful stimuli on central 5-HT content appeared to depend on the time sequence and the part of the brain in which the content was measured. The most extensive studies were done by Vermes and Telegdy (1975) who found that hypothalamic 5-HT content decreased between 0 and 10 min after application of stress, followed by an increase after 30 min. These changes were in negative correlation with the pituitary-adrenal activity. Similar short-term changes in 5-HT content were found by Palkovits et al. (1975), and in 5-HT turnover by Morgan et al. (1975).

Studies on the effect of corticosteroids on brain 5-HT also suggest a biphasic action. Corticosterone injection increases 5-HT content within 15 min, but after 1 h the content returns to normal or is even lower than normal (De Schaepdrijver et al., 1969; Millard et al., 1972; Telegdy and Vermes 1975). Glucocorticoids increase synaptosomal uptake of ^{3}H-tryptophan in vitro (Neckers and Sze 1975) and ^{3}H-5-HT (Vermes et al., 1977b).

The effect of 5-HT on CRF release in vitro from hypothalamic slices was found to be stimulatory by Hillhouse et al. (1975), but inhibitory by Smelik et al. (1976, see Fig. 5). The data generally suggest an inverse relationship between brain 5-HT and adrenocortical activity. This view has also been adopted by Azmitia and McEwen (1969, 1974) and by Vernikos-Danellis et al. (1973), in that a reciprocal control loop may exist between the central 5-HT system and the level of circulating glucocorticoids. This may have implications for the feedback control of the pituitary–adrenal system.

F.   GAMMA-AMINOBUTYRIC ACID (GABA)

There is accumulating evidence that GABA is an inhibitory neuro-transmitter in the brain (Curtis and Johnston, 1970; Kuhar, 1973; Bennett et al., 1974). It is present in hypothalamic nerve cells (Okada et al. 1971; Makara and Stark, 1975). Although Krieger and Krieger (1970) found a stimulatory effect of GABA injections into the cat hypothalamus, Makara and Stark (1974) observed an inhibitory

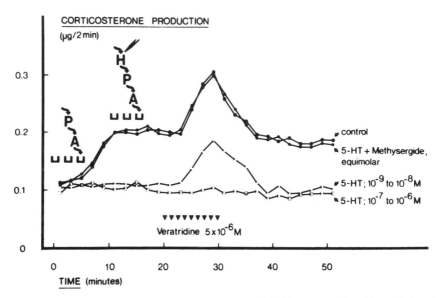

Fig. 5. Effect of serotonin (5-HT) on the release of CRF from superfused hypothalami. For the method, see Fig. 2. In hypothalami superfused with 10^{-8} or 10^{-9}M 5-HT during 15 min basal but not Veratridine induced CRF was blocked. Superfusion with 10^{-6} or 10^{-7}M 5-HT also prevented stimulated CRF release. The effect of 5-HT was prevented by equimolar concentrations of the 5-HT antagonist methysergide.

effect, whereas GABA antagonists appeared to be potent ACTH releasers. Burden *et al.* (1974) observed an inhibitory effect of GABA on CRF release from incubated hypothalamic slices.

G. OTHER PUTATIVE TRANSMITTERS

Histamine concentration in the hypothalamus is very high, especially in the median eminence region (Taylor *et al.*, 1972; Brownstein *et al.*, 1974). Although most of this substance is located within mast cells (Green, 1970; Garbarg *et al.*, 1973), there are some indications that it also occurs intraneuronally (Taylor *et al.*, 1972; Schwartz *et al.*, 1974). Whilst histamine has always been found to be a potent stimulator of ACTH release, its effect has been considered as unspecific (Guillemin and Rosenberg 1955; De Wied 1964; Lissak and Endröczi 1965). Similarly, quite high concentrations stimulated CRF release (Hillhouse *et al.*, 1975; Smelik *et al.*, 1976) and ACTH release (Smelik *et al.*, 1976). Its physiological role is still uncertain.

The amino acids glutamate and glycine have been studied recently because of their possible transmitter function. No evidence for a direct effect of these substances on CRF release has been reported (Makara and Stark 1974, Jones *et al.*, 1976a,b).

H. CONCLUSIONS

Although it is quite clear that the known neurotransmitters must be involved in neuroendocrine regulation, it appears to be very difficult to obtain clearcut evidence for a particular role. In our opinion this is mainly caused by the fact that the available techniques were not subtle and specific enough. Even with local application of drugs a number of unspecific or toxic effects cannot be excluded. Considering the multiple functions of a particular neurotransmitter system, impinging on a multitude of other neuronal elements on different levels, it is not surprising that one particular function cannot be singled out experimentally.

Summarizing the evidence collected during the past decade, it can be concluded that a cholinergic link in the activation of CRF neurons appears to have been shown with reasonable certainty. A noradrenergic inhibitory pathway may operate, but conclusive evidence is lacking. The serotonin system is probably involved in the control of ACTH secretion, but rather via corticosteroid feedback than via direct actions on the CRF neurons.

5. Extra-hypothalamic brain structures and pituitary–adrenal function

The basal hypothalamus can be regarded as the regulatory structure for pituitary function. Within the so-called "hypophysiotropic area" the neurosecretory elements are thought to be situated, which are capable of integrating several neural and humoral signals, resulting in the release of the hypophysiotropic factor. On the other hand, the hypothalamus can be considered as a nodal point in the circuitry of the brain stem, and it is to be expected that "higher" neural structures are capable of modulating the hypothalamic neuro-endocrine control system.

A logical experimental approach to the analysis of the role of extra-hypothalamic brain structures has been the isolation of the hypothalamus from the rest of the brain. The best technique for producing an "hypothalamic island" is deafferentation with a specially designed knife, controlled stereotaxically (Halász *et al.*, 1967a). After complete neural deafferentation the following results were obtained concerning pituitary–adrenal function.

In basal conditions pituitary ACTH content as well as that of the level of plasma corticosterone appeared to be increased, suggesting the existence of a tonic inhibitory influence of an extra-hypothalamic nature (Halász et al., 1967b; Greer and Rockie, 1968; Vermes et al., 1973). Circadian rhythm of pituitary–adrenal activity disappeared. The response of the system to "neurogenic" stresses (e.g. trauma, visual or acoustic stimuli, emotions) was abolished (Makara et al., 1969a; Feldman et al., 1970), the response to ether stress was somewhat reduced but still present (Halász et al., 1967a; Palka et al., 1969). Compensatory adrenal hypertrophy after unilateral adrenalectomy and dexamethasone-induced inhibition of ACTH production were both present, indicating that the feedback action of corticosteroids is not dependent on higher brain structures (Halász et al., 1967a; Feldman et al., 1973).

These data show that extra-hypothalamic CNS structures are essential for the integrated adaptive responses to environmental changes, but that the normal functioning of the system in basal conditions can be controlled at the hypothalamic level. Extrahypothalamic inputs to the system can be divided into two groups: (a) ascending tracts from the brain stem, entering the lateral hypothalamus mainly via the medial forebrain bundle, the periventricular tract and the tegmentomamillary tract; (b) descending fibre systems from limbic structures, mainly via the stria terminalis and the fimbria–fornix system.

The hypothalamus can be considered as an important relay centre in the midbrain–limbic circuitry (Nauta, 1963) and may act as an "endocrine motor system". Since the midbrain reticular formation system can be regarded as maintaining vigilance so that it responds to any stimuli, new or novel, demanding alertness and activity and the limbic system may incorporate incentives which operate at the level of recognition of earlier experiences of biologically significant stimuli (see Routtenberg, 1968) it may be expected that an adaptation system, such as that of the pituitary–adrenal axis, will be activated by both the midbrain and the limbic structures in case of noxious (nociceptive), threatening or emotional conditions.

A.   AFFERENTS FROM THE MIDBRAIN

Earlier work already suggested that the midbrain exerts a significant influence on the pituitary–adrenal system. Midbrain transections or lesions prevented activation of the adrenal, induced by stress (Royce and Sayers, 1958; Anderson et al., 1957; Egdahl, 1960; Martini et al., 1960), whereas electrical stimulation of the midbrain reticular formation increased the gland's activity (Mason, 1958; Okinaka *et al.*, 1960a; Endröczi and Lissak, 1960, 1962). Although these data suggest a stimulating influence of midbrain structures on pituitary–adrenal function, they should not be taken at face value. Indirect effects cannot be excluded; for instance, midbrain lesions may affect limbic structures profoundly and in this way modulate adrenocortical activity. Knigge and Hays (1963) found that hippocampal lesions prevented the inhibition produced by midbrain lesions. Moreover, the midbrain contains not only the reticular formation activating system, but also a "limbic midbrain area" (Nauta, 1963), comprising the ventral tegmentum and periaquaductal area, which may represent an inhibitory system. All these structures impinge heavily on forebrain areas, notably the septum–hippocampus complex.

B.   AFFERENTS FROM THE HIPPOCAMPUS

More than twenty years ago, Porter (1954) demonstrated that electrical stimulation of the uncus hippocampi inhibits the adrenocortical stress response in cats. Similar findings have been reported in monkeys (Mason, 1958; Smith and Root, 1971), in dogs, cats, rabbits and rats (Endröczi *et al.*, 1959) and in man (Mandell *et al.*, 1963; Rubin *et al.*, 1966). In an extensive and careful study, Kawakami *et al.* (1968) found that stimulation of the cornu ammoni hippocampi inhibited the response to immobilization stress in rabbits, an inhibition which could be prevented by dorsal fornix lesions, but not by those on the stria terminalis.

The importance of stimulation frequency was shown by Endröczi and Lissák (1962). They found that low frequency stimulation (12–36 c/sec) inhibited and high frequency stimulation (240 c/sec) potentiated the adrenal stress response. They also showed that hippocampal stimulation failed to alter adrenal responses if the animal was habituated to the stimulus. A similar finding was reported by Kawakami *et al.* (1971), who suggested that the hippocampus may play a role in the habituation to repeated emotional stimuli.

Thus, lesions in the hippocampus have been reported to increase the pituitary–adrenal stress response (Knigge 1961; Fendler *et al.*, 1961; Endröczi, 1972), although this has not necessarily been confirmed (Coover *et al.*, 1971; Wilson and Critchlow, 1973; Lanier *et al.*, 1975).

Septal lesions have also been found to remove inhibitory influences on pituitary–adrenal activity (Endröczi and Lissák, 1960; Bohus, 1961; Seggie, 1968). Resting levels of plasma corticosterone do not seem to be changed by septal lesions, but the stress response is more pronounced (Usher *et al.*, 1967; Montgomery and Berkut, 1969; Uhlir *et al.*, 1974; Seggie *et al.*, 1974), probably induced by a decrease in sensitivity threshold to emotional disturbances. It could be expected that interruption of the main connection between the septum–hippo-campus complex and the basal hypothalamus, the fornix bundle, may produce similar effects. However, experiments with fornix lesion or transection have given conflicting results (Moberg *et al.*, 1971; Lengvàri and Halász, 1973; Wilson and Critchlow, 1973), indicating the difficulty of isolating one type of modulation at that level.

C. AFFERENTS FROM THE AMYGDALA

In earlier studies the results were in good agreement, suggesting that the amygdaloid complex exerts a stimulatory influence on pituitary–adrenal activity. Amygdala lesions were reported to inhibit adrenal stress responses (Woods, 1956; Mason, 1958; Knigge, 1961; Ganong, 1963), whereas electrical stimulation of the amygdaloid complex was shown to facilitate stress responses (Mason, 1958; Endröczi *et al.*, 1959; Ganong and Goldfien, 1959; Okinaka *et al.*, 1960b; Mandell *et al.*, 1963, and many others). However, the different amygdaloid nuclei do not respond uniformly to electrical stimulation. Inhibition of ACTH release was reported to result from stimulation of anterior and medial nuclei, whilst increased release was induced by stimulation of the lateral and basal nuclei (Endröczi and Lissák, 1960; Bovard and Gloor, 1961). Moreover, the effect of amygdala stimulation appeared to depend on the existing plasma corticosteroid level, facilitation of ACTH production only being induced if the steroid levels were low before stimulation (Matheson *et al.*, 1971).

Kawakami *et al.* (1971) observed that the amygdala responded to emotional stress with a decrease in evolved potential amplitude but this response fades out after repeated stressful stimulation.

Electrical stimulation of the orbito-frontal cortex results in pituit-ary–adrenal activation (Endröczi and Nagy, 1951; Endröczi and Lissák, 1954; Porter, 1954; Okinaka *et al.*, 1960b). This activation can be prevented by bilateral amygdala lesions (Hall and Marr, 1975), suggesting that the amygdala may serve as a relay station between orbital cortex and hypothalamus.

It thus seems that the amygdaloid complex is involved in the response to neurogenic stimuli, and that it mainly facilitates pituit-ary–adrenal activation (Allen and Allen, 1974).

D. CONCLUSIONS

This review on experimental work concerning the role of those brain structures which may influence pituitary–adrenal activity, clearly shows that the hypothalamic controller can be modulated by a number of afferents, but at present we are still far from an analysis of how these impinge on the CRF neurons. It should be kept in mind that the techniques applied were rather crude (lesions and electrical stimulation) and the parameters used were indirect (steroid levels). Only a few studies took into account the behavioural conditions of the conscious animal, and these may be essential in studying the relationships between midbrain or limbic structures and the hypothalamic endocrine motor system. Another variable which may be of importance is the timing of the pituitary–adrenal response to single stimulation (Redgate, 1970) and the novelty of the stimulus (Kawakami et al., 1968). Nevertheless, the now classical idea of two main systems impinging on the hypothalamic controller—a midbrain system and a forebrain system—still seems to be valid. There are good indications that each system may exert a facilitatory and an inhibitory influence: for the midbrain, a facilitating action from the ascending reticular activating system (ARAS) and an inhibitory action from the limbic midbrain; for the forebrain, a facilitating influence from certain amygdaloid nuclei, and an inhibitory one from the hippocampus.

It remains to be elucidated how these modulating afferents function in the interplay between the adjustment of pituitary–adrenal activity and the adaptation of the organism to the ever-changing environmental conditions, including processes of learning and emotion.

6. Circadian rhythm of the pituitary–adrenal system

It has long been known that in many animals and in man rhythmic variations in adrenocortical activity occur. This rhythmicity is related to the dark and light variations, since nocturnal animals show a pattern which is almost exactly opposite to that of diurnal animals. Moreover, reversal of the day–night (diurnal) light cycle induces a reversal in endocrine rhythm. The daily fluctuations do not always coincide exactly with the 24-hour cycle. Therefore Halberg (1959) introduced the term "circadian rhythm", indicating rhythms with a cycle between 20 and 28 hours. Such rhythms are thought to be induced by an internal "clock" which may integrate a number of rhythmic external signals, or may be synchronized by an external

Zeitgeber such as light (Wurtman, 1966; Halberg, 1969; Retiene, 1970). Although other rhythms in adrenocortical activity have been found, either seasonal (Haus and Halberg, 1970; Weitzman *et al.*, 1975) or even very fast ("ultradian", Berson and Yalow, 1968; Krieger *et al.*, 1971), only the circadian rhythm will be discussed here in more detail.

A.   GLUCOCORTICOID RHYTHMICITY

The first observation on a daily rhythm in corticoid secretion rate was made in man by Pincus (1943), and has since been confirmed by numerous authors. The plasma cortisol level reaches a peak of 15–25 μg/100 ml between 06.00 and 08.00, and declines steadily to a nadir of less than 5 μg/100 ml at around 24.00 (Daly and Evans, 1974). A similar pattern is seen in monkeys (Migeon *et al.*, 1955) and dogs (Yates and Maran, 1974). In nocturnal animals including laboratory rats and mice, the peak in plasma corticosterone levels, ranging from 25–40 μg/100 ml, occurs just before the dark period, 17.00–20.00 and the nadir is found in the early morning between 04.00 and 06.00. The rhythm coincides with the hours of activity and of sleep reaching the maximum at the end of the rest period.

B.   POSTNATAL DEVELOPMENT OF CIRCADIAN RHYTHM

It takes about 2–3 years before a full circadian rhythm develops in children (Franks, 1967), thereafter it persists during the rest of human life (Silverberg *et al.*, 1968). In rats, the first appearance of a circadian rhythm occurs at 21–25 days (Allen and Kendall, 1967; Hiroshige and Sato, 1970) and is fully presented at day 30–32. Early handling or frequent stress advances the development of the rhythm (Ader, 1969). Since the maturation of the pituitary–adrenal system, including the stress responsiveness, occurs well before this period (Schapiro, 1962; Milkovic and Milkovic, 1968), it seems that development of the circadian rhythm is dependent on the maturation of some central nervous mechanism. The development of the rhythm can be delayed by administration of cortisol to neonatal rats (Krieger, 1974a). This suggests that a critical period exists, sensitive to corticosteroids, to provide central maturation.

C.   SEX DIFFERENCES IN CIRCADIAN RHYTHMS

Generally, the peak values during the rhythm are higher in the human female than in the male (Asfeldt, 1971) and this is true in other vertebrates (Halberg and Haus, 1960; Kitay, 1963; Critchlow *et al.*, 1963).

In rats it was found that gonadectomy decreased the afternoon peak in females, but not in males. Moreover the magnitude of the peaks in females varies with the oestrous cycle. This suggested that oestrogens play an important role in this sex difference. In man it has been shown that oestrogen administration increases cortisol peak levels (Daly and Elstein, 1972).

D. REGULATION OF THE CIRCADIAN RHYTHM

It has been suggested by many indirect indications, though often only by implication, that the circadian rhythm is controlled by the brain. The daily oscillations of the plasma corticoid levels are caused by synchronous variations in pituitary ACTH production (Retiene *et al.*, 1965; Krieger *et al.*, 1971) which persist after adrenalectomy (Cheifetz *et al.*, 1968).

Moreover, a circadian rhythm in hypothalamic CRF content has been demonstrated by several groups (David-Nelson and Brodish, 1969; Hiroshige and Sakakura, 1971; Takabe *et al.*, 1971) which is not suppressed by adrenalectomy or hypophysectomy.

Surgical isolation of the hypothalamus abolishes the circadian fluctuations as shown by Halász *et al.* (1967a); their investigations suggested that the anterior connections are crucial, since frontal deafferentation had the same effect. Other work supports this idea: Moore and Eichler (1972) found that destruction of the suprachiasmatic nucleus or small frontal cuts caudal but not rostral to the optic chiasma similarly obliterated and diurnal rhythm. Raisman (1975) showed that small suprachiasmatic lesions abolished a number of circadian rhythms, mainly behavioural activities.

If the suprachiasmatic region is indeed essential for the mainte- nance of circadian variations, it would be interesting to know which neuronal inputs might control its function. Fornix transection has been reported to abolish diurnal fluctuations of adrenal function (Mason *et al.*, 1957; Nakadate and De Groot, 1963; Moberg *et al.*, 1971), but others found that the effect was only temporary (Lengvári and Halász, 1973) or absent (Wilson and Critchlow, 1973). A similar discrepancy was apparent from studies with septal lesions (Seggie and Brown, 1971; Seggie *et al.*, 1974), or those of the hippocampus (Knigge, 1961; Fendler *et al.*, 1961; Endröczi, 1972; Lanier *et al.*, 1975).

In spite of these discrepancies, the evidence at present suggests that the limbic midbrain circuit may play an important role in the maintenance of the circadian rhythm *via* the suprachiasmatic nucleus, even when considering that the timing, localization and extent of the lesions varied considerably among the different experimental approaches.

E.   POSSIBLE ROLE OF PARTICULAR NEUROTRANSMITTER SYSTEMS

Since it has been shown that the content of neurotransmitters in the brain has a circadian rhythm (e.g. Quay, 1965; Reis *et al.*, 1968; Friedman and Walker, 1968), it could be anticipated that they may be involved in the regulation of the adrenocortical rhythm.

1.   *Acetylcholine*

Krieger *et al.* (1968) found, in cats, that the muscarinic blocking agent atropine is capable of blocking the daily rise in plasma corticosteroids, if administered just before the expected increase. This systemic injection of atropine did not alter pituitary–adrenal responsiveness to stress or adrenal sensitivity to ACTH. This suggested that, at least in cats, a cholinergic control of the circadian rhythm may exist.

2.   *Serotonin (5-HT)*

Shortly afterwards, the same group showed that drugs altering the activity of central serotonergic systems, but not that of other aminergic systems, also affect the adrenocortical rhythm. The possible involvement of the 5-HT system has attracted much attention since this original observation. A number of studies have shown that during the light period brain 5-HT is higher than during the dark period. This is true for many species as well as for several brain regions. Héry *et al.* (1972) observed that 5-HT synthesis and turnover in the rat brain stem and hypothalamus are maximal during the light period, and that 5-HT release is activated during the dark period. This would suggest that the 5-HT rhythm is reciprocal to the pituitary–adrenal rhythm.

   Interference with the activity of the 5-HT system in the brain was reported by many groups to abolish or at least blunt the circadian variations (Scapagnini *et al.*, 1971; Vernikos–Danellis *et al.*, 1973; Van Delft *et al.*, 1973; Vermes and Telegdy, 1975). It is not clear yet whether the 5-HT system is indispensable for the existence of a pituitary–adrenal rhythm. Krieger (1975) reported that in rats,

treated neonatally with central injections of the neurotoxic agent
5,6-dihydroxy–tryptamine, a normal adrenocortical rhythm developed.

3. *Noradrenaline (NA)*

Although the noradrenaline content in the brain shows a clear diurnal
rhythm, no relationship with pituitary–adrenal rhythmicity could be
demonstrated. The NA rhythm is inverse to that of 5-HT, and in
phase with the adrenal rhythm. Pretreatment with the synthesis
blocker α-m-PT did not modify the adrenocortical rhythm in cats
(Krieger and Rizzo, 1969) and in rats (Kaplanski and Smelik, 1973b);
similar results have been found after central administration of the
neurotoxic drug 6-hydroxydopamine (Ulrich and Yuwiler, 1973;
Kaplanski *et al.*, 1974).

In summary, the limbic cholinergic and serotonergic system may be
involved in the maintenance of the diurnal adrenocortical rhythm, but
there is no evidence as yet for any neurotransmitter system as an
obligatory link in the control of this rhythm. Any experimental
interference with brain structures or circuitry may alter the activity of
other rhythms like sleep, feeding behaviour, temperature, muscular
activity and sensory inputs, and in this way may affect the function of
the pituitary–adrenal system.

F. EXTERNAL SYNCHRONIZERS OF THE ADRENOCORTICAL RHYTHM

1. *Light*

Halberg was the first to show that changes in the light regimen could
influence adrenocortical function (Halberg and Vischer, 1950; Halberg, 1953). Blindness has been reported to abolish or disturb the
circadian rhythm in rats (Saba *et al.*, 1965; Haus *et al.*, 1967; Krieger,
1973) and in man (Orth and Island, 1969; Krieger and Rizzo, 1971).
A constant light environment also affects the rhythm to some extent
(Critchlow, 1963; Cheifetz *et al.*, 1968; Krieger, 1973), generally to the
extent that the nadir of the rhythm is elevated. A shift or reversal of
the light–dark cycle generally induces an adaptation of the adrenal
rhythm within about a week. This has been shown in animals
(Critchlow, 1963; Morimoto *et al.*, 1975) and man (Perkoff *et al.*, 1959;
Conroy *et al.*, 1968; Wegmann *et al.*, 1970).

Although a role of the pineal gland could be assumed in view of the
connections between pineal, light perception (Kappers, 1960) and

5-HT and melatonin rhythmicity (Wurtman, 1966), surgical or chemical pinealectomy did not affect the circadian rhythm of plasma corticosterone levels (Vermes *et al.*, 1974).

2. *Sleep*

A number of studies have been devoted to the relation of (monitored) sleep and adrenocortical activity in humans. Mandell *et al.* (1966) and Weitzman *et al.* (1968) found a correlation between periods of paradoxical or REM sleep and peaks of blood corticoid levels. However, this induction of episodic bursts of adrenal activity is independent of the diurnal rhythm, which persists during deprivation of REM sleep (Weitzman, 1974). Shifts in the sleep–wake pattern appeared to disturb the corticoid rhythm profoundly (Perkoff *et al.*, 1959; Orth *et al.*, 1967) and Krieger *et al.* (1969) showed that sleep is a more powerful synchronizer than light.

3. *Food and water intake*

Changes in the feeding regimen may alter adrenocortical rhythmicity (Halberg, 1953). Recently, Krieger (1974b) showed that restriction of food to a two-hour period during the resting phase caused a 12-hour shift in peaks of plasma corticosterone levels in rats, but other parameters were also altered: body temperature, running activity and brain monoamine content variations. Water deprivation also has been reported to modify adrenocortical rhythms (Coover *et al.*, 1971; Johnson and Levine, 1973), but here the constant and severe stress situation might explain the disappearance of the nadir in corticoid levels.

G.   SIGNIFICANCE OF THE CIRCADIAN RHYTHM

It may be obvious from the preceding section that there is no indication for the existence of a single "synchronizer". It seems that a number of environmental and internal stimuli, which "happen" to be presented periodically, are able to maintain the functioning of some kind of endogenous rhythmicity. This state of affairs has been interpreted as an adaptive mechanism (Sollberger, 1969; Aschoff, 1973). The organism maintains an internal oscillation in adapting itself to the temporally programmed environment, in which the diurnal and the seasonal rhythms are the most prominent oscillations with a fixed wave length.

Concerning the rhythmicity of the pituitary–adrenal system, it is tempting to speculate that it is in some way related to the important role of this system in adaptive processes. The activity of the adrenal cortex seems to follow the need for corticoids rather closely. Apart from the acute activation of the system by emotional and noxious stimuli, and the chronic hyperactivity during persistent stress situations, it is striking that the daily peak in adrenocortical activity coincides with the end of the sleep or rest period, as if the body has to be prepared again for action. This "arousal" function of the adrenal system (including the adrenal medullary outflow) has been recognized by several authors like Cannon (1939), Hess (1949), Selye (1952) and Gellhorn (1957). The critical point, however, in all theories has been that the physiological significance of the "need for corticoids" has not been clearly determined nor demonstrated.

7. The pituitary–adrenal stress response

It is a somewhat paradoxical situation that it is almost impossible to avoid the word "stress' when dealing with the functions of the pituitary–adrenal system, nevertheless "stress certainly is one of the most grandly imprecise terms in the lexicon of science" (Ganong, 1963). Coined by Selye (1936), it was meant to indicate a very basic and significant concept of interaction between the organism and its environment, but it has often been degraded to indicate any stimulus which activates the pituitary–adrenal system, an operational definition which actually is almost an offense to the founder of stress concept.

The development of the stress concept is in fact a quite interesting and important evolution in biological and medical thinking. It emerged from the ideas of Claude Bernard about the internal regulation of stability of the "milieu intérieur". This idea included the notion that the organism is capable of counteracting external stimuli which would threaten the maintenance of the internal balance.

Cannon introduced the concept of homeostasis as a closed system of processes which defend the dynamic equilibrium against environmental stimuli which tend to disturb the balance. He emphasized the role of the autonomic (sympathetic) system in the bodily defense mechanism. It was amply confirmed that adrenal medullary hormones are secreted as a reflex response to an endangering situation calling for either fight or flight ("emergency reaction"). Selye made his original discoveries in 1936 along completely different lines. When he found that a crude ovarian extract caused the same symptoms as any other toxic substance which he injected, the idea was born that the body reacts to any nonspecific noxious agent or tissue damage with the same syndrome: adrenal hypertrophy, atrophy of the thymus and gastric ulcers. It was soon found out that this syndrome was due to activation of adrenocortical secretion. His concept of the General Adaptation Syndrome included three stages, that of the alarm reaction, of resistance and eventually of exhaustion. The word "stress" was introduced to indicate the state of the organism when it has to respond to the noxious or threatening stimulus. The stimulus was then defined as the stressor agent which induces the stress response.

The notion that the adrenal cortex plays an important role in the stress response provided medical science with a comprehensive insight into the meaning of the manifold functions of the adrenocortical hormones, a new concept of adaptation, a common denominator for a group of non-specific harmful stimuli, and the basis for theories stretching far beyond this hormonal defense mechanism into the psycho-social realm ("stress and society", "everyday stress", etc.).

Although the impact of Selye's concept cannot be underestimated, it should be emphasized that the notion of Cannon's adrenomedullary emergency reaction was somewhat neglected. The adrenocortical response appeared to be only part of the stress reaction, and in fact represents the secondary reaction introducing the phase of restoration. The "alarm reaction" does not consist of secretion of cortical, but of medullary hormones, mediated by the sympathetic system. This results in arousal, muscular activity and a number of physiologic events (such as increased heart rate, glycogen breakdown, lipolysis) subserving the energy-demanding fight or flight reaction. The counterpart is the adrenocortical activation, starting a few minutes later, subserving a limitation of the consequences of the primary reflex by suppression of tissue reactions to damage and of the immune response to antigenic agents, of stimulation of protein breakdown and new formation of carbohydrates.

At the time it was not yet clear how a noxious stimulus would activate both the adrenal medulla and the adrenal cortex, and an elegant and unifying theory was advanced by Long. Starting from the observation that adrenaline is a powerful stimulus for ACTH secretion, he proposed that adrenaline released by the sympathetic reflex response to stress, induced ACTH secretion directly (Long, 1947; McDermott et al., 1950). However, it was soon found that removal of the adrenal medulla did not prevent the adrenocortical stress response (Gordon, 1950; Vogt, 1951).

The mechanism by which ACTH release is induced by stress, remained obscure for a long time. It was proposed by several groups that different kinds of stress stimuli may act via different pathways. Fortier (1951) was the first to offer a subdivision of stress stimuli, and proposed that "neural" stimuli like pain and fear would need an intact connection between pituitary gland and hypothalamus, whereas "systemic" stimuli like tissue damage or toxic agents would activate the pituitary directly (Fortier, 1952). Others followed this classification but deviated from Fortier's concept in that the pathways involved were different in their opinion. Mialhe-Voloss (1958) observed that neurogenic stress depletes ACTH occurring in the posterior lobe, whereas systemic stress depletes anterior lobe ACTH. This observation was confirmed by Rochefort et al. (1959), but not by Smelik (1960) and De Wied (1961a), although their work with posterior lobectomized rats showed that the neurogenic stress response is dependent on an intact posterior lobe. De Wied (1961a) demonstrated, however, that this response is restored if the posterior lobe hormone vasopressin is substituted by administration of Pitressin.

Work with hypothalamic lesions again suggested different pathways for neurogenic and for systemic stimuli. Smelik (1959) found that posterior hypothalamic lesions inhibited the adrenal response to painful and emotional stimuli, but enhanced the response to traumatic or systemic stimuli. More recent work confirmed that different pathways in the brain are involved in neurogenic and in systemic stress (Feldman et al., 1968; Makara et al., 1969a, 1971).

The growing conviction that the release of a hypothalamic CRF is the final common pathway for the pituitary ACTH secretion strengthened the opinion that all stress stimuli converge on the hypothalamic level, possibly with the CRF-producing neurons themselves as the integrating controller. These cells can be reached via the blood circulation by humoral factors or via different neural pathways by neurogenic stimuli.

The hypothalamic deafferentation technique, introduced by Halász and Pupp (1965), proved to be quite useful in demonstrating how stimuli can reach the hypothalamus. A number of stimuli, including humoral or metabolic factors, appeared to be capable of activating the system even after complete neural isolation of the hypothalamus. Among these stimuli are ether anaesthesia (Matsuda et al., 1964b; Halász et al., 1967a; Vermes et al., 1973), anoxia (Feldman et al.,

1970), immobilization (Palka *et al.*, 1969; Dunn and Critchlow, 1969), histamine, formalin, insulin and endotoxin (Makara *et al.*, 1970).

On the other hand, stimuli such as trauma (Greer and Rockie, 1968), sound and light (Feldman *et al.*, 1970, 1972), vibration, capsaicine injection and surgery (Makara *et al.*, 1970) or sciatic stimulation (Feldman *et al.*, 1975) all failed to provoke ACTH release in animals with hypothalamic deafferentation. On the basis of such results, a division in systemic and neurogenic stresses can be made (Stark *et al.*, 1970), although it should be noted that most stimuli may have both somatic or metabolic and neurogenic or emotional components (Smelik *et al.*, 1962).

In a series of experiments Greer and coworkers analysed the role of peripheral and central neural afferents in the mediation of traumatic stress. Damage to the hind leg did not stimulate ACTH release after spinal transection or interruption between the superior colliculus and the dorsal hypothalamic region (Matsuda *et al.*, 1964b). The neural pathway is contralateral in the spinal cord and up to the pons region, and then enters the hypothalamus with the medial forebrain bundle (Gibbs, 1969; Greer *et al.*, 1970).

Hind leg tourniquet appeared to activate the pituitary–adrenal system even in rats with complete hypothalamic deafferentation; nevertheless spinal cord transection blocked the response. This suggested that above the level of the thoracic segments the neural stimulus is converted into a humoral signal (Allen *et al.*, 1973, 1974).

Feldman and his colleagues studied extensively the neural pathways involved in the adrenal response to photic and acoustic stimuli. These stimuli appear to enter the hypothalamus from the lateral and posterior sides. Photic stimulation may reach the hypothalamus via the accessory optic tract, the medial forebrain bundle and the mammillary peduncle (Sarne and Feldman, 1971; Feldman *et al.*, 1972), whereas the propagation of acoustic stimuli would travel along the reticular formation and the medial forebrain bundle (Feldman *et al.*, 1971, 1972). Components of these stimuli may also reach the hypothalamus via other CNS structures, since interruption of one of these pathways never completely abolished the response. In similar studies, Makara *et al.* (1969b, 1970, 1972) also found that traumatic stimuli travel along the contralateral side of the spinal cord, then ascend in the region of the dorsal longitudinal fascicle and the medial forebrain bundle into the lateral hypothalamus.

The possibility cannot be excluded that certain stress stimuli can bypass the basal hypothalamus and stimulate the pituitary directly.

This was first suggested by Fortier (1951), and ever since reports have appeared which seem to confirm this possibility.

The *E. coli* endotoxin, when injected intraperitoneally, has been shown to stimulate ACTH secretion in rats bearing hypothalamic lesions or after pituitary stalk secretion (Makara *et al.*, 1971). It has no effect, however, on ACTH secretion from pituitaries *in vitro* nor on the adrenocortical secretion directly (Stark, 1972; Allen *et al.*, 1973). Another humoral mediator in the plasma of endotoxin-treated rats could not be demonstrated by Allen *et al.* (1973). The work of Brodish and co-workers about the existence of a "tissue CRF" has been referred to in an earlier section.

The original concept of the stress response as a result of any unspecific damage has been modified in several ways. In the first place, the majority of acute stress stimuli do not result in the classical triad of symptoms founded by Selye: adrenal hypertrophy, involution of the thymus and gastric ulcers. It is only the severe and chronic stress situation which causes this syndrome. In most studies on the effect of continuous stress such as immobilization, sound, cold, or water deprivation for several days, a biphasic adrenal response was found with an initial rise, a period of (sub)normal corticoid output generally between 4–12 hours after the onset, and a secondary rise for a longer period (Knigge *et al.*, 1959; Reck and Fortier, 1960; Bohus, 1969; Kawakami *et al.*, 1971; Sakellaris and Vernikoş-Danellis, 1974). The period of low corticoid levels can be prevented by posterior hypothalamic lesions (Knigge *et al.*, 1959; Feldman *et al.*, 1972), suggesting a neural suppressing mechanism. It may be, therefore, that Selye's syndrome rather represents maladaptation during chronic stress than successful adaptation. This may be relevant in view of permanent social and psychological stress conditions as seen in humans.

Moreover, when more sensitive and direct measurements of plasma corticoids levels became available, it appeared that even minor disturbances of the resting conditions like entrance of the experimenter in the animal room, sudden noise or the smell of ether vapour already activates the pituitary–adrenal system (Ader, 1969). In order to measure real basal values of corticoid levels it appeared to be necessary to take blood either quite rapidly from an undisturbed animal or during barbiturate anesthesia. The fact that the system responds promptly to minor emotional stimuli represents another challenge to the original stress concept. There appears to be a continuous scale of stimuli activating the pituitary–adrenal system, from very insignificant (but biologically meaningful) alerting signals to the full measure of massive tissue damage, intoxication or other somatic damage. This means that it is difficult, if even at all possible, to discern in every day life between stress and non-stress conditions. It would be preferable, therefore, to consider the pituitary–adrenal system as a servomechanism following closely the changing demands of the environment, rather than a homeostatic device subserving the constancy of the "milieu intérieur".

References

Abe, K. and Hiroshige, T. (1974). *Neuroendocrinology* **14,** 195–211.
Ader, R. (1969). *Ann. N.Y. Acad. Sci.* **159,** 791–805.
Allen, C F. and Kendall, J. W. (1967). *Endocrinology* **80,** 926–930.
Allen, J P. and Allen, C F. (1974). *Neuroendocrinology* **15,** 220–230.
Allen, J P., Allen, C. F. and Greer, M. A. (1974). *Neuroendocrinology* **13,** 246–254.
Allen, J P., Allen, C. F., Greer, M. A. and Jacobs, J. J. (1973). *In:* "Brain-Pituitary-Adrenal Interrelationships" (A. Brodish and E. S. Redgate, Eds), pp. 99–127, Karger Basel.
Anand, B. K. and Dua, S. (1955). *J. Physiol. Lond.* **127,** 153–156.
Anderson, E., Bates, R. W., Hawthorne, E., Haymaker, W., Knowlton, K., Rioch, D.McK., Spence, W. T. and Wilson, H. (1957). *Recent Prog. Horm. Res.* **13,** 21–66.
Arimura, A., Yamaguchi, T., Yoshimura, K., Imazeki, T. and Itoh, S. (1965). *Jap. J. Physiol.* **15,** 278–295.
Arimura, A., Bowers, C. Y., Schally, A. V., Saito, M. and Miller, M. C. (1969). *Endocrinology* **85,** 300–311.
Aschoff, J. (1973). *Aerospace Med.* **40,** 844–852.
Asfeldt, V. H. (1971). *Scand. J. clin. Lab. Invest.* **28,** 61–70.
Azmitia, E. F. and McEwen, B. S. (1969). *Science, N.Y.* **166,** 1274–1276.
Azmitia, E. F. and McEwen, B. S. (1974). *Brain Res.* **78,** 291–302.
Barchas, J. D. and Vernikos-Danellis, J. (1970). *In:* "Hormonal Steroids" (L. Martini, ed.) p 60. Excerpta Medica *I.C.S.,* **210,** Amsterdam.
Bennett, J. P., Mulder, A. H. and Snyder, S. H. (1974). *Life Sciences,* **15,** 1045–1056.
Berger, P. A., Barchas, J. D. and Vernikos–Danellis, J. (1974). *Nature, Lond.* **248,** 424–426.
Berson, S. A. and Yalow, R. S. (1968). *J. clin. Invest.* **47,** 2727–2751.
Bhargava, K. P., Bhargava, R. and Gupta, M. B. (1972). *Br. J. Pharmac.* **45,** 682–683.
Bhattacharya, A. N. and Marks, B. H. (1970). *Neuroendocrinology* **6,** 49–55.
Björklund, A., Falck, B., Hromek, F., Owman, Ch. and West, K. A. (1970). *Brain Res.* **17,** 1–23.
Björklund, A., Moore, R. Y., Nobin, A. and Stenevi, U. (1973). *Brain Res.* **51,** 171–191.
Björklund, A., Lindvall, O. and Nobin, A. (1975). *Brain Res.* **89,** 29–42.
Bliss, E. L., Aillon, J. and Zwanziger, J. (1968). *J. Pharmacol. exp. Ther.* **164,** 122–134.
Bock, R. (1966). *Histochemie* **6,** 362–369.
Bogdanove, E. M. (1963). *Endocrinology* **73,** 696–712.
Bohus, B. (1961). *Acta physiol. hung.* **20,** 373–377.

Bohus, B. (1969). *Acta physiol. hung.* **35**, 141–148.
Bohus, B. and Endröczi, E. (1964). *Acta physiol. hung.* **25**, 351–358.
Bohus, B., Nyakas, C. and Lissák, K. (1968). *Acta physiol. hung.* **34**, 1–8.
Bovard, E. W. and Gloor, P. (1961). *Experientia* **17**, 521–523.
Briggs, F. N. and Munson, P. L. (1955). *Endocrinology* **57**, 205–219.
Brodish, A. (1963). *Endocrinology* **73**, 727–735.
Brodish, A. (1964). *Endocrinology* **74**, 28–34.
Brodish, A. and Long, C. N. H. (1956). *Endocrinology* **59**, 666–676.
Brownstein, M. J., Saavedra, J. M., Palkovits, M. and Axelrod, J. (1974). *Brain Res.* **77**, 151–156.
Burden, J., Hillhouse, E. W. and Jones, M. T. (1974). *J. Physiol., Lond.* **239**, 116–117P.
Cannon, W. B. (1939). "The Wisdom of the Body". Norton New York.
Carlsson, A. and Lindquist, M. (1972). *J. Neural Transmission* **33**, 23–43.
Chan, L. T., Schaal, S. M. and Saffran, M. (1969). *Endocrinology* **85**, 644–651.
Cheifetz, P. N., Gaffud, N. T. and Dingman, J. F. (1968). *Endocrinology* **82**, 1117–1124.
Conroy, R. T. W. L., Hughes, B. D. and Mills, J. N. (1968). *Br. med. J.* **3**, 405–407.
Coover, G. D., Goldman, L. and Levine, S. (1971). *Physiol. Behav.* **7**, 727–732.
Corrodi, H., Fuxe, K. and Hökfelt, T. (1968). *Life Sciences* **7**, 107–112.
Critchlow, V. (1963). *In*: "Advances in Neuroendocrinology" (A. V. Nalbandov, Ed), pp. 377–402. Univ. of Illinois Press, Urbana Illinois.
Critchlow, V., Liebelt, R. A., Bar-Sela, M., Mountcastle, W. and Lipscomb, H. S. (1963). *Am. J. Physiol.* **205**, 807–815.
Csernus, V., Lengvári, I. and Halász, B. (1975). *Neuroendocrinology* **17**, 18–26.
Curtis, D. R. and Johnston, G. A. R. (1970). *In*: "Handbook of Neurochemistry" (A. Lajta, Ed), pp. 115–134. Plenum Press New York.
Dahlström, A. and Fuxe, K. (1964). *Acta physiol. scand.* **62**, Suppl. 232, 1–55.
Dahlström, A. and Fuxe, K. (1965). *Acta physiol. scand.* **64**, Suppl. 247, 1–36.
Daily, W. J. R. and Ganong, W. F. (1958). *Endocrinology* **62**, 442–454
Dallman, M. F. and Yates, F. E. (1968). *Mem. Soc. Endocr* **17**, 39–72.
Dallman, M. F. and Yates, F. E. (1969). *Ann. N.Y. Acad. Sci.* **156**, 696–721
Dallman, M. F., Jones, M. T., Vernikos-Danellis, J. and Ganong, W. F. (1972). *Endocrinology* **91**, 961–968.
Daly, J. R. and Elstein, M. (1972). *J. Obstet. Gynaec. Br. Common W.* **79**, 544–549
Daly, J. R. and Evans, J. I. (1974). *In*: "Advances in Steroid Biochemistry and Pharmacology" (M. H. Briggs and G. A. Christie, Eds), Vol. 4, pp. 61–110. Academic Press London and New York.
David-Nelson, M. A. and Brodish, A. (1969). *Endocrinology* **85**, 861–866.
Davidson, J. M. and Feldman, S. (1963) *Endocrinology* **72**, 936–946.
De Groot, J. and Harris, G. W. (1950). *J. Physiol. Lond.* **111**, 335–346.
Dekanski, J. (1952). *Br. J. Pharmac.* **7**, 567–572.
De Schaepdryver, A. F., Preziosi, P. and Scapagnini, U. (1969). *Arch. Int. Pharmacodyn. Thèr.* **180**, 11–18.
De Wied, D. (1961a). *Endocrinology* **68**, 956–970.
De Wied, D. (1961b). *Acta endocr. Copn.* **37**, 279–288.
De Wied, D. (1964). *J. Endocr.* **29**, 29–37.
De Wied, D., Bouman, P R. and Smelik, P. G. (1958). *Endocrinology* **62**, 605–613.
De Wied, D., Witter, A., Versteeg, D. H. G. and Mulder, A. H. (1969). *Endocrinology* **85**, 561–569.
De Wied, D., Bohus, B. and Van Wimersma Greidanus, Tj.B. (1976). *In*: "Cellular

and Molecular Bases of Neuroendocrine Processes" (E. Endröczi, Ed), Symp. Int. Soc. Psychoneuroendocrinol. Visegrád, pp. 547–553. Acad. Sci. Budapest.

Dhariwal, A. P. S., McCann, S. M., Taleisnik, S. and Tomatis, M. E. (1966). *Proc. Soc. exp. Biol. Med.* **121,** 996–998.

Dhariwal, A. P. S., Russell, S. M., McCann, S. M. and Yates, F. E. (1969). *Endocrinology* 84, 544–556.

Dixit, B. N. and Buckley, J. P. (1969). *Neuroendocrinology* **4,** 32–41.

Dordoni, F. and Fortier, C. (1950). *Proc. Soc. Soc. exp. Biol. Med.* **75,** 815–816.

Douglas, W. W. (1968). *Br. J. Pharmac.* **34** (3), 451–474.

Dunn, J. and Critchlow, V. (1969). *Brain Res.* **16,** 395–403.

Edwardson, J. A. and Bennett, G. W. (1974). *Nature, Lond.* **251,** 425–427.

Egdahl, R. H. (1960). *Acta endocr. Copnh. Suppl.* **51,** 49–50.

Egdahl, R. H. (1961). *Endocrinology* **68,** 574–581.

Egdahl, R. H. (1962). *Endocrinology* **71,** 926–935.

Endröczi, E. (1972). *In*: "Limbic System, Learning and Pituitary–Adrenal Function". Akadémiai Kiadó Budapest.

Endröczi, E. and Lissák, K. (1954). *Endokrinologie* **32,** 168–172.

Endröczi, E. and Lissák, K. (1960). *Acta physiol. hung.* **17,** 39–55.

Endröczi, E. and Lissák, K. (1962). *Acta physiol. hung.* **21,** 257–263.

Endröczi, E. and Nagy, D. (1951). *Acta physiol. hung.* **2,** 11–14.

Endröczi, E., Kovács, S. and Lissák, K. (1956). *Endokrinologie* **33,** 271–278.

Endröczi, E., Lissák, K., Bohus, B. and Kovács, S. (1959). *Acta physiol. hung.* **16,** 17–22.

Endröczi, E., Lissák, K. and Tekeres, M. (1961). *Acta physiol Hung.* **18,** 291–299.

Endröczi, E., Schreiberg, G. and Lissák, K. (1963). *Acta physiol hung.* **24,** 211–221.

Estep, H L., Mullinax, P. F., Brown, R., Blaylock, K. and Butts, E. (1967). *Endocrinology* **80,** 719–724.

Eto, S., Wood, J. M., Hutchins, M. and Fleischer, N. (1974). *Am. J. Physiol.* **226,** 1315–1320.

Feldman, S., Conforti, N., Chowers, I. and Davidson, J M. (1968). *Israel J. med. Sci.* **4,** 908–910.

Feldman, S., Conforti, N., Chowers, I. and Davidson, J. M. (1970). *Acta endocr. Copnh.* **63,** 405–414.

Feldman, S., Conforti, N. and Chowers, I. (1972). *J. endocr.* **51,** 745–749.

Feldman, S., Conforti, N., Chowers, I. and Davidson, J. M. (1973). *Neuroendocrinology* **10,** 316–323.

Feldman, S., Conforti, N., Chowers, I. and Davidson, J. M. (1975). *Acta endocrinol. Copnh.* **78,** 539–544.

Fendler, K., Karmos, G. and Telegdy, G. (1961). *Acta physiol. hung.* **20,** 293–297.

Fink, G., Smith, J. R. and Tibbals, J. (1971). *Nature, London.* **230,** 467–468.

Fink, G., Smith, G. C., Tibballs, J. and Lee, V. W. K. (1972). *Nature, Lond.* **239,** 57–59.

Fleischer, N., Donald, R. A. and Butcher, R. W. (1969). *Am. J. Physiol.* **217,** 1287–1291.

Fortier, C. (1951). *Endocrinology* **49,** 782–788.

Fortier, C. (1952). *In*: "Anterior pituitary secretion and hormonal influences in water metabolism" (G. E. W. Wolstenholme and M. P. Cameron, eds). Ciba Fdn. Colloq. Endocr., **4,** 124–136.

Franks, R. C. (1967). *J. clin. Endocrinol.* **27,** 75–78.

Friedman, A. H. and Walker, C. A. (1968). *J. Physiol. Lond.* **197**, 77–85.

Fuxe, K. (1964). *Z. Zellforsch. mikrosk. Anat.* **61**, 710–724.

Fuxe, K. (1965). *Acta physiol. scand.* **64**, Suppl. 247, 39–85.

Fuxe, K. and Hökfelt, T. (1966). *Acta physiol. scand.* **66**, 243–244.

Fuxe, K. and Hökfelt, T. (1969). *In*: "Frontiers in Neuroendocrinology" (W. F. Ganong and L. Martini, Eds), pp. 47–96. Oxford University Press New York.

Fuxe, K. and Hökfelt, T. (1970). *In*: "The Hypothalamus" (L. Martini, M. Motta and F. Fraschini, Eds), pp. 123–138. Academic Press, New York and London.

Fuxe, K., Corrodi, H., Hökfelt, T. and Johnsson, G. (1970). *Prog. Brain Res.* **32**, 42–56.

Fuxe, K., Hökfelt, T. and Johnsson, G. (1973). *In*: "Brain–Pituitary–Adrenal Interrelationships" (A. Brodish and E. S. Redgate, Eds), pp. 239–269. Karger Basel.

Fuxe, K., Schubert, J. Hökfelt, T. and Johnsson, G. (1974). *Adv. Biochem. Psychopharmacol.* **10**, 67–74.

Ganong, W. F. (1963). *In*: "Advances in Neuroendocrinology" (A. V. Nalbandov, Ed), pp. 92–149. Urbana: Univ. of Illinois Press.

Ganong, W. F. (1970). *In* "The Hypothalamus" (L. Martini, M. Motta and F. Fraschini, Eds), pp. 313–333. Academic Press New York and London.

Ganong, W. F. (1972). *Progr. Brain Res.* **38**, 41–54.

Ganong, W. F. and Goldfien, A. (1959). Program 41st Meeting, Endocrine Soc. Atlantic City, N. J., pp. 29.

Ganong, W. F. and Hume, D. M. (1954). *Endocrinology* **55**, 474–483.

Garbarg, M., Krishnamoorthy, M. S., Feeger, J. and Schwartz, J. C. (1973). *Brain Res.* **50**, 361–367.

Gellhorn, E. (1957). "Autonomic Imbalance and the Hypothalamus". University of Minneapolis Press, Minneapolis.

Gibbs, F. P. (1969). *Am. J. Physiol.* **217**, 84–88.

Gordon, M. L. (1950). *Endocrinology* **47**, 13–18.

Green, J P. (1970). *In* "Handbook of Neurochemistry" (A. Lajta, Ed), Vol. 4, pp. 221–250. Plenum Press New York.

Greer, M. A. and Rockie, C. (1968). *Endocrinology* **83**, 1247–1252.

Greer, M. A., Allen, C. F., Gibbs, F. P. and Gullickson, C. (1970). *Endocrinology* **86**, 1404–1409.

Guillemin, R. (1955). *Endocrinology* **56**, 248–255.

Guillemin, R. (1957). *Endokrinologie* **34**, 193–201.

Guillemin, R. (1964). *Recent Prog. Horm. Res.* **20**, 89–121.

Guillemin, R. and Hearn, W. R. (1955). *Proc. Soc. exp. Biol. Med.* **89**, 365–367.

Guillemin, R. and Rosenberg, B. (1955). *Endocrinology* **57**, 599–607.

Guillemin, R., Hearn, W. R., Cheek, W. R. and Housholder, D. E. (1956). *Fedn. Proc. Fedn Am. Socs exp. Biol.* **15**, 268.

Guillemin, R., Hearn, W. R., Cheek, W. R. and Housholder, D. E. (1957). *Endocrinology* **60**, 488–506.

Guillemin, R., Dear, W. E., Nichols, B. and Lipscomb, H. S. (1959). *Proc. Soc. exp. Biol.* **101**, 107–111.

Halász, B. and Pupp, L. (1965). *Endocrinology* **77**, 553–562.

Halász, B. and Szentágothai, J. (1959). *Z. Zellforsch, mikrosk. Anat.* **50**, 297–306.

Halász, B., Pupp, L. and Uhlarik, S. (1962). *J. Endocr.* **25**, 147–154.

Halász, B., Pupp, L., Uhlarik, S. and Tima, L. (1965). *Endocrinology* **77**, 343–355.

Halász, B., Slusher, M. A. and Gorski, R. A. (1967a). *Neuroendocrinology* **2**, 43–55.

Halász, B., Vernikos-Danellis, J. and Gorski, R. A. (1967b). *Endocrinology* **81,** 921–924.
Halberg, F. (1953). *The Lancet* **73,** 20–28.
Halberg, F. (1959). *Z. Vitamin. Hormon. Fermentforsch.* **10,** 235–296.
Halberg, F. (1969). *A. Rev. Physiol.* **31,** 675–725.
Halberg, F. and Haus, E. (1960). *Am. J. Physiol.* **199,** 859–862.
Halberg, F. and Visscher, M. B. (1950). *Proc. Soc. exp. Biol. Med.* **75,** 846–847.
Hall, R. E. and Marr, H. B. (1975). *Brain Res.* **93,** 367–371.
Harris, G. W. (1955). "Neural Control of the Pituitary Gland." Edward Arnold, London.
Haus, E. and Halberg, F. (1970). *Environmental Res.* **3,** 81–106.
Haus, E., Lakatha, D. and Halberg, F. (1967). *Exp. Med. Surg.* **25,** 7–45.
Hedge, G. A. (1971). *Endocrinology* **89,** 500–506.
Hedge, G. A. and Smelik, P. G. (1968). *Science, N.Y.* **159,** 891–892.
Hedge, G. A. and Smelik, P. G. (1969). *Neuroendocrinology* **4,** 242–253.
Hedge, G. A., Yates, M. B., Marcus, R. and Yates, F. E. (1966). *Endocrinology* **79,** 328–340.
Héry, F., Rouer, E. and Glowinski, J. (1972). *Brain Res.* **43,** 445–465.
Hess, W. R. (1949). *In* "Das Zwischenhirn: Syndrome, Lokalisationen, Funktionen". Schwabe Basel.
Hillhouse, E. W., Burden, J. and Jones, M. T. (1975). *Neuroendocrinology* **17,** 1–11.
Hinsey, J. C. and Markee, J. E. (1933). *Proc. Soc. exp. Biol. Med.* **31,** 270–271.
Hiroshige, T. and Abe, K. (1973). *In* "Neuroendocrine Control" (K. Yagi and S. Yoshida, Eds), pp. 205–228. Univ. Tokyo Press, Tokyo.
Hiroshige, T. and Sakakura, M. (1971). *Neuroendocrinology* **7,** 25–36.
Hiroshige, T. and Sato, T. (1970). *Endocrinology* **86,** 1184–1186.
Hokin, M. A., Hokin, L. E., Saffran, M., Schally, A. V. and Zimmermann, B. U. (1958). *J. biol.Chem.* **233,** 811–813.
Hume, D. M. and Nelson, D. H. (1957). *Proceedings of the Endocrine Society, U.S.A.* **39,** 98–99.
Hume, D. M. and Wittenstein, G. J. (1950). *In* "Proc. 1st Clin. ACTH Conf." (K. R. Mote, Ed), pp. 134–146. Blakiston, Philadelphia.
Ishii, S., Iwata, T. and Kobayashi, H. (1969). *Endocr. jap.* **16,** 171–177.
Itoh, S. (1964). *Gunma Symposia on Endocrinol.* **1,** 143–158.
Jacobowitz, D. M. and Marks, B. H. (1964). *Endocrinology* **75,** 86–88.
Jacobowitz, D. M. and Palkovits, M. (1974). *J. comp. Neurology* **157,** 13–28.
Johnson, J. T. and Levine, S. (1973). *Neuroendocrinology* **11,** 268–273.
Jones, M. T. and Tiptaft, E. W. (1977). *Br. J. Pharmac.* **59,** 35.
Jones, M. T., Brush, F. R. and Neame, R. L. B. (1972). *J. Endocr.* **55,** 489.
Jones, M. T., Hillhouse, E. W. and Burden, J. (1976a). *J. Endocr.* **69,** 1–11.
Jones, M. T., Hillhouse, E. W. and Burden, J. (1976b). *J. Endocr.* **69,** 34P.
Jones, M. T., Hillhouse, E. W. and Burden, J. L. (1977). *J. Endocr.* **73,** 405–417.
Kappers, J. A. (1960). *Z. Zellforsch. mikrosk. Anat.* **52,** 163–215.
Kaplanski, J. and Smelik, P. G. (1973a). *Acta Endocr Copnh.* **73,** 651–659.
Kaplanski, J. and Smelik, P. G. (1973b). *Res. Comm. Chem. Path. Pharm.* **5,** 263–271.
Kaplanski, J., Dorst, W. and Smelik, P. G. (1972). *Eur. J. Pharmac.* **20,** 238–240.
Kaplanski, J., Nyakas, C., Van Delft, A. M. L. and Smelik, P. G. (1974). *Neuroendocrinology* **13,** 123–127.
Kawakami, M., Sato, K., Terasawa, E., Yoshida, K., Miyamoto, T., Sekiguchi, M. and Hattori, Y. (1968). *Neuroendocrinology* **3,** 337–348.

Kawakami, M., Kimura, F., Ishida, S. and Yanase, M. (1971). *Endocr. jap.* **18**, 469–476.
Kendall, J. W. and Allen, C. (1968). *Endocrinology* **82**, 397–405.
Kitay, J. I. (1963). *Endocrinology* **73**, 253–260.
Knigge, K. M. (1961). *Proc. Soc. exp. Biol. Med.* **108**, 18–21.
Knigge, K. M. and Hays, M. (1963). *Proc. Soc. exp. Biol. Med.* **114**, 67–69.
Knigge, K. M., Penrod, C. H. and Schindler, W. J. (1959). *Am. J. Physiol.* **196**, 579–582.
Kobayashi, H. and Matsui, T. (1969). In "Frontiers in Neuroendocrinology" (W. F. Ganong and L. Martini, Eds), pp. 1–46. Oxford Univ. Press, New York.
Kraicer, J., Milligan, J. V., Gosbee, J. L., Conrad, R. G. and Branson, C. M. (1969). *Endocrinology* **85**, 1144–1153.
Krieger, D. T. and Krieger, H. P. (1964) In "Proceedings of the Second International Congress of Endocrinology" (S. Taylor, ed) 640–645. Excerpta Medica *I.C.S.* **83**, Amsterdam.
Krieger, D. T., Silverberg, A. I., Rizzo, F. A. and Krieger, H. P. (1968). *Am. J. Physiol.* **215**, 959–967.
Krieger, D. T., Kreuzer, J. and Rizzo, F. A. (1969). *J. clin. Endocr. Metab.* **29**, 1634–1638.
Krieger, D. T. (1973). *Endocrinology* **93**, 1077–1091.
Krieger, D. T. (1974a). *Neuroendocrinology* **16**, 355–363.
Krieger, D. T. (1974b). *Endocrinology* **95**, 1195–1201.
Krieger, D. T. (1975). *Neuroendocrinology* **17**, 62–74.
Krieger, D. T. and Krieger, H. P. (1970). *Am. J. Physiol.* **218**, 1632–1641.
Krieger, D. T. and Rizzo, F. A. (1969). *Am. J. Physiol.* **217**, 1703–1707.
Krieger, D. T. and Rizzo, F. A. (1971). *Neuroendocrinology* **8**, 165–179.
Krieger, D. T., Allen, W., Rizzo, F. A. and Krieger, H. P. (1971). *J. clin. Endocr. Metab.* **32**, 266–284.
Krieger, D. T., Liotta, A. and Brownstein, M. J. (1977). *Endocrinology* **100**, 227–237.
Krieger, H. P., Kolodny, H. and Krieger, D. T. (1964). *Fedn Proc. Fedn Am. Socs exp. Biol.*, **23**, 205.
Kuhar, M. J. (1973). *Life Sciences* **13**, 1623–1634.
Lang, R. E., Voigt, K. H., Fehm, H. L. and Pfeiffer, E. F. (1976). *Neuroscience Letters* **2**, 19–22.
Lanier, L. P., Van Hartesveldt, C., Weis, B. J. and Isaacson, R. L. (1975). *Neuroendocrinology* **18**, 154–160.
Laqueur, G. L., McCann, S. M., Schreiner, L. H., Rosemberg, E., Rioch, D.McK. and Anderson, E. (1955). *Endocrinology* **57**, 44–54.
Leeman, S. E., Glenister, D. W. and Yates, F. E. (1962). *Endocrinology* **70**, 249–262.
Lengvári, I. and Halász, B. (1973). *Neuroendocrinology* **11**, 191–196.
Lissák, L. and Endröczi, E. (1965). "The Neuroendocrine Control of Adaptation" Pergamon Press Oxford.
Long, C. N. H. (1947). *Bull. N.Y. Acad. Med.* **23**, 260–282.
Long, C. N. H. and Fry, E. G. (1945). *Proc. Soc. exp. Biol. Med.* **59**, 67–69.
Lowry, P. J. (1974). *J. Endocr.* **62**, 163–164.
Lutz, B., Koch, B., Briaud, B. and Mialhe-Voloss, C. (1975). *Exp. Brain Res.* **23**, 131.
Lymangrover, J. R. and Brodish, A. (1973). *Neuroendocrinology* **12**, 225–235.
Lymangrover, J. R. and Brodish, A. (1974). *Neuroendocrinology* **13**, 234–245.
Makara, G. B., Stark, E. and Mihály, K. (1969a). *Acta physiol. hung.* **35**, 331–333.
Makara, G. B. and Stark, E. (1974). *Neuroendocrinology* **16**, 178–190.

Makara, G. B. and Stark, E. (1975). *Neuroendocrinology* **18,** 213–216.

Makara, G. B. and Stark, E. (1976). *Neuroendocrinology* **21,** 31.

Makara, G. B., Stark, E., Palkovitis, M., Révész, T. and Mihály, K. (1969b). *J. Endocr.* **44,** 187–193.

Makara, G. B., Stark, E. and Mihály, K. (1970). *Acta physiol. hung.* **38,** 199–203.

Makara, G. B., Stark, E. and Mészáros, T. (1971). *Endocrinology* **88,** 412–414.

Makara, G. B., Stark, E., Marton, J. and Mészáros, T. (1972). *J. Endocr.* **53,** 389–395.

Mandell, A. J., Chapman, L. F., Rand, R. W. and Walter, R. D. (1963). *Science, N.Y.* **139,** 1212.

Mandell, A. J., Chaffey, B., Brill, P., Mandell, M. P., Rodnick, J., Rubin, R. T. and Sheff, R. (1966). *Science, N.Y.* **151,** 1558–1560.

Martini, L., Pecile, A., Saito, S. and Tani, F. (1960). *Endocrinology* **66,** 501–507.

Mason, J. W. (1958). *In* "Reticular Formation of the Brain" (H. H. Jasper, Ed), pp. 645–662. Little Brown Boston Massachusetts.

Mason, J. W., Harwood, C. T. and Rosenthal, N. R. (1957). *Am. J. Physiol.* **190,** 429–433.

Matheson, G. K., Branch, B. J. and Taylor, A. N. (1971). *Brain Res.* **32,** 151–167.

Matsuda, K., Duyck, C. and Greer, M. A. (1964a). *Endocrinology* **74,** 939–943.

Matsuda, K., Duyck, C., Kendall, J. W. Jr. and Greer, M. A. (1964b). *Endocrinology* **74,** 981–985.

McCann, S. M. (1953). *Am. J. Physiol.* **175,** 13–20.

McCann, S. M. and Brobeck, J. R. (1954). *Proc. Soc. exp. Biol. Med.* **87,** 318–324.

McCann, S. M. and Haberland, P. (1959). *Proc. Soc. exp. Biol.* **102,** 319–325.

McCann, S. M., Fruit, A. and Fulford, B. D. (1958). *Endocrinology* **63,** 29–42.

McCann, S. M., Antunes-Rodriques, J., Nallar, R. and Valtin, H. (1966). *Endocrinology* **79,** 1058–1064.

McDermott, W. V., Fry, E. G., Brobeck, J. R. and Long, C. N. H. (1950). *Yale J. Biol. Med.* **23,** 52–66.

McEwen, B. S., Weiss, J. M. and Schwartz, L. S. (1969). *Brain Res.* **16,** 227–241.

McEwen, B. S., Zigmond, R. E. and Gerlach, J. L. (1972). *In* "Structure and Function of the Nervous System", vol. 5, pp. 205–291. Academic Press, New York and London.

McKinney, W. T. Jr., Prange, A. J., Majchowicz, Jr. E. and Schlesinger, K. (1971). *Dis. Nerv. Syst.* **32,** 308–313.

Mess, B., Fraschini, F., Motta, M. and Martini, L. (1966). *In* "Hormonal Steroids" (L. Martini, F. Fraschini and M. Motta, eds) pp. 1004–1013. Excerpta Medica *I.C.S.* **132,** Amsterdam.

Mess, B., Zanisi, M. and Tima, L. (1970). *In* "The Hypothalamus" (L. Martini, M. Motta and F. Fraschini, Eds), pp. 259–277. Academic Press, New York and London.

Mialhe–Voloss, C. (1958). *Acta endocr. Copnh.* Suppl. **35,** 1–96.

Migeon, C. V., French, A. B., Samuels, L. T. and Bowers, J. Z. (1955). *Am. J. Physiol.* **182,** 462–468.

Milkovic, K. and Milkovic, M. (1968). *In* "Neuroendocrinology" (L. Martini and W. F. Ganong, Eds), Vol. 1, 371–405. Academic Press, New York and London.

Millard, S. A., Costa, E. and Gal, E. M. (1972). *Brain Res.* **40,** 545–551.

Miller, R. E. (1973). *In* "Brain–Pituitary–Adrenal Interrelationships" (A. Brodish and E. S. Redgate, Eds), pp. 328–331. Karger Basel.

Milligan, J. V. and Kraicer, J. (1971). *Endocrinology* **89,** 766–773.

Milligan, J. V. and Kraicer, J. (1974). *Endocrinology* **94,** 435–443.
Mirsky, I. A., Stein, M. and Paulisch, G. (1954). *Endocrinology* **55,** 28–39.
Moberg, G. P., Scapagnini, U., De Groot, J. and Ganong, W. F. (1971). *Neuroendocrinology* **7,** 11–15.
Montgomery, R. L. and Berkut, M. K. (1969). *Physiol. Behav.* **4,** 745–748.
Moore, R. Y. and Eichler, V. B. (1972). *Brain Res.* **42,** 201–206.
Morgan, W. W., Rudeen, P. K. and Pfeil, K. A. (1975). *Life Sciences* **17,** 143–150.
Morimoto, Y., Oishi, T., Arisue, K., Ogawa, Z., Tanake, F., Yano, S. and Yamamura, Y. (1975). *Acta endocr. Copnh.* **80,** 527–541.
Motta, M., Fraschini, F., Piva, F. and Martini, L. (1968). *Mem. Soc. Endocrinol.* **17,** 3–18.
Motta, M., Schaffini, O., Piva, F. and Martini, L. (1971). *In* "The Pineal Gland" (G. E. W. Wolstenholme and J. Knight, Eds), pp. 279–301. J. and A. Churchill, London.
Mulder, A. H., Geuze, J. J. and De Wied. (1970). *Endocrinology* **87,** 61–79.
Mulder, G. H. (1975). "Release of ACTH by Rat Pituitary Cells". Ph.D. Thesis, Free University Amsterdam.
Mulder, G. H. and Smelik, P. G. (1977). *Endocrinology* **100,** 1143–1152.
Mulder, G. H., Vermes, I. and Smelik, P. G. (1976). *Neuroscience Letters* **2,** 73–78.
Nagareda, C. S. and Gaunt, R. (1951). *Endocrinology* **48,** 560–567.
Nakadate, B. M. and De Groot, J. (1963). *Anat. Rec.* **145,** 338.
Naumenko, E. V. (1968). *Brain Res.* **11,** 1–10.
Naumenko, E. V. (1969). *Neuroendocrinology* **5,** 81–88.
Nauta, W. J. H. (1963). *In* "Neuroendocrinology" (A. V. Nalbandov, Ed), pp. 5–21. Univ. of Illinois Press Urbana Illinois.
Neckers, L. and Sze, P. Y. (1975). *Brain Res.* **93,** 123–132.
Nichols, B. and Guillemin, R. (1959). *Endocrinology* **64,** 914–919.
Nikitovitch-Winer, M. and Everett, J. W. (1958). *Endocrinology* **63,** 916–930.
Nikitovitch-Winer, M. and Everett, J. W. (1959). *Endocrinology* **65,** 357–368.
Okada, Y., Nitsch–Hassler, C., Kim, J. S., Bak., I. J. and Hassler, R. (1971). *Exp. Brain Res.* **13,** 514–518.
Okinaka, S., Ibayashi, H., Motohashi, K., Fujita, T., Yoshida, S., Ohsawa, N. and Marakawa, S. (1960a). *Acta endoc. Copnh.* **35,** Suppl. 51, 43–44.
Okinaka, S., Ibayashi, H., Motohashi, K., Fujita, T., Yoshida, S. and Ohsawa, N. (1960b). *Endocrinology* **67,** 319–324.
Orth, D. N. and Island, D. P. (1969). *J. clin. Endocr. Metab.* **29,** 479–486.
Orth, D. N., Island, D. P. and Liddle, G. W. (1967). *J. clin, Endocr. Metab.* **27,** 549–555.
Otsuka, K. (1966). *Tohoku J. exp. Med.* **88,** 165–168.
Palka, Y., Coyer, D. and Critchlow, V. (1969). *Neuroendocrinology* **5,** 333–349.
Palkovits, M. and Mitro, A. (1968). *Neuroendocrinology* **3,** 200–210.
Palkovits, M. and Stark, E. (1972). *Neuroendocrinology* **10,** 23–30.
Palkovits, M., Brownstein, M., Saavedra, J. M. and Axelrod, J. (1974a). *Brain Res.* **77,** 137–149.
Palkovits, M., Saavedra, J. M. and Brownstein, M. (1974b). *Brain Res.* **80,** 237–249.
Palkovits, M., Kobayshi, R. M., Kizer, J. S., Jacobowitz, D. M. and Kopin, I. J. (1975). *Neuroendocrinology* **18,** 144–153.
Papaikonomou, E. and Smelik, P. G. (1974). *Am. J. Physiol.* **227,** 137–143.
Pearlmutter, A. F., Rapino, E. and Saffran, M. (1974). *Neuroendocrinology,* **15,** 106–109.

Perkoff, G. T., Eik–Nes, K., Nugent, C. A., Fred, H. L., Nimer, R. A., Rush, L., Samuels, L. T. and Tyler, F. H. (1959). *J. clin. Endocr. Metab.* **19,** 432–443.

Pincus, G. (1943). *J. clin. Endocr. Metab.* **3,** 195–199.

Portanova, R. and Sayers, G. (1973). *Proc. Soc. exp. Biol. Med.* **143,** 661–666.

Portanova, R. and Sayers, G. (1974). *Biochem. biophys. Res. Commun.* **56,** 928–933.

Porter, J. C. and Jones, J. C. (1956). *Endocrinology* **58,** 62–67.

Porter, J. C. and Rumsfeld, H. W. (1959). *Endocrinology* **64,** 948–954.

Porter, J. C., Goldman, B. G. and Wilber, J. F. (1969). *In* "Hypophysiotropic Hormones of the Hypothalamus: Assay and Chemistry" (J. Meites, Ed), pp. 282–292. The Williams and Wilkins Co., Baltimore.

Porter, R. W. (1953). *Am. J. Physiol.* **172,** 515–519.

Porter, R. W. (1954). *Recent Prog. Horm. Res.* **10,** 1–27.

Preziosi, P., Scapagnini, U. and Nistico, G. (1968). *Biochem. Pharmacol.* **17,** 1309–1313.

Quay, W. B. (1965). *Life Sciences* **4,** 379–384.

Raisman, G. (1970). *In* "The Hypothalamus" (L. Martini, M. Motta and F. Fraschini, Eds), pp. 1–17. Academic Press, New York and London.

Raisman, G. (1975). *Exp. Brain Res.* **23,** 169.

Reck, D. C. and Fortier, C. (1960). *Proc. Soc. exp. Biol. Med.* **104,** 610–613.

Redgate, E. S. (1970). *Endocrinology* **86,** 806–823.

Reis, D. J., Weinbrein, M. and Corvelli, A. (1968). *J. Pharmacol. exp. Ther.* **164,** 135–145.

Retiene, K. (1970). *In* "The Hypothalamus" (L. Martini, M. Motta and F. Fraschini, Eds), pp. 551–568. Academic Press, New York and London.

Retiene, K., Espinoza, A., Marx, K. H. and Pfeiffer, E. F. (1965). *Klin. Wschr.* **43,** 205–207.

Réthelyi, M. (1975). *Neuroendocrinology* **17,** 330–339.

Rinne, U. K. (1960). *Acta endocr. Copnh.* **35,** Suppl. 57, 1–108.

Rinne, U. K. (1972). *In* "Brain–Endocrine Interaction. Median eminence: Structure and Function" (K. M. Knigge, D. E. Scott and A. Weindl, Eds), pp. 164–171. Karger Basel.

Rochefort, G. J., Rosenberger, J. and Saffran, M. (1959). *J. Physiol. Lond.* **146,** 105–116.

Rosecrans, J. A. (1968). *Fedn Proc. Fedn Am. Socs exp. Biol.* **27,** 540.

Routtenberg, A. (1968). *Psychol. Rev.* **75,** 51–80.

Royce, P. C. and Sayers, G. (1958). *Endocrinology* **63,** 794–800.

Royce, P. C. and Sayers, G. (1960). *Proc. Soc. exp. Biol. Med.* **103,** 447–450.

Rubin, R. T., Mandell, A. and Crandall, P. H. (1966). *Science, N.Y.* **153,** 767–768.

Rumsfeld, H. W. and Porter, J. C. (1962). *Endocrinology* **70,** 62–67.

Russell, S. M., Dhariwal, A. P. S., McCann, S. M. and Yates, F. E. (1969). *Endocrinology* **85,** 512–521.

Saavedra, J. M., Palkovits, M., Brownstein, M. and Axelrod, J. (1974a). *Brain Res.* **77,** 157–165.

Saavedra, J. M., Brownstein, M. and Palkovits, M. (1974b). *Brain Res.* **79,** 437–441.

Saba, P., Carnicelli, A., Saba, G. C., Maltinti, G. and Marescotti, V. (1965). *Acta endocr. Copnh.* **49,** 289–292.

Saffran, M. (1959). *Can. J. Biochem. Physiol.* **37,** 319–331.

Saffran, M. (1974). *In* "Handbook of Physiology" (Geiger, S. R., Ed), Sect. 7 (vol. 4) part 2, pp. 563–586. Waverly Press Baltimore.

Saffran, M. and Schally, A. V. (1955). *Can. J. Biochem. Physiol.* **33,** 408–415.

Saffran, M., Schally, A. V. and Benfey, B. G. (1955). *Endocrinology* **57**, 439–444.
Saffran, M., Pearlmutter, A. F., Rapino E. and Upton, G. V. (1972). *Biochem. biophys. Res. Commun.* **49**, 748–751.
Sakakura, M., Saito, Y., Takabe, K. and Ishii, K. (1976). *Endocrinology* **98**, 954–957.
Sakellaris, P. C. and Vernikos-Danellis, J. (1974). *Physiol. Behav.* **12**, 1067–1070.
Sarne, Y. and Feldman, S. (1971). *Electroenceph. clin. Neurophysiol.* **30**, 45–51.
Sato, T., Sato, M., Shinsako, J. and Dallman, M. F. (1975). *Endocrinology* **97**, 265–274.
Sayers, G. and Sayers, M. A. (1948). *Recent Prog. Horm. Res.* **2**, 81–106.
Scapagnini, U. and Preziosi, P. (1972). *Arch. int. Pharmacodyn. Thèr. Suppl.* **196**, 205–219.
Scapagnini, U. and Preziosi, P. (1973). *Neuropharmacology* **12**, 57–62.
Scapagnini, U., Van Loon, G. R., Moberg, G. P. and Ganong, W. F. (1970). *Eur. J. Pharmac.* **11**, 266–270.
Scapagnini, U., Moberg, G. P., Van Loon, G. R., De Groot, J. and Ganong, W. F. (1971). *Neuroendocrinology* **7**, 80–96.
Schally, A. V. and Guillemin, R. (1963). *Proc. Soc. exp. Biol. Med.* **112**, 1014–1017.
Schally, A. V., Saffran, M. and Zimmerman, B. (1958). *Biochem. J.* **70**, 97–103.
Schally, A. V., Andersen, R. N., Lipscomb, H. S., Long, J. M. and Guillemin, R. (1960). *Nature, Lond.* **188**, 1192–1193.
Schally, A. V., Lipscomb, H. S., Long, J. M., Dear, W. E. and Guillemin, R. (1962). *Endocrinology* **70**, 478–480.
Schally, A. V., Muller, E. E., Arimura, A., Bowers, C. Y., Saito, T., Redding, T. W., Sawano, S. and Pizzolato, P. (1967). *J. clin. Endocr. Metab.* **27**, 755–762.
Schapiro, S. (1962). *Endocrinology* **71**, 986–989.
Schiaffini, O., Motta, M., Piva, F. and Martini, L. (1970). In "Hormonal Steroids" (V. H. T. James and L. Martini, Eds), pp. 822–829. Excerpta Medica, *I.C.S.* **219**, Amsterdam.
Schwartz, J. C., Julien, C., Feger, I. and Garbarg, M. (1974). *Fedn Proc. Fedn Am. Socs exp. Biol.* **33**, 285.
Seggie, J. (1968). *J. comp. physiol. Psychol.* **66**, 820–822.
Seggie, J. and Brown, G. M. (1971). *Neuroendocrinology* **8**, 367–394.
Seggie, J., Shaw, B., Uhlir, I. and Brown, G. M. (1974). *Neuroendocrinology* **15**, 51–61.
Seiden, G. and Brodish, A. (1971). *Neuroendocrinology* **8**, 154–164.
Selye, H. (1936). *Br. J. exp. Path.* **17**, 234–248.
Selye, H. (1952). "The Story of the Adaptation Syndrome". Acta Inc. Med. Publ., Montreal, Canada.
Shute, C. C. D. (1970). In "The Hypothalamus" (L. Martini, M. Motta and F. Fraschini, Eds), pp. 167–179. Academic Press New York and London.
Shute, C. C. D. and Lewis, P. R. (1966). *Br. med. Bull.* **22**, 221–226.
Shute, C. C. D. and Lewis, P. R. (1967). *Brain* **90**, 497–520.
Silverberg, A., Rizzo, F. and Krieger, D. T. (1968). *J. clin. Endocr. Metab.* **25**, 1661–1663.
Smelik, P. G. (1959). "Autonomic Nervous Involvement in Stress-induced ACTH Secretion", 80 pp. N.V. Drukkerij/Born–Assen, Netherlands.
Smelik, P. G. (1960). *Acta endocr. Copnh.* **33**, 437–443.
Smelik, P. G. (1963a). *Acta endocr. Copnh.* **44**, 36–46.
Smelik, P. G. (1963b). *Proc. Soc. exp. Biol. Med.* **113**, 616–619.
Smelik, P. G. (1965). *Acta physiol. pharmac. néerl.* **13**, 370–371.
Smelik, P. G. (1967). *Neuroendocrinology* **2**, 247–254.

Smelik, P. G. (1969). *Neuroendocrinology* **5**, 193–204.
Smelik, P. G. and Sawyer, C. H. (1962). *Acta endocr. Copnh.* **41**, 561–570.
Smelik, P. G., Garaenstroom, J. H., Konijnendijk, W. and De Wied D. (1962). *Acta physiol. pharmac. néerl.* **11**, 20–33.
Smelik, P. G., Mulder, G. H. and Vermes, I. (1976). *In* "Cellular and Molecular Bases of Neuroendocrine Processes" (E. Endröczi, Ed). Symposium of International Society for Psychoneuroendocrinology. Visegrád, pp. 423–430. *Acad. Sci.* Budapest.
Sollberger, A. (1969). *Expl. Med. Surg.* **27**, 80–104.
Stark, E. (1972). *Acta med. hung.* **29**, 77–88.
Stark, E., Makara, G. B., Palkovits, M. and Mihály, K. (1970). *Acta physiol. hung.* **38**, 43–49.
Suzuki, T., Hirai, K., Yoshio, H., Kurovji, K. I. and Hirose, T. (1964). *J. Endocr.* **31**, 81–84.
Takabe, K., Sakakura, M., Horiuchi, Y. and Mashimo, K. (1971). *Endocr. jap.* **18**, 451–455.
Taylor, K. M., Gfeller, E. and Snyder, S. H. (1972). *Brain Res.* **41**, 171–179.
Telegdy, G. and Vermes, I. (1975). *Neuroendocrinology* **18**, 16–26.
Thierry, A. M., Javory, F., Glowinski, J. and Kety, S. S. (1968). *J. Pharmac. exp. Ther.* **163**, 163–171.
Uhlir, I., Seggie, J. and Brown, G. M. (1974). *Neuroendocrinology* **14**, 351–355.
Ulrich, R. S. and Yuwiler, A. (1973). *Endocrinology* **92**, 611–614.
Ungar, F. (1964). *Ann. N.Y. Acad. Sci.* **117**, 374–395.
Uotila, U. U. (1939). *Anat. Rec.* **76**, 183–189.
Upton, G. V. and Corbin, A. (1975). *Experientia* **31**, 249–250.
Usher, D. R., Kasper, P. and Birmingham, M. K. (1967). *Neuroendocrinology* **2**, 158–174.
Van Delft, A. M. L., Kaplanski, J. and Smelik, P. G. (1973). *J. Endocr.* **59**, 465–474.
Van Loon, G. R., Scapagnini, U., Cohen, R. and Ganong, W. F. (1971a). *Neuroendocrinology* **8**, 257–272.
Van Loon, G. R., Scapagnini, U., Moberg, G. P. and Ganong, W. F. (1971b). *Endocrinology* **89**, 1464–1469.
Van Loon, G. R. (1973). *In* "Frontiers in Neuroendocrinology" (W. F. Ganong and L. Martini, eds) pp. 209–247. Oxford University Press, New York.
Vermes, I. and Telegdy, G. (1972). *Acta physiol. hung.* **42**, 49–59.
Vermes, I. and Telegdy, G. (1973a). *Acta physiol. hung.* **43**, 105–114.
Vermes, I. and Telegdy, G. (1973b). *Acta physiol. hung.* **43**, 99–103.
Vermes, I. and Telegdy, G. (1975). *Expl. Brain Res.* **23**, 208.
Vermes, I., Molnár, D. and Telegdy, G. (1973). *Acta physiol. hung.* **43**, 239–245.
Vermes, I., Molnár, D., Dull, G. and Telegdy, G. (1974). *Acta physiol. hung.* **45**, 53–62.
Vermes, I., Mulder, G. H. and Smelik, P. G. (1977a). *Endocrinology* **100**, 1153–1159.
Vermes, I., Smelik, P. G. and Mulder, A. H. (1977b). *Life Sci.* **19**, 1719–1726.
Verney, E. B. (1947). *Proc. Roy. Soc. B.* **135**, 25–105.
Vernikos-Danellis, J. (1963). *Endocrinology* **72**, 574–581.
Vernikos-Danellis, J. (1964). *Endocrinology* **75**, 514–520.
Vernikos-Danellis, J. (1965). *Endocrinology* **76**, 122–126.
Vernikos-Danellis, J. and Harris, C. G. (1968). *Proc. Soc. exp. Biol. Med.* **128**, 1016–1019.
Vernikos-Danellis, J., Berger, P. and Barchas, J. D. (1973). *Prog. Brain Res.* **39**, 201–209.

Versteeg, D. H. G., Van der Gugten, J., de Jong, W. and Palkovits, M. (1976). *Brain Res.* **113,** 563–574.

Vogt, M. (1951). *J. Physiol. Lond.* **114,** 465–470.

Vogt, M. (1954). *J. Physiol. Lond.* **131,** 125–136.

Von Euler, U. S. (1946). *Acta physiol. scand.* **12,** 73–93.

Wegmann, H. M., Bruner, H., Jovy, D., Klein, K. E., Marberger, J. P. and Rimpler, A. (1970). *Aerospace Med.* **41,** 1003–1005.

Weiner, R. I., Shryne, J., Gorsky, R. and Sawyer, C. H. (1972). *Endocrinology* **90,** 867–873.

Weitzman, E. D. (1974). *In* "Chronobiological Aspects of Endocrinology" (Symposia Medica Hoechst), 9, pp. 169–184. F. K. Schattauer Verlag Stuttgart–N.Y.

Weitzman, E. D., Goldmacher, D., Kripke, D., MacGregor, P., Kream, J. and Hellman, L. (1968). *Trans Am. neurol. Assoc.* **93,** 153–157.

Weitzman, E. D., De Graaf, A. S., Sassin, J. F., Hansen, T., Godtlibsen, O. B., Perlow, M. and Hellman, L. (1975). *Acta endocr. Copnh.* **78,** 65–76.

Wiley, M. K., Pearlmutter, A. F. and Miller, R. E. (1974). *Neuroendocrinology* **14,** 257–270.

Wilson, M. and Critchlow, V. (1973). *Neuroendocrinology* **13,** 29–40.

Witorsch, R. J. and Brodish, A. (1972). *Endocrinology* **90,** 1160–1167.

Wittkowski, W. and Bock, R. (1972). *In* "Brain–Endocrine Interaction. Median Eminence: Structure and Function" (K. M. Knigge, D. E. Scott and A. Weindl, Eds), pp. 171–181. Karger Basel.

Woods, J. W. (1956) *Nature, Lond.* **178,** 869.

Wurtman, R. J. (1966). *In* "Neuroendocrinology" (L. Martini and W. F. Ganong, Eds), Vol. 2, 19–59. Academic Press New York and London.

Wurtman, R. J. (1970). *In* "The Hypothalamus" (L. Martini, M. Motta and F. Fraschini, Eds), pp. 153–166. Academic Press New York and London.

Yates, F. E., Leeman, S. E., Glenister, D. W. and Dallman, M. F. (1961). *Endocrinology* **69,** 67–80.

Yates, F. E., Russell, S. M., Dallman, M. F., Hedge, G. A., McCann, S. M. and Dhariwal, A. P. S. (1971). *Endocrinology* **88,** 3–15.

Yates, F. E. and Maran, J. W. (1974). *In* "Chronobiological Aspects of Endocrinology" (Symposia Medica Hoechst) 9, pp. 351–377. Schattauer Verlag Stuttgart–New York.

Yates, F. E. and Maran, J. W. (1975). *In* "Handbook of Physiology" (S. R. Geiger, Ed), Sect. 7, vol. 4, part 2, pp. 367–404. Waverly Press Baltimore.

Zimmerman, G. and Fleischer, N. (1970). *Endocrinology* **87,** 426–429.

2. Clinical Disorders Involving Adrenocortical Insufficiency and Overactivity

D. R. Cullen, J. P. D. Reckless† and E. H. McLaren*

The Royal Hallamshire Hospital, Sheffield, S10 2JF, England

* Consultant Physician, Stobhill General Hospital, Glasgow, Scotland.
† Consultant Physician, Royal United Hospital, Bath, England.

PART 1 ADRENOCORTICAL INSUFFICIENCY

1. Introduction

Thomas Addison (1795–1860) with wide enquiring clinical interests presented a paper, largely concerned with anaemia and entitled "On Anaemia: Disease of the Suprarenal Capsules", to the South London Medical Society in 1849. Much encouraged by S. Wilks, a fellow Senior Physician at Guy's Hospital, he published, in 1855, a monograph: "On the Constitutional and Local Effects of Disease of the Supra-Renal Capsules". Thereby he produced a classical book not only describing for the first time a disease due to defects of a ductless gland (see Introduction, Volume 1), which was later named Addison's Disease but also presenting its general characteristics in descriptive terms, little bettered today. He wrote "the leading and characteristic features of the morbid state to which I would direct attention are anaemia, general langour and debility, remarkable feebleness of the heart's action, irritability of the stomach and a peculiar change of colour in the skin, occurring in connection with a diseased condition of the supra-renal capsules". He thus described the classical features of primary adrenocortical insufficiency or Addison's Disease, namely: the slow onset, the asthenia, the gastro-intestinal symptoms, the hypotension, weight loss and pigmentation.

Adrenocortical insufficiency may be classified as (i) due to a "primary" failure giving hyposecretion from both adrenal cortices arising from their disease and (ii) "secondary" failure, consequent on reduced adenohypophysial adrenocorticotrophin (ACTH) secretion. The former condition is the major topic of this chapter though the second is considered. Another form of adrenal insufficiency, namely congenital adrenal hyperplasia, arising from genetic deficiency of hydroxylating enzymes, has been discussed by Sandor *et al.* (1976, Vol. 1), Idelman (1978, Vol. 2) and Kime *et al.* (this volume).

2. Aetiology and pathogenesis

A. PRIMARY ADRENOCORTICAL INSUFFICIENCY (ADDISON'S DISEASE)

At the beginning of this century, bilateral destruction of the adrenal glands by tuberculous infection was the main cause of Addison's disease, being found in 70–80% of autopsied cases (Conybeare and Millis, 1924; Barker, 1929; Guttman, 1930 a,b). Only a small proportion of patients were found to have idiopathic Addison's disease, that is adrenocortical atrophy. In a review of 566 cases, 17% were idiopathic and 70% were tuberculous (Guttman, 1930 a,b). These proportions have changed with the effective control of active tuberculosis so that the incidence of the idiopathic type predominates. For instance Stuart-Mason *et al.*, (1968), in a population survey in the London area, showed that 69% of cases had non-tuberculous disease of the adrenal glands whilst Nerup (1974) reported 66% of his cases to have idiopathic Addison's Disease with only 17% being classified as tuberculous in origin. Recently Irvine and Barnes (1972) in an international series of 224 patients with Addison's Disease reported 78% to be idiopathic and 21% to be tuberculous. However it is to be noted that the total number of cases has lessened, which correlates with the control of tuberculosis and resulting decreased mortality (Dunlop, 1963).

Idiopathic Addison's disease is associated with the presence of antibodies to the adrenal cortex in about 50% of this group. They are present in about 80% of females and 10% of males (Irvine *et al.*, 1967). Most patients with non-tuberculous Addison's disease may be considered to have autoimmune adrenalitis. Antibody titres are usually low and they are not related to the duration of the disease. In two large series (Irvine and Barnes, 1972; Nerup, 1974) adrenal antibodies were not found in any patients with unequivocal tuberculous Addison's disease so the presence of such antibodies would help to confirm the diagnosis of autoimmune adrenalitis, with the proviso that their absence does not positively confirm the tubercular type. Histologically, in autoimmune adrenalitis, the gland has increased adrenocortical fibrous tissue and lymphocyte filtration together with loss of glandular cells, all reminiscent of the thyroid gland in Hashimoto's disease. In comparison, tuberculous disease affects both the adrenal medulla and the cortex.

Very rarely, the patient with Addison's disease may have developed the disorder from destruction of the adrenal glands by infiltration from malignancy, amyloidosis, granulomas, fungal disease, haemochromatosis, Hodgkin's disease or from infarction secondary to atherosclerosis, periarteritis nodosa or systemic lupus erythematosis. Haemorrhage may be caused by anticoagulants, toxaemia of pregnancy or by meningococcal septicaemia (Rickards and Barrett, 1954; Soffer *et al.*, 1961; Symington, 1969; Irvine and Barnes, 1972).

B.   SECONDARY ADRENOCORTICAL INSUFFICIENCY

Disease of the hypothalamus or pituitary gland with consequent inadequate ACTH secretion leads to secondary adrenocortical failure. Lessened ACTH production is normally accompanied by diminished secretion of other adenohypophysial trophic hormones, which usually fail before ACTH. Thus chromophobe adenomas, craniopharyngiomas and postpartum haemorrhage initially lead to gonadotrophic and then somatotrophic loss before ACTH and TSH are affected. Rarely is ACTH secretion solely lost in an hypothalamic-pituitary disorder though it has been reported, perhaps secondary to an hypothalamic lesion (Odell *et al.*, 1960).

The most common impairment of ACTH secretion, singled out from the other anterior lobe hormones, occurs secondarily to hypothalamic-pituitary suppression by exogenous corticosteroids. During therapeutic corticosteroid administration, low ACTH secretory levels lead to atrophy of the adrenal cortex. This may be relatively easily reversed by exogenous ACTH but nevertheless, depression of the hypothalamic-pituitary axis may be much slower to recover after stopping corticosteroid therapy. Such patients remain susceptible to acute failure during periods of stress. Exogenous steroids are usually administered orally but some risk is associated with topical skin application, bronchial aerosols or with steroid enemata.

C.   INCIDENCE

Addison's disease is rare. An incidence of approximately one patient in 3,000–6,000 hospital admissions has been found (Rowntree and Snell, 1931; Bloodworth *et al.*, 1954; Soffer *et al.*, 1961; Ask-Upmark and Hull, 1972). The prevalence rate, in the age range 25–70 years in the London area, is estimated at 39 per million (Stuart Mason *et al.*, 1968) whilst for Danes of all age groups a figure of 60 per million is given (Nerup, 1974). The point has been made in Section 2A above that the number of cases of idiopathic Addison's disease (or autoimmune adrenalitis) has increased greatly from about 25% given in early studies (Conybeare and Millis, 1924; Barker, 1929; Guttman, 1930 a,b; Rowntree and Snell, 1931) to about 65–75% in later surveys

(Stuart Mason *et al.*, 1968; Maisey and Lessof, 1969; Irvine and Barnes, 1972; Nerup, 1974). However, these percentages are of a smaller number of cases as the mortality from tuberculous Addison's disease has sharply decreased (Guttman, 1930 a,b; Dunlop, 1963), together with little absolute change in the occurrence of autoimmune cases. The majority of cases of Addison's disease is found in the third to fifth decades of life but autoimmune cases may occur at all ages whilst those from tuberculosis are rare before 40 years of age (Thorn *et al.*, 1942; Maisey and Lessof, 1969; Irvine and Barnes, 1972; Nerup, 1974). Tuberculous adrenal disease is more common in males but the majority of patients with autoimmune adrenalitis are female (60–80% in surveys; Guttman, 1930 a,b; Nerup, 1974).

3.   Clinical symptoms and signs

A.   CHRONIC ADRENOCORTICAL INSUFFICIENCY

The original description by Addison (1855) has not been greatly altered over the subsequent century and does not differ materially between tuberculous and other causes. Weakness, increasing fatigue and weight loss are present in almost all patients. The symptoms develop insidiously and progressively. The weight loss may not be significant until the disease is well advanced. Gastrointestinal symptoms are common, the patient losing appetite and often complaining of nausea. Diarrhoea is frequent and may alternate with constipation. These symptoms vary in severity and may be such as to lead to a diagnosis of gastroenteritis, particularly in adrenal crisis.

While pigmentation may occur in many disease conditions, it is the most frequent sign that leads to the consideration of the diagnosis of Addison's disease (Fig. 1). Recently occurring pigmentation is obviously more suggestive of hypoadrenalism than longstanding changes. Pigmentation is seen most clearly in areas exposed to light such as the face, neck, elbow and knee joints, and the extensor aspects of the hands, or on pressure areas under belts or girdles. Pigmentation occurs in the skin creases of the palms of the hands, on the areola of the nipple, and also in newer but not old scars. Mucous membranes (conjunctival, buccal and vaginal) may also pigment (Fig. 2). The pigmentation results from increased β-melanocyte stimulating hormone secretion (βMSH) concurrent with increased ACTH release from the pituitary (Abe *et al.*, 1969). In some patients, pigmentation precedes the appearance of other signs of Addison's disease by a number of years. Associated with hyperpigmentation is the increased

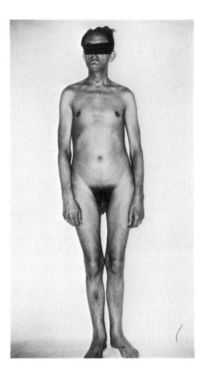

Fig. 1. A typical patient with Addison's disease.

occurrence of vitiligo, which occurs in approximately 15% of patients (Thorn, 1951). Vitiligo shows with increased frequency in other diseases with organ-specific autoimmune features (Cunliffe *et al.*, 1968).

Over 90% of patients with Addison's disease have hypotension in both supine and standing positions. It is uncommon to find systolic blood pressure above 110 mm Hg. There may also be an added postural hypotensive element, with failure of peripheral vasoconstriction and inadequate maintenance of blood pressure, leading to symptoms of dizziness and syncope. Electrolyte disturbance is common but not always present. The changes are those of hyponatraemia, hyperkalaemia and moderate uraemia. Occasionally hypercalcaemia may also occur. The loss of water associated with longstanding sodium loss may contribute to hypotension. Hypoglycaemic symptoms may be present particularly as a reactive feature after carbohydrate feeding, resulting from the effects of lack of glucocorticoids which limits gluconeogenesis and reduce insulin antagonism.

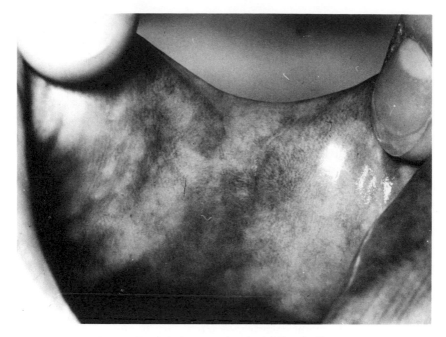

Fig. 2. Buccal pigmentation in Addison's disease.

Symptoms and signs referable to the aetiology of the disorder are often present. There may be evidence of tuberculosis, malignancy or other disease. In patients with autoimmune adrenalitis there may be associated failure of other glands. In female subjects, amenorrhoea may be present early in the disorder and there may be some loss of body hair since this is to a certain extent dependent on adrenal androgen secretion. Schmidt (1926) described thyroiditis in patients with idiopathic Addison's disease, and subsequent studies (Carpenter *et al.*, 1964; Irvine and Barnes, 1972; McHardy-Young *et al.*, 1972) have confirmed the association of autoimmune adrenal failure with Grave's disease, Hashimoto's thyroiditis and primary hypothyroidism. There are also associations between adrenal failure and diabetes mellitus (Beaven *et al.*, 1959; Irvine and Barnes, 1975), and particularly in young patients with idiopathic hypoparathyroidism (Irvine and Barnes, 1972).

In the original description by Addison, anaemia was a prominent feature of the disease. It would appear that Addison unwittingly recognized the now well established association between pernicious anaemia and adrenal failure. For instance Irvine and Barnes (1972) in a group of 174 patients with Addison's disease found ten patients with pernicious anaemia with malabsorption of vitamin B12, and lack of secretion of intrinsic factor, with the presence of gastric parietal cell, and intrinsic factor antibodies in the serum.

B.   ACUTE ADRENOCORTICAL INSUFFICIENCY

Acute adrenal failure is a medical emergency with a history of onset varying from hours to some days. The patient complains usually of vomiting, diarrhoea, nausea and extreme muscular weakness. Hypotension is present and is eventually followed by coma and death. Addison's disease may first present with acute failure of the gland, but most cases of acute adrenocortical failure occur in patients on steroid replacement therapy who have forgotten their medication, or who have not increased the dosage at times of stress such as intercurrent infection.

C.   SECONDARY ADRENOCORTICAL INSUFFICIENCY

The symptoms of secondary adrenocortical insufficiency are similar to those of Addison's disease with anorexia, lassitude and perhaps a past history of prostration following a minor illness. Symptoms of a pituitary tumour such as headache or visual field loss often leading to total loss of vision in one or both eyes may also be present. Physical examination may be normal and patients do not show the pigmentation characteristic of Addison's disease since ACTH secretion is defective. Signs of panhypopituitarism such as loss of body hair, testicular atrophy and pale skin would usually, however, be present. Hypotension may be found but is not as common as in primary hypoadrenocorticism because of the preservation of adrenal mineralocorticoid secretion. ACTH deficiency can also occur in the presence of functioning pituitary tumours so that the patient may appear acromegalic or have galactorrhoea.

4.   Diagnosis

The usual presenting symptoms of Addison's disease, fatigue and anorexia, are very common amongst the general population. The combination of these symptoms, however, with hypotension and pigmentation makes it more likely that the patient is suffering from

Addison's disease but because of the need for life-long replacement therapy, biochemical confirmation of the diagnosis should be made in every case. This can be after initiating therapy if the clinical condition warrants immediate treatment.

The diagnosis of Addison's disease is confirmed by the demonstration of inadequate secretion of corticosteroids, both basally and after stimulation with ACTH. It is then necessary to see whether this impairment is due to primary disease of the adrenal glands or is secondary to failure of pituitary ACTH secretion or atrophy following steroid therapy. Finally an attempt should be made to establish the aetiology of the disease in each particular case.

The following tests are available for the investigation of Addison's disease:

1. Basal levels of corticosteroids in blood and urine.
2. Basal levels of ACTH in blood.
3. Stimulation tests.
4. Tests of the peripheral action of corticosteroids.
5. Other non-endocrine investigations.

1. Basal levels of corticosteroids

Low plasma cortisol levels with loss of diurnal variation may be seen in severe adrenal insufficiency. Levels within the low normal range showing some diurnal variation are seen however in incomplete failure (Nabarro and Brook, 1975), but adequate responses to stress are not present. The cortisol secretion rate is normally reduced. Urinary oxogenic steroids and oxosteroids may be within or below the normal range. Basal measurements of plasma urinary corticosteroids are not, therefore, of much value in the diagnosis.

2. Basal levels of ACTH

Plasma ACTH levels as measured by either bioassay (Liddle et al., 1962) or radioimmunoassay (Besser et al., 1971) are elevated in Addison's disease (Fig. 3). In spite of the elevated levels a diurnal variation is still present (Besser et al., 1971). When measured at 0900 hours levels of 320–1200 ng/1 were found in patients with primary adrenal failure whilst patients with hypopituitarism or pituitary/ adrenal suppression secondary to steroid therapy showed levels from zero to the upper limit of normal (70 ng/1; Besser et al., 1971).

The combination of high plasma ACTH and low plasma cortisol

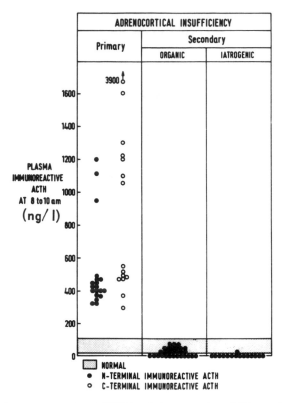

Fig. 3. Plasma immunoreactive ACTH levels between 8 and 10 a.m. in patients with adrenocortical insufficiency (from Besser *et al.*, 1971).

levels is particularly suggestive of primary adrenal failure, and measurement of these two parameters might replace the more conventional stimulation tests in the future. However, ACTH is very unstable in whole blood, and is destroyed by proteolytic enzymes in serum or plasma, before or after freezing or after thawing of the sample. Samples must, therefore, be taken in plastic syringes, kept frozen in plastic tubes, because ACTH is absorbed onto glass. In addition ACTH assays are complex and time consuming. Stimulation tests are therefore normally undertaken.

3.   *Stimulation tests*

Stimulation tests are used to test the reserve function of the adrenal glands. This will demonstrate impaired secretory function in those patients with Addison's disease, in whom basal levels may be normal. They also help to differentiate between primary and secondary failure of the adrenal glands. In these tests the synthetic preparation of the N-terminal 24 amino acids of ACTH, tetracosactrin (synacthen), is used, because of the lower incidence of allergic reactions with this preparation, together with its predictable and well established stimulatory effect on the adrenal cortex.

(a) Short synacthen test, or synacthen screening test (Wood *et al.*, 1965; Greig *et al.*, 1968). The plasma cortisol is measured in the resting, non-fasting patient before, and at thirty and sixty minutes after, the intramuscular injection of 250 μg of tetracosactrin (Fig. 4).

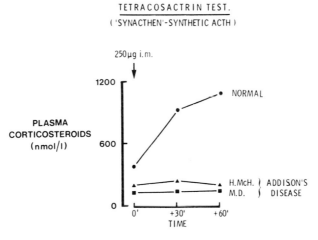

Fig. 4. The short tetracosactrin test (synacthen, CIBA; synthetic ACTH) in a normal subject and patients with Addison's disease.

The test is suitable for out-patient use, and is preferably performed at about 0900 hours. It may, however, be carried out later in the day. The initial level of plasma cortisol at 0900 hours should be between 160 and 660 nmol/l. With a normal adrenal response, plasma cortisol should rise by at least 200 nmol/l, and the peak value should be more than 550 nmol/l. All these criteria should be filled before the response is regarded as normal. Because of its simplicity this test is invaluable as a screening procedure. A normal result excludes adrenal insufficiency but as some patients with apparently normal adrenals give a subnormal response it is necessary to confirm the diagnosis with more prolonged ACTH stimulation. In addition the short synacthen test does not distinguish between primary and secondary adrenal insufficiency.

(b) Five hour depot synacthen test. Plasma cortisol levels are measured before and 5 h after the intramuscular injection of 1 mg of synacthen depot, which is tetracosactrin absorbed onto a zinc phosphate complex. It provides maximal stimulation of the adrenal cortex. In normal subjects the plasma cortisol level should double in one hour, and then rise more slowly to more than 1000 nmol/l at 5 h (Fig. 5). An absent or impaired cortisol response is seen in primary adrenocortical failure.

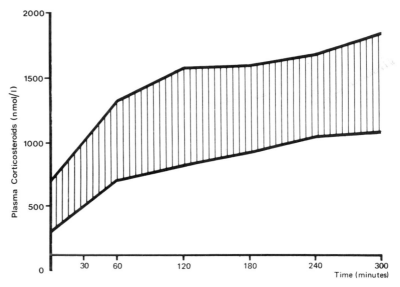

Fig. 5. Plasma corticosteroid response to a five hour synacthen test in normal subjects. Courtesy of Ciba Laboratories Ltd., Horsham, Sussex. "Hypothalamic–Pituitary–Adrenal Function Tests" by V. H. T. James and J. Landon.

(c) Long synacthen test. In long standing secondary adrenocortical insufficiency with atrophy of the adrenal cortices, an impaired response to the 5 h synacthen test may occur. 1 mg of tetracosactrin is therefore given intramuscularly at 1000 hours on each of three days. An incremental response in the plasma cortisol levels on each successive day will be seen in secondary adrenocortical insufficiency, but an absent response will still be seen in primary adrenocortical insufficiency.

In Europe the synacthen tests have largely superseded ACTH infusion but these are still used in the U.S.A. In the standard infusion test 40 units of ACTH are given by intravenous infusion over 8 h and the response is measured by the change in plasma or urinary corticosteroids (Jenkins *et al.*, 1955). For more prolonged stimulation the infusion is repeated on successive days, or a prolonged infusion over 48 h may also be used (Rose *et al.*, 1970).

It is necessary for practical as well as academic reasons to differentiate primary from secondary adrenocortical failure since differences in treatment are involved.

(d) Insulin-induced hypoglycaemia. In a patient in whom secondary adrenocortical insufficiency is suspected, and for whom ACTH radioimmunoassays are not readily available (see below), insulin-induced hypoglycaemia should be carried out. The normal response to hypoglycaemic stress is increased secretion of ACTH and therefore an increase in production of plasma cortisol. An adrenal gland normally responsive to ACTH *must* first be demonstrated as plasma cortisol is measured and the test is in effect a bioassay *in vivo*. The patient with hypopituitarism is sensitive to insulin and should not receive more than 0.1 units soluble insulin per kg body weight. The test should not be performed in subjects with myocardial ischaemia or epilepsy.

The patient is fasted overnight, and an intravenous line is established at least 30 minutes prior to the start of the test at 0900 hours. A medical practitioner must be present throughout the test and dextrose (25 or 50%) for intravenous use must be at hand. Blood for plasma cortisol and blood sugar is taken prior to intravenous administration of insulin, and at 30, 45, 60 and 90 minutes after insulin. The patient should sweat during the test (usually for about 10 minutes), and the blood sugar should fall either to below 2.2 mmol/l or to less than 50% of the fasting level, whichever is the lower. If sweating does not occur by 45 minutes the dose of insulin may be repeated and blood sampling is recommenced. Adequate hypoglycaemia may then be produced without the need to repeat the test with a higher insulin dose on a subsequent day. In normal subjects, plasma cortisol levels should rise by at least 166 nmol/l to a maximum of at least 550 nmol/l. An absence of a rise in the level of plasma cortisol in response to adequate hypoglycaemia is seen in adrenocortical insufficiency (Fig. 6). If hypoglycaemia is inadequate a lack of plasma cortisol rise is uninterpretable, though a normal steroid response in these circumstances is valid.

Rarely, the test has to be terminated early if hypoglycaemia is excessive. If sweating is profuse and continues for more than 20 minutes, or if loss of consciousness occurs or is impending, 25 ml of

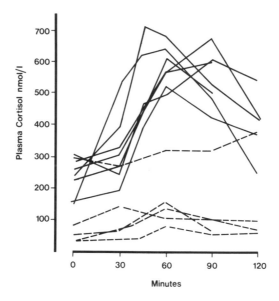

Fig. 6. Plasma corticosteroid response to insulin hypoglycaemia in five patients with secondary adrenocortical insufficiency (– – – –) compared to seven normal subjects (——).

25% dextrose should be given intravenously. Then the sampling should be continued as stress will have occurred. We have found it useful to estimate blood sugars at the bedside using a reflectance meter to ensure adequate hypoglycaemia has ensued.

(e) Metyrapone. Metyrapone influences the hypothalamic pituit-ary–adrenal axis and may be used as a test. The drug at 750 mg orally 4 hourly for 24 h is used to block the enzyme 11β-hydroxylase in the adrenal cortex. This is a final enzyme in cortisol synthesis and is discussed in Volume 1 (Sandor et al., 1976). Negative feedback on the hypothalamic–pituitary axis is reduced, and if the axis is normal ACTH secretion increases and a normal adrenal will produce further 11-deoxycortisol. Thus plasma fluorogenic corticosteroids fall, and urinary 17 oxogenic steroids rise during metyrapone administration and for the next 24 h. While insulin–hypoglycaemia tests the path-ways from CNS stress to the adrenal gland, metyrapone tests the hypothalamic–adrenal portion. In adrenocortical insufficiency however, the tests are effectively similar, and insulin hypoglycaemia is usually used. Metyrapone should not be used in primary adrenal failure, since it is likely to reduce to zero whatever endogenous cortisol secretion there is with catastrophic results.

(f) Lysine vasopressin (Gwinup, 1965). This hormone may be used to distinguish hypothalamic causes of secondary adrenal failure from those of pituitary origin. With hypothalamic lesions, or those involving the hypophysial portal system, the cortisol response to hypoglycaemic stress will be absent, but lysine vasopressin which stimulates the anterior pituitary will give a normal cortisol response. Lysine vasopressin is rarely used, as it is normally only required to establish if replacement treatment with cortisol is needed and this can be done with little discomfort to the patient using insulin hypoglycaemia. Lysine vasopressin causes vaso-constriction and should not be used in anyone with suspected myocardial ischaemia.

4.   *Tests involving the peripheral action of corticosteroids*

These tests are no longer used as there are now specific assays for corticosteroids. However, abnormal electrolyte concentrations in plasma may, in the first place, indicate the diagnosis.

(a) Plasma electrolytes. Due to aldosterone deficiency, patients with Addison's disease often show hyponatraemia, hyperkalaemia and mild uraemia but electrolyte concentrations may be normal (Fig. 7). In the series reported by Nerup (1974) 88% had a serum sodium

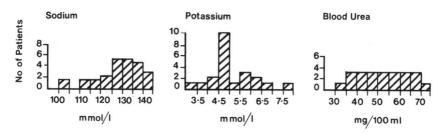

Fig. 7. Plasma electrolyte and blood urea concentrations in patients with Addison's disease (from Nabarro and Brook, 1975).

of less than 136 mmol/l, and 64% had a serum potassium greater than 4.9 mmol/l.

(b) Water excretion test. Patients with adrenocortical hormone deficiency are unable to excrete a water load as rapidly as normal. This used to be used as a test of adrenal function but is now obsolete.

5. *Other non-endocrine studies*

Evidence of active tuberculosis or a past history of extensive or disseminated disease would favour tuberculosis as the cause of adrenal failure, but it must be remembered that radiological evidence of old tuberculosis on the chest X-ray is common in the general population. Radiological calcification of the adrenals indicates a tuberculous aetiology but calcification may also be absent in this condition (Irvine and Barnes, 1972) (Fig. 8). As already discussed,

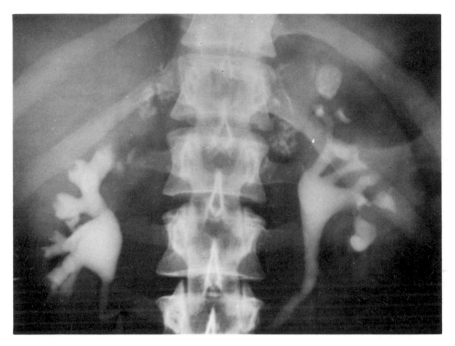

Fig. 8. An intravenous pyelogram showing adrenal calcification in a patient with tuberculous Addison's disease. Note the calcification in the upper and lower poles of the left kidney due to renal tuberculosis.

Addison's disease may rarely be caused by a number of other diseases which should be borne in mind in an atypical case (Nerup, 1974), and appropriate investigations undertaken. If autoimmune adrenalitis is diagnosed it is worthwhile performing screening tests to diagnose the more common of the diseases associated with this condition such as secretory failure of the thyroid, stomach, parathyroids and ovaries. In a group of patients studied by Irvine and Barnes (1972) 16% had thyroid disease, 6% pernicious anaemia, 10% diabetes mellitus and 7% had hypoparathyroidism (Fig. 9). Abnormal gonadal function was seen in 18%. Where antibodies to thyroid cytoplasm and

Disorder	Idiopathic	Tuberculous	Others
Thyroid			
Toxic	9 ⎫		
Hypothyroid	10 ⎪ 27	–	–
Hashimoto	4 ⎬		
Simple goitre	4 ⎭		
Pernicious anaemia	10	–	–
Diabetes mellitus	18	–	–
Hypoparathyroidism	12	–	–
Abnormal gonadal function	31	1	–
Vitiligo	16	1	–
No. of patients affected	83	2	–
Total no. of patients	174	46	4

Fig. 9. The principal clinical disorders associated with Addison's disease (from Irvine and Barnes, 1972).

thyroglobulin, to gastric parietal cells, to intrinsic factor and to gonadal cells are present in the absence of overt or subclinical disease, particular care should be taken in the follow up of these patients in case such diseases subsequently develop.

5. Treatment

A. PRIMARY ADRENOCORTICAL FAILURE

In primary adrenal insufficiency life-long glucocorticoid replacement therapy is mandatory to restore good health and to prevent the development of an Addisonian crisis. The vast majority of patients also require mineralocorticoid replacement but a few manage without this.

If the patient is markedly hypotensive, dehydrated or vomiting, parenteral therapy should be given at first as described in the treatment of Addisonian crisis. The majority of patients however, can be started on oral therapy in the first instance. Glucocorticoid replacement is given as cortisol or cortisone acetate in starting doses of 20 mg at 0800 hours and 10 mg at 1800 hours for cortisol and 25 mg and 12.5 mg for cortisone acetate. Cortisol is preferred because unlike cortisone acetate it does not have to be metabolized in the liver but this theoretical advantage is probably of little practical importance (Kehlet *et al.*, 1976). The correct maintenance dose should be determined for each individual patient by performing a steroid profile, since the levels achieved with standard doses vary widely in individual patients. Plasma corticosteroid levels peak about 30–60 min after an oral dose. Replacement therapy should be adjusted to give peak levels of 700–950 nmol/l after the morning dose, falling to 160–350 nmol/l before the evening dose, after which a peak of 400–550 nmol/l is reached (Besser and Edwards, 1972). Synthetic glucocorticoids, e.g. prednisolone, betamethasone, are useful for treating the patient during adrenal stimulation tests but for long term maintenance have the disadvantage that the exact dose cannot be determined since they are not measured in most routine glucocorticoid assays. Patients with primary adrenal disease normally have deficient aldosterone production. On cortisol or cortisone acetate alone, the blood pressure may be low, orthostatic hypotension may occur, or plasma urea or potassium levels may be high, but replacement is preferably undertaken even in the absence of these signs. Fludrocortisone is the steroid of choice, at a dose of 0.05 mg daily which may be increased if necessary to 0.1 to 0.2 mg per day. Excess of this hormone is manifest by the signs of sodium retention and potassium loss. An alternative is deoxycorticosterone (DOCA) but this has to be given by intramuscular injection. Patients who have hypothyroidism in addition to Addison's disease should not commence thyroxine before steroid replacement, since the increase in metabolism could otherwise precipitate an Addisonian crisis.

In tuberculous Addison's disease patients should be given antituberculous drugs for about a year in addition to steroid therapy. Those receiving sodium aminosalicylate (PAS) need higher than normal doses of steroids to keep well, because the drug induces hepatic microsomal enzymes with a resultant increase in cortisol catabolism (Edwards *et al.*, 1974). Theoretically, similar difficulties might arise with other enzyme-inducing drugs.

Patients with Addison's disease or secondary adrenal insufficiency are unable to increase their steroid output in response to the stress of infections, injury or operation. They should therefore always carry a card or wear a bracelet detailing their condition with instructions to give steroids if necessary. It is also useful for the patient to carry an ampoule of 100 mg of hydrocortisone sodium succinate or hydrocortisone sodium phosphate that can be given intravenously in an emergency.

Doubling the doses of hydrocortisone for a few days is usually sufficient to deal with mild intercurrent infections but the patient must be given clear instructions to seek help urgently if his condition deteriorates in spite of this or if associated vomiting makes it impossible to take the tablets. In these cases admission to hospital and parenteral hydrocortisone is necessary.

B. PREGNANCY IN ADDISON'S DISEASE

Fertility is usually normal in treated Addison's disease except where this may be complicated by a premature menopause but there is no contraindication to pregnancy.

Normal maintenance therapy is continued throughout pregnancy but delivery should be undertaken in hospital. During labour, hydrocortisone should be given intramuscularly every six hours until the patient is able to take her normal oral medication. If Caesarian section is needed replacement therapy should be as below.

C. SURGERY IN ADDISON'S DISEASE

Patients with Addison's disease should be admitted to hospital for all operations including minor procedures like dental extraction. For minor surgery it is usually sufficient to double the dose of hydrocortisone on the day of operation and the day afterwards. For more major procedures the patient should be admitted the day before operation to check that no fluid and electrolyte abnormalities are present. Hydrocortisone sodium succinate 100 mg should be given intra-muscularly with the premedication and 50 mg six hourly thereafter. Any unexplained hypotension should be regarded as an indication for further intravenous hydrocortisone. After operation it is advisable to continue with double the maintenance dose of hydrocortisone for a few days before gradually reducing it to normal. In emergency surgery, it is important to delay surgery if possible until adequate fluid, electrolyte and glucocorticoid replacement has been given.

In stressful situations, the maximum daily cortisol output from the normal adrenal is about 300 mg so it is reasonable to aim for this level of replacement in any emergency. The risks of overtreatment are much less than those of undertreatment so steroid therapy should, if anything, be over-generous.

Parenteral therapy must be given as hydrocortisone sodium succinate, since intramuscular cortisone acetate gives low inconstant plasma levels of corticosteroids (Plumpton *et al.*, 1969). In spite of adequate replacement therapy some patients with Addison's disease remain pigmented. This is probably due in some cases to higher levels of plasma cortisol being required to suppress ACTH and the dose used for replacement may not attain these levels (Holdway, 1973). This is at worst a minor inconvenience and is not an indication for increasing the dose of glucocorticoids, provided that adequate levels of plasma cortisol are maintained.

D. ADDISONIAN CRISIS

This condition is a medical emergency which can be fatal if untreated, though adequate treatment usually produces a rapid and dramatic improvement. The crisis is often precipitated by the stress of an infection or other illness in a patient with untreated Addison's disease or may occur in a previously diagnosed patient unable to take their medication because of vomiting. In addition to the immediate treatment of the crisis an attempt must be made to find the precipitating cause. It is usual to take blood for plasma cortisol, electrolytes, blood sugar and a blood culture but treatment is started without awaiting the results.

In Addisonian crisis there is a deficiency of both gluco- and mineralo-corticoids and the main needs are for replacement of cortisol, fluid, sodium ions and often glucose. Hydrocortisone sodium succinate is given intravenously in a dose of 100 mg immediately and 100 mg six hourly thereafter. Mineralocorticoid replacement is not necessary at this stage for successful management. Fluid replacement should consist of normal saline given according to the degree of dehydration but a half to one litre should be given as rapidly as possible. Intravenous dextrose may also be necessary to treat hypoglycaemia although saline replacement is more important. Intravenous fluids should be continued until the patient is able to tolerate fluids orally. The dose of cortisol should be reduced gradually over the next few days and mineralocorticoid replacement should be started once the patient is able to take oral medication. Any remedial cause of the crisis such as bacterial infections must be treated at the same time.

Sheridan and Mattingly (1975) have described a method for the simultaneous treatment and diagnosis of patients in Addisonian crisis. The crisis is treated with intramuscular prednisolone while the plasma 11-0H-CS are measured by fluorimetry before and after an injection of depot synacthen given at the same time.

E.   SECONDARY ADRENOCORTICAL FAILURE

1.   *Hypothalamic–pituitary disease*

The commonest cause of non-iatrogenic secondary adrenocortical insufficiency is a pituitary tumour, usually a chromophobe adenoma (Nabarro, 1972), though these are still uncommon causes of adrenal failure. The commonest hormone to be affected in pituitary tumours is growth hormone, with secretion of gonadotrophin, prolactin, ACTH and TSH affected in descending order of frequency (Nieman *et al.*, 1967). ACTH secretion was affected in 56% of cases in this series. The pattern of deficiency varies and prediction, in any particular case, is impossible.

If basal levels of cortisol are subnormal and there is an impaired response to insulin hypoglycaemia then long term steroid replacement therapy should be given as recommended for Addison's disease. Mineralocorticoid replacement is not required because adrenal production of aldosterone (which is not ACTH dependent, see Vol 2, Chapters 2 and 5) is normal. It is debatable whether steroid replacement therapy is indicated in those patients with normal basal cortisol levels and an impaired response to insulin hypoglycaemia. They should, of course be warned of their condition and carry a card or bracelet detailing it, but it is probably sufficient for them to have steroid cover merely at times of stress such as during operations or infections. Coma similar to that seen in Addisonian crisis does occur in hypopituitary patients but there is usually less evidence of fluid and electrolyte disturbance. In this situation urgent treatment with intravenous hydrocortisone and normal saline as well as dextrose is necessary.

The primary lesion should be treated as necessary in patients with secondary adrenocortical insufficiency. Evidence of a pituitary tumour should be sought by X-rays of the pituitary together with tomography of the pituitary fossa, computerized co-axial tomography (CAT scan), air encephalography and visual field charting. If these are all normal, it is necessary to postulate that the patient has a hypothalamic lesion or a microadenoma. Such patients should be kept under observation in case a previously undetected intrasellar lesion enlarges.

2.   *Post steroid therapy*

This iatrogenic disorder is the most common type of secondary adrenocortical insufficiency. High levels of circulating steroids from exogenous drug therapy cause a fall in ACTH levels by a direct action on the pituitary and probably also by reducing hypothalamic secretion of corticotrophin releasing factor (CRF). This in turn produces a secondary adrenal atrophy which has been noted in steroid-treated patients who died during operation (Fraser et al., 1952; Lewis et al., 1953; Salassa et al., 1953).

When steroids are withdrawn from a patient there is usually a rapid return of basal cortisol to normal levels. In many patients this takes place within 48 h (Robinson et al., 1962) but may take up to a month (Livanou et al., 1967). Sometimes, patients fail to show this rapid recovery but this is rare (Cope, 1966). A normal diurnal variation is seen as soon as four days after steroid withdrawal (Malone et al., 1970). In spite of normal basal levels, the cortisol response to hypoglycaemia is much reduced in patients on steroids (Daly et al., 1974) and the cortisol response may not become normal until a year after stopping therapy (Livanou et al., 1967). However, the cortisol response to major surgical stress was normal in patients who had been off steroids for as little as two months and some response, although impaired, was seen in patients currently receiving steroids (Plumpton et al., 1969). Although insulin hypoglycaemia is thought to be a good predictor of the ability to respond to stress, Janasi et al. (1968) found that those patients who had a poor response to the short synacthen test gave a sub-normal response to surgery and Plumpton et al. (1969) reported a few who responded to surgery but not to insulin-hypoglycaemia. Kehlet and Binder (1973) found that of 74 patients on steroid therapy at the time of operation, 39 had cortisol levels below that of normal patients one hour after the start of operation and 24 were similarly below normal at four hours. These authors showed that an impaired response was a function of both the dose and duration of treatment in that a dose of 12.5 mg prednisone for six months had the same effect as a dose of 7.5 mg for five years. There is general agreement that higher doses of steroids produce a greater suppression than lower doses (Livanou et al., 1967; Westerhof et al., 1972) and duration of treatment may be less important. However, there is no agreement about the minimum duration of treatment required to produce significant suppression of the hypothalamic–pituitary–adrenal (HPA) axis which certainly shows considerable individual variation in sensitivity to suppression in different patients (Livanou et al., 1967).

Topical steroids used both for skin lesions (Scoggins and Kilman, 1965; Feiwel et al., 1969), oral ulceration (Lehner and Lyne, 1969) and as an aerosol for asthma (Choo-Kang et al., 1972) can produce suppression of cortisol levels but there is some evidence of recovery of the HPA axis after changing from an oral steroid to a steroid aerosol (Bondarevesky *et al.*, 1976; Maberly *et al.*, 1973).

In spite of the well-documented effects on plasma steroid levels from therapeutic steroids, there is little correlation between low plasma levels and clinical symptoms. Unexplained hypotension during operation on patients who have received steroid therapy has been attributed to adrenal failure (Fraser *et al.*, 1952; Lewis *et al.*, 1953; Salassa *et al.*, 1953) but only a few cases have shown low plasma steroid levels and a convincing response to intravenous hydrocortisone (Sampson *et al.*, 1961; Janasi *et al.*, 1968). Kehlet and Binder (1973) found seven cases of unexplained hypotension in a series of 74 steroid-treated patients operated on without steroid cover. Four of these patients had a plasma cortisol response within the normal range and three abnormal. All recovered spontaneously without the use of parenteral steroids. In addition 33 patients had sub-normal cortisol levels without developing hypotension. Janasi *et al.* (1968) found one patient who collapsed with a cortisol level of 170 nmol/l, but two other patients had equally low levels and survived operation without trouble. Unfortunately there is no controlled trial showing whether patients given steroid cover during operation avoid even the low incidence of unexplained hypotension in those without steroid cover. These findings, nevertheless, support the statement by Cope (1966) that post-steroid-therapy collapse is usually associated with diagnoses other than adrenal failure.

In spite of this evidence, the majority of clinicians still give routine steroid cover for patients currently receiving steroids and undergoing surgery. Hydrocortisone sodium succinate 100 mg six hourly on the day of operation is probably adequate (Plumpton *et al.*, 1969). After the first day hydrocortisone is reduced to 50 mg I.V. six hourly, and subsequently to 20 mg orally six hourly, before a return to the normal maintenance therapy.

PART 2.   ADRENOCORTICAL OVERACTIVITY

1. Introduction

In 1912 Harvey Cushing, the Boston neurosurgeon, described the first case of the "peculiar polyglandular syndrome" which bears his name (Cushing, 1912). The patient was a 23 year old Russian Jewess named Minnie G. who had presented with increasing painful obesity, secondary amenorrhoea, headaches, nausea and vomiting, shortness of breath, palpitations and bruising of the skin. She also complained of a growth of facial hair with thinning of the scalp hair as well as increasing muscular weakness. On clinical examination Cushing described her as "an undersized kyphotic young woman . . . of most extraordinary appearance". She had a round face which was dusky and cyanosed with an abnormal growth of hair on the face. Her abdomen had the appearance of a full-term pregnancy, the breasts were hypertrophic and pendulous and there were pads of fat over the supraclavicular and posterior cervical regions which contrasted with her comparatively spare extremities. Her skin showed considerable pigmentation and there were subcutaneous ecchymoses over the lower limbs with numerous purple striae over the stretched skin of the lower abdomen, shoulders, breasts and hips. In addition her skin bruised easily and she was polycythaemic and hypertensive. Following this description, similar cases appeared in the literature (Turney, 1913; Oppenheimer and Fishberg, 1924; Weber, 1926). The clinical features and post-mortem pathological details of these and other cases were reviewed by Cushing (1932) who concluded that the primary abnormality in his syndrome was a basophil adenoma of the anterior pituitary gland producing adrenocortical hyperplasia. Subsequently Cushing's syndrome was reported in patients with a carcinoma or adenoma of one of the adrenal glands (Kepler *et al.*, 1934; Walters *et al.*, 1934). To make this syndrome even more complex an association between adrenocortical hyperactivity and "non-endocrine tumours" was recognized by Brown (1928) and established to be due to the ectopic secretion of ACTH by the tumour in the classical studies of Meador *et al.*(1962) and Mattingly *et al.* (1964). Although the clinical features of Cushing's syndrome are now well recognized, many advances have been made in recent years in our understanding of its pathophysiology, diagnosis and treatment.

2. Classification

Cushing's syndrome develops from prolonged exposure of the tissues to inappropriately elevated levels of circulating corticosteroids. The causes of the syndrome may be divided into those where the elevated corticosteroid levels are secondary to excess adrenocorticotrophic hormone secretion (ACTH dependent) or those where the primary disease lies in the adrenal gland and ACTH secretion is suppressed

(non-ACTH dependent). On this basis the syndrome may be classified as follows:

ACTH dependent
1. Pituitary ACTH hypersecretion causing bilateral adrenocortical hyperplasia. This is usually known as Cushing's disease.
2. Ectopic ACTH hypersecretion from a tumour of non-
   endocrine origin which also gives rise to bilateral adrenocortical hyperplasia.
3. Iatrogenic administration of exogenous ACTH or of its synthetic analogues over a prolonged period.

Non-ACTH dependent
1. Adenoma ⎫
   of the adrenal cortex
2. Carcinoma ⎭
3. Iatrogenic administration of pharmacological doses of corticosteroids in the treatment of chronic disease e.g. rheumatoid
   arthritis.

3.   Aetiology and pathogenesis

Bilateral adrenocortical hyperplasia is the commonest cause of
Cushing's syndrome (Fig. 10a, b, c, d) and is associated with
exaggerated adrenocortical secretory activity of the adrenal cortices.
This is characterized initially by loss of the normal circadian rhythm
of cortisol secretion so that plasma cortisol levels are as high in the
evening as in the early morning. As the condition becomes more
advanced, these levels eventually become elevated with concomitant
increase in the clinical symptoms and signs.

The pathological features of adrenocortical hyperplasia have been
reviewed by Neville and Symington (1967) in a series of 60 patients
with Cushing's syndrome due to pituitary-dependent adrenocortical
hyperplasia. They recognized two histological types of adrenocortical
hyperplasia: (a) simple hyperplasia, present in 50 patients, in which
there was a broad compact-cell zona reticularis occupying one-third
to one-half of the adrenal cortex divided from a wide clear-cell zona
fasciculata by an undulating line; and (b) adenomatous hyperplasia,
present in 10 patients, in which there were one or more adenomatous
nodules composed of fasciculata cells in both adrenal glands with
hyperplasia of the associated zona fasciculata (see Vol. 2, Idelman,
1978). This form of adrenal hyperplasia has been attributed to long
term exposure to ACTH but this has been refuted by Neville and

Fig. 10. (a) "Normal" human adrenal gland (low power) showing typical zonation. The gland was removed from a patient with breast carcinoma for therapeutic reasons. (b) Human adrenal gland from a patient with Cushing's syndrome (low power). The cortex shows marked broadening and consists mainly of compact cells which extend almost to the capsule and broad zona glomerulosa. (c) Higher power of "normal" zona fasciculata cells showing typical cytoplasmic arrangement. (d) Higher power of zona fasciculata in Cushing's syndrome showing compact cell conversion and virtual obliteration of the clear cells. (Material supplied by A. M. Neville.)

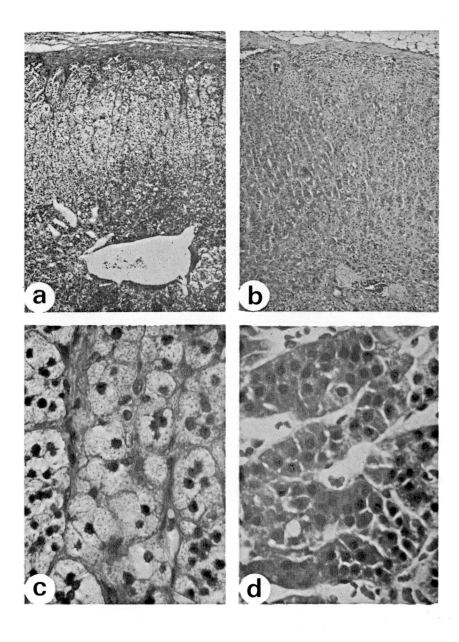

Symington (1967) and its presence in a 2-month-old child with Cushing's disease (O'Bryan *et al.*, 1964) is out of keeping with this hypothesis. The adrenal glands in these variants of hyperplasia also differed in weight, that of a single adrenal gland in "simple" hyperplasia lying between 6.0 and 12.0 g whereas in "adenomatous" hyperplasia it usually exceeded 12.0 g in contrast to an upper limit of 6.0 g for the normal gland (Studzinski *et al.*, 1963). It would appear, however, that simple and adenomatous hyperplasia are variants of the same disease process and do not signify a different aetiology. Occasionally the adrenal glands in Cushing's syndrome may be of normal weight and in the series reported by Neville and Symington (1967) seven patients had adrenal glands weighing less than 6.0 g although histologically they were hyperplastic having a compact cell zona reticularis which was wider than normal.

Bilateral adrenocortical hyperplasia has been shown, using both biological and immunological assay systems, to be due to overproduction of ACTH by the pituitary gland (Ney *et al.*, 1963; Nelson *et al.*, 1966; Besser and Landon, 1968). In normal subjects there is a circadian rhythm of secretion of ACTH which is maximum in the early morning and declines as the day progresses to reach a nadir in the late evening (Ney *et al.*, 1963; Hellman *et al.*, 1970). In early Cushing's disease, this circadian rhythm of secretion of ACTH is lost so that early morning ACTH values lie within or only slightly above the normal range and these levels are maintained throughout the 24 hours of the day. Subsequently plasma ACTH values become more markedly raised and this is reflected in increasing severity of the disease (Ney *et al.*, 1963; Besser and Landon, 1968). These plasma concentrations of ACTH in Cushing's disease have also been shown to correlate closely with adrenocortical activity as reflected by both plasma and urinary corticosteroid levels (Ney *et al.*, 1963; Besser and Landon, 1968).

Although it was suggested by Cushing that this excessive secretion of ACTH by the pituitary gland was due to the presence of a basophil adenoma in the pituitary gland, these tumours have been found in only 50% of cases of Cushing's disease at autopsy (Thompson and Eisenhardt, 1943; Plotz *et al.*, 1952). It is possible that the pituitary glands in other patients may contain microadenomata but the precise incidence of these is unknown. Although basophil adenomata usually remain small, they may grow large enough to expand the pituitary fossa and this may be seen radiologically in about 20% of patients with Cushing's disease on initial presentation. Occasionally Cushing's disease may be associated with a chromophobe or eosinophilic adenoma of the pituitary gland (Plotz *et al.*, 1952) and, rarely, with a carcinomatous pituitary tumour which may be locally invasive (Welbourne *et al.*, 1971). It is likely that in some cases of Cushing's disease the basic defect lies in the hypothalamus or even at a higher level, resulting in excessive secretion of corticotrophin releasing factor (CRF) which may be responsible for the development of a basophil adenoma. Evidence in favour of this hypothesis is obtained from the retention of the ability for endogenous cortisol and ACTH levels in Cushing's disease to be suppressed with a high dose of dexamethasone suggesting that the hypothalamic receptors which are sensitive to alteration in circulating cortisol levels are still functioning but are set to function at a higher level than normal (Cook *et al.*, 1976). Also degenerative changes have been reported on histological examination of the hypothalamus in four patients with Cushing's disease (Heinbecker, 1944).

The association of bilateral adrenocortical hyperplasia with 'non-endocrine" tumours such as a bronchogenic carcinoma was first recognized by Brown (1928). The initial theories advanced to explain this association varied from impairment of endogenous steroid conjugation to the stressful effects of malignancy on cortisol secretion and are well reviewed by Riggs and Sprague (1961). The syndrome was clearly defined, however, by Meador *et al.* (1962) who demonstrated elevated plasma ACTH activity, biologically active ACTH in the extracts of both primary tumours and their metastases and decreased pituitary ACTH. The observation by Mattingly *et al.* (1964) of the syndrome in an hypophysectomized patient provided further clinical evidence of the tumour source of ACTH. Subsequently, ACTH has been shown to be released by tumour cells grown in tissue culture and has been detected by both bioassay and radioimmunoassay in tumour extracts (Ratcliffe *et al.*, 1972; Orth, 1973; Orth *et al.*, 1973). It would appear that the ACTH secreted by tumours is biologically and immunologically very similar to pituitary ACTH although it has not been determined whether it has the same amino acid sequence.

Although bronchogenic carcinomas are most commonly associated with this syndrome, many other tumours have been reported to be associated with ectopic ACTH secretion. Tumours with ectopic ACTH secretion were categorized into four groups by Azzopardi and Williams (1968) as follows: (1) oat cell carcinomas of bronchus; (2) endocrine tumours of foregut origin including islet cell carcinomas of pancreas, carcinoid tumours of bronchus, stomach and pancreas, epithelial thymomas and medullary carcinomas of thyroid; (3) phaeochromocytomas and related tumours such as neuroblastoma and ganglioneuroblastoma; (4) certain ovarian tumours, e.g. arrheno-blastoma and adenocarcinoma. The syndrome now covers an even wider group of tumours which may synthesize ACTH, including carcinoma of the breast, oesophageal and gastric carcinoma and lymphoepitheliomata, renal carcinoma, bronchial adenocarcinoma and squamous carcinoma, and ventral mesothelioma (Liddle *et al.*, 1965, 1969; Ratcliffe *et al.*, 1972). It is thought that in these tumours ACTH is synthesized within intra-cellular granules which are similar to those found in typical peptide-secreting glands, having an electron dense core, a clear halo and a surrounding membrane (O'Neal *et al.*, 1968; Pimstone *et al.*, 1972).

The concentration of ACTH in the plasma of patients with Cushing's syndrome differs according to whether an adrenal tumour is present or if there is adrenal hyperplasia secondary to secretion of ACTH by the pituitary gland or from an ectopic tumour source (Fig. 11). Besser and Edwards (1972) found plasma immunoreactive ACTH levels in a series of patients with pituitary dependent adrenal hyperplasia to lie within the upper range of normal (12 to 80 ng/l) or to be slightly elevated up to 200 ng/ml, whereas the levels in patients with adrenal hyperplasia due to etopic ACTH secretion ranged from 100 to over 4000 ng/ml. Ratcliffe *et al.*(1972) reported that in patients with the ectopic ACTH syndrome, 83% of samples had plasma immunoreactive ACTH levels of between 100 and 1000 ng/ml. A characteristic feature of all the patients with Cushing's syndrome in both these series was the absence of the normal circadian rhythm of plasma ACTH. Ratcliffe *et al.* (1972) found no obvious relationship between the plasma concentration of ACTH and the histological tumour type but this does not apply to tumour concentrations of ACTH, e.g. bronchial carcinoids having a higher concentration than bronchial carcinomas. It is of interest that these workers also detected similar concentrations of bioactive and immunoreactive ACTH in control tumours not associated with the ectopic ACTH syndrome so it is possible that hormone synthesis is a common feature of many tumours with only certain tumours able to release the hormone to cause clinical symptoms.

Approximately 20% of cases of Cushing's syndrome are caused by an adenoma or carcinoma of the adrenal cortex. These tumours secrete glucocorticoids which suppress endogenous pituitary ACTH secretion so that the opposite adrenal cortex as well as that of the gland in which the tumour occurs, become atrophic. As the pituitary is suppressed, plasma ACTH levels are undetectable (Besser and Landon, 1968). In contrast to patients with adrenal hyperplasia they

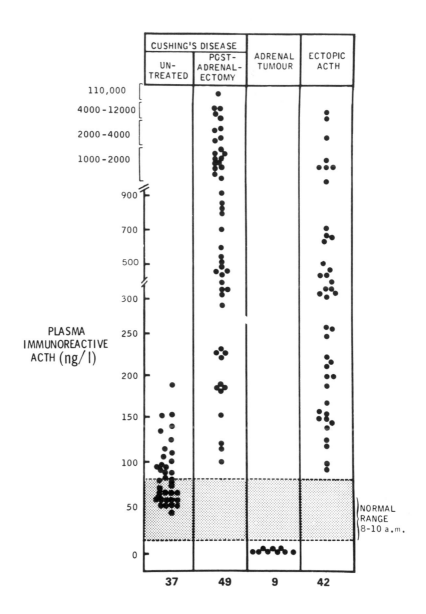

Fig. 11. Plasma immunoreactive ACTH levels in 137 cases of Cushing's syndrome (from Besser and Edwards, 1972).

are less likely to produce a pure syndrome of glucocorticoid excess since there are often associated features of virilization due to a concomitant hyper-secretion of androgens.

4. Incidence

Pituitary dependent adrenal hyperplasia occurs in about 1 per 5,000 of the population. It usually occurs in the female, with a female to male ratio of 8 : 1 (Ross *et al.*, 1966). Cushing's disease may occur at any age but it reaches its peak incidence in the third and fourth decades of life. On the other hand adrenal hyperplasia secondary to ectopic ACTH secretion is more common in males being found at post mortem in 2% of cases of bronchogenic carcinoma (Ross, 1966). Bilateral adrenocortical hyperplasia accounts for approximately 80% of all cases of Cushing's syndrome (Neville and Mackay, 1972).

Adrenocortical carcinoma and adenoma are responsible for the remaining 20% of cases of Cushing's syndrome. These tumours are also more common in females than in males in the ratio 2 : 1 (Hutter and Kayhoe, 1966). Their incidence in the population as a whole is difficult to assess but carcinoma of the adrenal cortex occurs in approximately 2 per one million of the population (Hutter and Kayhoe, 1966). Carcinoma of the adrenal cortex is the most common cause of Cushing's syndrome in children.

5. Clinical symptoms and signs

The most common presenting feature of Cushing's syndrome (Fig. 12) is obesity which may antedate the onset of other characteristic symptoms and signs. This is classically restricted to the face, neck and trunk giving the moon face, buffalo hump and truncal obesity with sparing of the extremities which is so typical of Cushing's syndrome. Ross *et al.* (1966) however found obesity to occur in this distribution in only 40% of their patients, the majority having a more generalized distribution of obesity. Although Cushing (1932) remarked that the obesity was painful, this is uncommon (Plotz *et al.*, 1952). The obesity of Cushing's syndrome is often more apparent than real. For instance, only one-third of patients in a series of 50 cases of Cushing's syndrome

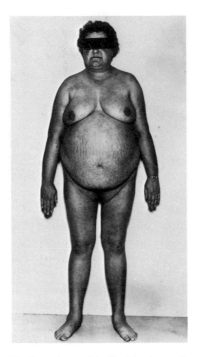

Fig. 12. A patient with Cushing's syndrome.

reported by Soffer *et al.* (1961) were more than 10% above standard weight and in only one-fifth were their weights 25% above normal. This appearance of obesity is probably due in most cases to a change in body fat distribution and some patients may lose height due to osteoporosis of the spine. The anti-anabolic effect of excessive cortisol secretion on protein (Albright, 1943) results in thinning of the skin which becomes dusky red in colour. Purple striae develop on the abdomen, thighs and breasts and the intensity of their colour may fluctuate from day to day. The same mechanism produces increased fragility of the small subcutaneous blood vessels and subsequent bruising of the skin. The plethoric face which is so characteristic of Cushing's syndrome is also the result of these factors and only infrequently is polycythaemia present (Ross *et al.*, 1966; Soffer *et al.*, 1961). Skin pigmentation may occur in some patients with Cushing's syndrome when plasma ACTH levels and possibly MSH levels are high as in pituitary dependent adrenal hyperplasia and particularly in the ectopic ACTH syndrome where pigmentation may be quite marked. The anti-inflammatory effects of excessive cortisol secretion also result in poor wound healing and increased susceptibility to skin infection so that severe cases of Cushing's syndrome may present with pyoderma or chronic fungus infections. Varying degrees of osteoporosis in Cushing's syndrome result from the effects of protein depletion on the bony matrix of the skeleton. Ross *et al.* (1966) found radiologically demonstrable osteoporosis in 50% of their patients with Cushing's syndrome and fractured ribs or crush fractures of the vertebrae were not uncommon. Lower back pain is a frequent symptom of osteoporosis in these patients and when bone disease is severe, loss of trunk height may occur. The anti-anabolic effect of excessive cortisol secretion on protein metabolism may also result in muscle weakness and wasting which is most marked in the proximal limb girdle muscles particularly of the lower limbs (Muller and Kugelberg, 1959). The histological changes in the affected muscles are consistent with a myopathy which is similar to that produced by exogenous steroid therapy (Perkoff *et al.*, 1959).

Some disturbances of carbohydrate metabolism are common in Cushing's syndrome as a result of increased gluconeogenesis from excessive secretion of cortisol. Soffer *et al.* (1961) found 84% of their patients to have laboratory evidence of disturbed carbohydrate metabolism but only 20% had overt diabetes mellitus. Ross *et al.* (1966) reported 15% of their series to have frank diabetes mellitus although many of their patients had an abnormal glucose tolerance curve. Despite this high incidence of disturbed carbohydrate metabolism in Cushing's syndrome symptomatic diabetes is unusual and when it occurs is often mild.

Hypertension may be a presenting symptom in Cushing's syndrome and its reported incidence is high: 85% (Plotz *et al.*, 1952); 90% (Sprague *et al.*, 1956); and 88% (Soffer *et al.*, 1961). Although it is usually benign, it may become malignant and be associated with retinopathy, albuminuria and renal damage. Hypertension, nevertheless, is largely responsible for the poor prognosis in untreated Cushing's syndrome by causing cardiac failure. Scholz *et al.* (1957) for instance in an autopsy study of 17 patients with Cushing's syndrome, all of whom had been hypertensive, found nine who had been in cardiac failure. The hypertension of Cushing's syndrome is probably due to a number of factors. The increased salt retention induced by excessive cortisol secretion may be partially responsible but plasma renin substrate levels have been found to be elevated in Cushing's syndrome which may contribute to increased angiotensin formation. In addition cortisol may increase the vascular responsiveness of the peripheral vessels to pressor agents (Crane and Harris, 1966; Schambelan *et al.*, 1971; Krakoff *et al.*, 1975).

Some degree of hirsutism which may be accompanied by acne is a common feature of Cushing's syndrome but this is only significant in approximately 25% of cases (Greenblatt, 1965). In most instances this consists simply of a downy growth of lanugo hair produced by the excessive cortisol secretion. Adrenocortical tumours, which may secrete high levels of adrenal androgens such as dehydroepiandrosterone, androsterone and testosterone in addition to cortisol, may present with more marked hirsutism sometimes associated with features of virilization such as scalp baldness, a male escutcheon of pubic hair, enlargement of the clitoris, a deep voice, severe acne and atrophy of the breasts. Sexual function is usually disturbed with loss of libido and amenorrhoea in women and impotence in men. A tendency for male patients to have soft testes and gynaecomastia has been observed by Ross *et al.* (1966).

The patient with Cushing's syndrome may also present with a wide range of psychiatric symptoms such as euphoria, anxiety, confusion, disorientation, depression, hypomania or as a psychotic state requiring institutional care. These symptoms are similar to those which may be produced in patients on exogenous steroid therapy and in similar manner usually disappear following effective treatment of the Cushing's syndrome.

The incidence of renal calculi is increased in Cushing's syndrome being reported as 5% (Sprague *et al.*, 1956); 24% (Wang and Robbins, 1956); 16% (Ross *et al.*, 1966) and 65% (Scholz *et al.*, 1957). These result from the excessive urinary secretion of calcium phosphate secondary to osteoporosis. In other more rare cases primary hyperparathyroidism may co-exist with Cushing's syndrome as part of the syndrome of multiple endocrine adenomatosis and may give rise to renal stones (Steiner *et al.*, 1968). Soffer *et al.* (1961) reported that in addition to the 20% of their patients with radio-opaque renal calculi there were a further 20% with radiologically demonstrable gall stones. Ross *et al.* (1966) found that 10% of their patients with Cushing's syndrome had gall stones, all of whom had renal calculi.

When it presents in childhood, Cushing's syndrome is invariably associated with short stature. Linear growth and skeletal maturation are both affected, possibly by the effects of excessive cortisol secretion diverting amino acids away from protein synthesis. Very rarely exophthalmos occurs in Cushing's syndrome due to an unknown mechanism.

6. Diagnosis

Laboratory confirmation of the diagnosis in even the most clinically obvious case of Cushing's syndrome is mandatory since treatment at present normally involves major surgery to the pituitary or adrenal glands. The clinician should also have a high index of suspicion for Cushing's syndrome since it is desirable that the diagnosis is made early in the course of the disease when symptoms and signs may not be florid.

The diagnosis of Cushing's syndrome is made in two stages. Firstly it is necessary to confirm the presence of adrenocortical hyperfunction. Secondly it is necessary to establish whether this is due to an adrenal tumour or secondary to excess secretion of ACTH either from the pituitary or an ectopic tumour source.

A. CORTICOSTEROID INVESTIGATIONS

(1) Basal adrenocortical function. The disease results from a sustained increase in the circulating free cortisol. This may be demonstrated by measurement of the plasma cortisol, urinary excretion of cortisol or its metabolites and the cortisol secretion rate.

(2) Suppression tests. Cushing's syndrome is characterized by an interruption of the negative feedback which high levels of circulating steroids normally exert on the hypothalamic pituitary axis. In practice this is shown by a failure of synthetic steroids to suppress endogenous steroid production.

(3) Stimulation tests. These measure the ability of the hypothalamic–pituitary–adrenal axis to respond to certain stimuli with an increase in steroid production.

1. *Tests of basal adrenocortical function*

(a) Measurement of plasma corticosteroids

The main circulating glucocorticoid in man is cortisol. This is usually measured by a competitive protein binding method, using cortisol-binding globulin (CBG) (Few and Cashmore, 1971). This method measures both free and protein bound plasma cortisol. In addition CBG binds corticosterone and 11-deoxycortisol so that the presence of these steroids may influence the assay. Elevated levels are found in conditions which increase the concentrations of CBG such as pregnancy and treatment with oestrogens. In these instances, however, the increase is mainly due to a higher level of protein bound cortisol and there is little elevation in the physiologically active free cortisol. It is also possible to measure plasma cortisol by radioimmunoassay using an anticortisol antibody (Fahmy *et al.*, 1975). This method is technically simpler than competitive protein binding and may replace it for routine use if sufficiently specific antisera become readily available. Steroids with an oxygen atom at C11 can be measured by the fluorescent method of Mattingly (1962). This method is reliable and simple but unfortunately subject to non-specific fluorescence and interference from drugs especially spironolactone, mepacrine and fucidic acid.

Normal subjects show a striking circadian rhythm of plasma cortisol concentrations with high levels of 160–600 nmol/l at 0900 hours declining to 0–200 nmol/l at 2400 hours. In Cushing's syndrome this normal circadian rhythm of plasma cortisol is absent and the high morning cortisol levels are maintained throughout the day. With increasing severity of the syndrome there is not only loss of the normal diurnal variation of plasma cortisol but levels become abnormally elevated. Unfortunately loss of the plasma cortisol diurnal variation also occurs in severe depression (Butler and Besser, 1968), congestive cardiac failure (Knapp *et al.*, 1967), severe infections or even under the stress of hospital admission. The late-evening cortisol concentration should be measured, therefore, in an unforewarned patient, woken from sleep. Should a well marked circadian rhythm be observed then the occurrence of Cushing's syndrome is unlikely though its absence is not, in itself, diagnostic.

(b) Concentration of ACTH in plasma

The introduction of radioimmunoassay techniques for the measurement of ACTH levels in plasma (Yalow *et al.*, 1964; Landon and Greenwood, 1968) has greatly enhanced the differential diagnosis of Cushing's syndrome (Besser and Landon, 1968). A circadian rhythm of plasma ACTH similar to that of cortisol occurs in the normal human with levels of 12 to 80 ng/l at 0800 hours and 5 to 30 ng/l at 2400 hours (Besser and Edwards, 1972; Besser, 1973). A characteristic feature of all patients with Cushing's syndrome is the absence of this normal circadian rhythm of plasma ACTH. Since many patients with the syndrome have normal morning ACTH levels this offers no advantage over conventional corticosteroid assays in its diagnosis. Plasma ACTH assays, however, have the practical value that they help to distinguish between the different causes of Cushing's syndrome (see Fig. 11). Thus the absence of circulating ACTH indicates the presence of an adrenal carcinoma or adenoma, whereas normal or elevated ACTH levels are found in adrenal hyperplasia due either to a pituitary or to an ectopic source of ACTH. Although ACTH levels in the ectopic ACTH syndrome overlap with those found in pituitary-dependent disease, levels greater than 200 ng/l make an ectopic source of ACTH very likely (Besser and Edwards, 1972).

(c) Basal urinary steroid excretion

Estimation of the urinary excretion of steroids and their metabolites gives a measure of the average hormone levels throughout the 24 hours.

(i) *Urinary free cortisol.* The unconjugated (free) cortisol in the urine is extracted and measured by the competitive protein binding technique (Few and Cashmore, 1971; Burke and Beardwell, 1973). The cortisol can also be measured by fluorimetry (Mattingly *et al.*, 1964) designated as urinary 11-OHCS. The free urinary cortisol is largely derived from renal ultrafiltration of the non-protein bound cortisol in plasma to give an excellent measure of the physiologically active plasma fraction. Since CBG is saturated at plasma cortisol levels of about 560 nmol/l, small increments of plasma cortisol above this level produce a marked elevation in the non-protein bound plasma cortisol which is reflected in the urinary free cortisol. Values for normal urinary free cortisol levels vary both with the laboratory involved and the assay method employed. There seems to be general agreement that a high level of cortisol gives a close correlation with the presence of Cushing's syndrome; thus this is one of the most useful tests for establishing the presence of adrenocortical hyperfunction (Burke and Beardwell, 1973; Eddy *et al.*, 1973; Fig. 13a). Other workers, however, have reported a small overlap with normal values (Nichols *et al.*, 1968). In particular the urinary free cortisol is normal in obesity and only advanced renal failure produces a decrease in the urinary levels. Urinary free cortisol is however increased by pregnancy, oestrogen treatment, severe stress and probably depression and cardiac failure (Burke and Beardwell, 1973).

(ii) *Urinary 17 oxogenic steroids (17 OGS).* This assay is so-called because it measures the urinary excretion of a group of C21 steroids which can be converted to 17 oxo steroids (17 OS). The endogenous 17 oxo steroids are first destroyed by borohydride reduction and the 17 OGS converted to 17 OS by periodate oxidation. The resulting 17 OS are measured by the Zimmerman reaction (Gray *et al.*, 1969). The steroids measured are mainly cortisol and its tetrahydro derivatives but include 17 hydroxyprogesterone, pregnanetriol, 11 deoxycortisol and cortisone (Kime, 1978). Many substances interfere with the

Zimmerman reaction. Glucose, chlordiazepoxide and digoxin cause a decrease in the level, while antibiotics and phenothiazines cause an increase. In the United States urinary steroids are usually assayed as

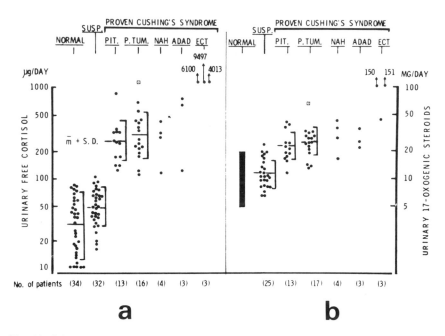

Fig. 13. (a) Cushing's syndrome urinary free cortisol levels compared with those of normal subjects. (b) Urinary 17-oxogenic steroid levels compared with those of normal subjects (from Burke and Beardwell, 1973).

Porter–Silber chromogens (known as 17 hydroxycorticosteroids or 17 OHCS) which measure cortisol, corticosterone and their tetrahydro derivatives but not pregnanetriol (Silber and Porter, 1954). Individual steroids can be measured by chromatography but this is mainly a research procedure at present. Basal levels of urinary 17 OGS or 17 OHCS are of little value in making or excluding the diagnosis of Cushing's syndrome since high levels are found in simple obesity and about 40% of cases of Cushing's syndrome have levels within the normal range (Burke and Beardwell, 1973) (Fig. 13b).

(iii) *Urinary 17-oxosteroids (17 OS)*. These are measured by the Zimmerman reaction after extraction of the urine. The steroids measured are mainly androgens derived from the adrenal cortex and include dehydroepiandrosterone, androstenedione and testosterone but small amounts are derived from cortisol metabolites (Gray et al., 1969). The normal range varies with the age and sex of the subject. Basal levels are of no value in establishing the diagnosis of Cushing's syndrome. Very high levels are often seen in adrenal carcinoma but not adrenal adenoma (Ernest, 1966) so this estimation very occasionally helps in establishing the aetiology of the disease.

(d) Cortisol secretion rate

This can be measured by an isotope dilution technique. ^{3}H labelled cortisol is given by mouth and the specific activity of a suitable urinary metabolite is determined (Cope and Black, 1958). The normal range is 16–110 μmol/24 h. Although raised levels are always found in Cushing's syndrome, high levels are often found in patients without the disease (Nichols et al., 1968). Since the test is difficult to perform it is not much used in routine practice.

2. *Suppression tests*

These tests are based on the fact that the pituitary–adrenal axis is more difficult to suppress with exogenous steroids in patients with Cushing's syndrome than in normal subjects.

(a) Short dexamethasone suppression test (Nugent et al., 1965)

Procedure
Day 1   0900 h   Basal plasma cortisol measured.
         2300 h   2 mg dexamethasone orally.
Day 2   0900 h   Basal plasma cortisol measured.

In normal subjects the post-dexamethasone cortisol level should be suppressed to 160 nmol/l or 30% of the basal levels. Nichols et al. (1968) reported that 97% of patients with Cushing's syndrome failed to suppress with this test. However a moderate number of hospital in-patients without the disease also failed to show suppression (Connolly et al., 1968) but fewer false positives are seen in out-patients. The test is simple to perform and the lack of false negative responses makes it very suitable for out-patient screening. A normal response is good evidence against the presence of Cushing's syndrome but an abnormal response needs further confirmation of the diagnosis.

(b) Long dexamethasone suppression test (Liddle, 1960)

Procedure

	Day	
Daily urinary 17 OGS	1⎱ 2⎰	Basal levels
or plasma cortisol	3⎱ 4⎰	Dexamethasone 0.5 mg 6 hourly
determination at 0900		
hours and 2400 hours	5⎱ 6⎰	Dexamethasone 2 mg 6 hourly

This test has been widely used both for the diagnosis and differential diagnosis of Cushing's syndrome. In normal subjects the urinary 17 OGS levels are suppressed to less than 18 μmol/24 h and the plasma cortisol to less than 160 nmol/l by the lower dose of dexamethasone. In pituitary-dependent Cushing's syndrome suppression of the urinary 17 OGS or plasma cortisol to less than 50% of the basal levels occurs with the higher but not the lower dose of dexamethasone (Fig. 14). Suppression is not usually seen with either dose in adrenal tumours or the ectopic ACTH syndrome (Fig. 15).

In practice, however, up to 40% of patients without Cushing's disease especially those with obesity (Burke and Beardwell, 1973) or severe depression (Butler and Besser, 1968) may fail to show suppres-

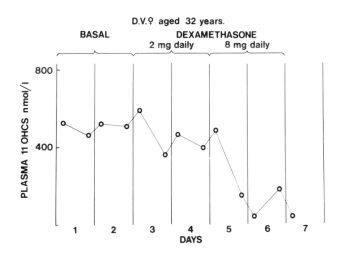

Fig. 14. Long dexamethasone suppression test in a patient with pituitary dependent Cushing's syndrome showing corticosteroid suppression with the high dose of dexamethasone.

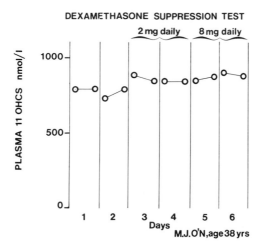

Fig. 15. Long dexamethasone suppression test in a patient with Cushing's syndrome due to an adrenocortical carcinoma showing lack of corticosteroid suppression with the high dose of dexamethasone.

sion on the low dose of dexamethasone. Reassessment of the urinary free cortisol levels increases the diagnostic precision but does not give any more information than estimation of basal urinary free cortisol levels (Burke and Beardwell, 1973). A normal suppression, however, with the low dose of dexamethasone excludes Cushing's syndrome. Use of the higher dose of dexamethasone in the differential diagnosis of Cushing's syndrome is useful but not entirely reliable. About 18% of patients with the pituitary-dependent disease fail to suppress on this dose while 9% of adrenal tumours and 27% of those with the ectopic ACTH syndrome show suppression (Nichols et al., 1968). Occasional cases of adrenal tumours (French et al., 1968; Epstein et al., 1973) and adrenal hyperplasia (James et al., 1965) show a paradoxical rise in steroid production following dexamethasone. In at least one case, however, the apparent paradoxical response was shown to be caused by an inherently cyclical pattern of adrenal secretion (Brown et al., 1973).

3.   Stimulation tests

These depend on the observation that patients with bilateral adrenal hyperplasia have a hyperactive response to ACTH as measured by urinary steroid estimations, not shown by normal subjects or those with autonomous adrenal tumours (Laidlaw *et al.*, 1955) in which ACTH secretion is suppressed.

(a) Metyrapone Test (Cope *et al.*, 1966)
    Procedure:

	Day 1	Basal levels determined.
Daily urine 17 OGS	2	Basal levels determined.
determination	3	Metyrapone 750 mg, 4 hourly.
	4	no drug.

Metyrapone is an 11β-hydroxylase inhibitor which blocks the final step in the synthesis of cortisol (Sandor *et al.*, 1976). The reduced cortisol level causes a rise in plasma ACTH which stimulates the adrenal to produce cortisol precursors including 11-deoxycortisol which can be measured as 17 OGS in the urine. In normal subjects a rise in 17 OGS levels of at least 35 nmol/24 h above basal levels is seen. It is most important because of its short half life that the drug is given 4 hourly otherwise maximal inhibition of 11β-hydroxylase may not occur. The effectiveness of inhibition can be checked by demonstrating a fall in plasma cortisol during the test. Plasma cortisol should be measured before, and hourly for four hours after, metyrapone and a fall of at least 50% should occur. When metyrapone is given, plasma cortisol should be measured by a fluorimetric method (Mattingly, 1962) as most radioimmunoassays show cross reactivity between cortisol and its precursor 11-deoxycortisol which accumulates after administration of metyrapone.

The value of this test in Cushing's syndrome lies in establishing the aetiology of the disease. In pituitary-dependent Cushing's syndrome, an exaggerated ACTH response is seen and the basal levels of 17 OGS should either double or rise by $70\,\mu$mol/24 h. In adrenal tumours or the ectopic ACTH syndrome, pituitary ACTH is suppressed and there should be no response. A diagnostic accuracy of 95% has been reported but occasional false positives and negatives are recorded (Burke and Beardwell, 1973).

(b) ACTH stimulation test
Although it has been suggested that exogenous ACTH stimulation will differentiate between adrenal hyperplasia and tumours, the results are so variable that this is a poor test in differentiating between the causes of Cushing's syndrome since adrenal adenomas and even

carcinomas have been known to produce a steroid response (Laidlaw et al., 1955; Rayfield et al., 1971).

(c) Lysine vasopressin test
In this test the adrenal steroid response to the infusion or intramuscular injection of lysine vasopressin, which has a corticotrophin releasing function, is measured. Although patients with bilateral adrenal hyperplasia have responded with increased adrenal responsiveness compared with adrenal tumours or carcinomas (Bettige et al., 1969), the lack of diagnostic precision limits its value (James et al., 1968).

(d) Insulin hypoglycaemia test
With very few exceptions a normal plasma cortisol response to insulin hypoglycaemia is not seen in Cushing's syndrome regardless of aetiology. It has been suggested (Besser and Edwards, 1972) that this effect results from the high circulating corticosteroid levels on the stress response, normally mediated by the hypothalamus. Since this is one of the few adrenal function tests which remains normal in obesity and depression, its use lies mainly in differentiating normal, obese or depressed patients from those with Cushing's syndrome.

B.   NON-CORTICOSTEROID INVESTIGATIONS

Patients with Cushing's disease may show glucose intolerance, polycythaemia, eosinopenia and a hypokalaemic alkalosis. The presence of these abnormalities may first raise the suspicion of adrenal hyperfunction but they cannot, of course, confirm the diagnosis. A marked hypokalaemic alkalosis is however characteristic of the ectopic ACTH syndrome.

Non-endocrine investigations are more useful in establishing the aetiology in a particular case of Cushing's syndrome. Evidence for a pituitary tumour may be found in a small proportion of patients by examination of the visual fields, radiology of the pituitary fossa or by CAT scan to show a suprasellar extension of a tumour. A plain X-ray of the abdomen may show an adrenal mass and fine calcification characteristic of adrenal carcinomas though this has also been reported in adenomas (Vermess et al., 1972). A soft tissue mass may be more clearly positioned by intravenous pyelography but either arterial angiography or retrograde venography (Reuter et al., 1967) are the best methods of visualizing an adrenal tumour (Fig. 16). Venography can be combined with venous sampling which may demonstrate differences in cortisol levels of adrenal venous blood,

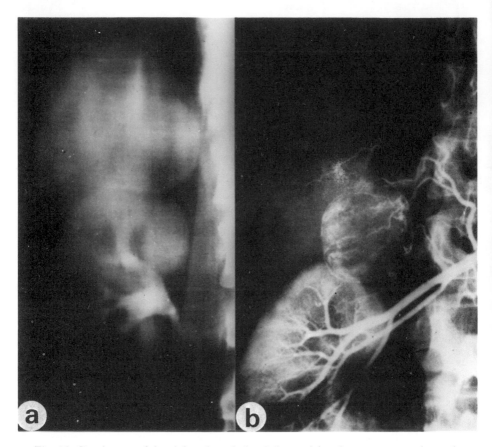

Fig. 16. Carcinoma of the right adrenal gland shown (a) at intravenous pyelography and (b) at subsequent arterial angiography.

thus helping to confirm the diagnosis. Occasionally this technique may unwittingly cure the disease by causing infarction of the adrenal tumour followed by an adrenal crisis because of coexisting suppression of the normal surrounding gland.

A recent development in the differential diagnosis of Cushing's syndrome is the use of isotope scanning with [131]I 19-iodocholesterol which is selectively taken up by functioning adrenal tissue. In pituitary-dependent Cushing's disease or the ectopic ACTH syndrome, intense bilateral activity is seen whilst with adrenal adenomas, there is unilateral uptake. In two cases of adrenal carcinoma no uptake was seen on either side (Moses *et al.*, 1974). This technique is said to be particularly valuable for visualizing remnants of adrenal tissue, missed at previous operation. CAT scanning of the adrenal glands is also proving of use in the diagnosis of adrenal tumours.

C. SUMMARY OF DIAGNOSIS OF CUSHING'S SYNDROME

In screening for suspected Cushing's syndrome in an out-patient, the short dexamethasone suppression test is probably the most useful test. When the patient is in hospital, the diurnal rhythm of plasma cortisol should be examined, the most important diagnostic level being an elevated cortisol concentration in the evening. If these tests are abnormal or equivocal or if clinical suspicion is high, then further investigation is necessary. The urinary free cortisol should be measured on two occasions, and if necessary an insulin hypoglycaemia test may help the diagnosis.

In establishing the exact aetiology of the disease in each case, measurement of the plasma ACTH level is most useful, but if not possible, tests of the long dexamethasone suppression or the metyrapone type may help. In pituitary-dependent Cushing's syndrome, ACTH is detectable in plasma and especially, if measured at midnight, may be elevated above normal. There is suppression of urinary 17 OGS or plasma cortisol levels with the high but not the low dose of dexamethasone, and a doubling of 17 OGS excretion is seen after metyrapone. It may be possible to demonstrate a pituitary tumour by radiological examination.

A patient with an adrenal adenoma should have no detectable ACTH in plasma, no suppression with either dose of dexamethasone, and a subnormal response to metyrapone. The urinary 17 OS are only moderately raised and it may be possible to demonstrate an adrenal tumour by angiography or retrograde venography. In adrenal carcinoma, plasma ACTH levels, and responses to dexamethasone or metyrapone, are similar to those seen with an adenoma. However, the 17 OS are often grossly raised, and fine calcification may be seen in the tumour. Nevertheless the distinction between adenoma and carcinoma is often not made until operation and even then the histology may be equivocal (Fig. 17). Patients with the ectopic ACTH syndrome have high levels of plasma cortisol and ACTH, are not suppressed by dexamethasone, and do not respond to metyrapone. If the ACTH level is greater than 200 ng/l then ectopic secretion is likely but levels overlapping with pituitary-dependent disease may be found. A marked hypokalaemic alkalosis is characteristic and it may be possible to demonstrate the primary or secondary tumour which is secreting ACTH.

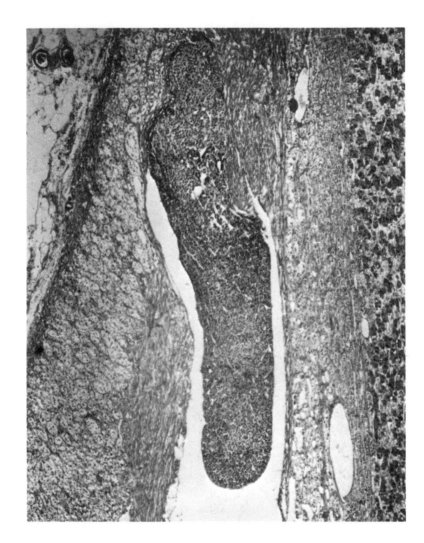

Fig. 17. Histology of an adrenal carcinoma seen to be invading a vein. This was thought to be an adenoma macroscopically.

Finally it must be remembered that paradoxical results have been recorded with virtually all the diagnostic tests used at present so that in cases of real doubt it may be necessary to perform all the available tests and take a consensus of the results.

7. Nelson's syndrome

Following adrenalectomy a proportion of patients with pituitary-dependent Cushing's disease develop a syndrome of hyperpigmentation, excessively high plasma ACTH levels and evidence of a pituitary tumour. This syndrome was first characterized by Nelson and now bears his name (Nelson et al., 1958; Nelson and Meakin, 1959; Nelson et al., 1960). The reported incidence of this complication varies from 15 to 30% (Welbourne et al., 1971; Nabarro and Brook, 1975; Nabarro, 1977). It is unusual to see it less than a year after adrenalectomy and it may take up to 14 years to become apparent.

A.   PATHOGENESIS

Plasma ACTH levels are elevated in all patients with Cushing's disease after adrenalectomy (Fig. 12) (Besser and Edwards, 1972) but in those who develop Nelson's syndrome, levels exceeding 1000 ng/l are seen. Cook et al. (1976) showed that in pituitary-dependent Cushing's syndrome, ACTH is suppressible by large, but not small, doses of dexamethasone. If this be the case then cure of excessively high cortisol levels by adrenalectomy would remove the suppressive effect of high dose steroids, allowing the plasma ACTH level to rise. It has also been postulated that high levels of circulating steroids might have a suppressive effect on the growth of a pituitary tumour, the subsequent fall in steroids allowing the pituitary to enlarge after adrenalectomy. It is not clear why only a proportion of patients develop this complication but it is possibly due to some cases of Cushing's disease being primarily hypothalamic rather than pituitary in origin. The pituitary tumours associated with this condition were originally described as chromophobe adenomas (Nelson et al., 1960) but are probably poorly granulated ACTH secreting tumours. ACTH has been demonstrated in the tumours by both immunofluorescence and electron microscopy (Cassar et al., 1976). Unlike most other pituitary adenomas these tumours are often locally invasive and distant metastases have been reported (Salassa et al., 1959).

B. DIAGNOSIS

The distribution of excess pigmentation is similar to that seen in Addison's disease but is usually much more intense and rapidly progressive. Visual field defects and cranial nerve palsy due to local invasion by the tumour may be seen. Radiological enlargement of the pituitary fossa is found in about 75% of cases but careful tomography shows abnormalities in nearly 100% (Weinstein et al., 1976). An CAT scan or air encephalogram should also be done to assess extrasellar extension. The diagnosis is confirmed by plasma ACTH levels exceeding 1000 ng/l but a progressive rise in ACTH levels following adrenalectomy should lead to earlier diagnosis.

C. PREVENTION AND TREATMENT

In view of the increasing incidence of this complication the treatment of pituitary-dependent Cushing's disease by adrenalectomy alone should be reconsidered. It is noteworthy that in the series reported by Orth and Liddle (1971) no cases of Nelson's syndrome were seen in 28 patients who received pituitary irradiation before adrenalectomy after an average follow-up of 9 years. Burke et al. (1973) also found no evidence of Nelson's syndrome in 31 cases of Cushing's disease with normal pituitary fossae treated by pituitary implants with or without adrenalectomy. However Wild et al. (1973) have reported the occurrence of Nelson's syndrome in two patients who received pituitary irradiation prior to adrenalectomy for Cushing's disease.

If total adrenalectomy alone is performed, patients must be carefully followed up with repeated measurements of plasma ACTH levels to detect early signs of the syndrome. Because of the invasive nature of the tumours seen in these cases, an aggressive policy of treatment is probably justified. Increasing pigmentation and rising ACTH levels without evidence of a pituitary tumour are best treated by external irradiation to the pituitary. Because of the relatively small numbers reported it is difficult to establish the best form of treatment for a definite pituitary tumour. Unfortunately, treatment is often unsatisfactory because of frequent extrasellar extension of the tumour and all attempts to eradicate it may fail (Clinicopathological Conference, 1969).

Surgical hypophysectomy supplemented by radiotherapy is the best generally available treatment, although it is not always successful (Nelson et al., 1960; Salassa et al., 1959; Welbourne et al., 1971). Microsurgical transnasal techniques have also been used but the results still need to be fully evaluated (Espinoza *et al.*, 1973). External irradiation alone has produced some cure but is less reliable (Nelson *et al.*, 1960; Salassa *et al.*, 1959). Pituitary implantation with yttrium-90 or gold-198 produced good results in 8 patients without extrasellar extension but this technique is not suitable where there is a large suprasellar extension, and in such cases a transfrontal craniotomy is indicated (Cassar *et al.*, 1976). Cryosurgery (Harrison *et al.*, 1970) and heavy particle therapy (Linfoot *et al.*, 1970) have also been used but these techniques are not generally available.

The aim of treatment should be eradication of the pituitary tumour, prevention of extrasellar extension and regression of the hyperpigmentation. It is probably not necessary to reduce the plasma ACTH levels to normal since of the patients who have had bilateral adrenalectomies, all have had raised plasma levels of ACTH in the absence of any evidence of Nelson's syndrome (Besser and Edwards, 1972).

8. Treatment of Cushing's syndrome

Effective treatment is essential in Cushing's syndrome whatever the aetiology since the mortality in inadequately treated cases is high. In the series reported by Plotz *et al.* (1952) 17 of 33 untreated patients died within five years of the known onset of the disease. Whilst spontaneous remissions have occurred in a small number of patients with Cushing's syndrome (Cushing, 1932; Pasqualini and Gurevich, 1956; Zondek and Leszynsky, 1956; Aber and Cheetham, 1961; Ross *et al.*, 1966), these are rare and treatment is indicated in all patients once the diagnosis is established. Ideally, treatment should leave the patient with physiological levels of circulating cortisol, and an intact stress response. In pituitary-dependent disease further growth of the pituitary tumour should be avoided. Unfortunately at present only a minority of patients achieve this ideal.

A. ACTH-DEPENDENT CUSHING'S SYNDROME

It is not yet known whether the abnormality in these patients lies mainly in the pituitary (Cushing, 1932) or in the hypothalamus (James *et al.*, 1968). Nevertheless treatment may be directed at the pituitary, the adrenal or both and it is still controversial which approach provides the best results.

1. *Methods of pituitary treatment*

(a) Conventional external irradiation

A dose of about 4–5 K rads is delivered to the pituitary by convention-al X-ray or Cobalt-60 irradiation. This produces cure of the disease in 23% of cases (Orth and Liddle, 1971) without significant hypopi-tuitarism. If this fails, a further proportion will be controlled using aminoglutethimide or metyrapone. In resistant cases, bilateral adre-nalectomy can then be performed and the previous pituitary irradia-tion appears to prevent the subsequent appearance of Nelson's syndrome (Orth and Liddle, 1971).

(b) Gold-198 or yttrium-90 implantation

This technique, which is only available in a few centres, delivers a dose of 5–10 K rads to the pituitary. There is a cure rate of about 60% but approximately 50% of patients are left with some evidence of hypopituitarism, especially hypogonadism (Burke *et al.*, 1973). There is also a small operative morbidity but this procedure also prevents the subsequent development of Nelson's syndrome if bilateral adre-nalectomy is necessary.

(c) Heavy particle irradiation

Use of a proton beam makes it possible to irradiate the pituitary with 10–12 K rads but is limited to those centres with access to a cyclotron. This produces a remission rate of about 60% but is complicated by some hypopituitarism (Lawrence *et al.*, 1976).

(d) Surgical hypophysectomy

Experience with this tecnhique, particularly by the trans-sphenoidal route, is limited, but cure rates of 50 to 60% have been reported (Fletcher *et al.*, 1971). In a more recent series of 13 patients, treated by trans-sphenoidal hypophysectomy (Carmalt *et al.*, 1977), 12 had complete remission and one partial, although there was one death later from endomyocardial fibrosis. There was also a high incidence of preservation of pituitary function post-operatively, with normal men-struation returning in 5 of the 7 pre-menopausal women, 3 of whom subsequently became pregnant. Hardy (1973) also reported good results with this technique though Nabarro (1977), on the contrary, has found trans-sphenoidal surgery to give disappointing results.

(e) Cryosurgery

A few patients have been treated by this method but the numbers are too small to make any objective assessment (Fraser, 1969).

2. *Methods of adrenal treatment*

(a) Subtotal adrenalectomy

This technique was devised in order to cure the high cortisol levels but to leave the patient with sufficient adrenal tissue to obviate the need for replacement therapy. In practice this ideal is difficult to achieve and in one series about 80% of patients treated in this way had impaired adrenocortical reserve (Welbourne *et al.*, 1971). In other patients the adrenal remnant may regenerate producing a recurrence of the disease.

(b) Bilateral adrenalectomy

This undoubtedly cures the high cortisol levels but leaves the patient dependent on glucocorticoid and mineralocorticoid replacement therapy. There is an operative mortality of about 4% (Welbourne *et al.*, 1971) but the major disadvantage of this procedure is the subsequent development of Nelson's syndrome in up to 30% of cases (Nabarro and Brook, 1975; Nabarro, 1977).

(c) Drug therapy

Metyrapone can be used to reduce cortisol levels but even in large doses it only produces a 90% enzyme inhibition so it is not always effective. Aminoglutethimide reduces cortisol synthesis by blocking the conversion of cholesterol to pregnenolone in the adrenal cortex. It is more effective than metyrapone but again its inhibitory effect may be overcome by increased ACTH secretion (Fishman *et al.*, 1967).

The adrenal blocking drugs can produce adverse effects which include rashes, depression and gastro-intestinal disturbances such as nausea, vomiting and diarrhoea. In addition aminoglutethimide may cause ataxia and somnolence as well as severe pruritus and it also has an antithyroid action. As can be anticipated from the site of its blocking action, aminoglutethimide suppresses the production of all steroid hormones. It is probably best prescribed, therefore, together with a small dose of fludrocortisone because of its effect on aldosterone synthesis.

Recent developments in the drug therapy of Cushing's disease are based on the theory that in some cases the primary defect lies in a failure of the hypothalamic regulation of ACTH secretion. This is thought to be caused by an imbalance of the various neurotransmitters found in the hypothalamus and attempts have been made to modify this with drug therapy. Miura *et al.* (1975) reported a high rate of cure in 18 patients using a combination of radiotherapy and reserpine, a dopamine agonist, and claimed the return of a normal nyctohemeral rhythm and dexamethasone suppressability. Kreiger *et al.* (1975) used cyproheptadine, an antihistamine with marked antiserotonin effects, on three patients and produced normal levels of urinary free cortisol and 17 OHCS. However, there was no restoration of the normal nyctohemeral variation and a paradoxical rise in steroid levels was seen in response to dexamethasone suppression. L-dopa has been found to be ineffective in the treatment of Cushing's disease (Krieger *et al.*, 1975). Work with the ergot alkaloid, bromocriptine or 2-bromo-α-ergokryptine-mesylate has shown that it produces marked clinical improvement in some patients with Cushing's disease to give lower levels of plasma ACTH and cortisol (Edwards and Jeffcoate, 1976; Lamberts and Birkenhager, 1976). It is uncertain whether this is due to altered serotonin activity in the hypothalamus or to the effects of suppression of prolactin on adrenal steroidogenesis. The role of bromocriptine in the treatment of Cushing's disease remains to be defined.

3. *Treatment of choice*

The ideal treatment for pituitary-dependent Cushing's disease remains controversial. Theoretically it would seem logical to treat the disease at a pituitary level. Conventional external radiation of the pituitary is effective, however, in only a small proportion of cases, whilst techniques which deliver a higher dose of irradiation to the pituitary are accompanied by an increased risk of hypopituitarism. Although experience with trans-sphenoidal hypophysectomy is still limited, it promises to be an effective method of securing remission in Cushing's disease. At present patients with clinically mild Cushing's disease are probably best treated by external pituitary irradiation in the first instance. If cure of the disorder, assessed clinically and biochemically, is not achieved within three to six months, then bilateral adrenalectomy should be performed. The previous pituitary irradiation then gives some protection against the subsequent appearance of Nelson's syndrome. In more florid disease where a rapid result is required, bilateral adrenalectomy should be performed in the first instance. This should be combined with pituitary irradiation to avoid later enlargement of the pituitary tumour, except when it is important to avoid hypogonadism in patients who have not completed their family. Careful follow up with measurement of ACTH levels is mandatory if pituitary irradiation is not given. In severe cases of Cushing's disease it is desirable to attempt to lower plasma cortisol levels with metyrapone or aminoglutethimide for a few weeks before operation in order to improve the clinical condition of the patient. The treatment of patients with Cushing's disease with drug therapy alone is still in the experimental stage and much greater experience will be required before this can be assessed adequately.

B.   NON-ACTH DEPENDENT CUSHING'S SYNDROME

1.   Adrenal adenoma

Surgical removal is the treatment of choice for this form of Cushing's syndrome but ideally both adrenals should be explored to exclude multiple adenomas. There is usually marked suppression of the hypothalamic–pituitary–adrenal axis and therefore of the contralateral adrenal gland because of the sustained high levels of circulating corticosteroids. A rise in plasma cortisol following tetracosactrin before operation should not be taken to indicate that the normal adrenal can respond to ACTH since the tumour itself may be ACTH responsive (Epstein et al., 1973). In view of this, steroid cover should be given in the same way as for bilateral adrenalectomy.

Although depot tetracosactrin 1 mg daily for 3–4 days will restore adrenal responsiveness to ACTH, the main problem is hypothalamic–pituitary suppression. For this reason prednisolone 5–10 mg daily which does not interfere with most methods of estimating endogenous corticosteroids is best used for replacement therapy. The prednisolone can then be gradually tailed off over a period of some weeks. The patient's plasma cortisol levels should be measured at intervals and when these are of the order of 300 nmol/l, then an insulin hypoglycaemia test should be performed and if a normal response is demonstrated, the prednisolone can be stopped completely.

2.   Adrenal carcinoma

Unfortunately, operative removal is rarely curative but worthwhile remissions may follow. Recurrence may be detected by rising levels of steroids in blood or urine. Occasionally, recurrence in the adrenal bed may be removed surgically or may respond to radiotherapy. If more distant metastases are present or it is impossible to remove the tumour, then the syndrome may be controlled using the drug ortho, para 'DDD (o,p'DDD). This isomer of the insecticide DDD causes selective necrosis of adrenocortical tissue and has been found to be effective in the treatment of some patients with adrenocortical carcinoma and its metastases. It is given in divided dosage, the starting daily dose of an adult being 2–6 gm which is then increased until the patient's endogenous steroid production is reduced to a minimum or until toxicity becomes a limiting factor. The most common toxic effects of o,p'DDD therapy are: gastro-intestinal disturbances such as anorexia, nausea or vomiting; neurological disturbances such as depression, lethargy, somnolence, dizziness, vertigo, confusion, weakness and paraesthesiae; and skin rashes. O,p'DDD therapy should be accompanied by steroid replacement therapy to prevent adrenal crises and either dexamethasone or prednisolone is best used to allow endogenous steroid levels to be continuously monitored. Since o,p'DDD induces liver enzymes which increase the breakdown of cortisol, a higher than average dose of steroid replacement is usually necessary (Fig. 18). The average survival time of patients with untreated adrenal carcinoma is very poor being 2.9 months (McFarlane, 1958). In a series of 107 inoperable patients the use of o,p'DDD therapy increased the mean survival time to 8.4 months with a range of 1 to 41 months (Lubitz *et al.*, 1973). Better results however have been reported in patients with operable disease where a combination

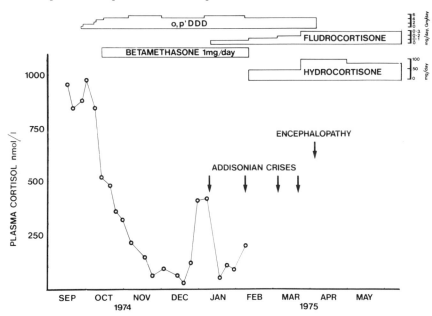

Fig. 18. The effect of o,p'DDD therapy on endogenous cortisol levels arising from metastatic deposits from an adrenocortical carcinoma in a patient in whom both adrenal glands had been removed surgically. Note the repeated episodes of Addisonian crisis necessitating an increase in steroid replacement therapy due to increased breakdown of corticosteroids by the induction of liver enzymes by o,p'DDD.

of surgery, radiotherapy and o,p'DDD therapy has been used (Hutter and Kayhoe, 1966).

3. *Ectopic ACTH syndrome*

Surgical removal of the tumour will effect a cure in these cases but this is rarely possible except in the case of bronchial carcinoid tumours (Olurin *et al.*, 1973). Often the associated tumour is highly malignant and death occurs within a few weeks. Attempts can be made to control the tumour with chemotherapy or radiotherapy and symptomatic relief can be obtained by reducing the patient's cortisol production with aminoglutethimide or metyrapone. O,p'DDD takes too long to work and is too toxic for most patients. Plasma ACTH levels are useful for monitoring the response to treatment of the tumour and often rise considerably before clinical relapse is evident (Besser and Edwards, 1972).

9. Conclusions

In both adrenal underactivity and overactivity, a high clinical index of suspicion is required to allow early diagnosis. These disorders require full biochemical investigation, to confirm the diagnosis, to localize precisely the endocrine defect, to allow the most appropriate therapy, and to monitor subsequent progress. These are conditions which need lifelong follow up, preferably in specialized units.

References

Abe, K., Nicholson, W. E., Liddle, G. W., Orth, D. N. and Island, D. P. (1969). *J. clin. Invest.* **48,** 1580–1585.

Aber, C. P. and Cheetham, H. D. (1961). *Br. med. J.* **1,** 336–338.

Addison, T. (1855). *In* "On the Constitutional and Local Effects of Disease of the Suprarenal Capsules" London, Samuel Highly VII–VIII, 1–39, plates I–XI. Reprinted by Dawsons of Pall Mall, 1968.

Albright, F. (1943). *The Harvey Lectures* **38,** 123–186.

Ask-Upmark, E. and Hull, R. (1972). *Acta. med. scand.* **192,** 445–446.

Azzopardi, J. G. and Williams, E. D. (1968). *Cancer* **22,** 274–286.

Barker, N. W. (1929). *Arch Path.* **8,** 432–450.

Beaven, D. W., Nelson, D. H., Renold, A. E. and Thorn, G. W. (1959). *New Engl. J. Med.* **261,** 443–454.

Besser, G. M. (1973). *Clin. Endocr.* **2,** 175–186.

Besser, G. M. and Edwards, C. R. W. (1972). *In:* "Clinics in Endocrinology and Metabolism". Vol. 1 No. 2 451–490. W. B. Saunders Co. Ltd., London, Philadelphia, Toronto.

Besser, G. M., Cullen, D. R., Irvine, W. J., Ratcliffe, J. C. and Landon, J. (1971). *Br. Med. J.* **1,** 374–376.

Besser, G. M. and Landon, J. (1968). *Br. med. J.* **ii,** 552–554.

Bettige, H., Bayer, J. M. and Winklemann, W. (1969). *Acta endocr. Copnh.* **60,** 47–59.

Bloodworth, J. M. B., Kirkendall, W. M. and Carr, T. L. (1954). *J. clin. Endocr. Metab.* **14,** 540–560.

Bondarevsky, E., Shapiro, M. S., Schey, G., Shahor, J. and Bruderman, I. (1976). *J. Am. med. Ass.* **236,** 1969–1971.

Brown, W. H. (1928). *Lancet* **ii,** 1022–1023.

Brown, R. D., Van Loon, G. R., Orth, D. N. and Liddle, G. W. (1973). *J. clin. Endocr. Metab.* **36,** 445–451.

Burke, C. W. and Beardwell, C. G. (1973). *Q. Jl. Med.* **165,** 175–204.

Burke, C. W., Doyle, F. H., Joplin, G. F., Arnot, R. N., Macerlean, D. P. and Russell Fraser, T. (1973). *Q. Jl. Med.* **42,** 693–714.

Butler, P. W. P. and Besser, G. M. (1968). *Lancet* **i,** 1234–1236.

Carmalt, M. H. B., Dalton, G. A., Fletcher, R. F. and Thomas Smith, W. (1977). *Q. Jl. Med.* **181,** 119–134.

Carpenter, C. C. J., Solomon, N., Silverberg, S. G., Bledsoe, T., Northcutt, R. C., Klinenbert, J. R., Bennet, I. L. and McGehee-Harvey, A. (1964). *Medicine (Baltimore)* **43,** 153–180.

Cassar, J., Doyle, F. H., Lewis, P. D., Mashiter, K., Van Noorden, S. and Joplin, G. F. (1976). *Br. med. J.* **2,** 269–272.

Choo-Kang, Y. F. J., Cooper, E. J., Tribe, A. E. and Grant, I. W. B. (1972). *Br. J. Dis. Chest* **66,** 101–106.

Clinicopathological Conference (1969). *Br. med. J.* **2,** 557–560.

Connolly, C. K., Gore, M. B. R., Stanley, N. and Wills, M. R. (1968). *Br. med. J.* **ii,** 665–667.

Conybeare, J. J. and Millis, G. C. (1924). *Guy's Hosp. Rep.* **74,** 369–375.

Cook, D. M., Kendal, J. W., Allen, J. P. and Lagerquist, L. G. (1976). *Clin. Endocr.* **5,** 303–312.

Cope, C. L. (1966). *Br. med. J.* **ii,** 847–853.

Cope, C. L. and Black, E. G. (1958). *Clin. Sci.* **17,** 147–163.

Cope, C. L., Dennis, P. M. and Pearson, J. (1966). *Clin. Sci.* **30,** 249–257.

Crane, M. G. and Harris, J. J. (1966). *J. clin. Endocr. Metab.* **26,** 1135–1143.

Cunliffe, W. J., Hall, R., Newell, D. J. and Stevenson, C. J. (1968). *Br. J. Derm.* **80,** 135–139.

Cushing, H. (1912). "The Pituitary Body and its Disorders" Lippincott Philadelphia and London.

Cushing, H. (1932). *Bull. Johns Hopkins Hosp.* **50,** 137–195.

Daly, J. R., Fletcher, M. R., Glass, D., Chambers, D. J., Bitensky, L. and Chayen, J. (1974). *Br. med. J.* **2,** 521–524.

Dunlop, D. M. (1963). *Br. med. J.* **2,** 887–891.

Eddy, R. L., Jones, A. L., Gilliland, P. F., Ibarra, J. D., Thompson, J. Q. and McMurray, J. F. (1973). *Am. J. Med.* **55,** 621–630.

Edwards, C. R. W. and Jeffcoate, W. J. (1976). *In:* "Pharmacological and Clinical Aspects of Bromocriptine (Parlodel)". Royal College of Physicians Symposium, May, 1976. (Eds. R. I. S. Bayliss, P. Turner and W. P. Maclay), pp. 42–51.

Edwards, O. M., Courtenay-Evans, R. J., Galley, J. M., Hunter, J. and Tait, A. D. (1974). *Lancet* **ii,** 549–551.

Epstein, S., McLaren, E. H. and Goldin, A. R. (1973). *Post Grad. Med. J.* **49,** 923–926.

Ernest, I. (1966). *Acta endocr. Copnh.* **51,** 511–525.

Espinoza, A., Nowakowski, H., Kautzky, R. and Ludecke, D. (1975). *Acta endocr. Copnh. Suppl.* **173,** 34.

Fahmy, D., Read, G. F. and Hillier, S. G. (1975). *Steroids* **26,** 267–280.

Feiwel, M., James, V. H. T. and Barnett, E. S. (1969). *Lancet* **i,** 485–487.

Few, J. D. and Cashmore, G. C. (1971). *Ann. clin. Biochem.* **8,** 205–209.

Fishman, L. M., Liddle, G. W., Island, D. P., Fleischer, N. and Kuchel, O. (1967). *J. clin. Endocr. Metab.* **27,** 481–490.

Fletcher, R., Dalton, G. A., Carmalt, M. H. B. and Smith, W. T. (1971). *Acta endocr. Copnh. Suppl.* **155,** 163.

Fraser, C. G., Preuss, F. S. and Bigford, W. D. (1952). *J. Am. med. Ass.* **149,** 1542–1543.

Fraser, T. R. (1969). *In:* "Progress in Endocrinology" (C. Gual, ed.) p. 1211. Excerpta Medica I.C.S. 184, Amsterdam.

French, F. S., Macfie, J. A., Baggett, B., Williams, T. F. and Van Wyk, J. J. (1968). *Am. J. Med.* **47,** 619–624.

Gray, C. H., Barn, D. N., Brooks, R. V. and James, V. T. H. (1969). *Lancet* **i,** 124–127.

Greenblatt, R. B. (1965). *In:* "The Hirsute Female" 75. C. C. Thomas, Springfield, Illinois.

Greig, W. R., Jasani, N. K., Boyle, J. A. and Maxwell, J. D. (1968). *Mem. Soc. Endocr.* **17,** 175–192.

Guttman, P. H. (1930a). *Archs Path.* **10,** 742–785.

Guttman, P. H. (1930b). *Archs Path.* **10,** 895–935.

Gwinup, G. (1965). *Lancet* **ii,** 572–573.

Hardy, J. (1973). *In:* "Diagnosis and Treatment of Pituitary Tumours". (Eds. P. O. Kohler and G. T. Ross). Excerpta Medica, Amsterdam.

Harrison, M. T., Jennett, W. B. and Cross, J. N. (1970). *Proc. Roy. Soc. Med.* **63,** 224–225.

Heinbecker, P. (1944). *Medicine (Baltimore)* **23,** 225.

Hellman, L., Weitzman, E. D., Roffwarg, H., Fukushima, D. K., Yoshida, K. and Gallacher, T. F. (1970). *J. clin. Endocr. Metab.* **30,** 686–689.

Holdway, I. N. (1973). *Clin. Endocr.* **2,** 37–41.

Hutter, A. M. and Kayhoe, D. E. (1966). *Am J. Med.* **41,** 572–580.

Idelman, S. (1978). *In:* "General, Comparative and Clinical Endocrinology of the Adrenal Cortex". (I. Chester Jones and I. W. Henderson, eds.) Vol. 2. pp. 1–199. Academic Press, London and New York.

Irvine, W. J. and Barnes, E. W. (1972). *Clinics in Endrocrinology and Metabolism* **1,** 549–594.

Irvine, W. J. and Barnes, E. W. (1975). *In:* "Clinical Aspects of Immunology". 130–154. Blackwells, Oxford.

Irvine, W. J., Stewart, A. G. and Scarth, L. (1967). *Clin. exp. Immunol.* **2,** 31–69.

James, V. H. T., Landon, J. and Wynne, V. (1965). *J. Endocr.* **33,** 515–524.

James, V. H. T., Landon, J., Wynne, V. and Greenwood, F. C. (1968). *J. Endocr.* **40,** 15–28.

Janasi, M. K., Freeman, P. A., Doyle, J. A., Reid, A. M., Diver, M. J. and Buchanan, W. W. (1968). *Q. Jl. Med.* **37,** 407–421.

Jenkins, D., Forsham, P. H., Laidlaw, J. G., Reddy, W. J. and Thorn, G. W. (1955). *Am. J. Med.* **18,** 3–14.

Kehlet, H. and Binder, C. (1973). *Br. J. Anaesth,* **45,** 1043–1048.

Kehlet, H., Binder, C. and Blichert-Toft, M. (1976). *Clin. Endocr.* **5,** 37–41.

Kepler, E. J., Kennedy, R. L. J., Davis, A. C., Walters, W. and Wilder, R. M. (1934). *Proc. Staff Meet. Mayo Clinic* **9,** 169–181.

Kime, D. E. (1978). *In:* "General, Comparative and Clinical Endocrinology of the Adrenal Cortex". (I. Chester Jones and I. W. Henderson, eds.). Vol. 2. pp. 265–290. Academic Press, London and New York.

Knapp, M. S., Keane, P. M. and Wright, J. G. (1967). *Br. med. J.* **2,** 27–30.

Krakoff, L., Nicholis, G. and Amsel, B. (1975). *Am. J. Med.* **58,** 216–220.

Krieger, D. T., Amorosa, L. and Linick, F. (1975). *New Engl. J. Med.* **293,** 893–896.

Laidlaw, J. C., Reddy, W. J., Jenkins, D., Haydan, N. A. Renold, A. E. and Thorn, G. W. (1955). *New Engl. J. Med.* **253,** 747–753.

Lamberts, S. W. J. and Birkenhager, J. C. (1976). *J. Endocr.* **70,** 315–316.

Landon, J. and Greenwood, F. C. (1968). *Lancet* **i,** 273–276.

Lawrence, J. H., Tobias, C. A., Linfoot, J. A., Born, J. L. and Chong, C. Y. (1976). *J. Am. med. Ass.* **235,** 2307–2310.

Lehner, T. and Lyne, C. (1969). *Br. med. J.* **4,** 138–141.

Lewis, L., Robinson, R. F., Yee, J., Hacken, L. A. and Eisen, G. (1953). *Ann. intern. Med.* **39,** 116–126.

Liddle, G. W. (1960). *J. clin. Endocr. Metab.* **20,** 1539–1560.

Liddle, G. W., Givens, J. R., Nicholson, W. E. and Island, D. P. (1965). *Cancer Res.* **25,** 1057–1061.

Liddle, G. W., Island, D. P. and Meador, C. K. (1962). *Recent Progr. Horm. Res.* **18,** 125–166.

Liddle, G. W., Nicholson, W. E., Island, D. P., Orth, D. N., Abe, K. and Lowder, S. C. (1969). *Recent Progr. Horm. Res.* **25,** 283–305.

Linfoot, J. A., Lawrence, J. H., Tobias, C. A., Born, J. L., Chong, C. Y., Lyman, J. T. and Monongian, E. (1970). *Trans. Am. Clin. Climatol. Ass.* **81,** 196–212.

Livanou, T., Ferriman, D. and James, V. H. T. (1967). *Lancet* **ii,** 856–859.

Lubitz, J. A., Freeman, L. and Okun, R. (1973). *J. Am. med. Ass.* **223,** 1109–1112.

Maberly, D. J., Gibson, G. J. and Butler, A. G. (1973). *Br. med. J.* **1,** 778–782.

Macfarlane, D. A. (1958). *Ann. R. Coll. Surg. Eng.* **23,** 155–164.

McHardy-Young, S., Lessof, M. H. and Maisey, M. N. (1972). *Clin. Endocr.* **1,** 45–56.

Maisey, M. N. and Lessof, M. H. (1969). *Guy's Hosp. Rep.* **118,** 362–372.

Malone, D. N. S., Grant, I. W. B. and Percy-Robb, I. W. (1970). *Lancet* **ii,** 733–735.

Mason, S. A., Meade, T. W., Lee, J. A. H. and Morris, J. N. (1968). *Lancet* **ii,** 744–747.

Mattingly, D. (1962). *J. clin. Path.* **15,** 374–379.

Mattingly, D., Keane, P. M., McCafthy, C. R. and Read, A. E. (1964). *Bristol Medicochirurgical Journal* **79,** 6–14.

Meador, C K., Liddle, G. W., Island, D. P., Nicholson, W. E., Lucas, C. P. Nuckton, J. G. and Luetscher, J. A. (1962). *J. clin. Endocr. Metab.* **22,** 693–703.

Miura, K., Aida, M., Nihara, A., Kato, K., Ojiman, M., Demura, R., Demura, H. and Okuyama, M. (1975). *J. clin. Endocr. Metab.* **41,** 511–526.

Moses, D. C., Schteingart, D. E., Sturman, M. F., Beierwaltes, W. H. and Ice, R. D. (1974). *Surgery Gynec. Obstet,* **139,** 201–204.

Muller, R. and Kugelberg, E. (1959). *J. Neurol. Neurosurg. Psychiat.* **22,** 314–319.

Nabarro, J. D. N. (1972). *Br. med. J.* **2,** 492–495.

Nabarro, J. D. N. and Brook, C. (1975). *Medicine (2nd series)* 351–369.
Nabarro, J. D. N. (1977). *J. R. Coll. Phys.* **11**, 363–375.
Nelson, D. H., Meakin, J. W., Dealy, J. B., Matson, D. D., Emerson, K. and Thorn, G. W. (1958). *New Engl. J. Med.* **259**, 161–164.
Nelson, D. H. and Meakin, J. W. (1959). *J. clin. Invest.* **38**, 1028–1029.
Nelson, D. H., Meakin, J. W. and Thorn, G. W. (1960). *Ann. intern. Med.* **52**, 560–569.
Nelson, D. H., Sprunt, J. G. and Mims, R. B. (1966). *J. clin. Endocr. Metab.* **26**, 722–728.
Nerup, J. (1974). *Danish Medical Bulletin* **21**, 201–217.
Neville, A. M. and Mackay, A. M. (1972). *Clinics in Endocrinology and Metabolism* 361–395.
Neville, A. M. and Symington, T. (1967). *J. Path. Bact.* **93**, 19–35.
Ney, R. L., Shimizu, N., Nicholson, W. E., Island, D. P. and Liddle, G. W. (1963). *J. clin. Invest.* **42**, 1669–1677.
Nichols, T., Nugent, C. A. and Tyler, F. H. (1968). *Am. J. Med.* **45**, 116–128.
Nieman, E. A., Landon, J. and Wynne, V. (1967). *Q. Jl. Med.* **36**, 357–392.
Nugent, C. A., Nichols, T. and Tyler, F. H. (1965). *Archs intern. Med.* **116**, 172–176.
O'Bryan, R. M., Smith, R. W., Fine, G. and Mellinger, R. C. (1964). *J. Am. med. Ass.* **187**, 257–261.
Odell, W. D., Green, G. M. and Williams, R. H. (1960). *J. clin. Endocr. Metab.* **20**, 1017–1028.
O'Neal, L. W., Kepnis, D. M., Luse, S. A., Lacy, P. E. and Jarrett, L. (1968). *Cancer* **21**, 1219–1232.
Olurin, E. D., Sofowara, E. D., Afonja, A. O., Kolawole, T. M. and Junaid, T. A. (1973). *Cancer* **31**, 1514–1519.
Oppenheimer, B. S. and Fishberg, A. M. (1924). *Archs intern. Med.* **34**, 631–644.
Orth, D. N. and Liddle, G. W. (1971). *New Engl. J. Med.* **285**, 243–247.
Orth, D. N. (1973). *Nature, Lond.* **242**, 26–28.
Orth, D. N., Nicholson, W. E., Mitchell, W. M. Island, D. P. and Liddle, G. W. (1973). *J. clin. Invest.* **52**, 1756–1759.
Pasqualini, J. R. and Gurevich, N. (1956). *J. clin. Endocr. Metab.* **16**, 406–411.
Perkoff, G. T., Silber, R., Tyler, F. H., Cartwright, G. E. and Wintrobe, M. M. (1959). *Am. J. Med.* **26**, 891–898.
Pimstone, B. L., Uys, C. J. and Vogelpoel, L. (1972). *Am. J. Med.* **53**, 521–528.
Plotz, C. M., Knowlton, A. I. and Ragan, C. (1952). *Am. J. Med.* **13**, 597–614.
Plumpton, F. S., Besser, G. M. and Cole, P. V. (1969). *Anaesthesia* **24**, 3–18.
Ratcliffe, J. G., Knight, R. A., Besser, G. M., Landon, J. and Stansfield, A. G. (1972). *Clin. Endocr.* **1**, 27–44.
Rayfield, E. J., Rose, L. I., Cain, J. P., Dluhy, R. G. and Williams, G. H. (1971). *New Engl. J. of Med.* **284**, 591–592.
Reuter, S. R., Blair, A. J., Schteingart, D. E. and Bookstein, J. J. (1967). *Radiology* **89**, 805–814.
Rickards, A. G. and Barrett, G. M. (1954). *Q. Jl. Med.* **23**, 403–423.
Riggs, B. L. and Sprague, R. G. (1961). *Archs intern. Med.* **108**, 841–849.
Robinson, B. H. B., Mattingly, D. and Cope, C. L. (1962). *Br. med. J.* **1**, 1579–1584.
Rose, L. I., Williams, C. M., Jagger, P. I. and Lauler, D. P. (1970). *Ann. intern. Med.* **73**, 49–54.
Ross, E. J. (1966). *Proc. Roy. Soc. Med.* **59**, 335–338.
Ross, E. J., Marshall-Jones, P. and Friedman, M. (1966). *Q. Jl. Med.* **138**, 149–192.

Rowntree, L. G. and Snell, A. M. (1931). "A Clinical Study of Addison's Disease, Mayo Clinic Monographs." W. B. Saunders Philadelphia and London.

Salassa, R. M., Bennet, W. A., Keating, F. R. and Sprague, R. G. (1953). *J. Am. med. Ass.* **152,** 1509–1515.

Salassa, R. M., Kearns, T. P., Kernohan, J. W., Sprague, R. G. and McCarthy, C. S. (1959). *J. clin. Endocr. Metab.* **19,** 1523–1539.

Sampson, P. A., Brooke, B. N. and Winstone, N. E. (1961). *Lancet* **i,** 337.

Sandor, T., Fazekas, A. G. and Robinson, B. H. (1976). *In:* "General, Comparative and Clinical Endocrinology of the Adrenal Cortex" (I. Chester Jones and I. W. Henderson, eds.). Vol. 1. pp. 25–142. Academic Press, London and New York.

Schambelan, M., Slaton, P. E. and Biglier, E. G. (1971). *Am. J. Med.* **51,** 299–303.

Schmidt, M. B. (1926). *Verh. dt. Path. Ges.* **21,** 212–220.

Scholz, D. A., Sprague, R. and Kernohan, J. W. (1957). *New Engl. J. Med.* **256,** 833–837.

Scoggins, R. B. and Kilman, B. (1965). *New Engl. J. Med.* **273,** 831–840.

Sheridan, P. and Mattingly, D. (1975). *Lancet* **ii,** 676–678.

Silber, R. H. and Porter, C. C. (1954). *J. biol. Chem.* **210,** 923–932.

Soffer, L. J., Dorfman, R. I. and Gabrilove, J. L. (1961). "The Human Adrenal Gland." Lea and Febiger, Philadelphia.

Soffer, L. J., Iannaccone, A. and Gabrilove, J. L. (1961). *Am. J. Med.* **30,** 129–146.

Sprague, R. G., Salassa, R. M., Randall, R. V., Scholz, D. A., Priestly, J. T., Walters, W. and Bulbulian, A. H. (1956). *Archs intern. Med.* **98,** 389–398.

Steiner, A. L., Goodman, A. D. and Powers, S. R. (1968). *Medicine (Baltimore)* **47,** 371–409.

Studzinski, G. P., Hay, D. C. F. and Symington, T. (1963). *J. clin. Endocr. Metab.* **23,** 248–254.

Symington, T. (1969). *In:* "Functional Pathology of the Adrenal Gland". E. and S. Livingstone, Edinburgh and London.

Thompson, K. W. and Eisenhardt, L. (1943). *J. clin. Endocr. Metab.* **3,** 445–452.

Thorn, C. W. (1951). *In:* "The Diagnosis and Treatment of Adrenal Insufficiency". Charles C. Thomas, Springfield Illinois.

Thorn, G. W., Dorrance, S. S. and Day, E. (1942). *Ann. intern. Med.* **16,** 1053–1064.

Turney, H. G. (1913). *Proc. Roy. Soc. Med.* **6,** 69–94.

Vermess, M., Schour, L. and Jaffe, E. S. (1972). *Br. J. Radiol.* **45,** 621–623.

Walters, W., Wilder, R. M. and Kepler, E. J. (1934). *Ann. Surg.* **100,** 670–688.

Wang, C. C. and Robbins, L. L. (1956). *Radiology* **67,** 17–24.

Weber, F. P. (1926). *Br. J. Derm.* **38,** 1–19.

Weinstein, M., Tyrell, B. and Newton, T. H. (1976). *Radiology* **118,** 363–365.

Welbourne, R. B., Montgomery, D. A. D. and Kennedy, T. L. (1971). *Br. J. Surg.* **58,** 1–16.

Westerhof, L., Van Ditmars, M. J., Der Kindersen, P. J., Thijssen, J. H. H. and Schwarz, F. (1972). *Br. med. J.* **2,** 195–197.

Wild, W., Nicholis, G. L. and Gabrilove, J. L. (1973). *Mt. Sinai J. Med. New York* **40,** 68–71.

Wood, J. B., Frankland, A. W., James, V. H. T. and Landon, J. (1965). *Lancet* **i,** 243–245.

Yalow, R. S., Gluk, S. M., Roth, J. and Berson, S. A. (1964). *J. clin. Endocr.* **24,** 1219–1225.

Zondek, H. and Leszynsky, H. E. (1956). *Br. med. J.* **1,** 197–200.

3. Metabolic Effects and Modes of Action of Glucocorticoids

Miguel Beato and Detlef Doenecke

Institut für Physiologische Chemie, Philipps-Universität, D 3550 Marburg/Lahn, F.R.G.

1. Introduction

For several decades, glucocorticoid hormones have been known to influence mammalian carbohydrate metabolism (Long *et al.*, 1940). The progress in enzyme research as well as in molecular and cell biology has greatly extended our knowledge on the mechanism of action of adrenocortical hormones. This review attempts to summarize current knowledge about the interactions of glucocorticoids with target cells to influence nucleic acid, protein and carbohydrate metabolism. In most cases, the effect of glucocorticoids on carbohydrate metabolism is preceded by actions on nucleic acid and protein synthesis, either in the same target organ or in other tissues. In other cases, the hormone evokes a complicated series of metabolic responses which eventually lead to cell differentiation. However, the general mechanisms eliciting these effects involve modulation of certain specific enzyme activities.

It should be mentioned that several steps in this process have still to be elucidated, and as will become obvious the physiological actions of glucocorticoid hormones must be viewed within a metabolic framework, in which several other hormones are simultaneously involved. Some of these hormones, unlike glucocorticoids, regulate enzyme activities by controlling the level of cyclic adenosine monophosphate in target cells (for reviews, see Robison *et al.*, 1968; Braun and Birnbaumer, 1975), but even in this case, it is known that certain cellular sensitivities to such hormones can be enhanced by adrenal steroids ("permissive action").

The metabolic pathways, under glucocorticoid control, are influenced by several variously acting hormones, and the concentration of metabolites resulting from hormonal actions may again affect enzyme activities in putative glucocorticoid target cells. It is beyond the scope of this review to describe the actions of hormones other than glucocorticoids nor will it be possible to mention all factors influencing the activity of enzymes themselves modulated by adrenocorticosteroids.

Any concepts concerning the mechanisms of glucocorticoid action must account for the apparent paradox that these hormones may be anabolic in some tissues, ultimately inducing the synthesis of macromolecules, and catabolic in others, inducing cellular breakdown and tissue regression. Both these effects, at least in part, contribute to enhanced gluconeogenesis, and both anabolic and catabolic actions are initiated by similar mechanisms in target cells.

Glucocorticoids participate in the regulation of gluconeogenesis, a metabolic pathway which is under the regulatory influence of several other signals and hormones, including glucagon which stimulates gluconeogenesis (Schimassek and Mitzkat, 1963; Schimassek, 1967; Ross *et al.*, 1967; Exton and Park, 1967, 1968; Menahan *et al.*, 1968; Williamson *et al.*, 1969) from lactate and several amino acids. This discussion concentrates on aspects of glucocorticoid-induced gluconeogenesis in relationship to the modulation of specific hepatic enzymes, rather than reviewing all the presently known facts about the regulation of gluconeogenesis. Current concepts in this field have been extensively reviewed (Landau, 1965; Seubert, 1967; Schimassek, 1967; Newsholme and Gevers, 1967; Hales, 1967; Weber, 1968; Scrutton and Utter, 1968; Exton et al., 1970; Söling and Willms, 1971; Exton, 1972).

A major advance in understanding the mechanisms of glucocorticoid action was the development of the concept of enzyme induction by steroid hormones (see Karlson et al., 1975). The discovery of such effects (Knox and Mehler, 1951; Knox and Auerbach, 1955; Knox et al., 1956; Lin and Knox, 1957; Greengard et al., 1963) and the formulation of theories for the mechanisms of action (Karlson, 1963) concentrated research on intracellular sites, where the synthesis of proteins is regulated (i.e. the cell nucleus carrying the genes for enzyme protein synthesis and the cytoplasmic site of protein synthesis). The fate of the hormones within the cell was thus studied in a wide variety of tissues known to be responsive to glucocorticoids (King and Mainwaring, 1974). The binding of the hormone to cytoplasmic proteins, its transfer to the cell nucleus, its binding to nuclear components and its influences on RNA and thence protein synthesis are all components of such studies. Several of these steps can be reproduced in vitro to permit analysis of the mechanisms of induction of specific enzymes which are known to be involved in gluconeogenesis.

Before detailing the initial steps of glucocorticoid hormone action, it is necessary to summarize their gross physiological or pharmacological effects on target cell metabolism. Such effects on gluconeogenic tissues have been recently reviewed (Manchester, 1968; Singer and Litwack, 1971; Baxter and Forsham, 1972; Thompson and Lipmann, 1974; King and Mainwaring, 1974; Leung and Munck, 1975; Karlson et al., 1975). Most authors find it necessary to distinguish "peripheral" from "gluconeogenic" tissues. It is tempting to describe the action in gluconeogenic tissues as anabolic, as opposed to the catabolic actions in the periphery; it will however be shown that even glucocorticoid-induced regression of tissues depends on the initial synthesis of RNA and proteins. In addition, glucocorticoids influence the differentiation of several tissues by inducing the synthesis of specific proteins. The following two sections give examples of the actions on gluconeogenesis, catabolism of peripheral tissues and target cell differentiation. In the last section the intracellular mode of action of glucocorticoids is discussed, focusing on the induction of specific hepatic enzymes.

2. Metabolic effects of glucocorticoids

A. EFFECTS OF GLUCOCORTICOIDS ON GLUCONEOGENESIS

1. *General considerations*

Gluconeogenesis describes the synthesis of glucose and glycogen from noncarbohydrate sources such as lactate, pyruvate, glycerol and several amino acids. The sites of the final pathway of gluconeogenesis, i.e. the cells where pyruvate is converted to glucose, are the liver and the kidney cortex, whereas metabolic processes in other organs build up a sequence of steps which supply signal molecules and substrates for the final common pathway (Weber, 1968). Exton (1972) suggested that the functions of gluconeogenesis include the supply of glucose under conditions of reduced carbohydrate intake, the reutilization of lactate and glycerol, the provision of NH_3 in the kidney to circumvent acidosis and finally—in the context of glucocorticoid function—the metabolism of amino acids either from alimentary sources or generated by catabolic processes in peripheral tissues.

The final sequence of reactions whereby glucose-6-phosphate is generated from noncarbohydrate precursors can be regarded as a reversal of the glycolytic pathway, insofar as most enzymatic reactions from pyruvate to glucose-6-phosphate are shared by glycolysis and gluconeogenesis. At several steps, however, other reactions circumvent glycolytic steps, which cannot be utilized for the synthesis of glucose from pyruvate.

The conversion of pyruvate to phosphoenolpyruvate is a key reaction in gluconeogenesis. It is initiated by the mitochondrial (Weiss *et al.*, 1974) enzyme *pyruvate carboxylase,* whose activity depends on the presence of acetyl CoA (Utter and Scrutton, 1969; Utter and Fung, 1970). Oxalacetate is formed and subsequently converted to phosphoenolpyruvate by a second key regulatory enzyme of gluconeogenesis, *phosphoenol–pyruvate carboxykinase.* Enzymes of the Embden–Meyerhof pathway then convert phosphoenolpyruvate to fructose-1, 6-diphosphate. The next step of the glycolytic pathway which must be circumvented during gluconeogenesis is the phosphofructokinase reaction. In gluconeogenesis, reversal of this step is achieved by 1,6 *diphosphofructose phosphatase.* Fructose-6-phosphate is converted by phosphohexose isomerase to glucose-6-phosphate which can be hydrolysed by *glucose-6-phosphatase* to raise free glucose levels.

Glucose yielded by gluconeogenesis can finally be stored as glycogen after stepwise conversion into glucose-1-phosphate and then UDP-glucose into glycogen. The final reaction is catalysed by glycogen synthetase, an enzyme which exists in active and inactive forms (Friedman and Larner, 1963; Bishop and Larner, 1967).

Besides hepatic cells, those of the renal cortex are actively gluconeogenic (for review, see Cohen and Barac-Nieto, 1973; Cahill and Aoki, 1975). A key factor affecting gluconeogenesis is metabolic acidosis (Goodman et al., 1966). Ammonia, which is excreted under these conditions, is mainly derived from glutamine and glutamate (Pitts, 1964). The product is α-ketoglutarate which, in rat kidney, is converted to glucose (Kamm and Strope, 1972) by the gluconeogenic pathway. Several models have been proposed for this relationship between gluconeogenesis and ammonia production (see Alleyne and Roobol, 1975). In each case, an enhanced activity of phosphoenol pyruvate carboxykinase is relevant to the gluconeogenic effect.

The supply of amino nitrogen from the liver, through the action of the induced amino acid metabolizing enzymes, leads to an increase in the synthesis of glutamic acid from α-ketoglutarate. After dehydrogenation, α-ketoglutarate is regenerated and ammonium ions are liberated, which, in turn, can become involved in purine biosynthesis. This sequence of events, postulated by Kornel (1973), relates the induction of amino acid metabolizing enzymes to carbohydrate and purine synthesis.

Pathways that increase blood glucose and glycogen deposition can be affected at each of the described levels by: changes in the supply, transport and uptake of substrates and metabolites; altered activities of several enzymes; and modified amounts of enzyme molecules as a result of induction of their biosynthesis. The regulated enzymes include not only gluconeogenic enzymes but also those which are involved in the supply of substrates.

Several reviews describe the current knowledge of the control of gluconeogenesis (Exton, 1972; Exton and Park, 1966, 1967; Exton et al., 1970; Hübener and Staib, 1965; Seubert, 1967; Söling and Willms, 1971).

2. Control by substrate supply

The supply of substrates for gluconeogenesis is a major regulatory factor and includes the supply of lactate (although the source of lactate is carbohydrate, it should be included in the scheme of gluconeogenesis, Krebs, 1964), pyruvate, glycerol and amino acids (Exton and Park, 1967; Exton et al., 1970).

The influence of substrates on the regulation of the opposing

metabolic processes, glycolysis and gluconeogenesis, has been examined in studies utilizing liver perfusion techniques (Schimassek,
1967). The role of glucose (Buschiazzo et al., 1970) as well as fructose
and lactate (Wimhurst and Manchester, 1973) in initiating adaptive
changes in glycolytic and gluconeogenic enzymes was thus established.

Using inhibitors of lipolysis, Froehlich and Wieland (1971) showed
that the supply of fatty acids is necessary for the full stimulatory effect
of glucagon on gluconeogenesis in perfused livers.

Amino acid mobilization in peripheral tissues appears to be a major
mechanism whereby glucocorticoids stimulate gluconeogenesis.
Those amino acids, converted to pyruvate or oxalacetate, can act as
substrates for gluconeogenic enzymes, and other amino acid degradation products can also be incorporated into the glucose molecule.

Amino acid metabolism not only furnishes substrate molecules for
gluconeogenesis, but several amino acids themselves influence the
activity of glycolytic and gluconeogenic enzymes. Friedrichs and
Schoner (1974), for example, have shown that alanine stimulates
gluconeogenesis by inhibiting pyruvate kinase. Thus, a stimulated
supply of certain amino acids increases hepatic gluconeogenesis both
as a substrate and via a catalytic action.

3.   Control of enzyme activity

The activity of the key gluconeogenic enzymes and glycogen synthetase has been tested under many metabolic conditions; adrenalectomy and administration of glucocorticoids are relevant to this
review.

Enzymes of the gluconeogenic pathway are influenced by glucocorticoids through activation and induction (see below). Pyruvate carboxylase present in liver, brain, adipose tissue and kidney is controlled by the substrate concentration, by activators, such as acetyl CoA
and by inhibitors, such as acetoacetyl-CoA (Utter and Scrutton, 1969;
Utter and Fung, 1970). In isolated liver cells a comparison between
glucagon-, cyclic AMP-, or adrenaline-stimulated glucose synthesis
and mitochondrial pyruvate carboxylase activity revealed the latter to
be a major point in glucagon and catecholamine regulation of
gluconeogenesis (Garrison and Haynes, 1975). Cortisol stimulates
pyruvate carboxylase (Henning et al., 1963; Seubert, 1967) and
phosphoenolpyruvate carboxykinase (Shrago et al., 1963; Foster et al.,
1966). Since this effect is in part inhibited by actinomycin D, the
induction of the respective enzymes or protein factors contributing to

their mode of action seem likely factors controlled by the hormone (see below).

On the other hand acetyl CoA, an activator of pyruvate carboxylase, increases with starvation (Wieland and Weiss, 1963; Bortz and Lynen, 1963) and after cortisol treatment (Seubert, 1967), the hormone having lipolytic actions. This process is again dependent on *de novo* protein synthesis (Fain, 1967; Seubert, 1967). Thus, the control of the acetyl CoA level seems to be a major site at which pyruvate carboxylase activity is regulated. In addition, cortisol can act by inhibiting glycolytic enzymes to alter metabolite concentrations (see Seubert, 1967).

Glycogen synthetase is also under glucocorticoid control. In this case, the actions of both glucocorticoids and glucose is intimately related to the reactions of glycogen phosphorolysis and the cAMP level which is controlling these reactions. de Wulf and Hers (1967a, b, 1968a, b) showed that cortisol, like glucose, provokes the conversion of the inactive glycogen synthetase *b* into the active *a* form by a phosphatase reaction which simultaneously inhibits glycogen degradation.

Glucagon, which acts through cAMP, counteracts this effect by activating the phosphorylase system (Rall and Sutherland, 1958) and inactivating glycogen synthetase (de Wulf and Hers, 1968b).

4. *Control by enzyme induction*

As noted above, the activities of key gluconeogenic enzymes can be partially inhibited by inhibitors of RNA and protein synthesis; such experiments will be discussed later. At this point, we suggest that key gluconeogenic enzymes and, as yet unidentified, control proteins are induced by glucocorticoid hormones.

The activities of glucose-6-phosphatase and fructose-1, 6-diphosphatase, both enzymes of the final gluconeogenic pathway, increase after glucocorticoid administration (Weber *et al.*, 1955). In a study of RNA synthesis using inhibitors of transcription, Weber *et al.* (1964) demonstrated that these increases resulted from *de novo* synthesis of enzymes, and could be completely blocked by actinomycin D.

Studies of the inductive mechanisms must consider the possibility that other substances might provoke the same effect as glucocorticoids through different modes of action. The synthesis of phosphoenolpyruvate carboxykinase in liver and hepatoma tissue culture cells, for example, has been conclusively shown (Krone *et al.*, 1974, 1975a, b) to be inducible by dibutyryl cAMP as well as by glucocorticoids. In chicken liver, where different forms of phosphoenolpyruvate carboxy-kinase occur in embryos or young chickens and in adult animals (Jo *et al.*, 1974a), application of gluconeogenic stimuli (such as predniso-lone) provoked marked changes in the relative proportions of the various forms of the newly synthesized enzyme (Jo *et al.*, 1974b).

Subsequent inhibitor experiments (Krone *et al.*, 1975b) clearly showed that only the glucocorticoid-induced enzyme synthesis in-volves the *de novo* synthesis of mRNA for the enzyme protein; the cyclic nucleotide acts at the translational level.

These findings were obtained in isolated liver cells, where the influence of other hormones can be controlled. Using whole animals, Gunn *et al.* (1975a) observed a decreased rate of hepatic phosphoenol-pyruvate carboxykinase synthesis after administration of triamcino-lone and cortisol to fed and starved rats. In diabetic rats, on the other hand, the rate of enzyme synthesis increased. It was concluded that insulin as the physiological de-inducer of the enzyme (PEP-CK) counteracted the glucocorticoid effect. Using hepatoma cells, Gunn *et al.* (1975b) found that in the absence of insulin, dexamethasone increases the *de novo* rate of phosphoenolpyruvate carboxykinase synthesis.

In rat kidney cortex both triamcinolone and experimental acidosis stimulate *de novo* phosphoenolpyruvate carboxykinase synthesis (Iynedjian *et al.*, 1975). This was shown for both the glucocorticoid and the acidotic stimulus by immuno-precipitation and inhibitor studies. This result does not agree with that of Flores and Alleyne (1971) which failed to reveal a sensitivity of the acidosis-mediated enzyme induction to actinomycin D. This discrepancy may reflect certain unfavourable experimental conditions applied by Flores and Alleyne (1971).

Thus glucocorticoid hormones have been conclusively shown to act directly on the *de novo* synthesis of gluconeogenic enzymes. In addition, they increase the supply of substrates from both peripheral tissues and target cells. This supply is guaranteed by enzymes involved with amino acid metabolism, such as tyrosine aminotrans-ferase, tryptophan oxygenase and alanine aminotransferase whose induction has been examined in relation to the mechanism of glucocorticoid action.

The mechanism of induction of specific enzymes will be discussed in detail in Section 4 H, but it is necessary to make initial mention of some facts relevant to the understanding of the hormonal regulation of gluconeogenesis.

The induction of hepatic tyrosine aminotransferase (Lin and Knox,

1957; Kenney and Flora, 1961) was one of the first examples in which
the link between RNA synthesis and enzyme induction was noted
(Greengard and Acs, 1962; Feigelson and Feigelson, 1966). The
enzyme exists, according to Johnson *et al.* (1973) in multiple forms, of
which one seems to be the initial product of translation after hormonal
induction, and it is then modified to various other forms. A similar
approach led Mertvetsov *et al.* (1973) to conclude that the enzyme
induction seems to be selective for certain isoenzymes; in rat liver,
cortisol stimulates the *de novo* synthesis of part of these isoenzymes,
whereas the levels of others remain unchanged. In contrast liver
regeneration after hepatectomy was accompanied by increases in all
known isoenzymes.

The induction of tyrosine aminotransferase, mostly studied in rat
liver and hepatoma tissue culture cells, can be linked to the presence
of glucocorticoid binding proteins in rat liver cytosol (Beato *et al.*,
1972a; Singer and Litwack, 1971; Litwack *et al.*, 1973), hepatoma
tissue culture cells (Gardner and Tomkins, 1969) and guinea pig liver
(Singer *et al.*, 1975).

The inducibility of enzymes by glucocorticoids as a model system
for gene regulation has been the basis of several theories on the
mechanism of hormone action (see Karlson *et al.*, 1975). The identification and localization of the genes involved in this regulation
remain to be established. One approach is the use of the cell
hybridization technique, in which cells of different origin are fused
and specific functions attributed to the presence or absence of specific
chromosomes in certain clones. Using this technique, Croce *et al.*
(1973) demonstrated that regulatory genes involved in the control of
tyrosine aminotransferase induction are located on the X chromosome. The cells fused were rat hepatoma cells and human fibroblasts.

The same technique was used to study the induction of alanine
aminotransferase (Sparkes and Weiss, 1973). The induction of this
enzyme is mediated by glucocorticoids (Knox, 1951; Rosen *et al.*,
1959; Segal and Kim, 1963; Lee and Kenney, 1970) and the mechanism was again studied by hybridizing inducible hepatoma cells and
non-inducible epithelial cells. It was shown that, in several hybrid
subclones, both the activity and inducibility of the enzyme were
reduced. Reexpression of the enzyme required the specific loss of
additional chromosomes. Thus, the genes for controlling factors could
be allotted to these chromosomes.

Tryptophan oxygenase was the first enzyme shown to be inducible
by glucocorticoids (Knox and Mehler, 1951). The enzyme is well
characterized (Schütz and Feigelson, 1972; Schütz *et al.*, 1972) and its

inducibility by glucocorticoids has been unequivocally established *in vitro* (Schütz *et al.*, 1973).

Tryptophan oxygenase can be induced by its substrate as well as by glucocorticoids (Staib *et al.*, 1969). In the former case, however, the induction of synthesis cannot be inhibited by inhibitors of transcription, whereas the induction of the enzyme by dexamethasone hemisuccinate was dependent on an intact RNA synthesis.

In addition to the enzymes described above, many other enzymes are induced by glucocorticoids in other tissues. In many cases, the induction of enzymes by glucocorticoids is intimately connected with characteristic steps of differentiation of the respective tissues.

5. *Conclusion*

It appears that the control of gluconeogenesis by glucocorticoids occurs at two main levels (Exton, 1972). Firstly the supply of substrates from peripheral tissues is enhanced and secondly gluconeogenesis is stimulated by the induction of enzymes involved both in the final gluconeogenic pathway, and in the supply of substrates. In most cases, the glucocorticoid effects depend on the intact synthesis of ribonucleic acid and proteins. As will become apparent, the catabolic actions of glucocorticoids in peripheral tissues result from similar steps that initiate the various types of reaction characteristic of the particular tissue.

B.   EFFECTS OF GLUCOCORTICOIDS ON PERIPHERAL TISSUES

1. *General considerations*

Glucocorticoids act on carbohydrate metabolism directly, by influencing enzyme activity and synthesis in gluconeogenic cells, and indirectly, by changing the supply of metabolites, chiefly amino acids, from the periphery. The characteristic effects of glucocorticoids on peripheral tissues have been extensively reviewed by Baxter and Forsham (1972), Thompson and Lipmann (1974) and Leung and Munck (1975).

Research on the catabolic and inhibitory actions of glucocorticoids stemmed from the early discovery of glucocorticoid induced lymphocytolysis (Selye, 1936; Dougherty and White, 1945). A key interpretative problem concerning the mechanism of hormone action is whether the data reported fit the general scheme of events at the molecular level (Section 4, below). Thus, it must be shown: (i) whether the cytoplasmic binding of glucocorticoid hormones by lymphoid and other hormone sensitive tissues involves receptor molecules; (ii) whether the hormone is transferred to the cell nucleus;

(iii) whether the known physiological effects depend on *de novo* RNA synthesis.

These physiological effects, eventually producing cell death, include inhibition of glucose, amino acid or nucleotide transport into target cells (Leung and Munck, 1975; Makman *et al.*, 1967). Subsequently, ATP synthesis is inhibited (Makman *et al.*, 1971; Hallahan *et al.*, 1973) and the cell dies.

Such inhibition of transport mechanisms by glucocorticoids occurs in lymphoid tissue (Munck, 1971), skin (decrease of glucose uptake: Ariyoshi *et al.*, 1973), muscle (altered amino acid transport: Kostyo, 1965), bone (decreased uptake of RNA precursors: Peck *et al.*, 1967), fibroblasts (decreased glucose uptake: Gabourel and Aronow, 1962; Gray *et al.*, 1971), leukocytes (decreased glucose uptake: Simonsson, 1972) and adipose tissue (decreased glucose uptake: Munck, 1961, 1962).

2. *Lymphoid cells*

In most examples cited macromolecular synthesis precedes the inhibitory action of the hormone. Whereas the synthesis of RNA and subsequently enzyme protein seems sufficient to account for enzyme induction by glucocorticoids, the action of these hormones on lymphoid tissues must depend on the induced synthesis of as yet hypothetical proteins which mediate the glucocorticoid effect on transport processes. Glucocorticoids are known to bind to cytoplasmic receptor proteins of thymocytes (Munck and Wira, 1971; Schaumburg and Bojesen, 1968) and to be transferred to the cell nucleus (Munck and Wira, 1971) where they interact with nuclear acceptors which are themselves essential for the transcriptional response to the glucocorticoids (van der Meulen *et al.*, 1972). Stimulation of RNA synthesis, an early characteristic, (Young, 1970) precedes the inhibition as observed after *in vivo* administration of the hormone (Fox and Gabourel, 1967) or after *in vitro* administration to either isolated thymocytes (Makman *et al.*, 1967) or thymic nucleic (Abraham and Sekeris, 1971). Actinomycin D inhibition studies on thymocytes revealed that the glucocorticoid effect depends on the RNA synthesized during the first minutes after hormone administration, so that this RNA may be messenger RNA for the proteins responsible for the catabolic response (Hallahan *et al.*, 1973). Studies with α-amanitin (van der Meulen *et al.*, 1972), a specific inhibitor of nucleoplasmic RNA polymerase, and the determination of the base composition of the RNA synthesized (Drews, 1969) suggest that the initial responses to glucocorticoid hormones include stimulation of mRNA synthesis and a subsequent decrease in ribosomal RNA synthesis.

The mechanism whereby glucocorticoids induce lymphocytolysis has been examined in cell fusion experiments. Gehring *et al.* (1972) formed hybrids between myeloma and lymphoma cell lines, both with cytoplasmic glucocorticoid receptors, but the lymphoma cells were resistant to the steroid. The lethal reaction to steroids occurred in the hybrid cells suggesting that the lymphoma cellular resistance does not reflect the presence of an inhibitor of steroidal actions.

Since the steroidal actions can be inhibited by actinomycin D and cycloheximide, Munck has postulated that RNA and protein synthesis are necessary for the inhibition of glucose transport into thymocytes; as a consequence of glucose deficiency, intracellular ATP decreases, protein and other synthetic processes decline and cellular death ensues (Munck, 1971; Hallahan *et al.*, 1973). Thompson and Lippman (1974) point out, however, that not all the catabolic effects are related to glucose uptake. Protein synthesis in muscle, for example, is inhibited by glucocorticoids although the transport of glucose is unchanged (see Section 2, B 4).

3. *Adipose tissue*

Adipose cells are the target of several hormones, including glucocorticoids, catecholamines, insulin and growth hormone. Fatty acid mobilization is enhanced by glucocorticoids (Jeanrenaud and Renold, 1960), but this effect seems mainly permissive in conjunction with catecholamines (Reshef and Shapiro, 1960). The mobilization of fatty acids does not, however, precede the onset of gluconeogenesis (Froehlich and Wieland, 1971), and the supply and oxidation of fatty acids limits hepatic gluconeogenesis.

The lipolytic action of glucocorticoids is intimately related to the action of growth hormone. The latter induces lipolysis with a two hour latent period, during which glucocorticoids are needed (Fain *et al.*, 1965). This permissive effect of glucocorticoids can be inhibited by actinomycin D and by cycloheximide. Glucocorticoids given after the latent period further enhance the growth hormone effect (Fain, 1967).

Glucose uptake by isolated fat cells (Fain, 1968) and intact adipose tissue (Munck, 1962) is inhibited by glucocorticoids. Methods which measure sugar uptake by adipocytes (Gliemann *et al.*, 1972; Livingstone and Lockwood, 1975) reveal that physiological concentrations of glucocorticoids inhibit the uptake of glucose and 3-0-methyl-glucose in a time dependent manner (Livingston and Lockwood,

1975). Such glucocorticoid actions on the sugar transport only occur in white fat cells (Fain, 1965) with a latency of two hours. Since it can be inhibited by actinomycin and cycloheximide (Czech and Fain, 1971; Livingston and Lockwood, 1975) glucocorticoids may be concluded to provoke the synthesis of messenger RNA coding for a protein involved in the inhibition of fat cell glucose uptake, or metabolism. An impaired glucose metabolism may thus be a major factor regulating fatty acid mobilization which impinges on other enzymes and hormones.

4. Muscle

Skeletal muscle displays catabolic responses to glucocorticoid hormones, although unlike in other tissues, glucocorticoids do not appear to impair glucose uptake by skeletal muscle (Munck and Koritz, 1962; Fishman and Reiner, 1972). On the other hand the uptake of amino acids by skeletal muscle decreases several hours after glucocorticoid administration (Kostyo, 1965; Kostyo and Redmond, 1966). This effect is closely related to the actions on target cell protein synthesizing capacity, since inhibitors of both protein and/or RNA synthesis (Kostyo and Redmond, 1966) give similar inhibitory effects on amino acid incorporation. The mechanisms involved are not understood.

While skeletal muscle does not display diminished glucose uptake after glucocorticoid administration, cardiac muscle shows a slight decrease (Morgan et al., 1961). There is also a marked inhibition of glucose-6-phosphate generation. Again, the regulatory role of glucocorticoids in target tissue metabolism cannot be clearly separated from other hormonal actions. Schaeffer et al. (1969) demonstrated that adrenalectomy blocks the ability of adrenaline or cyclic AMP to activate muscle glycogen phosphorylase. Cortisol restored this ability, and increased the enzyme concentration, indicating perhaps that glucocorticoids regulate the synthesis of muscle glycogen phosphorylase.

5. Conclusions

Glucocorticoid action on several non-gluconeogenic tissues involves the binding of the steroid to receptor molecules. This is followed, in most systems described, by a catabolic effect on the target cells, which depends on de novo synthesis of ribonucleic acids and proteins. The newly synthesized proteins may be involved in regulating cell membrane transport and intracellular metabolic processes. Such regulatory proteins are suggested by results of inhibitor studies. Future studies of the mechanisms of glucocorticoid-induced catabolism in target tissues must identify such proteins and focus on the transport processes involved in the hormonal response.

3. Role of glucocorticoids in differentiation

A. GENERAL CONSIDERATIONS

Hormones are well known to influence differentiation in a variety of organs and tissues; examples include many peptide hormones, thyroid hormones, androgens and oestrogens. Glucocorticoids, however, also regulate cytodifferentiation and organ development, and their actions are again correlated with the appearance of tissue-specific proteins. The increase in the apparent concentration of such proteins is preceded by changes in RNA transcription and translation. Finally, hormone receptor molecules occur in several tissues during embryology when cells become more or less sensitive to steroid hormones.

Three examples of glucocorticoid involvement in the regulation of development are given: the neural retina; the lung development and synthesis of pulmonary surfactant; and mammary tissues.

B. NEURAL RETINA

Glutamine synthetase in the embryonic neural retina of chickens follows a well-defined developmental pattern (Rudnick and Waelsch, 1955; Moscona and Hubby, 1963; Piddington and Moscona, 1965) associated with other differentiation patterns. A premature histological differentiation and a rise in glutamine synthetase in cultured retinal tissue occurs when the media are supplemented with adult animal sera (Moscona and Kirk, 1965). These changes in enzyme activity and cytological differentiation require functional RNA and protein synthesis (Moscona and Kirk, 1965; Kirk, 1965). The effective compound from adult serum was subsequently identified as cortisol, since cortisol provoked identical developmental and enzymatic changes both in organ cultures and *in situ* (Moscona and Piddington, 1966; Reif and Amos, 1966). Studies of steroid structure and inductive capacity revealed that the 11β-hydroxy-group is essential for the effect (Moscona and Piddington, 1967). This response to the hormone depends on histotypic associations and organization (Morris and Moscona, 1970, 1971). From actinomycin D experiments, Reif-Lehrer and Amos (1968) stated that cortisol is essential for the induction of glutamine synthetase at the transcriptional level (Raina and Rosen, 1968). Furthermore Wiens and Moscona (1972) observed an inhibition of glutamine synthetase induction using proflavine, a blocker of transcription. The inhibitory action of cordycepin in this system is especially interesting since this drug specifically inhibits the polyadenylate synthesis, an essential step in messenger RNA maturation (Sarkar *et al.*, 1973).

The effects of high concentrations of actinomycin D suggested that, in addition to the effect on the synthesis of RNA, the regulation by glucocorticoids might take place at the translational level (Moscona *et al.*, 1968), a phenomenon also described by Reif-Lehrer (1971) using lower concentrations. The enzyme induction thus requires RNA synthesis but this RNA would be particularly stable, so that additional RNA to regulate glutamine synthetase synthesis at translational levels was proposed (Alescio *et al.*, 1970). Such a model postulates that in addition to transcriptional regulation by glucocorticoids, two labile post-transcriptional repressors, i.e. a suppressor and a desuppressor, control glutamine synthetase synthesis (Moscona *et al.*, 1972).

The identification of the initial product of the glucocorticoid-induction sequence, the mRNA for glutamine synthetase of polysomes, was possible as a result of the high aspartic acid in glutamine synthetase (Sarkar and Moscona, 1971) combined with the application of antibodies against the purified enzyme (Schwartz, 1972; Sarkar and Moscona, 1973).

These studies on polysomal RNA after glucocorticoid application thus established an initial control of glutamine synthesis at transcriptional levels. On the other hand the validity of studies employing high actinomycin D concentrations, partly the basis for the model postulating an additional control step, were questioned by Schwartz (1973), who observed deleterious effects of actinomycin D on the oxidative respiration, the ATP pool size, the polyribosome content and the histological appearance of the retina. Thus, some of the effects described after administration of this inhibitor may have reflected the actions of a nonspecific metabolic poison.

The glucocorticoid effects on RNA and protein synthesis in the embryonic retina are consequent upon an uptake of hormone into cultured retinal cells (Reif-Lehrer and Chader, 1969). This accumulation is specific for those hormones which induce the enzyme (Chader and Reif-Lehrer, 1972). A soluble cytoplasmic receptor molecule was then demonstrated in chick embryonic retinas by gel filtration (Chader *et al.*, 1972). At 37°C, the uptake of cortisol was greater in the nuclear than in the cytosol fraction (Chader, 1973). This concurs with Horisberger and Amos (1970) who observed, that RNA induction and glutamine synthetase synthesis could only be achieved at 37°C.

In summary, steroid binding occurs in cytoplasm and nucleus and subsequently RNA and protein synthesis are affected. In this respect, the initial step seems to be at the level of transcription, but the high stability of the mRNA for glutamine synthetase and experiments based on this property suggest the presence of additional gene products which regulate the synthesis of glutamine synthetase post-transcriptionally.

C. MAMMARY GLAND

Both polypeptide and steroid hormones are necessary for lactation (for reviews see Lyons et al., 1958; Turkington and Kadohama, 1972). Lactogenesis is initiated by a dominance of prolactin and corticoids. A combination of insulin, cortisol and prolactin, added to a synthetic medium, stimulates the synthesis of casein by 300%–500% and induces histological changes in the alveolar epithelium of the mammary gland (Stockdale et al., 1966). Both mineralocorticoids and glucocorticoids may be involved in this process (Turkington et al., 1967). Glucocorticoids also increase DNA and RNA synthesis in the mammary gland (Kumaresan et al., 1967). Ultrastructural studies, performed by Mills and Topper (1970), showed that cortisol, when added after insulin which promotes cell proliferation (Stockdale and Topper, 1966), brings alveolar epithelium cells to an intermediate state of differentiation with an extensively increased rough endoplasmic reticulum (Oka and Topper, 1971). Such corticosteroid effects on the epithelial cells are prerequisites for prolactin actions; a permissive rather than an inductive influence on the synthesis of milk proteins may thus be present.

The hormonal stimulation of mammary glands produces both synthesis of milk proteins and induction of enzymes. Glucose-6-phosphate dehydrogenase synthesis is induced by insulin and prolactin (Rivera and Cummins, 1971; Leader and Barry, 1969) and prolactin and corticosteroids further maintain the elevated enzymatic activity. Actinomycin D and cycloheximide prevent this effect.

The concerted actions of the three hormones are also manifest in events in chromatin, the site of gene regulation. In cells which have undergone division after insulin stimulation, addition of prolactin initiates several changes in nuclei of alveolar epithelial cells, and this effect requires pretreatment of the alveolar cells with cortisol. Increased nuclear and ribosomal RNA synthesis, increased RNA polymerase activity and histone phosphorylation and, after a lag period, a rise in the synthesis of casein, lactalbumin and galactosyl transferase are all associated components of the response (Turkington and Kadohama, 1972).

The apparent analogy between this differentiation system and steroid hormonal induction of enzymes and the fact that glucocorticoids are rate limiting in the differentiation of the lactatory system (Lyons *et al.*, 1958), prompted several groups to search for glucocorticoid receptor molecules.

Tucker *et al.* (1971) demonstrated that the cytoplasm of bovine mammary cells cultured *in vitro* bound cortisol, which, after incubation at 37°C, entered the cell nucleus. In lactating murine mammary glands, Shyamala (1973) characterized a cytoplasmic glucocorticoid receptor which efficiently bound those steroid hormones that permit lactation. A distinct glucocorticoid binding protein has also been described in the lactating rat mammary gland (Gardner and Wittliff, 1973).

Another major action of glucocorticoids on mammary tissue is the maintenance of precancerous lesions and the development of tumors (Nandi, 1959). In spontaneous tumors, Shyamala (1974) detected cytoplasmic binding proteins with all the characteristics of specific glucocorticoid receptors. Subsequently Shyamala (1975) also demonstrated nuclear binding sites in mouse mammary tumors; in these, cortisol and corticosterone, but not oestradiol and progesterone, were translocated to the specific nuclear sites.

Dexamethasone stimulates the expression of murine mammary tumor virus, MMTV (Parks *et al.*, 1974; Ringold *et al.*, 1975a). The glucocorticoid induced MMTV production seems to be mediated by glucocorticoid receptor molecules (Young *et al.*, 1975a, b; Ringold *et al.*, 1975b). The latter studies further suggest that the synthetic rate of MMTV RNA increases after glucocorticoid application.

The role of glucocorticoids in mammogenesis and lactogenesis remains unclear. These hormones are rate limiting in their actions and permit prolactin effects. By analogy with other systems, the hormones are bound in the cytoplasm and interact with nuclear sites. In the murine mammary tumor virus system, these steps induce the synthesis of tumor virus.

D. FETAL LUNG

The fetal lung is another developmental system whose differentiation is influenced by glucocorticoids. Fetal steroid involvement in the maturation of fetal epithelial cells was proposed by Buckingham *et al.* (1968). Moreover, Liggins (1969) showed that lambs prematurely delivered as a result of dexamethasone infusion had partially inflated lungs and suggested that there was an accelerated appearance of lung surfactant activity. This observation has great clinical importance since surfactant deficiency is frequently a principal feature of the respiratory distress syndrome (see Avery, 1972). In infants with this syndrome, a cortisol deficiency occurs (Murphy, 1974). Measurements of pulmonary surface tension and pressure/volume relationships of twin fetal lambs given cortisol clearly demonstrated that cortisol is responsible for the premature appearance of pulmonary surfactant (De Lemos et al., 1969, 1970). Dexamethasone increased tracheal fluid secretion by fetal lambs by several hundred percent (Platzker et al., 1972). Similar effects on lung maturation were observed in fetal rabbits (Kotas et al., 1971; Motoyama et al., 1971) and Rhesus monkeys (DeLemos and McLaughlin, 1973). The accelerated epithelial cell maturation and surfactant production parallels an inhibited cell division (Carson et al., 1972). This was shown in rabbits, but an enhancement due to cortisol was observed in human fetal lung cells in monolayer culture (Smith et al., 1973). These latter cells converted cortisone to cortisol and this hormone enhanced the cell growth in culture.

Seeking specific steps at which the induction of pulmonary surfactant is controlled, pulmonary lipids have been examined after in utero decapitation of fetal rats; such "hypophysectomy" inhibited phospholipid, particularly lecithin, synthesis (Blackburn et al., 1972; Farrell and Blackburn, 1973). After administration of 9-fluoro-prednisolone, Farrell and Zachman (1973) observed a significant increase in lecithin concentration and choline phosphotransferase activity in fetal lungs. Experiments with actinomycin D and cycloheximide suggested that the mechanism of induction is principally at the translational level.

The relevance of these findings to specific therapies for respiratory distress syndromes was tested in controlled trials of antepartum (Liggins and Howie, 1972) and postpartum (Baden et al., 1972) glucocorticoid administration. Antepartum treatment resulted in no hyaline membrane disease or intra-ventricular cerebral haemorrhages when the mothers had received betamethasone for at least 24 hours before delivery. In contrast, glucocorticoids given to the children with respiratory distress syndrome had no obvious benefit.

The actual mechanisms whereby cell maturation and surfactant production are influenced are poorly understood. Fetal lung is certainly a target tissue for glucocorticoids in that receptor binding of the hormone and its transfer to the epithelial cell nucleus have been demonstrated (Ballard and Ballard, 1972, 1973, 1974; Ballard et al.,

1974) in rabbits (Ballard and Ballard 1972; Giannopoulos, 1973a, b), man (Ballard and Ballard, 1974), lamb (Ballard and Ballard, 1973), rats (Toft and Chytil, 1973) and guinea pigs (Giannopoulos, 1974). These receptors display the classical properties of steroid binding proteins, and can be transferred to a nuclear site in a temperature-dependent activation step.

In summary, glucocorticoid actions on fetal lung suggest that they accelerate lung differentiation and surfactant production. Lung development displays characteristics that clearly suggest the participation of steroidal hormones.

E.   CONCLUSION

The three examples chosen to illustrate glucocorticoid involvement in differentiation and development show that the action of these hormones is not restricted to metabolic processes such as gluconeogenesis and catabolism in several peripheral tissues. Like other steroid hormones, such as androgens, oestrogens and progestins, glucocorticoids actively participate in the control of development and differentiation.

4.   Intracellular mode of action of glucocorticoids

A.   GENERAL CONSIDERATIONS

Preceding sections have considered glucocorticoid actions on the metabolism of various types of cells. Glucocorticoid target cells probably include almost every mammalian tissue. The apparently paradoxical effects of these hormones are, in most cases, directed towards the same goal, the preservation or synthesis of glucose at the expense of proteins. The catabolic action results either from a decreased glucose uptake or from changes in amino acid utilization. These deficiencies produce cell death and tissue regression and increase the flow of substrates to gluconeogenic cells.

The hormonal actions both on amino acid metabolism and on the final gluconeogenic pathway are accompanied by the synthesis of proteins, as deduced from inhibitor studies. (The precise mechanisms of these effects on protein synthesis in target cells are described later in this Section).

The rat liver has been widely studied with the respect to the induction of certain enzymes such as tyrosine aminotransferase and tryptophan oxygenase. Evidence obtained from other steroid hormones, or from the interaction of glucocorticoids with other target organs, will be presented to illustrate specific aspects of the induction process which are less well understood in the hepatic system. In addition other systems exemplify the general mechanism of action of various steroid hormones.

Glucocorticoids clearly induce tyrosine aminotransferase and tryptophan oxygenase in the liver of the rat and other vertebrates. Phylogenetically, induction of both enzymes has been detected in mammals, birds and reptiles, but not in amphibia (Chan and Cohen, 1964; Spiegel, 1961; Ohisalo and Pispa, 1975). These two enzymes as well as alanine aminotransferase are induced as a result of *de novo* synthesis of enzyme proteins (Kenney, 1962; Segal and Kim, 1963; Schimke *et al.*, 1965; Granner *et al.*, 1968a, b). This section is concerned with the mechanisms whereby the hormone elicits the increased synthesis of these specific hepatic proteins. The description is chronological in that it will follow the events starting from the moment at which the hormone contacts the cell until the increase in protein synthesis can be detected.

B.   INTERACTION WITH SERUM PROTEINS

Glucocorticoids, like other steroid hormones, are not free in the blood serum but are partially bound to specific carrier proteins and to serum albumin (see Daughaday, 1967; Westphal, 1971). A specific glucocorticoid binding globulin, transcortin, has been purified to homogeneity from various species (Seal and Doe, 1961; Muldoon and Westphal, 1967; Chader and Westphal, 1968a, b; Rosner and Bradlow, 1971). The function of transcortin is unclear, however, and the question whether the free or the transcortin-bound steroid is the biologically active form remains (De Moor *et al.*, 1963; Keller *et al.*, 1969). Available evidence generally suggests that the free steroid is the active form (Blecher, 1966; Matsui and Plager, 1966; Westphal, 1971). Experiments using purified transcortin reveal no statistically significant effects of bound cortisol on blood lymphocyte numbers or on tyrosine aminotransferase induction in the liver (Rosner and Hochberg, 1972). Other findings suggest that transcortin protects the corticosteroids against the attack by metabolizing enzymes (Sandberg and Slaunwhite, 1963) to decrease their metabolic clearance rates (Koch *et al.*, 1970). Clearly transcortin is not essential for glucocorticoid induction of specific enzymes in the liver, since it has no detectable affinity for the synthetic glucocorticoids carrying a 9-α-fluoro atom (Florini and Buyske, 1961; Kolanowski and Pizzaro,

1969; Peets *et al.*, 1969), steroids which are very potent inducers of tyrosine aminotransferase and tryptophan oxygenase (Granner *et al.*, 1968a, b; Feigelson *et al.*, 1975).

There are other observations relevant to the hepatic function of transcortin. Firstly, transcortin has been detected intracellularly in liver and other glucocorticoid target cells (Beato *et al.*, 1972b; Werthamer *et al.*, 1973; Amaral *et al.*, 1974). Moreover transcortin is probably synthesized in the liver (Guidollet and Louisot, 1969a, b, 1970) so that any intrahepatic transcortin could either reflect its biosynthetic accumulation or intracellular migration of serum transcortin.

C.   UPTAKE INTO THE CELL

Relatively little is known about the mechanisms of glucocorticoid entry into the target cells. Plasma membranes of rat liver possess specific binding sites for natural glucocorticoids but not for dexamethasone (Suyemitsu and Terayama, 1975). Rao *et al.* (1976), using suspensions of rat hepatocytes, described a carrier-mediated transport mechanism across the cell membranes operating at concentrations of glucocorticoids below 10^{-7} M. At higher concentrations, transport across the membrane occurs by simple diffusion in both liver and hepatoma tissue culture cells, with the hormone being bound directly in the cytosol (Levinson *et al.*, 1972).

Protein–mediated glucocorticoid transport through the target cell membrane of cultured mouse L cells (Gross *et al.*, 1970) and mouse pituitary tumor cells (Harrison *et al.*, 1974, 1975) has been indicated. Treatment of these latter cells with neuraminidase or phospholipase A_2, which has no effect on the cytosol receptor, reduces cellular uptake of glucocorticoids while proteolytic enzymes, which destroy the receptor, have no effect. A similar protein-mediated entry of oestrogens into uterine cells has been postulated by Milgrom *et al.* (1973a).

On the basis of the sensitivity to sulfhydryl reagents, Jackson and Chalkley (1974a) have proposed a model in which the 4-S oestradiol receptor of the uterus is bound to the plasma membrane and picks up the steroid. After combining with the steroid the 4-S receptor is transformed to a 5-S form and leaves the membrane. In the cytosol, the oestradiol dissociates from the 5-S receptor, reverting to the 4-S form, and again binds to the plasma membrane. The basis for this model is that the 4-S form of the oestradiol receptor can bind to membranes, even in the absence of the hormone, in a temperature-dependent process which requires divalent cations (Jackson and Chalkley, 1974b).

D.   GLUCOCORTICOID BINDING PROTEINS OF THE CYTOSOL

1.   *Tissue specificity and subcellular distribution*

Unlike other steroid hormones which have a small number of well-defined target organs, glucocorticoids affect many tissues, and a clear definition of a target tissue is difficult. In the rat, glucocorticoid binding proteins are present in the cytoplasm of such diverse tissues as liver, thymus, skeletal muscle, heart, brain, pituitary, testes, kidney, spleen, lung, bone and stomach (Schaumburg, 1970; Shaler and McCarl, 1971; Beato and Feigelson, 1972; McEwen *et al.*, 1972; Munck *et al.*, 1972; Watanabe *et al.*, 1973; Kaiser *et al.*, 1973; Ballard *et al.*, 1974; Mayer *et al.*, 1974; Feldman *et al.*, 1975). In fetuses, the lung has higher concentrations of glucocorticoid receptor than the liver, reflecting the action of glucocorticoids on the development of pulmonary surfactant (Ballard and Ballard, 1972; Giannopoulos, 1973a; Toft and Chytill, 1973; Rooney *et al.*, 1975. See Section 3). The uterine cytosol of the rabbit and the mammary gland of mice also possess specific glucocorticoid binding macromolecules (Giannopoulos, 1973b; Shyamala, 1973; Gardner and Wittliff, 1973). The chick embryonic retina is another specific glucocorticoid target tissue with receptor proteins (Chader *et al.*, 1972; Lippman *et al.*, 1974a; Koehler and Moscona, 1975). Tissue culture has revealed numerous cell lines which bind glucocorticoids specifically; examples include various hepatoma lines, HeLa cells, fibroblasts, mammary cells and various types of lymphomas (Baxter and Tomkins, 1970; Melnykovich and Bishop, 1971; Hackney and Pratt, 1971; Tucker *et al.*, 1971; Kirkpatrick *et al.*, 1971; Baxter *et al.*, 1971; Lippman *et al.*, 1973).

   The present description focuses on studies of liver and hepatoma tissue culture cells, systems in which the induction of specific enzyme proteins takes place. Brief references will be made to the well-characterized lymphoid system and the chick embryonic retina.

   After administration of radioactive glucocorticoids to rats, entry into the liver is rapid and intracellular accumulation occurs (Bradlow *et al.*, 1954; Bellamy *et al.*, 1962; Bellamy, 1963; Bottoms *et al.*, 1969). The subcellular distribution of the radioactivity revealed that most is present in the higher speed supernatant fraction, the cytosol, and relatively small amounts are recovered in nuclear fractions (Litwack *et al.*, 1963; Dingman and Sporn, 1965; DeVenuto and Muldoon, 1968;

Yu and Feigelson, 1969a; Beato *et al.*, 1969a; Litwack *et al.*, 1973). However, whereas the hormone is extensively metabolized in the cytoplasm, the nuclear fraction contains mainly untransformed cortisol (DeVenuto and Westphal, 1961; Fiala and Litwack, 1966; Beato *et al.*, 1969b). It is interesting to note that the mitochondrial fraction also contains mainly unmetabolized cortisol (DeVenuto *et al.*, 1962; Beato *et al.*, 1969a, b).

A small fraction of the cytosolic radioactivity migrates with authentic cortisol on thin layer chromatography, and its proportion varies with the time elapsed after *in vivo* administration. This radioactive cortisol in the cytosol is bound to proteins as shown by Sephadex gel chromatography and enzyme digestion experiments (Beato *et al.*, 1969a). Column chromatography of these proteins on DEAE-Sephadex or DEAE-cellulose demonstrated at least three main protein fractions with affinity for active unmetabolized glucocorticoids in the cytosol (Beato *et al.*, 1972b; Beato and Feigelson, 1972; Snart *et al.*, 1970, 1972). In addition, two other proteins with affinity for anionic metabolites of glucocorticoids have been described and partially purified by Litwack and collaborators (Morey and Litwack, 1969; Litwack *et al.*, 1972a). The functional significance of a covalent complex of cortisol and a low molecular weight component of liver cytosol has yet to be determined (Morris and Barnes, 1966, 1967; Morris *et al.*, 1970).

2. *Identification of the receptor*

Double labelling experiments with blood serum and liver cytosol showed that one of the hepatic binding proteins for natural glucocorticoids co-chromatographed with serum transcortin and reacted with antibodies against this serum protein (Beato *et al.*, 1972b). The intracellular nature of this protein was verified by determining the degree of contamination of the cell preparation with blood using ^{14}C-inulin (Beato *et al.*, 1972b). Transcortin and another glucocorticoid binding protein from liver cytosol, called the A protein, only bind natural glucocorticoids and have an insignificant affinity for such potent inducers as dexamethasone and triamcinolone acetonide (Peets *et al.*, 1969; Beato and Feigelson, 1972; Beato *et al.*, 1972b). However, a third relatively unstable glucocorticoid binding protein present in the liver cytosol has an affinity for both natural and synthetic hormones (Beato and Feigelson, 1972; Koblinsky *et al.*, 1972; Litwack *et al.*, 1973). This protein exhibits many of the properties which may be expected from a receptor protein for glucocorticoids, and therefore will be referred to as the *glucocorticoid receptor*.

In hepatoma tissue culture cells (HTC) the situation is slightly simpler because there is a very slow metabolism of glucocorticoids, and cytosolic transcortin-like proteins have not been detected. Baxter and Tomkins (1970) demonstrated that at concentrations below 10^{-8}M the uptake of ^{3}H dexamethasone by HTC cells is a saturable process, involving some 10^5 binding sites per cell. The cytosol of HTC cells contains a high capacity and low specificity glucocorticoid binding protein (Gardner and Tomkins, 1969) as well as a high affinity and high specificity binding protein (Baxter and Tomkins, 1971). The latter protein co-chromatographs on DEAE-Sephadex with the presumptive hepatic glucocorticoid receptor (Singer *et al.*, 1973). This protein has been purified over 17 000-fold by affinity chromatography (Failla *et al.*, 1975).

The glucocorticoid receptor of rat liver can be purified over 1000 fold by a procedure involving double chromatography on phos-phocellulose, exploiting the different properties of the activated and inactivated receptor (Bugany and Beato, 1976; Climent *et al.*, 1976). Other reports on the purification of the liver glucocorticoid receptor are concerned with transcortin-like proteins (Beato *et al.*, 1970c; Gopalakrishnan and Sadgopal, 1972; Tu and Moudrianakis, 1973; Wong *et al.*, 1973).

The physico-chemical characterization of the glucocorticoid receptor protein in the form of a complex with dexamethasone, has been performed in crude preparations of liver cytosol (Koblinsky *et al.*, 1972). Under conditions of low ionic strength this complex partially sediments in the 7-S region of a sucrose gradient and is excluded from a Sephadex G-100 column, indicating a molecular weight of about 200 000. Increasing the ionic strength above 0.15 M changes the sedimentation behaviour from 7-S to 4-S, and the elution in gel chromatography suggests a molecular weight of 60 to 70 000. Similar results have been obtained from hepatoma tissue culture cells (Baxter and Tomkins, 1971). The complex of the hepatic receptor with cortisol sediments in the 4-S region of sucrose gradients even at very low ionic strength (Beato *et al.*, 1969a; Beato and Feigelson, 1972).

Unlike the transcortin-like proteins of liver cytosol, the binding of glucocorticoids to the receptor is very sensitive to sulfhydryl reagents such as p-chloromercuribenzoate or mercuriacetate, and shows a marked pH dependence with an optimum around 7.4 (Baxter and Tomkins, 1971; Koblinsky *et al.*, 1972). The receptor can also be distinguished from the transcortin-like proteins of the liver as a result of its very high sensitivity to elevated ionic strength and temperature (Koblinsky *et al.*, 1972). In both HTC-cells and liver the receptor can be partially stabilized against thermal denaturation by dexamethasone or triamcinolone acetonide binding (Rousseau *et al.*, 1972; Bugany and Beato 1976).

When a crude cytosol preparation is brought to 33% saturation with ammonium sulphate only the receptor precipitates, and the transcortin-like proteins remain in the supernatant (Koblinsky *et al.*, 1972). This provides a very simple procedure for testing the function of the separated glucocorticoid-binding proteins.

The temperature dependence of the steroid binding by hepatic cytosol proteins is such that at physiological temperatures the affinity of the receptor for glucocorticoids is considerably higher than that of the transcortin-like proteins although at low temperature all three proteins have apparent affinities of the same order of magnitude (K_a around $10^9 M^{-1}$) (Koblinsky *et al.*, 1972). Thermodynamically, the binding of cortisol to the transcortin-like proteins is accompanied by a negative change of entropy, while the interaction of dexamethasone with the receptor results in a positive entropic change; the binding reactions thus differ (Koblinsky *et al.*, 1972). Examination of the kinetics of association and dissociation of the steroid from the binding proteins give similar conclusions. Both rates are much faster for the interaction of glucocorticoids with the transcortin-like protein than they are for the binding of dexamethasone to the receptor. Furthermore, taking into account analysis of the stereochemical requirements that determine the affinity of various steroids for the binding site of the different proteins, these data strongly suggest that steroid binding to the receptor involves an interaction of almost every functional group in the steroid molecule within a deep groove in the protein molecule; the binding of glucocorticoids to the transcortin-like proteins on the other hand takes place in a shallower groove of the protein molecule and mainly involves the hydrophobic a-side of the steroid molecule (Koblinsky *et al.*, 1972, Koblinsky, 1974).

Very little is known about the quaternary structure of the receptor molecule. Chromatographic behaviour on Sephadex gels or characteristics on centrifugation in sucrose gradients depend on the ionic strength, suggesting that the receptor is composed of subunits, held together, among other forces, by ionic interactions.

3. *Correlation between binding and enzyme induction*

Before designating a glucocorticoid binding protein the receptor, it is essential that its mediation of the hormone's action is demonstrated. The evidence for this involvement is mainly circumstantial and based on the correlation between steroid binding to the receptor and the cellular response. This correlation has been shown in three ways: (i) correlation between the ability of different steroid derivatives to bind to the receptor, and their ability to evoke the specific hormonal response; (ii) correlation between the saturation of the receptor after different doses of hormone, and the extent of the biological response; and (iii) correlation between the sensitivity of different cell lines to the hormone and the cellular content of the receptor.

The binding of different steroids to the glucocorticoid receptor of HTC cells has been extensively investigated by Baxter and Tomkins (1971) and Rousseau *et al.* (1972). The ability of the steroids to induce tyrosine aminotransferase was assessed and steroids were classified as: "optimal inducers" (corticosterone, cortisol and dexamethasone); "suboptimal inducers" (5α-dehydrocortisol, 17β-hydroxy-progesterone, and 17α-hydroxy-progesterone); "anti-inducers" (cortisone, testosterone, oestradiol-17β, progesterone and 17α-methyl-testosterone), and "inactive steroids" (tetrahydrocortisol, epicortisol, 20β-hydroxy-cortisol, androstenedione, and 20α-hydroxy-cortisol) (Samuels and Tomkins, 1970). Only those steroids able to either induce tyrosine aminotransferase or to suppress the induction of the enzyme by the inducers had an affinity for the glucocorticoid receptor. In addition this affinity correlated with the biological activity (Baxter and Tomkins, 1971). With some minor variations, a very similar correlation has been demonstrated using the rat liver cytosol receptor, separated from the transcortin-like proteins by ammonium sulphate fractionation (Koblinsky, 1974). Such studies allow a series of conclusions about the chemical groups of the steroid molecule which donate binding characteristics and biological activity. Besides the Δ_4-3-keto function of the A-ring, the most important substitutions for binding and activity take place at the C-11 position. Addition of a keto-group at this position decreases both binding and biological activity, as does the addition of an hydroxy-group at 11α. The addition of an 11β-hydroxy-group generally increases binding and markedly potentiates biological activity. It is possible that the very pronounced biological activity of the synthetic derivatives with the 9α-fluor-atom results from an inductive effect on the 11β-hydroxy-group (Bush, 1962; Eger *et al.*, 1971). Many other substitutions at other positions reveal that both, the α- and β-sides of the steroid molecule, as well as the side-chains, affect receptor interactions.

Samuels and Tomkins (1970) proposed an allosteric model of the receptor, involving an active and an inactive form, to explain the differences in biological activities of inducers and anti-inducers. Anti-inducers would bind to the inactive configuration, whereas the binding of inducers favours the active configuration. Rousseau et al. (1972) demonstrated that anti-inducers compete with inducers for the receptor. Differences in the kinetics of binding and the stabilization of the receptor by inducers and anti-inducers support the allosteric model, and suggest that the 11β-hydroxy group is a prerequisite for changing the configuration of the receptor, compatible with biological activity. Steroids such as progesterone, without substituents at C-11, bind to the receptor but do not convert it to the active configuration. Competition between the anti-inducer cortexolone and the inducer cortisol in the thymus has been demonstrated (Munck et al., 1972; Turnell et al., 1974).

Other evidence that implicates receptors in the biological action of the hormone is the correlation between the dose-dependent saturation of the receptor and the extent of the hormonal response, as observed in HTC cells (Baxter and Tomkins, 1970) and in rat liver after cortisol administration in vivo (Beato et al., 1972a). In the latter system the effect of the steroid-receptor complex on enzyme induction is cumulative; that is both the degree and the time of saturation of the receptor must be considered when correlating the extent of enzyme induction.

The interpretation of receptor saturation data is influenced by certain theoretical considerations. If the cellular response to the steroid is assumed to be graded, then the number of occupied receptor molecules and the cellular response must correlate. If, on the other hand, the target cell responds as a quantal unit when the number of occupied receptor sites exceeds a given threshold, then the receptor saturation must correlate with the number of responding cells (Rodbard, 1973).

4. Sensitive and resistant cell lines

The third line of evidence suggesting that the receptor is involved in the action of glucocorticoids comes from the receptor contents of different cell lines displaying different sensitivities to the hormone. The best experimental evidence in this area comes from lymphoid tumors and different lines of lymphoma cells, whose growth can be inhibited by glucocorticoids. Glucocorticoid-sensitive and -resistant cell lines have been derived from the P1798 mouse lymphosarcoma

(Lampkin and Potter, 1958). Both natural and synthetic glucocorticoids reveal a correlation between the presence of specific cytoplasmic binding proteins and responsiveness to the hormone (Hollander and Chiu, 1966; Kirkpatrick *et al.*, 1971). Cell lines derived from a murine lymphoma (Horibata and Harris, 1970; Sibley and Tomkins, 1973) give similar results: glucocorticoid resistant cell sublines have lower receptor concentrations in the cytosol (Baxter *et al.*, 1971; Rosenau *et al.*, 1972; Kondo *et al.*, 1975). In fibroblast cell lines a correlation of this type has also been demonstrated (Aronow and Gabourel, 1963; Hackney *et al.*, 1970).

In all cases, however, cell lines with receptor activity, but which are insensitive to the hormone, occur (Gehring and Tomkins, 1974). In this respect it is interesting to compare guinea pig and rat livers with regard to the inducibility of tyrosine aminotransferase and the receptor content. Although the guinea pig liver is 6 times less responsive to cortisol than that of the rat, both show the same amount of binder II, which according to Singer *et al.* (1975) is the glucocorticoid receptor. Guinea pig liver, however, has much less activity of binders III and IV, a finding which suggests that these binders may be very relevant to glucocorticoid action, an assumption further supported by the fact that both binders III and IV activities are reduced in immature rat liver, where tyrosine aminotransferase induction is also very poor (Singer and Litwack, 1971).

5.   *Ontogeny of the receptor*

Greengard (1970) has shown that the ontogeny of rat hepatic enzymes is a complex, coordinated process, For instance, the induction of tryptophan oxygenase by glucocorticoids is only evident three weeks after birth (Greengard, 1970). Tyrosine aminotransferase inducibility is absent from fetal livers but can be observed 2 days after birth (Sereni *et al.*, 1959; Cake *et al.*, 1973).

Careful titrations of the cellular content of the hepatic glucocorticoid receptor during development have shown the levels to be very low at birth, but to increase rapidly to reach the adult concentrations before the end of the first week (Cake *et al.*, 1973; Feldman, 1974; Giannopoulos, 1975a). In the neonate receptor activity cannot be detected in the cytosol (Cake *et al.*, 1973), but this is probably due to the high levels of circulating corticosteroids (Milkovic and Milkovic, 1963). In fact, the nuclei of newborn animals incorporate radioactive dexamethasone quite efficiently, suggesting that the cytosol receptor has been translocated into the nucleus (Giannopoulos, 1975b; see below).

Even during late fetal life (18–20th day of pregnancy), receptors occur in fetal liver (Giannopoulos, 1975a, b). It is therefore not surprising that, in explants of liver from 16-day-old fetuses, cortisol induces glycogen synthetase activity (Eisen *et al.*, 1973). These findings confirm the previous statement that the receptor is a prerequisite for induction, so that in no case is induction detected in the absence of receptor, but its presence is insufficient for a specific response.

Similar results have been obtained from ontogenetic studies of the chicken neural retina (Lippman *et al.*, 1974a). In this system the inducibility of glutamine synthetase increases during development. During the first 7 to 8 post-natal days, glucocorticoids double the enzyme activity, while on days 12 to 14 induction is 20-fold. The concentration of glucocorticoid receptors on day 6, however, is as high as at the time of maximal induction, indicating that factors, operative in steps following receptor binding, are responsible for the final appearance of maximal inducibility.

As previously noted, the action of the hormone–receptor complex involves intranuclear migration, interaction with the chromatin and modulation of specific transcription. The lack of induction observed during early development may thus be due to the failure of the receptor to migrate into the nucleus and bind to chromatin. This is not so however, since in both the newborn rat liver and the chick embryonic retina, the receptors indeed migrate to the nucleus (Feldman, 1974; Lippman *et al.*, 1974a). In this respect it is important to note that the intracellular distribution of the receptor molecules depends on the concentration of circulating glucocorticoids. In the adult rat liver there are some 50–70 000 receptor molecules per cell (Beato *et al.*, 1974). Following adrenalectomy, the majority of these receptor molecules occur in the cytosol, whereas in the intact animal some 24% of the total population are intranuclear (Beato *et al.*, 1974). Furthermore, cortisol injected into adrenalectomized animals produces a redistribution of the receptor and subsequently an accumulation of intranuclear binding to levels of intact animals.

The factors that regulate the total cellular content of glucocorticoid receptors are poorly understood. After adrenalectomy there is a transient elevation in hepatic cytosolic receptor (Beato *et al.*, 1974; Giannopoulos, 1975a). In other systems the receptor levels are under hormonal control. For example in rats or guinea pigs oestrogens increase the concentrations of the progesterone and the oestradiol receptors in the endometrium (Milgrom *et al.*, 1971; Mester *et al.*, 1974; Sarff and Gorski, 1971; Cidlowski and Muldoon, 1974), while progesterone rapidly reduces the concentrations of its own receptor

(Milgrom *et al.*, 1973c). Although the effect of oestrogen can be prevented by actinomycin D or cycloheximide, the progesterone effect appears to be independent of RNA and protein synthesis.

E.   ACTIVATION OF THE CYTOSOL RECEPTOR

After the interaction of the steroid with the receptor *in vivo* the steroid–receptor complex migrates into the nucleus, as shown by titrating the receptor molecules present in the cytosol and in the nucleus in adrenalectomized animals before and after cortisol administration (Beato *et al.*, 1974). In HTC cells and thymocyte suspensions this intranuclear translocation of the steroid–receptor complex is temperature dependent (Baxter and Tomkins, 1970; Wira and Munck, 1970). At low temperatures the receptor-bound hormone can only be detected in the cytosol; the cells can then be washed with non-radioactive solutions and, upon warming to 37°C, most of the radioactivity bound to the receptor migrates into the nuclear fraction.

The temperature-dependent nuclear binding of glucocorticoids can be reproduced in cell-free systems, reconstituted from cytosol and purified nuclei. In these systems the steroid–receptor complex, rather than the nuclei, undergo a temperature-dependent transformation (Baxter *et al.*, 1972; Wira and Munck, 1974). Such findings confirm the initial observations of a "two-step" mechanism of oestradiol binding to uterine nuclei (Jensen *et al.*, 1968). Nuclear binding experiments with rat liver demand the use of adrenalectomized animals, in which the nuclei are depleted of receptor molecules (Beato *et al.*, 1974). If the cytosol receptor is labelled with radioactive steroid and incubated with isolated nuclei from adrenalectomized rats, there is a temperature-dependent nuclear uptake of radioactivity (Kalimi *et al.*, 1973). The process can be divided into three steps: (i) forming the steroid–receptor complex at low temperature; (ii) pre-warming the steroid–receptor complex above 20°C for a brief period; (iii) incubating the activated steroid–receptor complex with isolated nuclei at low temperatures. Under these conditions, the nuclear uptake is quite efficient, indicating that it is the activation of the steroid–receptor complex and not the nuclear uptake which is dependent on higher temperature. Activation can be induced both by elevated temperature, by increasing the ionic strength and by adding low concentrations of divalent cations, in particular Ca^{2+} (Kalimi *et al.*, 1975; Milgrom *et al.*, 1973c; Higgins *et al.*, 1973b). Theophylline and other methylxanthines appear to activate the steroid–receptor complex at 0°C (Cake and Litwack, 1975).

Little is known about the chemical nature of the transformation or activation process. In contrast to the uterine oestradiol receptor, the glucocorticoid receptor displays no change in sedimentation behaviour upon activation (Kalimi *et al.*, 1975). A conformational change, however, is suggested since the activated and non-activated complexes have different isoelectric points and different affinities for phosphocellulose and other anionic ion-exchangers (Milgrom *et al.*, 1973c; Climent *et al.*, 1976). The inactivated complex of receptor and steroid has no affinity for phosphocellulose, while the activated form binds to phosphocellulose at neutral pH and can only be eluted with salt concentrations above 0.25 M. This property of the activated steroid–receptor complex allows the purification of the transformed receptor by simple chromatography on two consecutive phosphocellulose columns; a crude cytosolic preparation containing the non-activated steroid–receptor complex is passed through an excess of phosphocellulose to absorb other cytosolic proteins and the collected unabsorbed steroid–receptor complex is activated by increasing both ionic strength and temperature and it is then re-applied to a second phosphocellulose column. The bound receptor is then eluted with buffer containing 0.5 M NaCl and bovine serum albumin. This very simple procedure purifies over 1000-fold (Climent *et al.*, 1976).

An interesting observation is that even under optimal conditions of activation, only about 50% of steroid–receptor complexes formed in the cytosol can be transformed into the active form, which is translocatable to the nucleus (Higgins *et al.*, 1973b; Milgrom *et al.*, 1973c). In liver and HTC cells the hormone is absolutely essential for transformation of the receptor. In the case of the oestradiol receptor, however, some reports suggest that the receptor can be activated in the absence of the steroid (DeSombre *et al.*, 1972; Chatkoff *et al.*, 1974). In this case the mechanisms of activation may be different, as a change in the sedimentation coefficient of the steroid–receptor complex occurs after activation (Jensen *et al.*, 1971). The non-activated oestradiol–receptor complex sediments at 4-S in sucrose gradients run at physiological ionic strength, while after activation it sediments as a 5-S complex. This 4-S to 5-S transformation, studied by Yamamoto and Alberts (1972), may involve an increase in the molecular weight, which cannot be explained in terms of a conformational change. It was suggested that this transformation may result from the addition of a new subunit to the oestradiol binding component of the receptor (Yamamoto, 1974). Another possibility is that the 4-S to 5-S transformation involves a dimerization of the oestradiol binding subunit of the receptor. A receptor transforming factor which catalyses this conversion has been described in calf uterus (Puca *et al.*, 1972). The dimerization hypothesis is supported by the recent findings that the transformation of the receptor can take place even after various purification steps (Nielsen and Notides, 1975). Very careful kinetic analyses of the transformational process have revealed a bimolecular reaction with a high energy of activation, which may explain its temperature dependence (Notides *et al.*, 1975; Notides and Nielsen, 1975). These findings support a dimerization model, probably accompanied by conformational changes and exposure of positively charged areas of the protein molecule which will account for the nuclear affinity of the activated receptor.

Whatever the nature of the activation process it is important to note that in HTC cells only those steroids which are biologically active provoke the transformation of the receptor (Rousseau *et al.*, 1972). Progesterone, which has a relatively high affinity for the receptor, cannot induce its transformation, probably reflecting the absence of an 11-hydroxyl group. An allosteric model of the steroid–receptor complex which can fluctuate between two different conformations is supported therefore. The equilibrium between these conformations is affected by the type of ligand steroids and these can be considered as allosteric modulators of receptor activity. Recent measurements of the kinetics of glucocorticoid binding to the receptor of mouse fibroblasts are compatible with a two-step mechanism involving a slow transformation of a weak steroid–receptor complex to a tight active form (Pratt *et al.*, 1975).

F. NUCLEAR BINDING

The activated steroid–receptor complex formed *in vivo* migrates into the cell nucleus. The mechanism of passage across the nuclear membrane is poorly understood. Jackson and Chalkley (1974b) have postulated a high affinity binding site for oestradiol in the nuclear membrane of bovine endometrium. The hepatic glucocorticoid receptor, however, seems unlikely to employ the nuclear membrane for intranuclear binding of the steroid, since no significant differences in binding existed between nuclei purified through hypertonic sucrose and those submitted to detergent treatment (Beato *et al.*, 1969b; Kalimi *et al.*, 1973) Giannopoulos (1975b), however, claims that around 20% of the nuclear binding sites in adult rat liver can be eliminated by Triton-X-100, and takes this to indicate binding on the nuclear membrane.

1. *Specificity and functional significance*

The tissue specificity of the nuclear binding of the steroid–receptor complex is a controversial issue. In this regard it is important to note that the hormonal status of the animal may markedly affect the results. The intracellular distribution of the receptor is of course partially determined by the concentration of circulating hormones. Therefore, nuclei prepared from animals not previously deprived of the hormone contain a large proportion of the cellular receptor molecules, and can incorporate free steroid by exchange (Brecher and Wotiz, 1969; Anderson *et al.*, 1973b; Beato *et al.*, 1974; Shaskas and Bottoms, 1974). Hepatic nuclei from adrenalectomized rats bind very little free steroid but incorporate considerably larger amounts of the steroid–receptor complex than those nuclei of intact animals or nuclei prepared from kidney or spleen of adrenalectomized rats (Kalimi *et al.*, 1973). This partial specificity of the nuclear acceptor sites has been confirmed in both purified chromatin (Hamana and Iwai, 1973; Bugany and Beato, 1976), and in hepatoma tissue culture cells (Higgins *et al.*, 1973a, c). Lippman and Thompson (1973) have shown that although isolated nuclei from HTC cells and fibroblasts incorporate both the homologous and the heterologous dexamethasone receptor complexes, the acceptor sites for the receptors differ. Nuclei saturated with the homologous dexamethasone receptor complex still bind the heterologous complex, so that the nuclear acceptor site distinguishes between the two receptors. In all these studies only a partial specificity has been demonstrated, however, in that nuclei from non-target tissues can incorporate small amounts of the steroid–receptor complex (Kalimi *et al.*, 1973).

The specificity of the nuclear binding of the oestradiol receptor is also controversial (Shyamala, 1972; Chamness *et al.*, 1973, 1974; Jackson and Chalkley, 1974b). The functional significance of the nuclear binding of the oestradiol receptor complex has indeed been questioned (Chamness *et al.*, 1974). Two lines of evidence indicate, however, that nuclear binding is relevant to the uterine actions of this hormone.

An initial effect of oestradiol on the endometrium is to induce the synthesis of a cytoplasmic protein, termed the induced protein or I.P., which is detectable thirty minutes after hormonal administration (Katzenellenbogen and Gorski, 1972). The extent of induction of this protein by oestradiol, oestrone or oestriol correlates with the extent of their nuclear binding (Katzenellenbogen and Gorski, 1972; Ruh *et al.*, 1973). The inducibility of the I.P. during the oestrous cycle also correlates with the nuclear accumulation of oestradiol (Iacobelli, 1973; Katzenellenbogen 1975). On the other hand, the uterotropic response to oestradiol is clearly a function of the nuclear accumulation of the oestradiol–receptor complex (Anderson *et al.*, 1972, 1973a, b, 1974, 1975; Luck *et al.*, 1973).

These observations together give functional significance to the nuclear binding of oestradiol. It is possible, however, that there are fundamental differences between the mechanisms whereby oestradiol and glucocorticoids are bound to the nuclei. Thus Higgins *et al.* (1973c) have demonstrated that whereas the acceptor sites for the glucocorticoid receptor, both in HTC cell nuclei and in uterine nuclei, are destroyed by treatment with deoxyribonuclease I, those in uterine nuclei, for the oestradiol receptor, are unaffected (Higgins *et al.*, 1973c). In addition, no acceptor sites for the oestradiol receptor could be detected in the nuclei of HTC cells, and the oestradiol receptor did not compete with the glucocorticoid receptor for nuclear acceptor sites.

The biological significance of the binding of the steroid–receptor complex to the nucleus is further emphasized in the studies of lymphoma and hybrid cell lines. Screening lymphoma sub-lines, resistant to glucocorticoid treatment, Gehring and Tomkins (1974) found that most of them have low concentrations of cytoplasmic receptors otherwise similar to those found in the wild type. In some resistant lines, however, transfer of the steroid–receptor complex into the nucleus does not take place, suggesting that those properties of the receptor which are required for the nuclear binding are altered. In hybrids of HTC cells and mouse fibroblasts, both hepatoma and fibroblast receptors are present, and the nuclear acceptor sites distinguish between the two types of receptor (Lippman and Thompson, 1974). These findings again support the specificity of the nuclear acceptor sites.

2.   *Role of the cytosolic receptor*

Other debatable questions concern the involvement of the cytosolic receptor in the nuclear binding of the steroid and the relationship between the nuclear and the cytoplasmic receptor. In isolated rat liver nuclei using the glucocorticoid binding proteins from liver cytosol, the receptor is the factor responsible for nuclear binding of the hormone, and the transcortin-like proteins of the cytosol are not required for the specific binding of glucocorticoids to the nuclei *in vitro* (Beato *et al.*, 1973). Moreover, during the incorporation of radioactive steroids into the nucleus, the concentration of cytosolic receptors decreases in proportion to the amounts of radioactivity taken up by the nucleus (Beato *et al.*, 1973). This phenomenon also occurs in the oestradiol system (Jensen *et al.*, 1971) and may be interpreted in two ways: either the receptor transfers the steroid to another protein in the nucleus and is inactivated during this process, or the steroid–receptor complex moves from the cytoplasm into the nucleus. Extraction of the nuclei with high salt buffers after incubation with the cytosol receptor, solubilizes a steroid–protein complex which, on sucrose gradient or gel filtration, is indistinguishable from the cytosolic receptor (Beato *et al.*, 1973). This, and the fact that the intracellular receptor sites are redistributed after cortisol injection into adrenalectomized animals, suggests that the steroid–receptor complex is transferred as a whole from the cytosol into the nucleus.

The optimal conditions for the transfer of the steroid–receptor complex into the nucleus has been studied in reconstituted cell-free systems, composed of partially purified steroid–receptor complexes from cytosol and isolated nuclei. In these systems the transfer process is dependent on the ionic strength, the concentration of divalent cations and temperature (Baxter *et al.*, 1972; Kalimi *et al.*, 1973). The magnesium effects appear to be mediated by structural changes in the chromatin, whereas the effect of temperature is mainly attributable to the transformation of the steroid–receptor complex into the active form.

When nuclear binding is measured as a function of the concentration of steroid–receptor complex in the incubation medium, there is a saturation curve which reaches a half maximum at about 10^{-9} M (Kalimi *et al.*, 1973). The form of this saturation curve is however controversial, and in some cases no saturation has been observed *in vitro* (Higgins *et al.*, 1973a). It is important to note in this respect, that nuclear saturation is only detectable under very precise ionic conditions, after subtraction of non-specific binding. Suitable conditions will produce a saturation curve with both purified chromatin and nuclei, although partially purified receptor preparations bind to the chromatin as a linear function of receptor concentration (Bugany and Beato, 1976).

In the reconstituted cell-free system, composed of cytosol and nuclei, the number of receptor molecules bound per nucleus at saturation, is between 6 and 8000, which is in agreement with the number reported for other target tissues (Jensen *et al.*, 1971; Baxter *et al.*, 1972). The significance of this high number of acceptor sites is not yet clear, but it should be noted that after *in vivo* administration of the hormone more steroid molecules are localized within the nucleus (Jensen *et al.*, 1971). It is conceivable that only a few of these thousands of molecules are actually interacting with the biologically significant acceptor sites within the nucleus, and that the rest are only required to increase the local concentration of receptor or to facilitate the specific interactions (See Section 3 below).

There may be direct interaction between the steroid hormone and nucleic acids and in particular with DNA (T'so and Lu, 1964; Goldberg and Atchley, 1966; Sunaga and Koide, 1967a; Cohen and Kidson, 1969; Cohen, *et al.*, 1969; Kidson *et al.*, 1970; Chin and Kidson, 1971; Arya and Yang, 1975). The affinity of the steroids for nucleic acids is, however, relatively low compared with the biologically active concentrations, and no correlation exists between the binding affinities of various steroids and biological activity. These data indicate that the direct interaction of the hormones with nucleic acids is probably not relevant to enzyme induction.

3. *Nature of the nuclear acceptor sites*

The chemical nature of the nuclear acceptor sites for the steroid–receptor complex may now be considered. It is necessary to decide whether the receptor binds to the nucleolus or to the extra-nucleolar chromatin in the first instance. In the cell-free, reconstituted system from rat liver, the interaction takes place mainly with extranucleolar chromatin and very little receptor is bound to the nucleolus (Beato *et al.*, 1973) There are two possible mechanisms within the chromatin: (i) the receptor interacts with available DNA sites recognizing specific sequences; or (ii) other structural components of the chromatin, including chromosomal proteins and possibly RNA, are part of the acceptor site.

Deoxyribonuclease effects on isolated nuclei suggest that DNA is part of the chromatin acceptor site for the receptor (Baxter *et al.*, 1972; Beato *et al.*, 1973). The activated glucocorticoid receptors of rat liver and hepatoma tissue culture cells bind to both native and denatured DNA *in vitro* (Baxter *et al.*, 1972; Beato *et al.*, 1973). This binding, however, lacks certain properties displayed by isolated nuclei. For instance, in contrast to nuclei, no saturation was detected in the nanomolar range of receptor concentration (Rousseau *et al.*, 1975; Bugany and Beato, 1976). In addition, using DNA coupled to cellulose, no specificity in relation to the source of the DNA was found. Thus the binding to rat liver, calf thymus and *E. coli* DNA was very similar, and the differences were only quantitative (Bugany *et al.*,

1976.) This suggests that DNA may not be the only component of the acceptor site in chromatin and that other chromosomal components are involved in the binding of the receptor.

The role of the DNA in the acceptor function of chromatin has been further elucidated from the binding kinetics of the steroid–receptor complex to DNA and chromatin at different salt concentrations (Bugany and Beato, 1976). Such data suggest that the vast majority of receptor molecules, bound to chromatin, interact with available DNA sites, although it is possible that a small population of receptor molecules is stabilized by other forces in its interaction with the chromatin. This additional interaction may involve chromosomal protein or secondary structural features of the DNA unique to chromatin.

It is possible that the lack of specificity in the binding of the receptor to DNA reflects methodological limitations of the techniques used to characterize the interaction of the receptor with the DNA *in vitro,* so that small populations of specific acceptor sites are indistinguishable from the background non-specific binding. It should be emphasized that the "non-specific" binding to DNA occurs in all DNA binding proteins, even when a specific recognition of a particular sequence is present (von Hippel *et al.*, 1974); even "non-specific" interaction with DNA is a prerequisite for the recognition of the specific sequences. In the case of *lac* operon of *E. coli,* the non-specific binding of the repressor to DNA impinges on both the basal activity of the *lac* operon and the kinetics of induction (von Hippel *et al.*, 1974). In higher organisms, with much more nuclear DNA, greater non-specific interaction might be expected, magnifying the practical difficulties of detecting a small population of molecules specifically bound to particular sequences on the DNA (Yamamoto and Alberts, 1975).

Other evidence in favour of an interaction of the receptor with DNA associated with the nuclear binding of the hormone is that with HTC cells grown in the presence of bromodeoxyuridine the binding of the receptor to the nuclei is much tighter than in normally grown cells (Rousseau *et al.*, 1974). Moreover, in several glucocorticoid resistant lymphoma cell lines, a defective binding of the receptor to the nucleus occurs. The ability of the receptor of these cell lines to bind to purified DNA is decreased in parallel with the loss of nuclear binding, indicating that the same mutation affects both functions (Gehring and Tomkins, 1974; Yamamoto *et al.*, 1974). The latter finding emphasizes the functional significance of the interaction of the receptor with DNA.

The chemical nature of the interaction of the receptor with DNA is poorly understood. The fact that the activated receptor also binds to phosphocellulose and CM-cellulose suggests ionic forces. In addition, N-ethylmaleimide or iodoacetomide prevents receptor binding to DNA invoking the involvement of sulfhydryl groups (Young et al., 1975a, b).

The role of chromosomal proteins in the nuclear binding of the glucocorticoid receptor is equivocal. Reports concerning the interaction of radioactive glucocorticoids with histones (Sekeris and Lang, 1965; Sluyser, 1966, 1969), should be considered with caution, since the radioactive glucocorticoids may have been contaminated with 21-dehydrocortisol and this could account for the observed binding (Monder and Walker, 1970). Tsai and Hnilica (1971) have also shown that contamination of the histone fraction with non-histone proteins is probably responsible for the observed binding of cortisol to arginine-rich H3 histones *in vitro* (Sunaga and Koide, 1967b). Interactions between radioactive glucocorticoids and non-histone nuclear proteins have also been described (Dastugue et al., 1971; Defer et al., 1974; Lavrinenko et al., 1971), but their biological significance is unclear.

Other steroid hormones are also inconclusive in this respect. The uterine oestradiol receptor, for instance, binds purified DNA (Musliner and Chader, 1971, 1972; Clemens and Kleinsmith, 1972; King and Gordon, 1972; Yamamoto and Alberts, 1972; André and Rochefort, 1973), but an interaction with non-histone chromatin proteins has also been postulated (Maurer and Chalkley, 1967; King and Gordon, 1967; Alberga et al., 1971; Puca et al., 1974, 1975). In the well-described chick oviduct, the progesterone receptor appears to be composed of two different subunits, one of which interacts with purified DNA and the other has affinity for a specific fraction of the non-histone chromatin proteins (Schrader et al., 1972; O'Malley et al., 1972). A definitive answer to this question may come from chromatin reconstitution experiments which will show whether the function of the chromosomal proteins reflects a passive masking of certain sequences on the DNA, or whether they play an active role in binding of the receptor to the chromatin.

The interactions between the oestradiol–receptor complex and RNA polymerase (Arnaud et al., 1971; Müller et al., 1974), as well as with ribonucleoprotein particles of the nucleus (Liang and Liao, 1972, 1974), are relevant to steroidal actions on transcription (see below).

The data discussed thus suggest that the cytoplasmic receptor, after binding the steroid, migrates, as an activated complex, to the cell nucleus, where it binds to chromatin. Virtually nothing is known about the later fate of the hormone. The very interesting question of whether the steroid and/or the receptor are inactivated or recycled has received little attention. In thymocytes and fibroblasts the steroid does not seem to be metabolized in the chromatin (Munck and Brinck–Johnson, 1974), and the receptor itself may recycle in a process which is dependent on energy and temperature (Munck *et al.*, 1972; Ishi *et al.*, 1972; Bell and Munck, 1973).

G. REGULATION OF TRANSCRIPTION

1. *Rate of RNA synthesis*

An initial effect of cortisol on hepatic metabolism is the stimulation of protein and RNA synthesis (Clark, 1953; Silber and Porter, 1953; Korner, 1960; Feigelson *et al.*, 1962a; Leon *et al.*, 1962; Kenney and Kull, 1963; Lang and Sekeris, 1964a; Garren *et al.*, 1964b; Greenman *et al.*, 1965; Barnabei *et al.*, 1966; Leon, 1966; Koike *et al.*, 1968; Angelov and Richter, 1969). The increased protein synthesis accompanies an accumulation of active polysomes (Cammarano *et al.*, 1968; Cox and Mathias, 1969; Enwonwu and Munro, 1971). The stimulated RNA synthesis cannot be explained solely by changes in the specific activity of the precursor pool (Yu and Feigelson, 1969a). The latter observation, alongside the fact that actinomycin D blocks the induction of specific enzyme activity (Greengard and Acs, 1962) suggested that RNA synthesis was involved in the mechanism of enzymatic induction, Kenney and Kull (1963) had shown stimulation of nuclear RNA synthesis after a latent period of about 30 minutes, and also that the accumulation of induced tyrosine aminotransferase began about 30 minutes after the effect on RNA synthesis. The RNA synthesized after hormonal administration includes both ribosomal precursor RNA and rapidly labelled heterogeneous nuclear RNA (Greenman *et al.*, 1965; Venkov *et al.*, 1967; Koike *et al.*, 1967; Drews and Brawerman, 1967; Brossard and Nicole, 1969). Yu and Feigelson (1969b) showed that during the first hour after hormonal injection the induced RNA was rich in uracyl, whereas after the second hour a G-rich RNA was synthesized. In addition shortly after glucocorticoid administration, the rate of precursor incorporation into the RNA of nuclear ribonucleoprotein particles, which are transported into the cytoplasm, is increased (Niessing and Sekeris, 1970;

Lund–Larsen and Berg, 1973). Characterization of a hormone-specific class of RNA in sucrose gradients, polyacrylamide gels or counter-current distribution has been imperfect (Kidson and Kirby, 1964; Finkel *et al.*, 1966; Leader and Barry, 1967; Jackson and Sells, 1968; MacGregor and Mahler, 1969), largely because the induced enzymes represent a very low percentage of cellular protein synthesis.

Inhibition of hormonal enzyme induction by actinomycin D or aflatoxin B_1 can only occur during the two hours after hormonal administration (Wogan and Friedman, 1968), suggesting that the RNA synthesized during this time includes the messenger RNA for the inducible enzymes (Feigelson *et al.*, 1975). In hybridization competition experiments, Drews and Brawerman (1967) found new species of mRNA in the liver after cortisol administration, a finding confirmed by Doenecke and Sekeris (1970), who also showed that the cortisol induced a RNA which preferentially hybridized with slowly renaturing DNA; cortisol may thus directly influence transcription of specific genes. In addition, base analysis of the RNA induced by glucocorticoids in isolated hepatic nuclei has confirmed Yu and Feigelson (1970a), in that the initial stimulation of RNA synthesis by cortisol mainly concerns U-rich RNA (Beato *et al.*, 1970b).

In hepatoma tissue culture cells, glucocorticoids do not affect the rate of synthesis of total cellular RNA (Gelehrter and Tomkins, 1967), but induce the synthesis of one of the two isoaccepting species of phenylalanine tRNA (Yang *et al.*, 1974; Lippman *et al.*, 1974b). Although this effect is quite rapid, occurring one hour after hormonal administration, the relationship to the hormonal induction of tyrosine aminotransferase is not established. In the liver, glucocorticoids do not significantly change the pattern of tRNA species (Agarwal *et al.*, 1969; Agarwal and Hanoune, 1970), although the appearance of a new leucyl-tRNA isoaccepting species has been reported (Altman *et al.*, 1972).

Furthermore there is a very rapid effect of glucocorticoids on mitochondrial RNA synthesis (Yu and Feigelson, 1970b), although how this is related to the induced nuclear RNA synthesis is unclear (Mansour and Nass, 1974).

The glucocorticoid effects on nucleic acid synthesis are of interest in relation to their ability to suppress somatic growth of developing animals (Henderson *et al.*, 1971; Loeb and Yeung, 1973). The latter is also seen in the liver of rapidly growing rats, where low doses of glucocorticoids quickly inhibit DNA synthesis (Kimberg and Loeb, 1971). Glucocorticoids also inhibit DNA synthesis and mitosis in regenerating liver (Webb and Wozney, 1968; Raab and Webb, 1969; Rizzo et al., 1971; Dorman and Webb, 1974). Loeb et al. (1973) found that glucocorticoids markedly suppress DNA synthesis and cell proliferation in a line of rat hepatoma cells in vitro. The inhibition of DNA synthesis in hepatoma cells in contrast to lymphoid cells, is accompanied neither by cell lysis, nor inhibition of RNA synthesis. In addition, this hepatoma cell line did not display tyrosine aminotransferase induction, and exemplifies the dissociation between two glucocorticoid-sensitive phenomena: suppression of DNA synthesis and enzyme induction.

2.   DNA-dependent RNA polymerase

The mechanisms of glucocorticoid stimulation of hepatic nuclear RNA synthesis have been extensively studied and two possibilities emerge: either the activity of RNA polymerase is stimulated, or a structural change in the chromatin results in increased template activity. Indeed an increased RNA polymerase activity is an early effect of cortisol on rat hepatic nuclei (Barnabei et al., 1966; Lang and Sekeris, 1964b). Nucleolar RNA polymerase activity is also increased by cortisol (Yu and Feigelson, 1971). It is the activity, not amount, of nucleolar RNA polymerase which is increased after cortisol administration (Jacob et al., 1969; Blatti et al., 1970; Sajdel and Jacob, 1971; Schmid and Sekeris, 1975). Yu and Feigelson (1973) found that glucocorticoid stimulation of nucleolar RNA polymerase activity is inhibited when either α-amanitin or cycloheximide are given with the hormone. Although these findings are controversial (Benecke et al., 1973), it may be that the hormone induces the synthesis of a mRNA for a factor or a subunit of the nucleolar RNA polymerase, itself responsible for the increased activity of the enzyme. This hypothetical factor will be induced as a result of the extranucleolar RNA polymerase activity, so explaining the sensitivity to α-amanitin. Attempts to measure a direct increment in the activity, or the amount, of the α-amanitin-sensitive RNA polymerase following cortisol administration have failed (Jacob et al., 1969; Benecke et al., 1973). Thus in the liver the enhancement of nucleolar RNA synthesis by cortisol is secondary to an effect on extranucleolar RNA synthesis, although a direct interaction of the oestradiol receptor with nucleolar RNA polymerase has been postulated (Arnaud et al., 1971; Andress et al., 1974; Müller et al., 1974). There is no evidence of a direct interaction with the polymerase molecule.

3. *Chromatin template activity*

Hormonal effects on RNA synthesis may result from an interaction with the chromatin template. In 1965, Dahmus and Bonner demonstrated that cortisol increased hepatic chromatin template activity. This activation, detected 4 h after cortisol injection, has been confirmed (Vorob'ev and Konstantinova, 1972; Earp, 1974), and may indeed occur earlier (Sekeris *et al.*, 1970a). Similar effects on RNA synthesis, RNA polymerase activity and chromatin template activity have been reported for almost all the steroid hormones (see King and Mainwaring, 1974).

Some of the effects of glucocorticoids on RNA synthesis also occur *in vitro*. Isolated rat liver nuclei incubated with cortisol *in vitro* display an increased RNA synthesis and increased RNA polymerase activity (Dukcs and Sekeris, 1965; Lukács and Sekeris, 1967; Beato *et al.*, 1968), an effect confirmed with isolated nuclei (Ohtsuka and Koide, 1969; Umemura and Sakano, 1973) as well as with purified chromatin (Beato *et al.*, 1970d; Stackhouse *et al.*, 1968). Although high concentrations of the steroid are required under such circumstances, addition of hepatic cytosol macromolecules to isolated nuclei allows stimulation of RNA polymerase activity at lower cortisol concentrations (Beato *et al.*, 1970a). This partial requirement for cytosol proteins which was later confirmed (Bottoms *et al.*, 1972) is interesting since Beato *et al.* (1969b) showed that rat liver nuclei isolated from intact animals can incorporate radioactive cortisol because they contain about 25% of the total receptor capacity of the hepatocyte (Beato *et al.*, 1974). These nuclei can thus respond to free cortisol *in vitro* with an increased RNA synthesis.

In other steroid target tissues this *in vitro* effect on nuclear RNA synthesis has not been observed, possibly as a result of the particular experimental conditions. For example, there is no oestradiol effect on the nuclear RNA synthesis by isolated uterine nuclei obtained from ovariectomized animals, since such nuclei contain no receptor protein. In this example the cytosol receptor was an absolute requirement for an *in vitro* effect on RNA synthesis (Raynaud–Jammet and Baulieu, 1970; Mohla *et al.*, 1972). Such considerations may also be relevant to the difficulties encountered in reproducing the effect of glucocorticoids on isolated nuclei (Drews and Bondy, 1966; Monder and Walker, 1970; Dahmus and Bonner, 1965).

The *in vitro* effect of cortisol on nuclear RNA synthesis can be elicited in thymocyte nuclei from intact animals (Abraham and Sekeris, 1971), but after adrenalectomy the *in vitro* inhibition of RNA synthesis is only observed in the presence of the nuclear glucocorticoid receptor (Van der Meulen *et al.*, 1972).

A direct effect of glucocorticoids on chromatin was suggested by changes in certain physical and chemical properties of the chromatin after cortisol administration (Chakrabarti and Darzynkiewicz, 1973). For example hepatic chromatin binds more toluidin blue or actinomycin D one hour after cortisol administration (Salganik *et al.*, 1969; Beato *et al.*, 1970d).

Cortisol may also modify chromosomal proteins to give an increased thiol content and increased acetylation and phosphorylation of histones (Allfrey, 1966; Sekeris *et al.*, 1968; Murphy *et al.*, 1970; Doenecke *et al.*, 1972). A few hours after cortisol administration the synthesis of nuclear acidic proteins, including some specific fractions, detected in polyacrylamide gels, appears to be stimulated (Shelton and Allfrey, 1970; Buck and Schauder, 1970; Schauder and Buck, 1971); these are, however, later events and probably not related to the initial mechanism of enzyme induction.

In summary, available evidence supports the hypothesis that glucocorticoids interact with liver chromatin and influence nuclear RNA synthesis. The relationship between this effect on transcription and the observed enzyme induction is not yet clear (See next Section). However, it is interesting that glucocorticoids stimulate the transcription of certain RNA species while probably inhibiting thr synthesis of other RNA classes. This assumption is based on the observation that actinomycin D given prior to hormone injection, inhibits RNA synthesis to give levels well below those observed with actinomycin D alone (Homoki *et al.*, 1968). Although this finding needs further clarification, it may suggest that the steroid–receptor complex acts both as a positive and as a negative modulator of transcription.

H. MECHANISMS OF ENZYME INDUCTION

1. *General considerations*

One of the most extensively studied and most rapid effects of glucocorticoids is the induction of specific enzymes such as tryptophan oxygenase and tyrosine aminotransferase in rat liver. The mechanisms of induction have been studied in great detail (Knox, 1951; Knox and Auerbach, 1955; Lin and Knox, 1957, 1958; Civen and Knox, 1959; Kenney and Flora, 1961), and models, attempting to explain the regulation of enzyme induction by steroid hormones at the transcriptional level, have been introduced (Karlson, 1963; for review see Karlson *et al.*, 1975). Tryptophan oxygenase can be induced by its substrate tryptophan, and although the hormonal induction can be inhibited by actinomycin D, the substrate induction is not (Feigelson *et al.*, 1962b; Rosen and Milholland, 1963; Staib *et al.*, 1969). Adrenalectomy, however, decreases this hepatic response to tryptophan, and non-inducing doses of cortisol restore this capacity, suggesting that the hormone plays a "permissive" role in substrate induction (Lee and Baltz, 1962).

The direct action of the glucocorticoids on enzyme induction was demonstrated in the perfused rat liver (Goldstein *et al.*, 1962; Kim and Miller, 1969), and in liver slices (Liberti *et al.*, 1971). The glucocorticoid mediated induction of both tyrosine aminotransferase and tryptophan oxygenase may also occur in primary cell suspensions of the liver (Haung and Ebner, 1969; Berg *et al.*, 1972), as well as in established cell lines derived from hepatocytes (Gerschenson *et al.*, 1970, 1974; Sellers and Granner, 1974). The induction of tyrosine aminotransferase by glucocorticoids in a hepatoma tissue culture cell line, HTC cells, has been considered in relation to the receptor studies (Thompson *et al.*, 1966).

Many metabolic and endocrine factors influence the inducibility of enzymes by glucocorticoids. For instance, glucose pre-treatment, or noradrenaline, markedly reduce cortisol induction of these enzymes (Yuwiller *et al.*, 1970; Sitaraman and Ramasarma, 1974). Insulin also induces tyrosine aminotransferase, independently of the effect of glucocorticoids (Kenney *et al.*, 1967; Reel *et al.*, 1970; Seglen, 1971). Hypophysectomy enhances the hepatic enzyme response to glucocorticoids, but a single dose of growth hormone markedly inhibits the effect (Schapiro, 1968; Csanyi and Greengard, 1968; Liberti *et al.*, 1970). Factors in serum, other than insulin, also induce tyrosine aminotransferase (Gelehrter and Tomkins, 1969; Lee and Kenney, 1971). Moreover progesterone and deoxycorticosterone can partially interfere with the induction by glucocorticoids, and are classified as anti-inducers (Samuels and Tomkins, 1970; Braidman and Rose, 1970, 1971).

Before considering the details of the induction mechanism it should be noted that other hepatic enzymes and proteins are probably also controlled by glucocorticoids (See Section 2). Alanine aminotransferase (Lee and Kenney, 1970), phenylalanine hydroxylase (Haggerty *et al.*, 1973), phosphoenolpyruvate carboxykinase (Wicks *et al.*, 1972), and even serum albumin (Bancroft *et al.*, 1969) have all been hormonally induced. The ontogeny of rat hepatic enzymes is such that tyrosine aminotransferase and tryptophan oxygenase are inducible at different times after birth (Franz and Knox, 1967; Greengard, 1970). Whereas tyrosine aminotransferase can be induced by cortisol in 2- to 3-day-old rats, hormonal and substrate induction of tryptophan oxygenase appears during the second or third postnatal week. This delayed appearance of inducibility is not due to the lack of the cytosol receptor, which is present from the beginning of the postnatal life (Cake *et al.*, 1973; Feldman, 1974; Giannopoulos, 1975a). Cortisol, given at day four after birth, and twenty four hours before the administration of tryptophan, allows the substrate induction of tryptophan oxygenase, which usually can only be detected two weeks postnatally (Greengard and Dewey, 1971). The effect of cortisol but not that of tryptophan is sensitive to actinomycin D; the hormone plays a "permissive" role which involves RNA synthesis (Greengard *et al.*, 1963).

Explants of fetal rat liver react very poorly to the hormone during the first 12 to 24 h of culture, but thereafter a good response occurs (Wicks, 1971a, b). The lack of response initially could be due to the fact that such fetal liver explants metabolize cortisol very rapidly to tetrahydrocortisol, whereas after a preincubation this metabolism slows down (Coufalik and Monder, 1974). Furthermore the induction of tyrosine aminotransferase in this and other systems is complex in that multiple forms of the enzyme exist, and apparently only some of these isoenzymes are induced by glucocorticoids (Holt and Oliver, 1969a; Miller and Litwack, 1969; Litwack *et al.*, 1972b; Johnson *et al.*, 1973; Mertvetsov *et al.*, 1973, see Section 2).

Another aspect of hormonal induction of enzymes worthy of comment is the role played by cyclic AMP. The latter can induce tyrosine aminotransferase in fetal and neonatal liver, in perfused adult rat liver and in hepatoma tissue culture cells, (Holt and Oliver, 1969b; Wicks, 1971; Barnett and Wicks, 1971b; Krone *et al.*, 1974). The effects of cyclic AMP and glucocorticoids are synergistic rather than additive (Wicks *et al.*, 1969), the latter playing a "permissive" role for the effect of the former. Tyrosine aminotransferase induction by dibutyryl cAMP in adult liver is impaired by adrenalectomy (Wicks *et al.*, 1974), and cortisol enhances the absolute degree of enzyme induction in responsive hepatoma tissue culture cells (Butcher *et al.*, 1971; Stellwagen, 1972). In addition, whereas the effect of cyclic AMP on both tyrosine aminotransferase and phosphoenolpyruvate carboxykinase can be inhibited by cycloheximide but not by actinomycin D or cordycepin, the glucocorticoid effect is prevented by these inhibitors of RNA synthesis (Holt and Oliver, 1969b; Krone *et al.*, 1974, 1975a and b). Moreover, cortisol and cAMP induce different isoenzymes of tyrosine aminotransferase (Holt and Oliver, 1969b). Glucocorticoids thus appear to act on the induction process at the level of transcription, whereas cAMP operates at a post-transcriptional protein synthetic level. cAMP is not an absolute requirement for hormonal enzyme induction since low levels of cAMP exist in HTC cells which do not respond to the nucleotide, but which are very sensitive to glucocorticoid induction of tyrosine aminotransferase (Granner *et al.*, 1968a; Butcher *et al.*, 1971). Recently Granner (1976) has shown that HTC cells, normally nonresponsive to cAMP, acquire the response after incubation with dexamethasone. This is thus an excellent model for the study of the "permissive" action of glucocorticoids.

2.   *The superinduction phenomenon*

The application of specific antibodies against the purified enzymes have helped elucidate some mechanisms of induction. Such antibodies, used to examine the rate of incorporation of radioactively labelled amino acids into the specific enzyme protein, revealed that enzyme induction is accompanied by accumulation of enzyme molecules in the cytoplasm (Feigelson and Greengard, 1962) and further, that an increased rate of incorporation of amino acids into the protein occurred after hormone administration (Kenney, 1962; Segal and Kim, 1963; Schimke *et al.*, 1965; Granner *et al.*, 1968). The hormonal induction of these enzymes can be blocked by both actinomycin and α-amanitin, so that continued RNA synthesis is involved in the induction (Greengard and Acs, 1962; Sekeris *et al.*, 1970b).

Actinomycin D administered four or more hours after the hormone causes an additional increase in enzyme activity (Garren *et al.*, 1964a). This phenomenon, called superinduction, has been confirmed and demonstrated in HTC cells (Reel and Kenney, 1968; Tomkins *et al.*, 1966, 1972; Auricchio *et al.*, 1969; Thompson *et al.*, 1970). The interpretation of this phenomenon is however controversial. Tomkins and his collaborators claim that under the conditions employed, actinomycin D does not affect the degradation of tyrosine aminotransferase, while Kenney and his collaborators have clearly shown that, under other experimental conditions, the half life of the enzyme is prolonged after administration of actinomycin D, and that the synthetic rate of the enzyme is not increased by the antibiotic (Auricchio *et al.*, 1969; Lee *et al.*, 1970; Auricchio and Ligucri, 1971; Barker *et al.*, 1971; Herschko and Tomkins, 1971; Kenney *et al.*, 1973). Based on the former interpretation of the superinduction phenomenon, and on the observation that inducibility of tyrosine amino-transferase in synchronized HTC cells is only observed in late G_1 and in S-phase (Martin *et al.*, 1969a, b; Martin and Tomkins, 1970), a mechanism for hormonal induction has been proposed that involves an unstable translational repressor, the synthesis or activity of which is blocked by the hormone (Tomkins *et al.*, 1969). The level of messenger RNA for tyrosine aminotransferase would thus increase after hormone induction because the hormone interferes with the production, or the activity, of the labile translational repressor; the latter usually inactivates the messenger RNA, and favors its degradation. If this model is correct, actinomycin D should cause an accumulation of the mRNA for tyrosine aminotransferase, and therefore an increased rate of enzyme synthesis. Such predictions have not been forthcoming (Kenney *et al.*, 1973; Killewich *et al.*, 1975). Moreover, other inhibitors of RNA synthesis, such as camptothecin, do not cause superinduction (Bushnell *et al.*, 1974), and certain drugs which interfere with RNA synthesis have different effects on enzyme induction. Studies with purine and pyrimidine analogues have led to the postulate that there is a posttranscriptional control of the hepatic tyrosine aminotransferase induction by glucocorticoids (Levitan and Webb, 1969a, b, 1970a, b; Levitan *et al.*, 1971). A re-evaluation of the superinduction phenomenon, and of the kinetics of deinduction of tyrosine aminotransferase suggests that actinomycin D also stabilizes the enzyme protein, and that the steroids probably act by accelerating the rate of messenger RNA production (Tomkins, 1974; Steinberg *et al.*, 1975a, b).

3. *Titration of specific mRNAs*

The mechanisms involved in enzyme induction and superinduction will only be elucidated when the messenger RNAs of the specific enzymes are titrated, and their rates of synthesis studied under various conditions. An assay for the mRNA of tryptophan oxygenase has been developed, and its accumulation during hormonal induction has been directly demonstrated (Schütz *et al.*, 1973). Moreover there is a very good correlation between the enzyme activity and the cellular content of the specific mRNA for tryptophan oxygenase at different times after the injection of various doses of hormone (Schütz *et al.*, 1975). The levels of tryptophan oxygenase mRNA under conditions of superinduction do not increase as would be expected if the translation repressor theory applied (Killewich *et al.*, 1975). Similar titration experiments for the mRNA of tyrosine aminotransferase have been handicappped by difficulties in purifying the enzyme and obtaining sufficient mRNA. However, direct measurements of tyrosine aminotransferase synthesis in cell-free extracts of hepatoma tissue culture cells and measurements of the concentration of growing enzyme molecules in polysomes reveal that the level of tyrosine aminotransferase mRNA is also increased 10 to 18 times following hormone induction (Beck *et al.*, 1972; Scott *et al.*, 1972). Recently the mRNA for hepatic tyrosine aminotransferase has been translated in a wheat germ, cell-free system, and an accumulation of this mRNA after cortisol administration was demonstrated (Roewekamp *et al*, 1977).

4.   *Cell hybridization*

A genetic approach to the mechanisms of enzyme induction by hormones is possible by using cell hybridization techniques. Heterokaryons from inducible and non-inducible rat cells reveal that the control of tyrosine aminotransferase induction is probably negative, as the heterokaryons have low enzyme activity and are not inducible (Thompson and Gelehrter, 1971; Schneider and Weiss, 1971). However, when heterokaryons lose some chromosomes, tyrosine aminotransferase and alanine aminotransferase induction reappear independently of each other (Weiss and Chaplan, 1971; Sparkes and Weiss, 1973). The independent re-expression of alanine aminotransferase and tyrosine aminotransferase inducibility suggests that they are regulated at separate chromosomal loci. In rat–human hybrid cells the presence of the human X chromosome correlates with the suppression of tyrosine aminotransferase inducibility although the presence of the glucocorticoid receptor is independent of the X chromosome (Croce *et al.*, 1973).

Hybrids of inducible rat hepatoma cells and non-inducible mouse fibroblasts are non-inducible although they contain a cytosol dexamethasone receptor able to transfer the steroid into the nucleus (Croce *et al.*, 1974). In other hybrid lines the situation is reversed. Hybrids of glucocorticoid-sensitive lymphoma and insensitive myeloma cells exhibit glucocorticoid receptors and are killed by the hormones (Gehring *et al.*, 1972).

5.   General conclusions

This discussion has emphasized enzyme inductive mechanisms and largely ignored other possible glucocorticoid effects on cells. The mechanisms described appear very similar to those applicable to other steroid hormones which induce the synthesis of specific proteins. In blood the hormones are partially bound to specific proteins and penetrate cell membranes by a poorly understood mechanism. Within the cytoplasm the hormone combines with specific receptor molecules, which undergo distinct structural transformations to donate a nuclear affinity, resulting in an hormonal accumulation in the nucleus and thence binding to chromatin. Within the nuclei, the interaction of the steroid–receptor complex with the chromatin results in chemical modifications of the template and an increased rate of RNA synthesis. Thereafter the amounts of mRNAs specific for the inducible enzymes in the cytoplasm increase and the levels of these mRNAs correlate with enzyme activities. The relationship between the accumulation of mRNA and the induced nuclear RNA synthesis remains unclear. Procedures to detect specific sequences of the induced proteins in the nuclear RNA are now essential. In the liver this is a formidable task, considering that after induction the cellular protein synthesized as a result of the induced enzymes only reflects about a 0.1% increase. Therefore the purification of induced mRNA and the detection of specific sequences in nuclear RNA is much more difficult than in other systems where the induced proteins account for up to 50% of the total protein synthesized.

The induction of RNA tumor virus in cell cultures by glucocorticoids circumvents some of these difficulties (McGrath, 1971; Paran *et al.*, 1973; Dickson *et al.*, 1974; Parks *et al.*, 1974, 1975; Ringold *et al.*, 1975a). This effect is mediated by hormone receptors similar to those involved in enzyme induction (Young *et al.*, 1975a, b; Ringold *et al.*, 1975b), and the glucocorticoid effect is dependent on RNA synthesis (Ringold *et al.*, 1975b). Obviously, this system is appropriate to a more detailed study of the mechanism whereby the hormone regulates the transcription of specific genes, and may even allow development of regulated cell-free transcriptional systems. The hepatic system, however, has certain advantages as regards mechanisms of interaction between receptor and chromatin structures, another fascinating aspect of enzyme induction by hormones.

References

Abraham, A. D. and Sekeris, C. E., (1971). *Biochim. biophys. Acta* **247**, 562–569.
Agarwal, M. K. and Hanoune, J. (1970). *Biochem. J.* **118**, 31P.
Agarwal, M. K., Hanoune, J., Yu, F. L., Weinstein, I. B. and Feigelson, P. (1969). *Biochemistry, N.Y.* **8**, 4806–4812.

Alberga, A., Massol, N., Raynaud, J. P. and Baulieu, E.-E. (1971). *Biochemistry, N.Y.* **10**, 3835–3843.
Alescio, T., Moscona, M. and Moscona, A. A. (1970). *Exptl. Cell Res.* **61**, 342–346.
Alleyne, G. A. O. and Roobol, A. (1975). *Med. Clins. N. Am.* **59**, 781–795.
Allfrey, V. G. (1966). *Cancer Res.* **26**, 2026–2039.
Altman, K., Southren, A. L., Uretsky, S. C., Zabos, P. and Acs, G. (1972). *Proc. natn. Acad. Sci. U.S.A.* **69**, 3567–3569.
Amaral, L., Lin, K., Samuels, A. J. and Werthamer, S. (1974). *Biochim. biophys. Acta* **362**, 332–345.
Anderson, J. N., Clark, J. H. and Peck, E. J. Jr. (1972). *Biochem. biophys. Res. Commun.* **48**, 1460–1468.
Anderson, J. N., Peck, E. J. Jr. and Clark, J. H. (1973a). *Endocrinology* **92**, 1488–1495.
Anderson, J. N., Peck, E. J. Jr. and Clark, J. H. (1973b). *Endocrinology* **93**, 711–717.
Anderson, J. N., Peck, E. J. Jr. and Clark, J. H. (1974) *Endocrinology* **95**, 174–178.
Anderson, J. N., Peck, E. J. Jr. and Clark, J. H. (1975) *Endocrinology* **96**, 160–167.
André, J. and Rochefort, H. (1973). *FEBS Letters* **29**, 135–140.
Andress, D., Borgna, J.-L., Cazaubon, C. and Mousseron–Canet, M. (1974). *J. Steroid Biochem.* **5**, 895–903.
Angelov, E. Z. and Richter, G. (1969). *Acta biol. med. germ.* **22**, 499–507.
Ariyoshi, Y., Plager, J. and Matsui, N. (1973). *Acta Endocr. Copnh.* **74**, 723–731.
Arnaud, M., Beziat, Y., Borgna, J. L., Guilleaux, J. C. and Mousseron–Canet, M. (1971). *Biochim. biophys. Acta* **254**, 241–254.
Aronow, L. and Gabourel, J. D. (1963). *Proc. Soc. exp. Biol. Med.* **111**, 348–349.
Arya, S. K. and Yang, J. T. (1975). *Biochemistry, N.Y.* **14**, 963–969.
Auricchio, F. and Liguori, A. (1971). *FEBS Letters* **12**, 329–332.
Auricchio, F., Martin, D. Jr. and Tomkins, G. (1969). *Nature, Lond.* **224**, 806–808.
Avery, M. E. (1972). *Pediatrics, Springfield* **50**, 513–514.
Baden, M., Bauer, C. R., Colle, E., Klein, G., Taeusch, H. W. and Stern, L. (1972). *Pediatrics, Springfield* **50**, 526–534.
Ballard, P. L. and Ballard, R. A. (1972), *Proc. natn. Acad. Sci. U.S.A.* **69**, 2668–2672.
Ballard, P. L. and Ballard, R. A. (1973). *Pediatric. Res.* **7**, 308.
Ballard, P. L. and Ballard, R. A. (1974). *J. clin. Invest.* **53**, 477–486.
Ballard, P. L., Baxter, J. D., Higgins, S. J., Rousseau, G. G. and Tomkins, G. M. (1974). *Endocrinology* **94**, 998–1002.
Bancroft, F. C., Levine, L. and Tashjian, A. H. Jr. (1969). *Biochem. biophys. Res. Commun.* **37**, 1028–1035.
Barker, K. L., Lee, K.-L. and Kenney, F. T. (1971). *Biochem. biophys. Res. Commun.* **43**, 1132–1138.
Barnabei, O., Romano, B., di Bitonto, G., Tomasi, V., and Sereni, F. (1966). *Archs. Biochem. Biophys.* **113**, 478–486.
Barnett, C. A. and Wicks, W. D. (1971). *J. biol. Chem.* **246**, 7201–7206.
Baxter, J. D. and Forsham, P. H. (1972). *Am. J. Med.* **53**, 573–589.
Baxter, J. D. and Tomkins, G. M. (1970). *Proc. natn. Acad. Sci. U.S.A.* **65**, 709–715.
Baxter, J. D. and Tomkins, G. M. (1971). *Proc. natn. Acad. Sci. U.S.A.* **68**, 932–937.
Baxter, J. D., Harris, A. W., Tomkins, G. M. and Cohn, M. (1971). *Science, N.Y.* **171**, 189–191.
Baxter, J. D., Rousseau, G. G., Benson, M. C., Garcea, R. L., Ito, J. and Tomkins, G. M. (1972). *Proc. natn. Acad. Sci. U.S.A.* **69**, 1892–1896.
Beato, M. and Feigelson, P. (1972). *J. biol. Chem.* **247**, 7890–7896.

Beato, M., Homoki, J., Lukacs, I. and Sekeris, C. E. (1968). *Hoppe–Seyler's Z. Physiol. Chem.* **349,** 1099–1104.

Beato, M., Biesewig, D., Braendle, W. and Sekeris, C. E. (1969a). *Biochim. biophys. Acta* **192,** 494–507.

Beato, M., Homoki, J. and Sekeris, C. E. (1969b). *Expl. Cell Res.* **55,** 107–117.

Beato, M., Braendle, W., Biesewig, D. and Sekeris, C. E. (1970a). *Biochim. biophys. Acta* **208,** 125–136.

Beato, M., Homoki, J., Doenecke, D. and Sekeris, C. E. (1970b). *Experientia* **26,** 1074–1076.

Beato, M., Schmid, W., Braendle, W. and Sekeris, C. E. (1970c). *Steroids* **16,** 207–216.

Beato, M., Seifart, K. H. and Sekeris, C. E. (1970d). *Archs. Biochem. Biophys.* **138,** 272–284.

Beato, M., Kalimi, M. and Feigelson, P. (1972a). *Biochem. biophys. Res. Commun.* **47,** 1464–1472.

Beato, M., Schmid, W. and Sekeris, C. E. (1972b). *Biochim. biophys. Acta* **263,** 764–774.

Beato, M., Kalimi, M., Konstam, M. and Feigelson, P. (1973). *Biochemistry N.Y.* **12,** 3372–3379.

Beato, M., Kalimi, M., Beato, W. and Feigelson, P. (1974). *Endocrinology* **94,** 377–387.

Beck, J.-P., Beck, G., Wong, K. Y. and Tomkins, G. M. (1972). *Proc. natn. Acad. Sci. U.S.A.* **69,** 3615–3619.

Bell, P. A. and Munck, A. (1973). *Biochem. J.* **136,** 97–107.

Bellamy, D. (1963). *Biochem. J.* **87,** 334–340.

Bellamy, D., Phillips, J. G., Chester Jones, I. and Leonard, R. A. (1962). *Biochem. J.* **85,** 537–545.

Benecke, B. J., Ferencz, A. and Seifart, K. H. (1973). *FEBS Letters* **31,** 53–58.

Berg, T., Boman, D. and Seglen, P. O. (1972). *Expl. Cell Res.* **72,** 571–574.

Bishop, J. S. and Larner, J. (1967). *J. biol. Chem.* **242,** 1354–1356.

Blackburn, W. R., Travers, H., Kelley, J. S. and Rhoades, R. A. (1972). *Fedn. Proc. Fedn. Am. Soc. exp. Biol.* **31,** Abstract Number 156.

Blatti, S. P., Ingles, C. J., Lindell, T. Y., Morris, P. W., Weaver, R. F., Weinberg, F. and Rutter, W. J. (1970). *Cold Spring Harb. Symp. quant. Biol.* **35,** 649–655.

Blecher, M. (1966). *Endocrinology* **79,** 541–546.

Bortz, W. M. and Lynen, F. (1963). *Biochem. Z.* **339,** 77–82.

Bottoms, G. D., Stith, R. D. and Burger, R. O. (1969). *Proc. Soc. exp. Biol. Med.* **132,** 1133–1136.

Bottoms, G. D., Stith, R. D. and Roesel, O. F. (1972). *Proc. Soc. exp. Biol. Med.* **140,** 946–949.

Bradlow, H. L., Dobriner, K. and Gallagher, T. F. (1954). *Endocrinology* **54,** 343–352.

Braidman, I. P. and Rose, D. P. (1970). *Biochem. J.* **118,** 7P–8P.

Braidman, I. P. and Rose, D. P. (1971). *Endocrinology* **89,** 1250–1253.

Braun, T. and Birnbaumer, L. (1975). *In* "Comprehensive Biochemistry" (M. Florkin and E. H. Stotz, eds.) **25,** pp. 65–106 Elsevier, Amsterdam.

Brecher, P. I. and Wotiz, H. H. (1969). *Endocrinology* **84,** 718–726.

Brossard, M. and Nicole, L. (1969). *Can. J. Biochem.* **47,** 226–229.

Buck, M. D. and Schauder, P. (1970). *Biochim. biophys. Acta* **224,** 644–646.

Buckingham, S., McNary, W. F., Sommers, S. C. and Rothschild, J. (1968). *Fed. Proc. Fedn. Am. Socs. exp. Biol.* **27,** 328.

Bugany, H. and Beato, M. (1976). *Molec, cell. Endocr.* **7,** 49–66.

Bush, I. E. (1962). *Pharmac. Rev.* **14,** 317–445.

Feigelson, M., Gross, P. R. and Feigelson, P. (1962b). *Biochim. biophys. Acta* **55**, 495–504.
Feigelson, P., Beato, M., Colman, P., Kalimi, M., Killewich, H. A. and Schütz, G. (1975). *Rec. Prog. Horm. Res.* **31**, 213–242.
Feldman, D. (1974). *Endocrinology* **95**, 1219–1227.
Feldman, D., Dziak, R., Koehler, R. and Stern, P. (1975). *Endocrinology* **96**, 29–36.
Fiala, E. M. and Litwack, G. (1966). *Biochim. biophys. Acta* **124**, 260–266.
Finkel, R. M., Henshaw, E. C. and Hiatt, H. H. (1966). *Molec. Pharmac.* **2**, 221–226.
Fishman, R. A. and Reiner, M. (1972). *J. Neurochem.* **19**, 2221–2224.
Flores, H. and Alleyne, G. A. O. (1971). *Biochem. J.* **123**, 35–39.
Florini, J. R. and Buyske, D. A. (1961). *J. biol. Chem.* **236**, 247–251.
Foster, D. O., Ray, P. D. and Lardy, H. A. (1966). *Biochemistry, N.Y.* **2**, 555–562.
Fox, K. E. and Gabourel, J. D. (1967). *Molec. Pharmac.* **3**, 479–486.
Franz, J. M. and Knox, W. E. (1967). *Biochemistry, N.Y.* **6**, 3464–3471.
Friedman, D. L. and Larner, J. (1963). *Biochemistry, N.Y.* **2**, 669–675.
Friedrichs, D. and Schoner, W. (1974). *Biochim. biophys. Acta* **343**, 341–355.
Froehlich, J. and Wieland, O. (1971). *Eur. J. Biochem.* **19**, 557–562.
Gabourel, J. D. and Aronow, L. (1962). *J. Pharmac. exp. Ther.* **136**, 213–221.
Gardner, R. S. and Tomkins, G. M. (1969). *J. biol. Chem.* **244**, 4761–4767.
Gardner, D. G. and Wittliff, J. L. (1973). *Biochim. biophys. Acta* **320**, 617–625.
Garren, L. D., Howell, R. R. and Tomkins, G. M. (1964a). *J. molec. Biol.* **9**, 100–108.
Garren, L. D., Howell, R. R., Tomkins, G. M. and Crocco, R. M. (1964b). *Proc. natn. Acad. Sci. U.S.A.* **52**, 1121–1129.
Garrison, J. C. and Haynes, R. C. (1975). *J. biol. Chem.* **250**, 2769–2777.
Gehring, U. and Tomkins, G. M. (1974). *Cell* **3**, 301–306.
Gehring, U., Mohit, B. and Tomkins, G. M. (1972). *Proc. natn. Acad. Sci. U.S.A.* **69**, 3124–3127.
Gelehrter, T. D. and Tomkins, G. M. (1967). *J. molec. Biol.* **29**, 59–76.
Gelehrter, T. D. and Tomkins, G. M. (1969). *Proc. natn. Acad. Sci. U.S.A.* **64**, 723–730.
Gerschenson, L. E., Anderson, M. and Okigaki, T. (1970). *Science, N.Y.* **170**, 859–861.
Gerschenon, L. E., Davidson, M. B. and Anderson, M. (1974). *Eur. J. Biochem.* **41**, 139–148.
Giannopoulos, G. (1973a). *J. biol. Chem.* **248**, 3876–3883.
Giannopoulos, G. (1973b). *Biochem. biophys. Res. Commun.* **54**, 600–606.
Giannopoulos, G. (1974). *Endocrinology* **94**, 450–458.
Giannopoulos, G. (1975a). *J. Steroid Biochem.* **6**, 623–641.
Giannopoulos, G. (1975b). *J. biol. Chem.* **250**, 5847–5851.
Gliemann, Østerlind, K., Vinten, J. and Gommeltoft, S. (1972). *Biochim. biophys. Acta* **286**, 1–9.
Goldberg, M. L. and Atchley, W. A. (1966). *Proc. natn. Acad. Sci. U.S.A.* **55**, 989–996.
Goldstein, L., Stella, E. J. and Knox, W. E. (1962). *J. biol. Chem.* **237**, 1723–1726.
Goodman, A. D., Fuisz, R. E. and G. F. Cahill Jr. (1966). *J. clin. Invest.* **45**, 612–619.
Golpalakrishnan, T. V. and Sadgopal, A. (1972). *Biochim. biophys. Acta.* **287**, 164–186.
Granner, D. K. (1976). *Nature, Lond.* **259**, 572–573.
Granner, D. K., Chase, L., Aurbach, G. D. and Tomkins, G. M. (1968a). *Science, N.Y.* **162**, 1018–1020.
Granner, D. K., Hayashi, S., Thompson, B. and Tomkins, G. M. (1968b). *J. molec. Biol.* **35**, 291–301.

Gray, J. G., Pratt, W. B. and Aronow, L. (1971). *Biochemistry, N.Y.* **10**, 277–284.
Greengard, O. (1970). *In* "Biochemical Action of Hormones" (G. Litwack, Ed.) Vol. I, p. 53–87. Academic Press, New York and London.
Greengard, O. and Acs, G. (1962). *Biochim. biophys. Acta* **61**, 652–653.
Greengard, O. and Dewey, H. K. (1971) *Proc. natn. Acad. Sci. U.S.A.* **68**, 1698–1701.
Greengard, O., Smith, M. A. and Acs, G. (1963). *J. biol. Chem.* **238**, 1548–1551.
Greenman, D. L., Wicks, W. D. and Kenney, F. T. (1965). *J. biol. Chem.* **240**, 4420–4426.
Gross, S. R., Aronow, L. and Pratt, W. B. (1970). *J. Cell Biol.* **44**, 103–114.
Guidollet, J. and Louisot, P. (1969a). *Clinica chim. Acta* **23**, 121–132.
Guidollet, J. and Louisot, P. (1969b). *Acta Endocr., Cpnh.* **62**, 468–476.
Guidollet, J. and Louisot, P. (1970). *Clinica chim. Acta* **27**, 133–138.
Gunn, J. M., Hanson, R. W., Meyuhas, O., Reshef. L. and Ballard, F. J. (1975a). *Biochem. J.* **150**, 195–203.
Gunn, J. M., Tilghman, S. M., Hanson, R. W., Reshef, L. and Ballard, F. J. (1975b). *Biochemistry, N.Y.* **14**, 2350–2357.
Hackney, J. F. and Pratt, W. B. (1971). *Biochemistry, N.Y.* **10**, 3002–3008.
Hackney, J. F., Gross, S. R., Aronow, L. and Pratt, W. B. (1970). *Molec. Pharmac.* **6**, 500–512.
Haggerty, D. F., Young, P. L., Porják, G. and Carnes, W. H. (1973). *J. biol. Chem.* **248**, 223–232.
Hales, C. N. (1967). *Essays in Biochemistry* **3**, 73–98.
Hallahan, C., Young, D. A. and Munck, A. (1973). *In* "Endocrinology" (R. O. Scow, ed.), pp. 415–420. Excerpta Medica I.C.S. **273**. Amsterdam.
Hamana, K. and Iwai, K. (1973). *Gunma Symp. Endocr.* **10**, 77–88.
Harrison, R. W., Fairfield, S. and Orth, D. N. (1974). *Biochem. biophys. Res. Commun.* **61**, 1262–1267.
Harrison, R. W., Fairfield, S. and Orth, D. N. (1975). *Biochemistry, N.Y.* **14**, 1304–1307.
Haung, Y. L. and Ebner, K. E. (1969). *Biochim. biophys. Acta* **191**, 161–163.
Henderson, I. C., Fischer, R. E. and Loeb, J. N. (1971). *Endocrinology* **88**, 1471–1476.
Henning, H. V., Seiffert, I. and Seubert, W. (1963). *Biochim. biophys. Acta* **77**, 345–348.
Hershko, A. and Tomkins, G. M. (1971). *J. biol. Chem.* **246**, 710–714.
Higgins, S. J., Rousseau, G. G., Baxter, J. D. and Tomkins, G. M. (1973a). *Proc. natn. Acad. Sci. U.S.A.* **70**, 3415–3418.
Higgins, S. J., Rousseau, G. G., Baxter, J. D. and Tomkins, G. M. (1973b). *J. biol. Chem.* **248**, 5866–5872.
Higgins, S. J., Rousseau, G. G., Baxter, J. D. and Tomkins, G. M. (1973c). *J. biol. Chem.* **248**, 5873–5879.
Hollander, N. and Chiu, Y. W. (1966). *Biochem. biophys. Res. Commun.* **25**, 291–297.
Holt, P. G. and Oliver, I. T. (1969a). *Biochemistry, N.Y.* **8**, 1429–1437.
Holt, P. G. and Oliver, I. T. (1969b). *FEBS Letters* **5**, 89–91.
Homoki, J., Beato, M. and Sekeris, C. E. (1968). *FEBS Letters* **1**, 275–278.
Horibata, H. and Harris, A. W. (1970). *Expl. Cell Res.* **60**, 61–77.
Horisberger, M. and Amos, H. (1970). *Biochem. J.* **117**, 347–353.
Hübener, H. J. and Staib, W. (1965). "Biochemie und Klinik" (Weitzel G. and Zöllner, N. eds.) pp. 1–253. Thieme, Stuttgart.
Iacobelli, S. (1973). *Nature New Biology* **245**, 154–155.
Ishii, D. N., Pratt, W. B. and Aronow, L. (1972). *Biochemistry, N.Y.* **11**, 3896–3904.

Iynedjian, P. B., Ballard, F. J. and Hanson, R. W. (1975). *J. biol. Chem.* **250,** 5596–5603.
Jackson, V. and Chalkley, R. (1974a). *J. biol. Chem.* **249,** 1627–1636.
Jackson, V. and Chalkley, R. (1974b). *J. Biol. Chem.* **249,** 1615–1626.
Jackson, C. D. and Sells, B. H. (1968). *Biochim. biophys. Acta* **155,** 417–423.
Jacob, S. T., Sajdel, E. M. and Munro, H. N. (1969). *Eur. J. Biochem.* **7,** 449–453.
Jeanrenaud, B. and Renold, A. E. (1960). *J. Biol. Chem.* **235,** 2217–2223.
Jensen, E. V., Suzuki, T., Kawashima, T., Stumpf, W. E., Jungblut, P. W. and De Sombre, E. R. (1968). *Proc. natn. Acad. Sci. U.S.A.* **59,** 632–638.
Jensen, E. V., Numata, M., Brecher, P. I. and De Sombre, E. R. (1971). In "The Biochemistry of Steroid Hormone Action" (R. M. Smellie, Ed.). Biochemical Society Symposium No. 32. pp. 133–159. Academic Press, London.
Jo, J. S., Ishihara, N. and Kikuchi, G. (1974a). *Archs Biochem. Biophys.* **160,** 246–254.
Jo, J. S., Ishihara, N. and Kikuchi, G. (1974b). *FEBS Letters* **43,** 345–348.
Johnson, R. W., Roberson, L. E., and Kenney, F. T. (1973). *J. biol. Chem.* **248,** 4521–4527.
Kaiser, N., Milholland, R. J. and Rosen, F. (1973). *J. biol. Chem.* **248,** 478–483.
Kalimi, M., Beato, M. and Feigelson, P. (1973). *Biochemistry, N.Y.* **12,** 3365–3371.
Kalimi, M., Colman, P. and Feigelson, P. (1975). *J. biol. Chem.* **250,** 1080–1086.
Kamm, D. E. and Strope, G. L. (1972). *J. clin. Invest.* **51,** 1251–1263.
Karlson, P. (1963). *Perspectives in Biology and Medicine* **6,** 203–214.
Karlson, P., Doenecke, D. and Sekeris, C. E. (1975). In "Comprehensive Biochemistry" (M. Florkin and E. H. Stotz, eds.), Vol. 25, pp. 1–63. Elsevier, Amsterdam, London, N.Y.
Katzenellenbogen, B. S. (1975). *Endocrinology* **96,** 289–297.
Katzenellenbogen, B. S. and Gorski, J. (1972). *J. biol. Chem.* **247,** 1299–1305.
Keller, N., Richardson, U. I. and Yates, F. E. (1969). *Endocrinology* **84,** 49–62.
Kenney, F. T. (1962). *J. biol. Chem.* **237,** 1610–1614.
Kenney, F. T. and Flora, R. M. (1961). *J. biol. Chem.* **236,** 2699–2702.
Kenney, F. T., and Kull, F. J. (1963). *Proc. Natl. Acad. Sci. U.S.A.* **50,** 493–499.
Kenney, F. T., Holten, D. and Albritton, W. L. (1967). *Nat. Cancer Inst. Monogr.* **27,** 315–323.
Kenney, F. T., Lee, K.-L., Stiles, C. D. and Fritz, J. E. (1973). *Nature New Biology* **246,** 208–210.
Kidson, Ch. and Kirby, K. S. (1964). *Nature, Lond.* **203,** 599–603.
Kidson, Ch., Thomas, A. and Cohen, P. (1970). *Biochemistry, N.Y.* **9,** 1571–1576.
Killewich, L., Schütz, G. and Feigelson, P. (1975). *Proc. natn. Acad. Sci. U.S.A.* **72,** 4285–4287.
Kim, J. H. and Miller, L. L. (1969). *J. biol. Chem.* **244,** 1410–1416.
Kimberg, D. V. and Loeb, J. N. (1971). *Biochim. biophys. Acta* **246,** 412–420.
King, R. J. B. and Gordon, J. (1967). *J. Endocr.* **39,** 533–542.
King, R. J. B. and Gordon, J. (1972). *Nature New Biology* **240,** 185–187.
King, R. J. B. and Mainwaring, W. I. P. (1974). "Steroid-Cell Interactions" Butterworths & Co., London.
Kirk, D. L. (1965). *Proc. natn. Acad. Sci. U.S.A.* **54,** 1345–1353.
Kirkpatrick, A. F., Milholland, R. J. and Rosen, F. (1971). *Nature New Biology* **232,** 216–218.
Knox, W. E. (1951). *Br. J. exp. Path.* **32,** 462–469.
Knox, W. E. and Auerbach, V. U. (1955). *J. biol. Chem.* **214,** 307–313.
Knox, W. E. and Mehler, A. H. (1951). *Science, N.Y.* **113,** 237–241.

Knox, W. E., Auerbach, V. H. and Lin, E. C. C. (1956). *Physiol. Rev.* **36,** 164–254.
Koblinsky, M. (1974) "Glucocorticoid Binding to Liver Cytosol" Ph.D. Thesis, College of Physicians and Surgeons, Columbia University, New York.
Koblinsky, M., Beato, M., Kalimi, M. and Feigelson, P. (1972). *J. biol. Chem.* **247,** 7897–7904.
Koch, B., Lutz, B., Schmitt, G. and Mialhe, C. (1970). *Horm. Metab. Res.* **2,** 292–297.
Koehler, D. E. and Moscona, A. A. (1975). *Arch. Biochem. Biophys.* **170,** 102–113.
Koike, K., Otaka, T. and Okui, S. (1967). *J. Biochem., Tokyo* **61,** 709–715.
Koike, K., Otaka, T. and Okui, S. (1968). *J. Biochem., Tokyo* **63,** 709–715.
Kolanowski, J: and Pizzaro, M. A. (1969). *Annls Endocr.* **30,** 177–197.
Kondo, H., Kikuta, A. and Noumura, T. (1975). *Expl Cell Res.* **90,** 285–297.
Kornel, L. (1973). *Acta endocr. Suppl.* **178,** 5–45.
Korner, A. (1960). *J. Endocr.* **21,** 177–189.
Kostyo, J. L. (1965). *Endocrinology* **76,** 604–613.
Kostyo, J. L. and Redmond, A. F. (1966). *Endocrinology* **79,** 531–540.
Kotas, R. V., Fletcher, B. D., Torday, J. and Avery, M. E. (1971). *Pediatrics* **47,** 57–64.
Krebs, H. A. (1964). *Proc. Roy. Soc.* Series B. **159,** 545–564.
Krone, W., Huttner, W. B., Seitz, H. J. and Tarnowski, W. (1974). *FEBS Letters* **46,** 158–161.
Krone, W., Marquardt, W., Seitz, H. J. and Tarnowski, W. (1975a). *FEBS Letters* **57,** 64–67.
Krone, W., Huttner, W. B., Seitz, H. J. and Tarnowski, W. (1975b). *FEBS Letters* **52,** 85–89.
Kumaresan, P., Anderson, R. R. and Turner, C. W. (1967). *Endocrinology* **81,** 658–661.
Lampkin, J.McC. and Potter, M. (1958). *J. Nat. Cancer Inst.* **20,** 1091–1112.
Landau, B. R. (1965). *Vitamins and Hormones* **23,** 1–59.
Lang, N. and Sekeris, C. E. (1964a). *Hoppe–Seyler's Z. physiol. Chem.* **339,** 238–248.
Lang, N. and Sekeris, C. E. (1964b). *Life Sci.* **3,** 391–393.
Lavrinenko, I. A., Morozova, T. M. and Yuschkova, L. F. (1971). *Mol. Biol.* (Rus.) (*Molekulyarnaya Biologiya*) **5,** 12–16.
Leader, D. P. and Barry, J. M. (1967). *Nature, Lond.* **215,** 1374–1375.
Leader, D. P. and Barry, J. M. (1969). *Biochem. J.* **113,** 175–182.
Lee, N. D. and Baltz, B. E. (1962). *Endocrinology* **70,** 84–87.
Lee, K.-L. and Kenney, F. T. (1970). *Biochem. biophys. Res. Commun.* **40,** 469–475.
Lee, K.-L. and Kenney, F. T. (1971). *J. biol. Chem.* **246,** 7595–7601.
Lee, K.-I., Reel, J. R. and Kenney, F. T. (1970). *J. biol. Chem.* **245,** 5806–5812.
Leon, H. A. (1966). *Endocrinology* **78,** 481–486.
Leon, H. A., Arrhenius, E. and Hultin, T. (1962). *Biochim. biophys. Acta* **63,** 423–433.
Leung, K. and Munck, A. (1975). *A. Rev. Physiol.* **37,** 245–272.
Levinson, B. B., Baxter, J. D., Rousseau, G. G. and Tomkins, G. M. (1972). *Science N.Y.* **175,** 189–191.
Levitan, I. B. and Webb, T. E. (1969a). *Biochim. biophys. Acta* **182,** 491–500.
Levitan, I. B. and Webb, T. E. (1969b). *J. biol. Chem.* **244,** 4684–4688.
Levitan, I. B. and Webb, T. E. (1970a). *J. molec. Biol.* **48,** 339–348.
Levitan, I. B. and Webb, T. E. (1970a). *Science, N.Y.* **167,** 283–285.
Levitan, I. B., Morris, H. P. and Webb, T. E. (1971). *Biochim. biophys. Acta* **240,** 287–295.
Liang, T. and Liao, S. (1972). *Biochim. Biophys. Acta* **277,** 590–594.

Liang, T. and Liao, S. (1974). *J. biol. Chem.* **249**, 4671–4678.
Liberti, J. P., Longman, E. S. and Navon, R. S. (1970). *Endocrinology* **86**, 1448–1450.
Liberti, J. P., DuVall, C. H. and Wood, D. M. (1971). *Can. J. Biochem.* **49**, 1357–1361.
Liggins, G. C. (1969). *J. Endocr.* **45**, 515–523.
Liggins, G. C. and Howie, R. N. (1972). *Pediatrics* **50**, 515–525.
Lin, E. E. C. and Knox, W. E. (1957). *Biochim. biophys. Acta* **26**, 85–88.
Lin, E. E. C. and Knox, W. E. (1958). *J. biol. Chem.* **233** 1186–1189.
Lippman, M. E. and Thompson, E. B. (1973). *Nature* **246**, 352–354.
Lippman, M. E. and Thompson, E. B. (1974). *J. biol. Chem.* **249**, 2483–2488.
Lippman, M., Halterman, R., Perry, S., Leventhal, B. and Thompson, E. B. (1973). *Nature New Biology* **242**, 157–158.
Lippman, M. E., Wiggert, B. O., Chader, G. J. and Thompson, E. B. (1974a). *J. biol. Chem.* **249**, 5916–5917.
Lippman, M. E., Yang, S. S. and Thompson, E. B. (1974b). *Endocrinology* **94**, 262–266.
Litwack, G., Sears, M. L. and Diamondstone, T. I. (1963). *J. biol. Chem.* **238**, 302–305.
Litwack, G., Lichtash, E. and Diamondstone, T. I. (1972a). *Nature New Biology,* **237**, 149–150.
Litwack, G., Morey, K. S. and Ketterer, B. (1972b). *In* "Effects of Drugs on Cellular Control Mechanisms" (B. R. Rabin and R. B. Freedman, eds.) pp. 105–130. Macmillan Press, London, New York.
Litwack, G., Filler, R., Rosenfield, S. A., Lichtash, N., Wishman, C. A. and Singer, S. (1973). *J. biol. Chem.* **248**, 7481–7486.
Livingston, J. N. and Lockwood, D. H. (1975). *J. biol. Chem.* **250**, 8353–8360.
Loeb, J. N. and Yeung, L. L. (1973). *Proc. Soc. exp. Biol. Med.* **143**, 502–507.
Loeb, J. N., Borek, C. and Yeung, L. L. (1973). *Proc. natn. Acad. Sci. U.S.A.* **70**, 3852–3856.
Long, C. N. H., Katzin, B. and Frye, E. G. (1940). *Endocrinology* **26**, 309–344.
Luck, D. N., Gschwendt, M. and Hamilton, T. H. (1973). *Nature New Biology* **245**, 24–25.
Lukács, I. and Sekeris, C. E. (1967). *Biochim. biophys. Acta.* **134**, 85–90.
Lund–Larsen, T. and Berg, T. (1973). *Hoppe–Seyler's Z. physiol. Chem.* **354**, 1334–1338.
Lyons, W. R., Li, C. H. and Johnson, R. E. (1958). *Rec. Prog. Horm. Res.* **14**, 219–254.
MacGregor, R. R. and Mahler, H. R. (1969). *Biochemistry, N.Y.* **8**, 3036–3049.
Makman, M. H., Nakagawa, S. and White, A. (1967). *Rec. Prog. Horm. Res.* **23**, 195–227.
Makman, M. H., Dvorkin, B. and White, A. (1971). *Proc. natn. Acad. Sci. U.S.A.* **68**, 1269–1273.
Manchester, K. L. (1968). *In* "The Biological Basis of Medicine" (E. E. Bittar and N. Bittar, eds.), pp. 221–255. Academic Press, London and New York.
Mansour, A. M. and Nass, S. (1974). *Acta Endocr. Copnh.* **77**, 298–309.
Martin, D. W. and Tomkins, G. M. (1970). *Proc. Natl. Acad. Sci. U.S.A.* **65**, 1064–1068.
Martin, D. W., Tomkins, G. M. and Bresler, M. A. (1969a). *Proc. natn. Acad. Sci. U.S.A.* **63**, 842–849.
Martin, D. W., Tomkins, G. M. and Granner, D. (1969b). *Proc. Natl. Acad. Sci. U.S.A.* **62**, 248–255.
Matsui, N. and Plager, J. E. (1966). *Endocrinology* **78**, 1159–1164.

Maurer, H. R. and Chalkley, G. R. (1967). *J. molec. Biol.* **27**, 431–441.
Mayer, M., Kaiser, N., Milholland, R. J. and Rosen, F. (1974). *J. biol. Chem.* **249**, 5236–5240.
McEwen, B. S., Magnus, C., and Wallach, G. (1972). *Endocrinology* **90**, 217–226.
McGrath, C. M. (1971). *J. natn. Cancer Inst.* **47**, 455–467.
Melnykovych, G. and Bishop, C. F. (1971). *Endocrinology* **88**, 450–455.
Menahan, L. A., Ross, B. O. and Wieland, O. (1968). *Biochem. biophys. Res. Commun.* **30**, 38–44.
Mertvetsov, N. P., Chesnokov, V. N. and Salganik, R. I. (1973). *Biochim. biophys. Acta* **315**, 61–65.
Mester, I., Martel, D., Psychoyos, A. and Baulieu, E.-E. (1974). *Nature, Lond.* **250**, 776–778.
Milgrom, E., Atger, M. and Baulieu, E.-E. (1971). *Steroids* **16**, 741–750.
Milgrom, E., Atger, M. and Baulieu, E.-E. (1973a). *Biochim. biophys. Acta* **320**, 267–283.
Milgrom, E., Thi, L., Atger, M. and Baulieu, E.-E. (1973b). *J. biol. Chem.* **248**, 6366–6374.
Milgrom, E., Atger, M. and Baulieu, E.-E. (1973c). *Biochemistry, N.Y.* **12**, 5198–5205.
Milkovic, K. and Milkovic, S. (1963). *Endocrinology* **73**, 535–539.
Miller, J. E. and Litwack, G. (1969). *Biochem. biophys. Res. Commun.* **36**, 35–41.
Mills, E. S. and Topper, Y. J. (1970). *J. Cell Biol.* **44**, 310–328.
Mohla, S., De Sombre, E. R. and Jensen, E. V. (1972). *Biochem. biophys. Res. Commun.* **46**, 661–667.
Monder, C. and Walker, M. C. (1970). *Biochemistry, N.Y.* **9**, 2489–2497.
Morey, K. S. and Litwack, G. (1969). *Biochemistry, N.Y.* **8**, 4813–4821.
Morgan, H. E., Regen, D. M., Henderson, M. J., Sawyer, T. K. and Park, C. R. (1961). *J. biol. Chem.* **236**, 2162–2168.
Morris, F. J. and Barnes, F. W. Jr. (1966). *Fedn. Proc. Fedn. Am. Socs, exp. Biol.* **25**, Abstract Number 514.
Morris, D. J. and Barnes, F. W. Jr. (1967). *Biochim. biophys. Acta.* **136**, 67–78.
Morris, D. J., Sarma, M. H. and Barnes, F. W. (1970). *Endocrinology* **87**, 486–493.
Morris, J. E. and Moscona, A. A. (1970). *Science, N.Y.* **167**, 1736–1738.
Morris, J. E. and Moscona, A. A. (1971). *Devl. Biol.* **25**, 420–444.
Moscona, A. A. and Hubby, J. L. (1963). *Devl. Biol.* **7**, 192–206.
Moscona, A. A. and Kirk, D. L. (1965). *Science, N.Y.* **148**, 519–521.
Moscona, A. A. and Piddington, R. (1966). *Biochim. biophys. Acta* **121**, 409–411.
Moscona, A. A. and Piddington, R. (1967). *Science, N.Y.* **158**, 496–497.
Moscona, A. A., Moscona, M. H. and Saenz, N. (1968). *Proc. natn. Acad. Sci. U.S.A.* **61**, 160–167.
Moscona, M., Frenkel, N. and Moscona, A. A. (1972). Devl. Biol. **28**, 229–241.
Motoyama, E. K., Orzalesi, M. M., Kikkawa, Y., Kaibara, M.., Wu, B., Zigas, C. J. and Cook, C. D. (1971). *Pediatrics* **48**, 547–555.
Muldoon, T. G. and Westphal, U. (1967). *J. biol. Chem.* **242**, 5636–5643.
Müller, W. E. G., Totsuka, A. and Zahn, R. K. (1974). *Biochim. biophys. Acta.* **366**, 224–233.
Munck, A. (1961). *Endocrinology* **68**, 178–180.
Munck, A. (1962). *Biochim. biophys. Acta* **57**, 318–326.
Munck, A. (1971). *Perspectives in Biology and Medicine* **14**, 265–289.
Munck, A. and Brinck–Johnson, T. (1974). *J. Steroid Biochem.* **5**, 203–205.
Munck, A. and Koritz, S. B. (1962). *Biochim. biophys. Acta* **57**, 310–317.

Munck, A. and Wira, C. (1971). *In* "Advances in the Biosciences" **7,** 301–327 Schering Workshop on Steroid Hormone Receptors (G. Raspé, ed.), Pergamon Press, Oxford.

Munck, A., Wira, C., Young, D. A., Mosher, K. M., Hallahan, C. and Bell, P. A. (1972). *J. Steroid Biochem.* **3,** 567–578.

Murphy, B. E. P. (1974). *J. Clin. Endocrinol. Metab.* **38,** 158.

Murphy, L. D., Pradhan, D. S. and Sreenivasan, A. (1970). *Biochim. biophys. Acta* **199,** 500–510.

Musliner, T. A. and Chader, G. J. (1971). *Biochem. biophys. Res. Commun.* **45,** 998–1003.

Musliner, T. A. and Chader, G. J. (1972). *Biochim. biophys. Acta* **262,** 256–263.

Nandi, S. (1959). *University of California, Berkeley. Publ. Zool.* **65,** 1 (cit. by Shyamala, 1975).

Newsholme, E. A. and Gevers, W. (1967). *Vitamins and Hormones* **25,** 1–87.

Nielsen, S. and Notides, A. C. (1975). *Biochim. biophys. Acta* **381,** 377–383.

Niessing, J. and Sekeris, C. E. (1970). *Hoppe–Seyler's Z. physiol. Chem.* **351,** 1161–1163.

Notides, A. C. and Nielsen, S. (1975). *J. Steroid Biochem.* **6,** 483–486.

Notides, A. C., Hamilton, D. E. and Awer, H. E. (1975). *J. biol. Chem.* **250,** 3945–3950.

Ohisalo, J. J. and Pispa, J. P. (1975). *Biochim. biophys. Acta* **397,** 94–100.

Ohtsuka, E. and Koide, S. S. (1969). *Biochim. biophys. Res. Commun.* **35,** 648–652.

Oka, T. and Topper, Y. J. (1971). *J. biol. Chem.* **246,** 7701–7707.

O'Malley, B. W., Spelsberg, T. C., Schrader, W. T., Chytil, F. and Steggles, A. W. (1972). *Nature, Lond.* **235,** 141–144.

Paran, M., Gallo, R. C., Richardson, L. S. and Wu, A. M. (1973). *Proc. natn. Acad. Sci. U.S.A.* **70,** 2391–2395.

Parks, W. P., Scolnick, E. M. and Kozikowski, E. H. (1974). *Science N.Y.* **184,** 158–160.

Parks, W. P., Ransom, J. C., Young, H. A. and Scolnick, E. (1975). *J. biol. Chem.* **250,** 3330–3336.

Peck, W. A., Brandt, J. and Miller, I. (1967). *Proc. natn. Acad. Sci. USA.* **57,** 1599–1606.

Peets, E. A., Staub, M. and Symchowicz, S. (1969). *Biochem. Pharmac.* **18,** 1655–1663.

Peterkofsky, B. and Tomkins, G. M. (1968). *Proc. natn. Acad. Sci. U.S.A.* **60,** 222–228.

Piddington, R. and Moscona, A. A. (1965). *J. Cell Biol.* **27,** 247–252.

Pitts, R. F. (1964). *Am. J. Med.* **36,** 720–742.

Platzker, A. C. G., Kittermann, J. A., Clemend, A. and Tooley, W. H. (1972). *Pediatrics Res.* **6,** 406/146 (abstract).

Pratt, W. B., Kaine, J. L. and Pratt, D. V. (1975). *J. biol. Chem.* **250,** 4584–4591.

Puca, G. A., Nola, E., Hibner, U., Cicala, G. and Sica, V. (1975). *J. biol. Chem.* **250,** 6452–6459.

Puca, G. A., Nola, E., Sica, V. and Bresciani, F. (1972). *Biochemistry, N.Y.* **11,** 4157–4165.

Puca, G. A., Sica, V. and Nola, E. (1974). *Proc. natn. Acad. Sci. U.S.A.* **71,** 979–983.

Raab, K. H. and Webb, T. E. (1969). *Experimentia* **25,** 1240–1242.

Raina, P. N. and Rosen, F. (1968). *Biochim. biophys. Acta* **165,** 470–475.

Rall, T. W. and Sutherland, E. W. (1958). *J. biol. Chem.* **232,** 1065–1091.

Rao, M. L., Rao, G. S., Höller, M., Breuer, H., Schattenberg, P. J. and Stein, W. D. (1976). *Hoppe–Seyler's Z. Physiol. Chem.* **357,** 573–584.

Raynaud–Jammet, C. and Baulieu, E.-E. (1970). *Annls. Endocr.* **31,** 775–783.

Reel, J. R. and Kenney, F. T. (1968). *Proc. Natl. Acad. Sci. U.S.A.* **61**, 200–206.
Reel, J. R., Lee, K.-L., and Kenney, F. T. (1970). *J. biol. Chem.* **245**, 5800–5805.
Reif, L. and Amos, H. (1966). *Biochem. biophys. Res. Commun* **23**, 39–48.
Reif–Lehrer, L. (1971). *J. Cell Biol.* **51**, 303–311.
Reif–Lehrer, L. and Amos, H. (1968). *Biochem. J.* **106**, 425–430.
Reif–Lehrer, L. and Chader, G. J. (1969). *Biochim. biophys. Acta* **192**, 310–317.
Reshef, L. and Shapiro, B. (1960). *Metabolism* **9**, 551–555.
Ringold, G. M., Lasfargues, E. Y., Bishop, J. M. and Varnus, H. E. (1975a). *Virology* **65**, 135–147.
Ringold, G. M., Yamamoto, K. R., Tomkins, G. M., Bishop, J, M. and Varmus, H. E. (1975b). *Cell.* **6**, 299–305.
Rivera, E. M. and Cummins, E. P. (1971). *Gen. comp. Endocr.* **17**, 319–326.
Rizzo, A. J., Heilpern, P. and Webb, T. E. (1971). *Cancer Res.* **31**, 876–881.
Robison, G. A., Butcher, R. W. and Sutherland, E. W. (1968). *A. Rev. Biochem.* **37**, 149–174.
Rodbard, D. (1973). *In* "Receptors for Reproductive Hormones" (B. W. O'Malley and A. R. Means, eds.) pp. 342–364. Plenum Press, New York and London.
Roewekamp, W. G., Hofer, E. and Sekeris, C. E. (1977) *Eur. J. Biochem.* **70**, 259–268.
Rooney, S. A., Gross, I., Gassenheimer, L. N. and Motoyama, E. K. (1975). *Biochim. biophys. Acta* **398**, 433–441.
Rosen, F. and Milholland, R. J. (1963). *J. biol. Chem.* **238**, 3730–3735.
Rosen, F., Roberts, N. R. and Nichol, C. A. (1959) *J. biol. Chem.* **234**, 476–480.
Rosenau, W., Baxter, J. D., Rousseau, G. G. and Tomkins, G. M. (1972). *Nature New Biology* **237**, 20–24.
Rosner, W. and Bradlow, H. L. (1971). *J. Clin. Endocr. Metab.* **32**, 193–198.
Rosner, W. and Hochberg, R. (1972). *Endocrinology* **91**, 626–632.
Ross, B. D., Hems, R. and Krebs, H. A. (1967). *Biochem. J.* **102**, 942–951.
Rousseau, G. G., Baxter, J. D. and Tomkins, G. M. (1972). *J. molec. Biol.* **67**, 99–115.
Rousseau, G. G., Higgins, S. J., Baxter, J. D. and Tomkins, G. M. (1974). *J. Steroid Biochem.* **5**, 935–939.
Rousseau, G. G., Higgins, S. J., Baxter, J. D., Gelfand, D. and Tomkins, G. M. (1975). *J. biol. Chem.* **250**, 6015–6021.
Rudnick, D. and Waelsch, H. J. (1955). *Exp. Zool.* **129**, 309–326.
Ruh, T. S., Katzenellenbogen, B. S., Katzenellenbogen, J. A. and Gorski, J. (1973). *Endocrinology* **92**, 125–134.
Sajdel, E. M. and Jacob, S. T. (1971). *Biochem. biophys. Res. Commun.* **45**, 707–715.
Salganik, R. I., Morozova, T. M. and Zakharov, M. A. (1969). *Biochim. biophys. Acta* **174**, 755–757.
Samuels, H. H. and Tomkins, G. M. (1970). *J. molec. Biol.* **52**, 57–74.
Sandberg, A. A. and Slaunwhite, W. R. (1963). *J. clin. Invest.* **42**, 51–54.
Sarff, M. and Gorski, J. (1971). *Biochemistry, N.Y.* **10**, 2557–2563.
Sarkar, P. K. and Moscona, A. A. (1971). *Proc. natn. Acad. Sci. U.S.A.* **68**, 2308–2311.
Sarkar, P. K. and Moscona, A. A. (1973). *Proc. natn. Acad. Sci. U.S.A.* **70**, 1667–1671.
Sarkar, P. K., Goldman, B. and Moscona, A. A. (1973). *Biochem. biophys. Res. Commun.* **50**, 308–315.
Schaeffer, L. D., Chenoweth, M. and Dunn, A. (1969). *Biochim. biophys. Acta* **192**, 304–309.
Schapiro, S. (1968). *Endocrinology* **83**, 475–478.
Schauder, P. and Buck, M. D. (1971). *Biochim. biophys. Acta* **240**, 151–153.
Schaumburg, B. P. (1970). *Biochim. biophys. Acta* **214**, 520–532.

Schaumburg, B. P. and Bojesen, E. (1968). *Biochim. Biophys. Acta* **170**, 172–188.
Schimassek, H. (1967). *In:* "Wirkungsmechanismen der Hormone" (18. Coll. Ges. Physiol. Chem.), pp. 33–50. Springer, Berlin, Heidelberg, New York.
Schimassek, H. and Mitzkat, H. J. (1963). *Biochem. Z.* **337**, 510–518.
Schimke, R. T., Sweeney, E. W. and Berlin, C. M. (1965). *J. biol. Chem.* **240**, 322–331.
Schmid, W. and Sekeris, C. E. (1975). *Biochim. biophys. Acta* **402**, 244–252.
Schneider, J. A. and Weiss, M. C. (1971). *Proc. natn. Acad. Sci. U.S.A.* **68**, 127–131.
Schrader, W. T., Toft, D. O. and O'Malley, B. W. (1972). *J. biol. Chem.* **247**, 2401–2407.
Schütz, G. and Feigelson, P. (1972). *J. biol. Chem.* **247**, 5327–5332.
Schütz, G., Chow, W. and Feigelson, P. (1972). *J. biol. Chem.* **247**, 5333–5337.
Schütz, G., Beato, M. and Feigelson, P. (1973). *Proc. Natl. Acad. Sci. U.S.A.* **70**, 1218–1221.
Schütz, G., Killewich, L., Chen, G. and Feigelson, P. (1975). *Proc. natn. Acad. Sci. U.S.A.* **72**, 1017–1020.
Schwartz, R. J. (1972). *Nature New Biology* **237**, 121–125.
Schwartz, R. J. (1973). *J. biol. Chem.* **248**, 6426–6435.
Scott, W. A., Shields, R. and Tomkins, G. M. (1972). *Proc. natn. Acad. Sci. U.S.A.* **69**, 2937–2941.
Scrutton, M. C. and Utter, M. F. (1968). *A. Rev. Biochem.* **37**, 249–302.
Seal, U. S. and Doe, R. P. (1961). *Fedn. Proc. Fedn. Am. Socs. exp. Biol.* **20**, 179.
Segal, H. L. and Kim, Y. S. (1963). *Proc. natn. Acad. Sci. U.S.A.* **50**, 912–918.
Seglen, P. O. (1971). *Biochim. biophys. Acta* **230**, 319–326.
Sekeris, C. E. and Lang, N. (1965). *Hoppe–Seyler's Z. physiol. Chem.* **340**, 92–96.
Sekeris, C. E., Beato, M., Homoki, J. and Congote, L. F. (1968). *Hoppe–Seyler's Z. physiol. Chem.* **349**, 857–858.
Sekeris, C. E., Beato, M. and Seifart, K. H. (1970a). Proc. IVth Internat. Congr. Pharmacol. 292–299. Schwabe & Co. Publ., Basel.
Sekeris, C. E., Niessing, J. and Seifart, K. H. (1970b). *FEBS Letters* **9**, 103–107.
Sellers, L. and Granner, D. (1974). *J. Cell. Biol.* **60**, 337–345.
Selye, J. (1936). *Br. J. exp. Path.* **17**, 234–248.
Sereni, F., Kenney, F. T. and Kretchmer, N. (1959). *J. biol. Chem.* **234**, 609–612.
Seubert, W. (1967). *In* "Wirkungsmechanismus der Hormone" (18. Coll. Ges. Physiol. Chem.) pp. 158–191. Springer, Berlin, Heidelberg and New York.
Shaler, R. C. and McCarl, R. L. (1971). *J. Cell Biol.* **49**, 205–209.
Shaskas, J. R. and Bottoms, G. D. (1974). *Proc. Soc. exp. Biol. Med.* **147**, 232–238.
Shelton, K. R. and Allfrey, V. G. (1970). *Nature, Lond.* **228**, 132–134.
Shrago, E., Lardy, H. A., Nordlie, R. C. and Foster, D. O. (1963). *J. biol. Chem.* **238**, 3188–3192.
Shyamala, G. (1972). *Biochem. biophys. Res. Commun.* **46**, 1623–1630.
Shyamala, G. (1973). *Biochemistry, N.Y.* **12**, 3085–3090.
Shyamala, G. (1974). *J. biol. Chem.* **249**, 2160–2163.
Shyamala, G. (1975). *Biochemistry N.Y.* **14**, 437–444.
Sibley, C. H. and Tomkins, G. M. (1973). *Genetics* **74**, s253–s254.
Silber, R. H. and Porter, C. C. (1953). *Endocrinology* **52**, 518–525.
Simonsson, B. (1972). *Acta physiol. scand.* **86**, 398–409.
Singer, S. and Litwack, G. (1971). *Endocrinology* **88**, 1448–1455.
Singer, S., Becker, J. E. and Litwack, G. (1973). *Biochem. biophys. Res. Commun.* **52**, 943–950.
Singer, S. S., Gebhart, J. and Krol, J. (1975). *Eur. J. Biochem.* **56**, 595–604.

Sitaraman, V. and Ramasarma, T. (1974). *Biochem. biophys. Res. Commun.* **59,** 578–583.

Sluyser, M. (1966). *J. molec. Biol.* **19,** 591–595.

Sluyser, M. (1969). *Biochim. biophys. Acta* **182,** 235–244.

Smith, B. T., Torday, J. S. and Giroud, C. J. P. (1973). *Pediatrics Res.* **7,** 308.

Snart, R. S., Sanyal, N. N. and Agarwal, M. K. (1970). *J. Endocrinology* **47,** 149–158.

Snart, R. S., Shepherd, R. E. and Agarwal, M. K. (1972). *Hormones* **3,** 293–312.

Söling, H. D. and Willms, B., eds. (1971). "Regulation of Gluconeogenesis", Thieme, Stuttgart.

Sparkes, R. S. and Weiss, M. C. (1973). *Proc. natn. Acad. Sci. U.S.A.* **70,** 377–381.

Spiegel, M. (1961). *Biol. Bull. mar biol. Lab., Woods Hole,* **121,** 547–553.

Stackhouse, H. L., Chetsanga, C. J. and Tan, C. H. (1968). *Biochim. biophys. Acta* **155,** 159–168.

Staib, R., Thienhaus, R., Ammedick, U. and Staib, W. (1969). *Eur. J. Biochem.* **11,** 213–217.

Steinberg, R. A., Levinson, B. B. and Tomkins, G. M. (1975a). *Cell* **5,** 29.

Steinberg, R. A., Levinson, B. B. and Tomkins, G. M. (1975b). *Proc. natn. Acad. Sci. U.S.A.* **72,** 2007–2011.

Stellwagen, R. H. (1972). *Biochem. biophys. Res. Commun.* **47,** 1144–1150.

Stockdale, F. E. and Topper, Y. J. (1966). *Proc. natn. Acad. Sci. U.S.A.* **56,** 1283–1289.

Stockdale, F. E., Juergens, W. G. and Topper, Y. J. (1966). *Devl. Biol.* **13,** 266–281.

Sunaga, K. and Koide, S. S. (1967a). *Steroids* **9,** 451–456.

Sunaga, K. and Koide, S. S. (1967b). *Arch. Biochem. Biophys.* **122,** 670–673.

Suyemitsu, T. and Terayama, H. (1975). *Endocrinology* **96,** 1499–1508.

Thompson, E. B. and Gelehrter, T. D. (1971). *Proc. natn. Acad. Sci. U.S.A.* **68,** 2589–2593.

Thompson, E. B. and Lippman, M. E. (1974). *Metabolism* **23,** 159–202.

Thompson, E. B., Tomkins, G. M. and Curran, J. F. (1966). *Proc. natn. Acad. Sci. U.S.A.* **56,** 296–303.

Thompson, E. B., Granner, D. K. and Tomkins, G. N. (1970). *J. molec. Biol.* **54,** 159–175.

Toft, D. and Chytil. F. (1973). *Archs Biochem. Biophys.* **157,** 464–469.

Tomkins, G. M. (1974). *Harvey Lectures* **68,** 37–65.

Tomkins, G. M., Gelehrter, T. D., Granner, D., Martin, D. Jr., Samuels, H. H. and Thompson, E. B. (1969). *Science N.Y.* **166,** 1474–1480.

Tomkins, G. M., Levinson, B. B., Baxter, J. D. and Dethlefsen, L. (1972). *Nature New Biology* **239,** 9–14.

Tomkins, G. M., Thompson, E. B., Hayashi, S., Gelehrter, T., Granner, D. and Peterkofsky, B. (1966). *Cold Spring Harb. Symp. quant. Biol.* **31,** 349–360.

Tsai, Y.-H. and Hnilica, L. S. (1971). *Biochim. biophys. Acta* **238,** 277–287.

Ts'o, P. O. P. and Lu, P. (1964). *Proc. natn. Acad. Sci. U.S.A.* **51,** 17–24.

Tu, A. S. and Moudrianakis, E. N. (1973). *Biochemistry* **12,** 3692–3700.

Tucker, H. A., Larson, B. L. and Gorski, J. (1971). *Endocrinology* **89,** 152–160.

Turkington, R. W. and Kadohama, N. (1972). *In* "Gene Transcription in Reproductive Tissue". Karolinska Symposium Number 5, pp. 346–368. (E. Diczfalusy, ed.) Karolinska Institute, Stockholm.

Turkington, R. W., Juergens, W. G. and Topper, Y. J. (1967). *Endocrinology* **80,** 1139–1142.

Turnell, R. W., Kaiser, N., Milholland, R. J. and Rosen, F. (1974). *J. biol. Chem.* **249,** 1133–1138.

Umemura, Y. and Sakano, Y. (1973). *Gunma Symp. Endocr.* **10,** 63–75.
Utter, M. E. and Fung, C. H. (1970). Hoppe–Seyler's *Z. physiol. Chem.* **351,** 284–285.
Utter, M. F. and Scrutton, M. C. (1969). *Current Topics in Cell Regulation* **1,** 253–296.
Van der Meulen, N., Abraham, A. D. and Sekeris, C. E. (1972). *FEBS Letters* **25,** 116–122.
Van der Meulen, N., Marx, R., Sekeris, C. E. and Abraham, A. D. (1972). *Expl. Cell Res.* **74,** 606–610.
Venkov, P. V., Angelov, E. Z., Valeva, L. I. and Hadjiolov, A. A. (1967). *Nature Lond.* **213,** 807–809.
von Hippel, P. H. Revzin, A., Gross, C. A. and Wang, A. C. (1974). proc. natn. Acad. Sci. U.S.A. **71,** 4808–4812.
Vorob'ev, V. I. and Konstantinova, I. M. (1972). *FEBS Letters* **21,** 169–172.
Watanabe, H., Orth, D. N. and Toft, D. O. (1973). *J. biol. Chem.* **248,** 7625–7630.
Webb, T. E., and Wozney, L. R. (1968). *Arch biochem. Biophys.* **125,** 69–75.
Weber, G. (1968). *In* "The Biological Basis of Medicine" (E. E. Bittar and N. Bittar, eds.). Vol. 2, pp. 263–307. Academic Press, New York and London.
Weber, G., Allard, C., De Lamirande, G. and Cantero, A. (1955). *Biochim biophys. Acta* **16,** 618–619.
Weber, G., Srivastava, S. K. and R. L. Singhal (1964). *Life Sci,* **3,** 829–837.
Weiss, M. C. and Chaplan, M. (1971). *Proc. natn. Acad. Sci U.S.A.* **68,** 3026–3030.
Weiss, G., Lamartinière, C. A., Müller–Ohly, B. and Seubert, W. (1974). *Eur. J. Biochem.* **43,** 391–403.
Werthamer, S., Samuels, A. J. and Amaral, L. (1973). *J. biol. Chem.* **248,** 6398–6407.
Westphal, U. (1971). "Steroid-Protein Interactions". Springer–Verlag, Berling, Heidelberg, New York.
Wicks, W. D. (1971a). *J. biol. Chem.* **243,** 900–906.
Wicks, W. D. (1971b). *J. biol. Chem.* **246,** 217–223.
Wicks, W. D., Kenney, F. T. and Lee, K. L. (1969). *J. biol. Chem.* **244,** 6008–6013.
Wicks, W. D., Lewis, W. and McKibbin, J. B. (1972). *Biochim. biophys. Acta* **264,** 177–185.
Wicks, W. D., Barnett, C. A. and McKibbin, J. B. (1974). *Fedn. Proc. Fedn. Am. Socs, exp. Biol.* **33,** 1105–1111.
Wieland, O. and Weiss, L. (1963). *Biochem. biophys. Res. Commun.* **10,** 333–339.
Wiens, A. W. and Moscona, A. A. (1972). *Proc. natn. Acad. Sci. U.S.A.* **69,** 1504–1507.
Williamson, J. R., Browning, E. T. and Scholz, R. (1969) *J. biol. Chem.* **244,** 4607–4616.
Wimhurst, J. M. and Manchester, K. L. (1973). *Biochem. J.* **134,** 143–156.
Wira, C. and Munck, A. (1970). *J. biol. Chem.* **245,** 3436–3438.
Wira, C. R. and Munck, A. (1974). *J. biol. Chem.* **249,** 5328–5336.
Wogan, G. N. and Friedman, M. A. (1968). *Arch Biochem. Biophys.* **128,** 509–516.
Wong, K. C., Kornel, L., Bezkorovainy, A. and Murphy, B. E. P. (1973). *Biochim. biophys. Acta* **328,** 133–143.
Yamamoto, K. R. (1974). *J. biol. Chem.* **249,** 7068–7075.
Yamamoto, K. R. and Alberts, B. M. (1972). *Proc. natn. Acad. Sci. U.S.A.* **69,** 2105–2109.
Yamamoto, K. R. and Alberts, B. (1975). *Cell* **4,** 301–310.
Yamamoto, K. R., Stampfer, M. R. and Tomkins, G. M. (1974). *Proc. natn. Acad. Sci. U.S.A.* **71,** 3901–3905.
Yang, S. S., Lippman, M. E. and Thompson, E. B. (1974). *Endocrinology* **94,** 254–261.
Young, D. A. (1970). *Fedn. Proc. Fedn. Am. Socs. exp. Biol.* **29,** Abstract Number 3006.

Young, H. A., Parks, W. P. and Scolnick, E. M. (1975a). *Proc. natn. Acad. Sci. U.S.A.*
   72 3060–3064.
Young, H. A., Scolnick, E. M. and Parks, W. P. (1975b). *J. biol. Chem.* **250,** 3337–3343.
Yu, F.-L. and Feigelson, P. (1969a). *Archs. Biochem, Biophys.* **129,** 152–157.
Yu, F.-L. and Feigelson, P. (1969b). *Biochem. biophys. Res. Commun.* **35,** 499–504.
Yu, F.-L. and Feigelson, P. (1970a). *Arch. Biochem. Biophys.* **141,** 662–667.
Yu, F.-L. and Feigelson, P. (1970b). *Biochim. biophys. Acta* **213,** 134–141.
Yu, F.-L. and Feigelson, P. (1971). *Proc. natn. Acad. Sci. U.S.A.* **68,** 2177–2180.
Yu, F.-L. and Feigelson, P. (1973). *Biochem. biophys. Res. Commun.* **53,** 754–760.
Yuwiler, A., Wetterberg, L. and Geller, E. (1970). *Biochim. biophys. Acta* **208,** 428–433.

4.   Adrenal–Gonad Relationships

D. E. Kime,[1] G. P. Vinson,[2] Patricia W. Major[3] and R. Kilpatrick[4]

[1]Department of Zoology, University of Sheffield, Sheffield S10 2TN, England; [2]Department of Biochemistry and Chemistry, The Medical College of St. Bartholomew's Hospital, Charterhouse Square, London EC1M 6BQ, England; [3]Department of Pharmacology and Therapeutics, University of Sheffield, Sheffield S10 2TN, England; [4]School of Medicine, University of Leicester, Medical Sciences Building, University Road, Leicester LE1 7RH, England.

1.   Introduction

It is inevitable that most of the information about adrenal/gonad relationships should be based on man and a few species of eutherian mammals. The interplay of these two endocrine structures may be reflections of their controlling bodies. These include the hypothalamus (e.g. Sawyer, 1975; MacKinnon, 1978; Döhler, 1978) and may be related to the secretion of steroids by the embryo (e.g. Price *et al.*,

1975). There may be, also, a type of "competition" between the cells
and the anterior lobes of the pituitary in respect of, for example, the
corticotrophs and gonadotrophs (Nowell and Chester Jones, 1957).
This chapter deals primarily with the mammalian adrenal cortex
which, besides production of corticosteroids, secretes small amounts
of sex steroids. These may modulate, directly or indirectly, gonadal
function. It is possible, therefore, to envisage the following possible
effects of the adrenal on gonadal function:

   (i) Provision of precursors (e.g. progesterone, dehydroepiandros-
       terone, etc.) for gonadal steroid biosynthesis.
  (ii) Direct effects of adrenal corticosteroids and sex hormones on
       gonadal steroid producing enzymes.
 (iii) Effects of adrenal corticosteroids and sex hormones on pituitary
       ACTH and gonadotrophin production or release.
  (iv) Effects of adrenal corticosteroids and sex hormones on the
       hepatic catabolism of gonadal steroids.
   (v) Effects of hepatic adrenal steroid catabolites on the gonads
       either directly or via the pituitary.
  (vi) Effects of adrenal steroids on the protein binding of gonadal
       hormones.
 (vii) Effects of adrenal steroids on the structure of the gonads.

Similarly the following effects of the gonads on adrenal function may
be envisaged:

   (i) Provision of precursors (e.g. progesterone) for adrenal steroid
       biosynthesis.
  (ii) Direct effects of gonadal steroid hormones on adrenal steroid
       producing enzymes.
 (iii) Effects of gonadal hormones on pituitary ACTH and gonado-
       trophin production or release.
  (iv) Effects of gonadal hormones on the hepatic catabolism of
       corticosteroids.
   (v) Effects of gonadal hormones on protein binding of adrenal
       steroids.
  (vi) Effect of gonadal hormones on the structure of the adrenal.

These relationships are depicted schematically in Fig. 1.
   It is clear that the normal functioning of both gonad and adrenal is
dependent upon the balance and interplay of these effects. Most of our
information is derived from such conditions as castration,
hypophysectomy, adrenal carcinoma and enzyme deficiencies which
upset normal interrelationships.

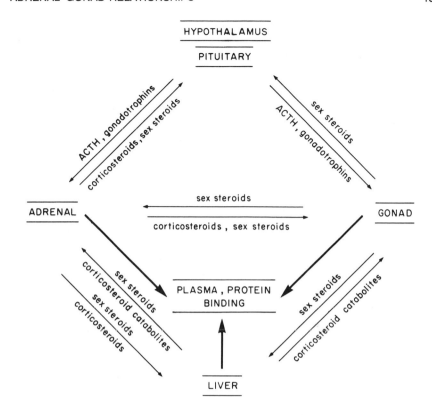

Fig. 1. Diagram to show the main possible interrelationships in the hypothalamus–pituitary–adrenal–gonad system.

In the normal adult the gonad and adrenal each have their specific function though each influences the other. During the last decade, however, it has been realized that the fetal adrenal has very different characteristics to that of the adult, both in function and in steroid hormone secretion. The concept of the feto-placental unit (Diczfalusy, 1964) as the complete endocrine organ has helped to elucidate the important role of the fetal adrenal in the production of oestrogens in pregnancy and in the initiation of parturition and lactation.

2.   Secretion of sex steroids by the adrenal cortex

A.   SECRETION OF ANDROGENS

In man it has long been recognized that the adrenal cortex is a source of C_{19} steroids, although their physiological significance is largely obscure. Some, such as testosterone and androstenedione are produced by other tissues, notably the gonads, but at least two, dehydroepiandrosterone (and its sulphate) and 11β-hydroxyandrostenedione seem to be produced exclusively by the adrenal cortex. Excessive adrenal androgen production may lead to virilization as in the adrenogenital syndrome, in which enzymatic defects in the biosynthesis of steroids by the adrenal cortex leads to increased androgen production due to loss of negative feedback to the pituitary. This is discussed more fully by Sandor *et al.* (1976).

1.   *Dehydroepiandrosterone and its sulphate*

It has long been appreciated that, quantitatively, dehydroepiandrosterone and its sulphate are the most important C_{19} steroids; less widely realized, perhaps, is that they are by far the most prominent products of adrenocortical origin in circulating plasma, at perhaps 10–20 times the level of cortisol. Studies on dehydroepiandrosterone are for some reason less popular than studies on, for example, cortisol or aldosterone, with the bizarre result that the most prominent adrenocortical product is of unknown physiological significance. Strangely, however, no evidence exists for the secretion of dehydroepiandrosterone or the sulphate in species other than man and a few primates. Although the role of dehydroepiandrosterone and its sulphate in the adult primate is not known, it is perhaps not without significance that it is only within the Anthropoidea that the adrenal synthesis of these compounds plays such an important role in the maintenance of pregnancy (see Section 6, C).

Dehydroepiandrosterone was first isolated by direct extraction of normal and abnormal human adrenal tissue by Bloch *et al.* (1956) and was soon confirmed by other workers (Plantin *et al.*, 1957; Keller *et al.*, 1958; Revol *et al.*, 1960; Baulieu, 1960; Cohn and Mulrow, 1961). Later it was found to be secreted into the adrenal vein, and to be present in both adrenal vein and circulating plasma largely as the sulphate (Oertel *et al.*, 1963a, b; Wieland *et al.*, 1963a; 1965; Baulieu 1965). Using a method for the determination of overall secretion rate, Roberts *et al.* (1961) obtained a value of 110 mg per 24 h in normal subjects. Its major importance as an adrenal secretory product is, for example, given by the data of Wieland *et al.* (1963b) giving values for adrenal vein blood concentrations of 165 μg/100 ml for dehydroepiandrosterone sulphate, and 70 μg per 100 ml for the free form, compared with only 288 μg per 100 ml for total 17-hydroxycorticosteroids. In circulating plasma, mean values for dehydroepiandrosterone were

$1.28 \pm 0.47\,\mu g$ per 100 ml in men, and 1.04 ± 0.38 in women, and in most subjects values obtained at 4 p.m. were perhaps 20% lower than at 8 a.m., suggesting diurnal variation (Kirschner *et al.*, 1965). Considerably higher values have been obtained in circulating plasma for the sulphate. Wang *et al.* (1968) showed that the concentration of dehydroepiandrosterone sulphate climbed abruptly at about age 7, reaching a peak at about 20–30 years of approximately $200\,\mu g$ per 100 ml: it thereafter declined slightly. In women, the pattern was similar but the values lower, with the maximum concentration at about 100–120 μg per 100 ml. In samples obtained from hirsute women and infertile men, Nieschlag *et al.* (1973) obtained values essentially similar to those reported by Kirschner *et al.* (1965) and Wang *et al.* (1968), with about $0.8\,\mu g$ per 100 ml for dehydroepiandrosterone and 50–200 μg per 100 ml for its sulphate in women, and 200–300 μg per 100 ml in men. The differences between peripheral and adrenal vein blood for the free compound were up to two orders of magnitude, but the adrenal vein content of the sulphate was only about twice that of circulating plasma. Together with the very high levels of the sulphate in circulating plasma of perhaps 10–20 times the value for cortisol (a much higher ratio than in adrenal vein blood, see above) this may indicate that dehydroepiandrosterone sulphate, like testosterone sulphate (Wang *et al.*, 1967) has a lower metabolic clearance rate than the free steroid. It is also possible that sulphurylation of dehydroepiandrosterone may take place in extra-adrenal sites, such as the liver. Although the production of dehydroepiandrosterone may also occur in gonadal tissue (Neher and Wettstein, 1960a, b, c; Eik-Nes and Hall, 1962; Nieschlag *et al.*, 1973), the major source is generally agreed to be the adrenal cortex (Nieschlag *et al.*, 1973; De Jong *et al.*, 1974), and this is supported by studies on the mode of control of its secretion (see Section 3). In addition, its secretion may be increased in cases of virilizing adenoma or Cushing's syndrome (Ibayashi and Yamaji, 1968; Saez *et al.*, 1970; 1971) and in the case of virilism this may result from decreased Δ^5-3β-hydroxysteroid dehydrogenase activity (Neville *et al.*, 1969a, b; Sandor *et al.*, 1976; this volume, chapter 2).

In species other than man, the production of dehydroepiandrosterone and its sulphate has been reported in certain primates *in vivo*, including *Macacus rhesus*, *Papio hamadryas*, *Cercopithecus aethiops* and *Erythrocebus patas* (Goncharov *et al.*, 1969; 1971), and also *in vitro* using homogenates of *Macaca mulatta* glands with pregnenolone and pregnenolone sulphate as added precursors (Gorwill *et al.*, 1970). No unequivocal evidence has been provided for the possibility that these two compounds are secreted by the adrenal cortex in other, non-primate species, despite particular efforts in some cases, notably in the guinea-pig (Deshpande *et al.*, 1971), pig and dog (Heap *et al.*, 1966; Wilroy *et al.*, 1968).

The pathways for the formation of dehydroepiandrosterone seem clear. In human glands, the major route for the formation of cortisol is *via* cholesterol side chain cleavage to pregnenolone, which is then hydroxylated at 17α prior to oxidation to 17α-hydroxyprogesterone (Weliky and Engel, 1963; Whitehouse and Vinson, 1968). It seems likely that 17-hydroxypregnenolone is also an intermediate in the formation of dehydroepiandrosterone (Deshpande *et al.*, 1970; Jensen *et al.*, 1971). Sulphurylation may also take place early in the pathway; at the point of pregnenolone formation (Gorwill *et al.*, 1970) but most of the dehydroepiandrosterone sulphate arises from direct sulphurylation of the free dehydroepiandrosterone (Doouss *et al.*, 1975); in man and guinea pig adrenals, this action takes place predominantly in the zona reticularis (Jones and Griffiths, 1968; Cameron *et al.*, 1969; and see Chapter 2, Volume 2).

2.   *11β-Hydroxyandrostenedione*

11β-hydroxyandrostenedione is another C_{19} steroid which appears to be an exclusive product of the adrenal cortex. It was apparently first described in rat adrenal vein blood (Bush, 1953) and later in human adrenal vein blood (Pincus and Romanoff, 1955; Romanoff *et al.*, 1953; Lombardo *et al.*, 1959; Hirschmann *et al.*, 1960) and adrenal tissue (Bloch *et al.*, 1956; Baulieu, 1960; Cohn and Mulrow, 1961). The values obtained by Lombardo *et al.* (1959) with adrenal vein blood suggest a concentration similar to that of cortisol, but much lower values were obtained by other authors. Values obtained for example by Hirschmann *et al.* (1960) suggest that 11β-hydroxyandrostenedione is only secreted into the adrenal vein in small amounts, giving concentrations of 3–8 μg per 100 ml plasma, compared with 3–5 μg for androstenedione, 8–47 μg for dehydroepiandrosterone and 115–300 μg for cortisol. Values found by Pincus and Romanoff (1955) were significantly higher, 84 μg per 100 ml compared with 850 μg for cortisol. These discrepancies may arise, of course, from deficiencies in the methods used for measuring steroids. However it may also be due to variation in the extent of dilution of the adrenal vein blood by contributions from vessels other than the adrenal vein: a notorious difficulty with human adrenal venous drainage. Furthermore, the effects of surgical stress on ACTH secretion may obviously be variable.

Unlike dehydroepiandrosterone, there is very good evidence that 11β-hydroxyandrostenedione is produced in the adrenals of a wide variety of species, including cattle (Bloch et al., 1954; Bryson and Sweat, 1962; Hudson et al., 1974), dog (Hechter et al., 1955; Oertel and Eik-Nes, 1962; Heap et al., 1966), pig (Wettstein and Anner, 1954; Heap et al., 1966; Ling and Loke, 1966; Holzbauer and Newport, 1969), cat (Bush, 1953), sheep (Bush and Ferguson, 1953), guinea pig (Deshpande et al., 1971), mouse (Bloch et al., 1960; Lucis and Lucis, 1970) and baboon (Axelrod, et al., 1973).

The biosynthesis of 11β-hydroxyandrostenedione is of interest, since there are at least two major pathways in mammalian adrenals. The most important is via 11β-hydroxylation of C_{19} substrates, and androstenedione has been shown to be a good substrate for 11β-hydroxylation (Deshpande et al., 1970; Axelrod et al., 1973). Dehydroepiandrosterone may also be converted to 11β-hydroxy-androstenedione, presumably through prior conversion to androstenedione (Neville et al., 1969a, b; Jones et al., 1970; Lucis and Lucis, 1970). However, an alternative route may involve the side chain scission of cortisol (Deshpande et al., 1970; Hudson and Killinger, 1972; Hudson et al., 1974) but this would seem to be relatively less important (Hudson and Killinger, 1972; Axelrod et al., 1973), and in general the C_{21} substrates are not converted so effectively as C_{19} substrates in vitro (Lucis and Lucis, 1970; Jones et al., 1971; Hudson and Killinger, 1972).

3.   *Androstenedione*

Other C_{19} products of the adrenal cortex are also formed in the gonads. One is androstenedione, which has been isolated from human adrenal tissue (Bloch et al., 1956; Baulieu, 1960; Cohn and Mulrow, 1961; Borgstede et al., 1963) and adrenal vein blood (Romanoff et al., 1953; Hirschmann et al., 1960; Wieland et al., 1963a, b; 1965; Weinheimer et al., 1966; Deshpande et al., 1967; Baird et al., 1969a, b). Concentrations of androstenedione in adrenal venous plasma given by Baird et al., (1969b) vary from about 0.5 μg per 100 ml (a male without endocrine disease) to 7.6 μg per 100 ml (female with congenital adrenal hyperplasia) compared with two values of 77.6 and 34.3 ng per 100 ml for circulating plasma in two other female subjects with endocrine disease: in the first case the amount of androstenedione in the adrenal venous plasma was about 1% of that of cortisol, while in the second, as may be expected, the value was much higher, about 25%.

Androstenedione has also been shown to be produced by adrenal tissue in numerous other species, including cattle (Bloch *et al.*, 1954; Bryson and Sweat, 1962; Ewald *et al.*, 1964; Cheatum *et al.*, 1967; Neville *et al.*, 1968b; Kowal *et al.*, 1964a, b; Munro *et al.*, 1969), pig (Ichii *et al.*, 1965; Gower and Ahmad, 1967; Holzbauer and Newport, 1969), sheep (Ward and Engel, 1966), *Macaca mulatta* (Sharma and Gabrilove, 1969), *M. rhesus, Papio hamadryas, Cercopithecus aethiops, Erythrocebus patas* (Goncharov *et al.*, 1969), in marsupials the brush tailed possum, *Trichosurus vulpecula* (Vinson *et al.*, 1971; Weiss, 1975) and the koala, *Phascolarctus cinereus* (Weiss and Richards, 1970). The major pathway for androstenedione formation is via conversion of dehydroepiandrosterone in human and bovine adrenals (Neville *et al.*, 1969a, b; Munro *et al.*, 1969; Jones et al., 1971) although an alternative pathway via 17-hydroxyprogesterone can also occur (Deshpande *et al.*, 1970; Jones *et al.*, 1971). This may very well not be true for other species, in which dehydroepiandrosterone is not readily demonstrated (see above), although the possibility that the transformation is extremely rapid, and does not allow dehydroepiandrosterone accumulation cannot be overlooked. In *Trichosurus vulpecula*, no evidence for the so-called Δ^5 pathway (i.e. involving 17-hydroxypregnenolone and dehydroepiandrosterone as intermediates following pregnenolone) was found, and it would appear that androstenedione was formed from 17-hydroxyprogesterone (Vinson *et al.*, 1971).

4.   *Testosterone*

The importance of the adrenal cortex as a source of testosterone is less clear. Testosterone was extracted from abnormal human tissue by Anliker *et al.* (1956) and Borgstede *et al.* (1963), while the transformation of labelled progesterone to testosterone, in low yields, was shown in human adrenal preparations by Kase and Kowal (1962), Ward and Grant (1963), Axelrod and Goldzieher (1967), Goldzieher *et al.* (1968) and Bryson *et al.* (1968). Pregnenolone may also be converted to testosterone (Dorfman *et al.* 1965; Axelrod *et al.*, 1965; 1969; Villee *et al.*, 1967; Bryson *et al.*, 1968) and the formation of testosterone sulphate from progesterone and pregnenolone has also been shown (Dixon *et al.*, 1965; Griffiths *et al.*, 1968). It is probable, however, as in the case of androstenedione formation, that the major pathway for testosterone synthesis is from pregnenolone, via 17-hydro-xypregnenolone, dehydroepiandrosterone and androstenedione as intermediates (Neville *et al.*, 1969a, b; Jones *et al.*, 1971).

It should be remembered of course that much of the tissue used in the studies *in vitro* was abnormal, and that in any case *in vitro* studies with radioactive precursors give little quantitative data that can be interpreted in physiological terms. Secretion of testosterone by the human adrenal into the adrenal vein is slight. Wieland *et al.* (1965) compared the adrenal vein plasma content of various C_{19} steroids. They found that the difference between peripheral plasma and adrenal vein plasma testosterone concentrations were minimal in two male subjects, values of 0.1 and 0.3 μg per 100 ml being obtained when the peripheral values of about 1 μg are subtracted from the adrenal vein levels. This offers a marked contrast to androstenedione [for in the same study the values for androstenedione were 6–12 μg per 100 ml in three female subjects (with undetectable peripheral values) and up to 28 μg per 100 ml in two males] and also to dehydroepiandrosterone sulphate (highest value: nearly 300 μg per 100 ml). Clear evidence for secretion of testosterone was obtained in two female subjects by Baird *et al.* (1969b), who obtained values of 145 and 21 ng per 100 ml adrenal vein plasma, compared with 23.2 and 9.5 ng per 100 ml peripheral plasma in the same subjects. The amounts are undeniably slight, however, even when compared with androstenedione (6954 and 646 ng per 100 ml adrenal vein, and 77.6 and 34.3 ng per 100 ml peripheral plasma) in the same subjects as for testosterone estimation (see also Kirschner and Bardin, 1972). Burger *et al.* (1964) have also reported a markedly higher level of testosterone in adrenal vein plasma (8400 ng/100 ml) than in peripheral plasma (640 ng/100 ml) but in only one of the six patients (with idiopathic hirsutism) that they examined. From studies on hormone dynamics, it appears that in the human female most of the circulating testosterone arises only indirectly from the adrenal cortex, through peripheral conversion of androstenedione (Horton and Tait, 1966; Baird *et al.*, 1969a). Horton (1967) concludes that this is also the same source of testosterone in cases of adrenal hyperplasia, although some testosterone may be secreted directly by the adrenal in cases of adrenogenital syndrome: the adrenal cortex is the major original source of the steroid in both cases. These views are supported by studies on the control of circulating C_{19} steroids (see Section 3).

The evidence for testosterone production by the adrenals in other species is patchy. In mouse adrenals incubated *in vitro*, testosterone production has been demonstrated both from labelled 17-hydroxy-progesterone (Hofmann and Christy, 1961) and from proges-terone, pregnenolone and dehydroepiandrosterone (Rosner et al., 1966). In pig adrenal homogenates testosterone is also formed from 17-hydroxyprogesterone and $17\alpha,20\alpha$-dihydroxypregn-4-en-3-one (Ichii et al., 1965). In rats, Koref et al. (1971) found no formation of any C_{19} steroids from radioactive cholesterol, pregnenolone or 17-hydroxyprogesterone. On the other hand, Askari et al. (1970) found small yields of testosterone from pregnenolone of from 0.03 to 0.14%, compared with 0.22 to 0.52% for androstenedione and 6.6 to 12.9% for corticosterone in incubated rat adrenal quarters. These authors found higher yields of corticosterone in female glands, in accordance with the general literature (see Section 4, B) and lower yields of testosterone when compared with the males. Milewich and Axelrod (1972a, b) also found that testosterone was produced from pregneno-lone, together with androstenedione, adrenosterone, dehydroepian-drosterone, 16α-hydroxyandrostenedione, and, in the presence of metapyrone, 11β-hydroxyandrostenedione and androsterone. Yields of testosterone were 1–2% of the radioactivity recovered, compared with similar amounts for androstenedione, 38–50% for 11-deoxycorticosterone, and 26–40% for corticosterone with a slight difference depending on sex (cf Askari et al., 1970). More recently, Vinson et al. (1976) obtained yields of 0.01–0.04% of testosterone from radioactive progesterone in female adrenal preparations. This compares with yields of 1–4% of corticosterone in the same incuba-tions. The results of Askari et al. (1970), Milewich and Axelrod (1972a) and Vinson et al. (1976) are thus reasonably consistent with regard to the relative amounts of testosterone and corticosterone produced by rat glands in vitro. In addition, Vinson et al. (1976) measured the production of testosterone from endogenous precursors and, depending on conditions of stimulation etc. (see Section 3), obtained values for testosterone which were of the order of 0.8% of those of corticosteroid measurable by a competitive protein binding method based on dog transcortin (i.e. largely corticosterone in this case). On this occasion, higher values for testosterone of endogenous origin were obtained in the females of the Wistar strain of rats used. Values for testosterone in adrenal vein plasma are not different from peripheral plasma in intact animals; however, after castration adrenal vein plasma still contains testosterone, whereas circulating levels are undetectable (Bardin and Peterson, 1967; Vinson and Phillips, 1976). In addition, strong indirect evidence for the production of testosterone by the rat adrenal cortex in vivo was suggested in the work of Kniewald *et al.* (1971) who found that circulating levels of testosterone were elevated in the period immediately after castration, and only later decreased to the very low values normally found. On the other hand, circulating testosterone levels were rapidly reduced by adrenalectomy. As well as indicating the possible importance of the adrenal cortex as a source of testosterone, at least under certain conditions, these findings hint at a possibility that an adrenal precursor might be required for testicular testosterone production. This possibility is also suggested by the work of Wasserman and Eik-Nes (1969) who found that both testosterone and androstenedione output are increased if the adrenal vein content is diverted to the testicular artery in dogs. They considered it likely that the adrenal production of progesterone could act as a substrate for testicular enzymes. However, using rat tissue, Vinson *et al.* (1976) found that the addition of rat adrenal tissue to incubations of interstitial tissue from the testis gave no further increment to overall testosterone production than would be expected by the simple summation of testosterone produced by each tissue type.

It is likely that there are considerable species differences in adrenal testosterone production. One striking example is the very much greater capacity for androstenedione and testosterone production seen in adrenal tissue from various marsupial species. In *Trichosurus vulpecula*, *Macropus eugenii* and *Didelphis marsupialis* the capacity of the tissue to synthesize testosterone or androstenedione from a radioactive precursor such as ^{14}C progesterone is comparable with its capacity to synthesize cortisol (Vinson *et al.*, 1971; Vinson and Renfree, 1975; Catling and Vinson, 1976). In all cases, the production of testosterone from endogenous precursors is also prominent, and depending on the degree of stimulation it may amount to about 5% of the value for total corticosteroid as measured by a competitive protein binding assay (Vinson, 1974a, b; Vinson and Renfree, 1975; Catling and Vinson, 1976). In *Trichosurus* its secretion may be associated with the presence of a discrete adrenocortical zone found only in the adult female. The range of variation in amount of this zone is demonstrated in Fig. 2a, b, c. It may be sparse or absent in the young female but during pregnancy it can occupy as much as 44% of the whole gland. However in both this and other species, some testosterone production occurs in the adrenals of both sexes, even, in *Macropus eugenii*, in pouch young animals aged 20 days or less (Catling and Vinson, 1976). In *Didelphis*, circulating testosterone levels are unaffected by ovariectomy, and in both *Trichosurus* and *Didelphis* females the effects of stimulation support the view that the adrenal cortex is the major source of circulating testosterone (Vinson, 1974a, b; Vinson and Renfree, 1975; and see Section 3).

5. *Androsterone*

Androsterone is another C_{19} steroid which may be of adrenocortical origin at least in man. It has been isolated from adrenal tissue (Keller *et al.*, 1958) and apparently from adrenal vein blood (Bush *et al.*, 1956) in which the amount present in a 30 ml sample was estimated at 12 μg, compared with 68 μg for cortisol, and 28 μg each for dehydroepiandrosterone and 11β-hydroxyandrostenedione. It is present in circulating plasma chiefly as the sulphate, like dehydroepiandrosterone, and is increased by ACTH stimulation (Baulieu, 1960). Mean circulating levels for androsterone sulphate are similar in men and women (Vihko, 1966) and again like dehydroepiandrosterone sulphate, they reach a maximum in both sexes at age about 30–40, in this case of about 50 μg per 100 ml plasma (Wang *et al.*, 1968). It has also been shown to be secreted into the adrenal vein in some primates (Goncharov *et al.*, 1969).

6. *Other adrenal adrogens*

Further C_{19} steroids may be produced by the adrenal cortex, at least under some conditions, but their importance as possible secretory products has not been studied. In man, these include androstenetrione (Reichstein, 1936a, b; Bloch *et al.*, 1954; Korus *et al.*, 1959; Chang *et al.*, 1963) which is also secreted in the pig and dog (Holzbauer and Newport, 1969). Various hydroxylated derivatives of the C_{19} steroids already described have been isolated from the adrenals of various species, including 11β-hydroxyandrosterone together with 11β-hydroxyepitestosterone from pig glands, 11β-hydroxyepiandrosterone from bovine tissue (von Euw and Reichstein, 1941), 6β-hydroxyandrostenedione (Meyer *et al.*, 1955), 6α-hydroxyandrostenedione and 19-hydroxyandrostenedione (Meyer *et al.*, 1955). In human adrenals too, 16α-hydroxylation may occur yielding 16α-hydroxydehydroepiandrosterone (Shahwan *et al.*, 1967, 1968; Palacios *et al.*, 1968; Villee *et al.*, 1967) especially in the fetus (Colas and Heinrichs, 1965; Easterling *et al.*, 1966; Simmer *et al.*, 1966) (see also Section 6, C). 16α-hydroxytestosterone and 16α-hydroxyandrostenedione can be formed from progesterone (Axelrod *et al.*, 1969). Testosterone may be 11β-hydroxylated (Chang *et al.*, 1963; Engel and Dimolene, 1963), androstenedione may be hydroxylated at

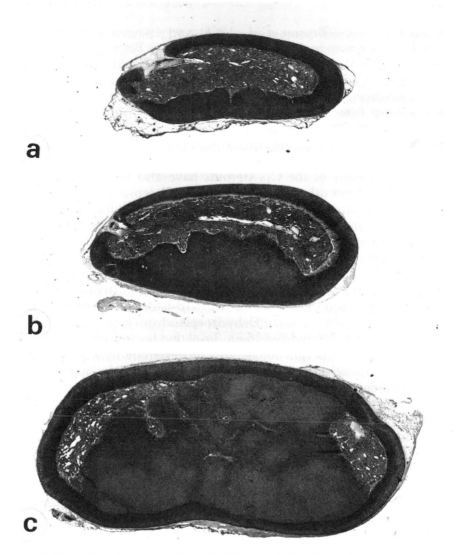

Fig. 2. Examples of the adrenal gland of the female brush tailed possum *Trichosurus vulpecula;* sectioned at 7μ, stained with Ehrlich's haematoxylin and eosin. Magnification × 10. (Material supplied by G. B. Sharman; processed and measured by W. Mosely). (a) Non-pregnant. The corpus luteum was removed on day 10 of the experiment and killed 8 days later. The special zone is just visible at the bottom of the section and occupies 7.15% of the overall gland. (b) Pregnant. The special zone is more obvious and occupies 18.5% of the overall gland. The experiment on this animal consisted of removing the corpus luteum on day 7 with progesterone injections on the following 3 days and killed on day 15. (c) Pregnant. Untreated and killed on day 15 of pregnancy. The special zone occupied 44% of the overall gland. The figures for the day are calculated by calling oestrus as equal to day 0.

larger than that of the male. Similar differences have been reported for other species, e.g. man (Swinyard, 1940); mouse (Chester Jones, 1955); woodchuck (*Marmota monax*) (Christian, 1962). In other species the weights are equal in males and females: cat, dog, guinea-pig, rabbit (Chester Jones, 1957), various species of voles (Delost, 1955, 1956; Delost and Delost, 1954, 1955). A full table is given in Idelman (1978, Volume 2 of this work). Changes in adrenal weights often occur in the female reproductive period of various species (Parkes, 1945; Parkes and Deanesly, 1966; Idelman, 1978). Although the sexual dimorphism in the non-pregnant and pregnant animal may be related to the effect of gonadal secretion, some caution must be exercised in identifying the rise in adrenal weight with sexual activity and increased corticosteroid secretion in those species which exhibit an annual breeding cycle.

Christian *et al.* (1965) have shown that in the woodchuck, the increase in adrenal weight begins during the period of sexual activity in the spring but continues for much longer. It appears it is related to external stress such as social pressure and intraspecific aggression (see Discussion in Nowell, Chapter 6). In a study of reproduction of the bank vole (*Clethrionomys glareolus*) in southern and northern Sweden it was shown that the relative adrenal weights of the female adrenal increase during the reproductive season. The males always had lower relative adrenal weights which did not change during the reproductive season; that of non-reproducing females had intermediate values (Fig. 3; Meurling, Gustafsson and Anderson, personal communication).

Direct evidence that sex steroids secreted by the gonads are able to affect adrenal weight and activity in laboratory animals has been provided during extensive studies by Kitay (1968). In the intact rat in which the female adrenal is larger than the male, testosterone treatment decreases the weight of the female gland but does not affect that of the male: conversely oestrogen treatment while increasing the weight of the male adrenal does not affect the female gland (Table I). Orchiectomy causes an increase in the weight of the male adrenal which is reversed by testosterone administration. Ovariectomy only slightly decreased the adrenal weight in females, but androgen therapy of the castrate causes a pronounced drop in weight. In the hamster, where the normal sexual dimorphism in adrenal weight is reversed, prepubertal gonadectomy results in a decrease in adrenal weight in both sexes, testosterone or oestradiol replacement therapy restoring the weights to control values (Peczenik, 1944; Holmes, 1955; Chester Jones, 1955; Gaskin and Kitay, 1970).

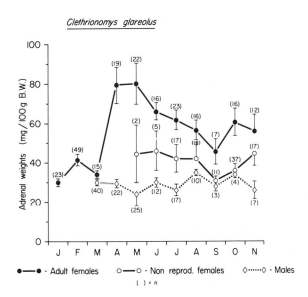

Fig. 3. Adrenal weights of the wild bank vole (*Clethrionomys glareolus*) from monthly samples over a year in Sweden. The adult females breed from April (A) to August (A) or September (S). During this season, the reproductive females are probably continuously pregnant as there is a post-partum oestrus. In the cases of non-reproductive females and males the adrenal weights remain steady (from unpublished observations of P. Meurling, T. Gustaffson and B. Andersson).

TABLE I

The effect of sex hormones on the adrenal weights of the mature rat.

	MALE 1			MALE 2			MALE 3			FEMALE 1			FEMALE 2			FEMALE 3		
	n	BW (g)	AW (mg)	n	BW (g)	AW (mg)	n	BW (g)	AW (mg)	n	BW (g)	AW (mg)	n	BW (g)	AW (mg)	n	BW (mg)	AW (g)
Intact control	12	222	33.2	12	263.0	38.0	19	220.1	36.6	12	175	53.2	12	185.5	55.3	23	164.5	48.6
+ oestrogen	12	195	48.4	9	210.4	43.7	9	189.2	51.8[a]	12	148	56.3	10	173.0	45.5	16	178.6	55.1[a]
+ testosterone	12	224	29.8	9	244.8	39.2	8	197.1	31.1[c]	12	188	40.5	10	194.8	41.9	14	166.7	32.9[c]
Gonadectomized	17	234	46.0	12	249.3	46.0	11	195.4	41.2	20	213	52.0	13	213.1	50.1	13	173.0	43.3
+ oestrogen	8	—	49.4	10	207.4	46.6	12	187.0	58.4[a]	8	—	66.7	12	192.2	48.7	10	136.9	54.6[b]
+ testosterone	8	—	33.2	11	248.7	40.8	9	199.8	31.1[c]	8	—	38.8	11	219.9	39.9	13	195.7	32.5[c]

Range of standard errors: Body weight ± 3 to ± 9; Adrenal weight ± 1 to ± 3. n = number of animals: BW = Body weight: AW = Absolute adrenal weight. 1. Kitay (1963b, c, d): polyoestradiol phosphate, 2 mg/100 mg body wt; testosterone phenylacetate, 5 mg/100 g body wt. Single injection 14 days before sampling. 2. Carter (1954): 100 μg oestradiol dipropionate; 1.5 mg testosterone propionate. Injections given daily for 22 days. 3. Greep and Chester Jones (1950): [a] 0.17 μg oestradiol benzoate; [b] 0.83 μg oestradiol benzoate; [c] 50 μg testosterone propionate. Injections given daily for 45 days. (Recalculated from Chester Jones, 1955).

Studies at the cellular level in rats have shown that these sex related changes in adrenal weight are due to an increase in the cytoplasmic volume of the individual cells and not to an increase in their number. Orchiectomy increased the volume of the cytoplasm in both the fascicular and reticular zone cells, the effect being reversed by adminstration of testosterone. Neither ovariectomy nor oestradiol replacement affected the cytoplasmic volume (Malendowicz, 1974a). Similarly, karyometric studies have shown that the nuclei in the female zona fasciculata cells are approximately 18% larger than those of males, but only slight sex differences (approx. 2%) are observed in the zona reticularis. Orchiectomy resulted in enlargement of cell nuclei in all zones of the adrenal cortex, but testosterone replacement reversed this. The nuclear volume is a sensitive parameter of cell activity and in the adrenal cortex is dependent on ACTH levels and reflects corticosteroid secretion of these cells (Malendowicz, 1974b). Mäusle (1971) has also reported larger zona fasciculata cells in the female rat and has noted that this sex also has larger mitochondria and smaller and more dispersed liposomes. Dhom et al. (1971) have described changes in the ultrastructure of the rat adrenal cortex with age. Beginning at the fourth week of life, the parenchymal cells in the zona fasciculata of the female are characterized by a light, finely granular cytoplasm without distinct lipid vacuoles, whereas in the male, the cells show a dense compact cytoplasm with the clearly visible vacuoles increasing in size.

B.   GONADAL EFFECTS ON THE FUNCTION OF THE ADRENAL CORTEX

The work initially was stimulated by the observation that in the rat the adrenal gland of the female is much larger than that of the male and that circulating levels of corticosterone are also higher and show a more sensitive response to ACTH (Kitay, 1961a, b; Cortes et al., 1963). In addition, corticosterone secretion varies with the stage of the oestrous cycle in females, and is highest during pro-oestrus (Dean et al., 1959; Critchlow et al., 1963; Hinsull and Crocker, 1970; Raps et al., 1971; Holzbauer and Godden, 1974; Phillips and Poolsanguan, 1978). In vivo, ovariectomy results in decreased levels of circulating cortico-sterone, (Kitay, 1963a, b, c, d) and there is decreased output of corticosterone by subsequently incubated adrenal slices in vitro (Kitay, 1961a, b; 1965a, b). Treatment of ovariectomized animals with oestradiol restores the secretion of corticosterone in vivo (Kitay, 1963c, d) and also in vitro, but is inhibitory at higher doses (see also Rennels and Singer, 1970, Kitay, 1963d, 1965b) although treatment of intact animals with oestradiol has no effect (Kitay, 1963b, c, d). In males gonadectomy results in unchanged levels of circulating corticosterone (Kitay, 1963b, c, d) although adrenal vein content of corticosterone and the secretion of corticosterone *in vitro* was reduced. Treatment with testosterone or oestradiol *in vivo* tended to restore these parameters to the control condition. The findings suggest that gonadal hormones can affect the secretion of corticosteroids in a number of ways through action at various loci, including the hypothalamus, the liver, and the adrenal itself, or by changes in the plasma protein binding of corticosteroids.

C.   EFFECTS INDIRECTLY MEDIATED

The results of various groups supports the view that oestrogen stimulates ACTH synthesis. Gemzell (1952) found that treatment of intact male rats with oestradiol increased blood ACTH content, and Barrett (1960) found that circulating ACTH levels after stress were higher in females than in males, and that the difference was greatest in females in pro-oestrus, and lowest in dioestrus. The differences between males and females was abolished after ovariectomy. Kitay (1963b, c, d) found evidence for the increased synthesis of ACTH in both males and females under oestrogen treatment; however, whereas in males the increased ACTH was released by the pituitary, as judged by consequent adrenal hypertrophy and unchanged pituitary ACTH content, in the females adrenal weight was unchanged while pituitary ACTH was increased. In the males, adrenal RNA was increased while DNA was unchanged (reflecting hypertrophy rather than hyperplasia) but in the females adrenal RNA and DNA were reduced, and since adrenal weight was unchanged this indicates a combination of hypoplasia with hypertrophy of the remaining tissue. (In addition to the pituitary effects, the results also support the hypothesis of a direct oestrogen effect on the adrenal, see below.) Testosterone treatment produced adrenal hyperplasia in both sexes (see also Mietkiewski *et al.*, 1969) although pituitary ACTH content was reduced only in the females: thus ACTH synthesis was reduced in both sexes with greater reduction in the females.

The converse effects of testosterone and oestradiol on ACTH secretion are also suggested by gonadectomy. Castration of male rats is followed by increased pituitary ACTH and adrenal hyperplasia, whereas ovariectomy results in diminished ACTH secretion. In both sexes treatment with the appropriate gonadal steroid restores the control condition (Nowell and Chester Jones, 1957; Kitay,

1963b, c, d; Ramirez *et al.*, 1965). While it is possible that changes in pituitary ACTH could arise indirectly through feedback from adrenal steroids, at least a partial direct effect on the pituitary is indicated by the result of Kitay (1963b, c, d) who found that oestradiol enhanced and testosterone reduced ACTH synthesis in gonadectomized adrenalectomized animals.

Further evidence for the effects of oestradiol on ACTH secretion were given by Coyne and Kitay (1969). Plasma concentrations of ACTH in unstressed adrenalectomized rats were decreased, following ovariectomy, to 76% and in stressed animals to 57% of control values. Release of ACTH by incubated whole pituitary glands *in vitro* did not differ from control values with or without oestradiol replacement, but their response to CRF was significantly decreased in the ovariectomized group, with partial reversal after treatment with polyoestradiol phosphate. The hypothalamic content of CRF was similar in intact and gonadectomized rats, but reduced in gonadectomized rats treated with oestradiol. The results therefore suggest that the hyposecretion of ACTH by ovariectomized animals results from decreased sensitivity of the pituitary to CRF, as well as a decreased capacity to synthesize ACTH. Moreover, the fact that ovariectomy induced a further decrease in ACTH levels in adrenalectomized animals suggests that the effects of oestradiol are to some extent at least independent of corticosteroid feedback. Similar experiments in male animals (Coyne and Kitay, 1971) showed that plasma concentrations of ACTH were increased to 197% of control values following castration of unstressed adrenalectomized animals, and to 159% of control values in stressed animals with intact adrenals. Release of ACTH by incubating whole pituitary glands did not change, but addition of pituitary stalk/median eminence extract to the pituitary incubations resulted in greater release of ACTH by the castrate group in both adrenalectomized and adrenal-intact groups. Partial restoration of control conditions was obtained in animals treated with testosterone, although testosterone added *in vitro* was ineffective. Hypothalamic content of CRF was apparently unaffected by castration or testosterone replacement, and therefore it appears that as in the case of the female/oestrogen results, the effects are at least partially mediated by changes in pituitary responsiveness to CRF, and are independent of concomitant changes in adrenal function or conditions related to steroid feedback.

A sex difference in the circadian periodicity of CRF activity in the rat hypothalamus has been demonstrated (Hiroshige *et al.*, 1973). The CRF content of the female hypothalamus is greater in the morning than in the afternoon, whereas the reverse is the case in the male. In the female a peak CRF content was found at 8 a.m. and a precipitous fall ensued about noon concomitant with a sharp rise in plasma corticosteroid levels. Ovariectomy reduced the morning levels but did not raise those of the afternoon. A similar result was obtained by studying the effect of the oestrous cycle on the hypothalamic content of the rat (Hiroshige and Wada-Okada, 1973). During pro-oestrus and oestrus, the CRF content was significantly greater at 9 a.m. than at 4 p.m. but during dioestrus no essential difference was observed between morning and afternoon levels. It thus appears that oestrogens may affect both the hypothalamic CRF content and the responsiveness of the pituitary to this factor.

Further experiments by Perklev (1971) indicate the possibility that other mechanisms may contribute to the overall effects. A single injection of polydiethylstilboestrol (200 μg/100 g body weight) into male rats results in elevated plasma corticosterone levels, adrenal weights and levels of adrenal Δ^5-3β-hydroxysteroid dehydrogenase 11–14 days later. These effects, including the enzyme activation were consistent with mediation of the oestrogen action through increased ACTH secretion (see above, also Surina, 1970). The effects were reversed by LH or long acting testosterone ester treatment. In castrated animals plasma corticosterone levels were normal (see above) but adrenal weight and steroid dehydrogenase activities were increased. In this case the effects were reversed by testosterone ester treatment but not by LH. In view of earlier results on the concomitant decrease in pituitary gonadotrophin secretion induced by this oestrogen (Perklev and Groning, 1969), Perklev (1971) suggested that one of the mechanisms by which oestrogens exert their effects is by regulating LH secretion and thereby testosterone secretion. It is known that corticosteroid binding globulin activity increases after a fall in testosterone levels (Gala and Westphal, 1965a) and it is probable that if the increased binding capacity resulted in a reduction in unbound corticosterone, there would be decreased negative feedback of corticosterone on pituitary ACTH secretion (bound corticosterone may be relatively inactive in this respect; Kawai and Yates, 1966). This does not exclude direct effects of oestrogens on the pituitary of course, especially in view of the fact that oestrogen given to castrates results in greater enlargement of the adrenals than seen in castrate control animals. Also castration does not result in an increase in circulating levels of corticosterone, although other adrenal parameters are stimulated. Despite the conclusions of Perklev (1971) the possibility of direct effects of gonadotrophins on the adrenals themselves cannot be eliminated, and the mechanism may account for some anomalies (see Section 3).

Phillips and Poolsanguan (1978) have shown that intravenous injection of NIH-LH (ovine) to female rats in the morning of metoestrus results in a significant increase in corticosterone output by the adrenal within 1–2 min of injection, suggesting a direct effect of the hormone at the adrenal level. Since no stimulation of corticosterone secretion occurs until 3 min after ACTH injection (Richardson and Schulster 1972) it was suggested that LH may act in a different manner possibly as a releasing hormone which may alter the permeability of membranes and increase the release of stored corticosterone. Gonadotrophins may well act upon the adrenal in several ways since when hCG or PMSG were administered 12 h before sampling a marked suppression of corticosterone was found in adrenal vein plasma (Poolsanguan, 1975).

There are obviously species variations in these parameters, but perhaps one of the most interesting at this point is the comparison of the rat with the hamster. In many respects the adrenal gland of the hamster appears to be unique, and compared with those of other animals, its adrenals contain very little cholesterol (Agate, 1952; Marks et al., 1958) or other lipids (Deane and Greep, 1946; Alpert, 1950; Deane and Lyman, 1954; Zieger et al., 1974). The sexual dimorphism in the size of the glands is the reverse of that seen in the rat, the male having the larger gland (Peczenik, 1944; Parkes, 1945; Keyes, 1949; Snyder and Wyman, 1951). Spaying has little effect on adrenal size, and the adrenal of the nulliparous female possibly contains an X zone, which that of the male does not have at any stage of development (Holmes, 1955). Selye (1941) suggested that androgens, in contrast to their action in the rat, stimulate ACTH secretion by the hamster pituitary. The differences obviously merit attention at the steroidogenic level and Gaskin and Kitay (1970, 1971) found that prepubertal gonadectomy resulted in decreased adrenal weight in both sexes, and testosterone or oestradiol treatment restored adrenal weight to the control level. As might be expected from the differences in adrenal weights, male hamsters have higher plasma corticosteroids, and higher in vivo and in vitro secretion of corticosteroids than the females. Hepatic clearance of cortisol is also greater in the male, resulting in a shorter half life for circulating cortisol; prepubertal castration results in decreases of secretion and of circulating levels of corticosteroids, and the effects are reversed by testosterone treatment. In contrast, the functional parameters of the female adrenal are unaffected by ovariectomy, and oestradiol has no effect on the secretion of corticosteroids (Gaskin and Kitay, 1970). Administration of ACTH abolished the decreases in adrenal weight and steroid secretion seen in castrated males. Testosterone treatment failed to promote steroid secretion *in vitro* in tissue from castrated hypophysectomized animals. It appeared that testosterone acts through stimulating ACTH secretion, and thus its actions in some ways parallel the actions of oestrogens in the rat (Gaskin and Kitay, 1971): spaying and oestrogen treatment in the females resulted in no clear changes.

The result of treating rats with sex steroids may depend in part on their age. Hacik (1968) found that a single injection of testosterone propionate inhibited adrenal and testis growth in 10 day old rats, whereas it did not affect the adrenal of 45 or 120 day rats. Oestradiol propionate given in a single dose to intact 3 day old female rats resulted in increased secretion and plasma levels of corticosterone over the following two weeks, indicating the sustained changes which this treatment may induce (Hacik, 1969; see also Ghraf *et al.*, 1975).

The effects of progesterone on ACTH secretion are less well documented. Givner and Rochefort (1972) have reviewed the data concerning the effects of synthetic progestogens on adrenal function in female rats. The most intensively studied of these, medroxyprogesterone, produces adrenal atrophy, decreased plasma corticosterone and decreased *in vitro* corticosterone production, effects which have been ascribed to a block in pituitary ACTH secretion. Similar results were obtained with implants or injection of progesterone (Rodier and Kitay, 1974).

D.   DIRECT EFFECTS OF GONADAL HORMONES ON THE ADRENAL ENZYMES

While the effects of gonadectomy may be countered by treament with the appropriate sex steroids, NADPH or ACTH alone are ineffective (Kitay, 1965a, b; Kitay *et al.*, 1965). Furthermore adrenal corticosterone production is enhanced in castrated hypophysectomized rats by gonadal hormone administration (Kitay *et al.*, 1965, 1966; Hirai *et al.*, 1968). In part at least it is possible that this may be attributed to the presence of a factor which inhibits steroidogenesis in ovariectomized animals (Kitay, 1965a). However later work shows that the sex steroids themselves may have a direct effect on the adrenal, and Kitay *et al.* (1966, 1970) found that as measured by acid fluorescence or ultra-violet light absorption *in vitro* production of steroids was decreased in castrated rats of either sex, but that when a blue tetrazolium reduction method (specific for the α-ketol arrangement in the side chain, but unaffected by the ring A configuration) was used no changes in steroid secretion were detected. The total corticosteroid output, it may be concluded, is unchanged with gonadectomy, but there is a decrease in the proportion of the total formed by Δ^4-3-ketones. From this it may be deduced that the sex steroids inhibit a reaction which forms a metabolite of corticosterone. Kitay *et al.* (1970) showed that this metabolite was $3\beta,5\alpha$-tetrahydrocorticosterone. The data were consistent with the hypothesis that adrenal 5α-reductase activity is increased in both sexes after castration, and decreased by replacement with oestradiol or with testosterone (Colby and Kitay, 1972; Maynard and Cameron, 1972). It is also possible that this effect is mediated by ACTH. Lantos *et al.* (1966; 1967) showed that the formation of putative $3\alpha,5\alpha$-tetrahydrocorticosterone from labelled progesterone, deoxycorticosterone or corticosterone by male rat adrenal slices was inhibited when ACTH was added to the incubation medium. Colby *et al.* (1973) have demonstrated a synergism between ACTH and gonadal hormones in the control of adrenal 5α-reductase. Cortisone administration was found to suppress corticosterone secretion and total corticosteroid production but to increase 5α-reductase activity. This was unaffected by ovariectomy but orchiectomy synergistically enhanced the action of cortisone to increase 5α-reductase activity and lower corticosterone production without affecting total steroid output. Later work of Kitay *et al.* (1971a) confirmed that the *in vitro* findings on adrenal reductase activities in gonadectomized rats also applied to *in vivo* conditions. Kitay *et al.* (1971b) found that hypophysectomy increased 5α-reductase activity (after 24 h, a more rapid effect than gonadectomy) and ACTH administration restored the normal condition in adrenal homogenate incubations. Indeed, using the same techniques for measuring "total" corticosteroid (i.e. blue tetrazolium reducing steroid, which does not include the 18-hydroxysteroids which are prominent components of the rat adrenocortical secretion) and for corticosterone measurement which they used in their studies on the effects of gonadectomy, they found that total steroid production was not affected by hypophysectomy in male rats, and only minimally in females, whereas corticosterone was greatly reduced. ACTH administration again did not affect the production of the total steroids, whereas corticosterone was specifically affected. While the possibility of specific effects of the pituitary on individual enzymes is of great interest, it does seem that the lack of effect of hypophysectomy on total steroid production is not compati-

ble with the bulk of the evidence which points to the side chain
scission of cholesterol as a major site of ACTH action in the
biosynthesis pathway. No explanation is immediately obvious. One
possibility is that the blue tetrazolium method on crude lipid extracts
is not as specific a method as these authors would like to believe.
Nevertheless the finding certainly deserves closer inspection. Furthermore the additional finding that testosterone administration to
hypophysectomized male animals inhibited the reductase activity and
increased the secretion of corticosterone, whereas oestradiol was
without effect (Kitay et al., 1971a, b; see also Zizine, 1970), further
supports the general hypothesis. It is difficult to understand the
underlying mechanisms when two stimulants as unrelated as testosterone and ACTH have similar effects on the adrenal cortex. Conceiv-
ably the ACTH effect is indirect, through stimulation of adrenal
testosterone production (see Section 3), and testosterone itself acts as
a local regulator of corticosteroid output. It is also possible that the
gonadal hormones may exert their influence both by long and by short
term effects.

There is evidence that during the neonatal period, the gonadal
hormones may irreversibly "imprint" or "programme" the activities
of the adrenal and hepatic enzyme systems for adult life (Goldman et
al., 1974, Begue et al., 1973; Simmons et al., 1973; Ghraf et al., 1975).
Neonatal injection with 300 μg oestradiol benzoate gave a greatly
increased steroid 5α-reductase activity in both sexes when the animals
were killed on day 75. Other enzymes, including 3α, 3β and 17β-
hydroxysteroid dehydrogenases, were unaffected. Testosterone propionate had no effect. Neonatal castration gave a higher 5α-reductase
activity in the adult than testectomy at 14 days of age. 11β-
Hydroxylase activity is also subject to androgenic inhibition in two
ways: (1) irreversible imprinting neonatally and (2) reversible suppression in later life (Goldman et al., 1974). Direct evidence for this
second type of interaction is found in the work of Dorfman et al. (1966)
who showed that in bovine adrenal enzyme preparations, 11β-
hydroxylation of deoxycorticosterone was inhibited by the addition of
microgram quantities of androstenedione, testosterone, dehydroepiandrosterone or dehydroepiandrosterone sulphate to the incubations media. Testosterone, dehydroepiandrosterone or androstenedione could also inhibit 21-hydroxylation of $\Delta^5,3\beta$-hydroxy-
steroids (pregnenolone or 17-hydroxypregnenolone) although 21-
hydroxylation of the corresponding $\Delta^4,3$-ketones (progesterone and
17-hydroxyprogesterone) was unaffected. In the experiments of
Fragachan et al. (1969) saturated solutions of dehydroepiandrosterone

in 5% glucose (3 mg/ml) perfused through *in situ* adrenal glands in anaesthetized dogs inhibited 11β-hydroxylation of labelled deoxycorticosterone and deoxycortisol, and subsequent incubation of adrenal tissue following dehydroepiandrosterone perfusions also gave decreased yields of aldosterone from endogenous precursors. Both the studies of Dorfman *et al.* (1966) and Fragachan *et al.* (1969) appear to suffer from the disadvantage that the amounts of dehydroepiandrosterone used are very large, and since there is no evidence that such concentrations are found in the adrenal of the intact animal under normal conditions the possibility that such mechanisms may play a role in the moment to moment control of adrenocortical secretion appears to be remote, although it is still possible that longer term effects may be mediated in this way. In human tissue on the other hand, Jensen *et al.* (1972) find that 17-hydroxyprogesterone and dehydroepiandrosterone itself are competitive inhibitors of the side chain scission of 17-hydroxypregnenolone to dehydroepiandrosterone, and that in this case the concentrations of these inhibitors normally found in the tissue are sufficient to allow this interaction to occur under normal circumstances. Their view is that the production of dehydroepiandrosterone may be regulated in this way.

Some more detailed studies on the mechanism of the androgen effects on the rat adrenal cortex were studied by Brownie *et al.* (1970). In this case the authors were examining the production of both corticosterone and 18-hydroxydeoxycorticosterone from progesterone. After treatment with several synthetic androgens, the yields of the two products were greatly reduced, and this was associated with a reduction in the level of adrenal mitochondrial cytochrome P-450. Neither the fall in the levels of cytochrome P-450 nor the *in vitro* production of corticosterone could be prevented by simultaneous administration of ACTH, although the adrenal weight was maintained. (These effects are hence correlated with decreases in hydroxylase activities, and are not due to increases in 5α-reductase as described above.) The effects of the androgens appeared to be direct and, since cytochrome P-450 levels are unaffected by metapyrone, are not necessarily correlated with their potential as inhibitors of steroid hydroxylases. One of the androgens used (11β-hydroxymethyltestosterone) is not an 11β-hydroxylase inhibitor when added directly to incubation media. Roy and Mahesh (1964) found that treatment of rats with testosterone propionate also reduced adrenal corticosterone content, whether or not ACTH was administered at the same time.

In female rats, acute intravenous injection of as little as 1 μg of oestradiol results in increased hydroxylation of deoxycorticosterone by adrenal homogenates after 30 minutes. Larger doses, for example $10\ \mu g$ gave less effect, illustrating again the biphasic nature of the response to oestrogen (Ruhmann-Wennhold *et al.*, 1970). Larger amounts of oestradiol added to adrenal homogenates *in vitro* (10^{-5}M) stimulated succinate supported hydroxylation, although only when succinate was supplied in rate limiting amounts. The effect of oestrogen pretreatment on hydroxylation of DOC was only observed in females, and had no effect in males. Castration of males did increase deoxycorticosterone metabolism, although testosterone administration was without effect (Ruhmann-Wennhold and Nelson, 1970; Fonzo and Nelson, 1970).

In mice, 20α-hydroxysteroid dehydrogenase can be detected histochemically (using 20α-hydroxy-4-pregnen-3-one as substrate) and by a spectrometric assay in the 33 000 g supernatant fraction of the adult or weanling female mouse, but is not similarly detectable in the male (Stabler and Ungar, 1970). Injection of oestradiol ($1\ \mu g$ per day for 1 week) into males induced the activity, while injection of testosterone into the females reduces it. Addition of oestrogen *in vitro* had no effects but ACTH, which stimulated the 20α-hydroxysteroid dehydrogenase activity only to the same extent as the $\Delta^5,3\beta$-hydroxysteroid dehydrogenase activity, could not account for all the effects since oestrogen treatment specifically stimulated the 20α-dehydrogenase.

The wealth of data on gonadal hormone effects on the adrenal while indicative in some senses are nevertheless baffling in their complexity. It seems as though a clear understanding of the mechanisms involved is as remote as ever. One problem is that little or no knowledge has been acquired of the sequence of events in the adrenal which follow gonadectomy, or sex hormone administration. Thus which events are primary and which secondary? Is the increase in 5α-reductase activity caused by ovariectomy the cause of decreased corticosterone production, or its result? The fact that both ACTH administration and oestrogens stimulate corticosterone output, and at the same time inhibit 5α-reductase can really be interpreted in either light, indeed it is simpler to conceive the change in 5α-reductase as *resulting* from changes in corticosterone production, rather than postulate a common mode of action for two hormones as dissimilar as ACTH and oestrogens. The fact too that under different conditions of dosage, timing of administration etc., oestrogens can have completely opposite effects (cf. Kitay 1963d, 1965b; Hacik, 1968; Ghraf *et al.*, 1975) is confusing, and does not lend credence to the view that the ultimate causes of sex differences in adrenocortical function in the rat have been discovered. For example while the work of Kitay *et al.* suggests that oestrogen treatment will enhance corticosterone output by inhibiting 5α-reductase, injection of neonates with oestrogen increases 5α-reductase activity, while at the same time increasing adrenal weight, at least in males (Ghraf *et al.*, 1975) and presumably also increasing corticosterone output (Hacik, 1968). It is also a pity that the possibility of interaction between ACTH and gonadotrophin effects on the adrenal have not been examined more closely. It is certainly possible that gonadotrophins, particularly LH may support the weight of the adrenal, and also affect corticosteroid secretion (Section 4, C). It is possible that some sex differences may depend on this interaction—it would seem quite likely that increases in corticosterone secretion in female rats in pro-oestrus may depend on gonadotrophin secretion, indeed there would at present seem to be a few other possibilities.

The experiments of Ogle and Kitay (1979) are relevant to the many problems discussed here. They followed the observation that 5α-reductase activity increases ten- to fifteen-fold within 24 h of hypophysectomy of the rat and the enhancement of adrenal capacity to produce DHB and THB at the expense of corticosterone output [reduced metabolites to corticosterone: 11β,21-dihydroxy-5α-pregnane-3,20, dione (DHB) and 3β,11β,21-trihydroxy-5α-pregnane-20-one (THB)]. They then showed that prolactin played a significant role in maintenance of corticosterone secretion, principally as a potent inhibitor of 5α-reductase. This was found when administered alone to hypophysectomized rats and in combination with ACTH. ACTH, on the other hand, seemed to stimulate precursor availability as well as inhibit 5α-reductase activity.

E.   GONADAL EFFECTS ON THE CATABOLISM OF ADRENAL STEROIDS

1.   *Differences in the hepatic clearance rates of steroids*

In rats sex differences are observed in the rates of steroid clearance, and a significantly more rapid clearance time in females has been shown (Glenister and Yates, 1961; Kitay, 1961a). Curiously, oestradiol treatment appears to decrease the biological half-life of corticosterone in male rats, but lengthens it in females (another example of a paradoxical result which suggests primary causes are still hidden) and these changes are also reflected in the capacity of the liver to metabolize corticosterone *in vitro*. This capacity is greater in the intact female than in the intact male, but following oestradiol treatment the values decreased in the female and increase in the male, and cease to be significantly different (Kitay, 1961a; 1963b, c, d; 1968). Testosterone had no effect on steroid clearance in the male, but prolonged it in females. Gonadectomy reduces the half life of corticosterone in males and increases it in females. There seems to be an inconsistency between these results and those obtained by direct measurement of circulating steroids and secretory capacity of the gland. Thus, in males gonadectomy results in unchanged levels of circulating corticosterone, although the adrenal vein corticosterone content and the secretion of corticosterone *in vitro* were both reduced. With a shorter half life, it is difficult to see how the circulating levels of steroid are maintained in view of the decreased secretion of steroid. Possibly the results of Linet and Lomen (1971) are more accurate here, they show an increased half life of corticosterone after androgen treatment: this certainly fits better with the other data. In hamsters, which show a sexual dimorphism in the adrenal which is the reverse of that seen in the rat (see Section 4, A), the biological half life of corticosterone in the male is shorter than in the female, and the capacity of the liver to metabolize cortisol is greater. Castration results in a prolonged half life, and a decreased capacity of the liver to metabolize cortisol. Testosterone replacement reverses or prevents the effects of orchiectomy. Thus, as in the case, of the secretory capacity of the pituitary to produce ACTH, and the adrenal to produce cortisol, the effects of testosterone in the hamster are similar to those of oestradiol in the rat. Oestradiol has no effect on the half life parameters in the hamster (Gaskin and Kitay, 1970).

Interestingly, similar studies on biological half life and hepatic clearance of cortisol have been made on a teleost fish, the sockeye salmon (*Oncorhynchus nerka*). Apparently, the volume of distribution of cortisol, the metabolic clearance rate and the cortisol secretion rate increase in both males and females during maturation. The changes were reversed in fish that were gonadectomized just before reaching sexual maturity and allowed to recuperate for 2 or 8 weeks (Donaldson and Fagerlund, 1970). As in rats, oestrogen and androgen replacement restores the normal conditions (Donaldson and Fagerlund, 1969; Fagerlund and Donaldson, 1969). The similarity between these effects and those described in mammals are very striking, and since the groups are so divergent in an evolutionary sense, the basic vertebrate pattern of gonads and adrenals may have been closely interconnected from an early stage in evolutionary history.

2. *Differences in the nature of hepatic catabolites*

In addition to the sex related differences in the metabolic clearance rates of adrenal steroids, there are very significant differences in the nature of the hepatic catabolites between the sexes reflecting the differences in enzyme activities (for review see Schriefers, 1967). Of the two rat liver enzymes responsible for reduction of the 4-ene function in 4-en-3-keto-steroids, the cytoplasmic Δ^4-5β-hydrogenase and the microsomal Δ^4-5α-hydrogenase (Forchielli and Dorfman, 1956), only the 5α-enzyme is sex dependent (Forchielli *et al.*, 1958), being more active in the female than the male. Yates *et al.* (1958) have shown this to be due to a difference in the quantity of microsomal enzyme protein rather than to a difference in the availability of NADPH or the presence of activators or inhibitors. Rubin and Strecker (1957, 1961) have shown that the 3β-hydroxysteroid dehydrogenase activity in rat liver may be up to ten times more active in the male than the female. Of especial importance in the regulation of corticosteroid activity is the finding that male liver is approximately twice as active as the female in the conversion of cortisone into the biologically more active cortisol (Koerner and Hellman, 1964). Similarly, hydrogenation of the C-20 ketone in the 17α-hydroxycorticosteroids is greater in male rat liver than in female tissue (Troop, 1959; Hagen and Troop, 1960; Schriefers *et al.*, 1962), although there appears to be no sex difference in the rate of metabolism of corticosterone by this organ (Kitay, 1961b). In the intact animal there is no sex difference in the Δ^4-5α-hydrogenase activity up to 36 days of age, but then the activity continues to increase in the female but declines in the male, whereas the reverse occurs with 20 ketoreductase activity. Castration of the male leads to an increase in Δ^4-5α-reductase activity to 3 to 4 times that of the control values (see Schriefers, 1967) but a decrease in 20-ketoreductase to normal female control values. However, ovariectomy only delayed the increase in Δ^4-5α-hydrogenase activity to control values and did not affect 20-ketoreductase activity. Oestradiol administered to female rats on day 25 and examined on day 60 showed no influence on Δ^4-5α-reductase activity but oophorectomy decreased the activity of this enzyme only if carried out on immature animals. Yates *et al.* (1958) have interpreted this as indicating that in female rats the Δ^4-5α-hydrogenase is under the influence of non-gonadal factors. This may be related to the work of Denef and de Moor (1972) and Gustafsson and Stenberg (1974) who showed that the masculine pattern of hepatic enzymes is imprinted by the gonads during neonatal life.

Oxidative metabolism is also sex related, and in the C_{19} steroids at least is more active in the male (see Schriefers, 1967). A neonatal sex differentiation of corticosterone metabolites in the bile of rats has been observed (Begue *et al.*, 1973). Neonatal testosterone in this case suppressed activity of the 15α-hydroxylase enzyme in adult life.

It is clear that these sex-linked differences in enzyme activities will lead to different catabolites in males and females. In the rat, cortisol is catabolized by liver homogenates to $3\beta,5\alpha$-tetrahydro derivatives with some $3\alpha,5\beta$-tetrahydro product in the male, but only to $3\alpha,5\alpha$-tetrahydrocortisol in the female. With C_{19} steroids, male liver produces more hydroxylated derivatives and less reduced products than the female.

Conjugate formation may also be affected by sex differences. In the rat, glucuronide formation is higher in males than females during incubations of progesterone with liver homogenates. The reverse occurs with the hamster, but no sex difference was observed with rabbits and guinea pigs (Rao and Taylor, 1965). The relation of these results to the adrenal weight (Table I) is striking (see below). Hepatic cortisol sulphotransferase activity in the rat is also affected by gonadal hormones (Singer and Sylvester, 1976), testosterone again appearing to act as a suppressor to activity. The reverse appears to be the case with sulphatase activity which was lowered by castration of males and raised by testosterone replacement (Burstein, 1968).

3.   *Control of adrenal function by the liver*

It is clear from the above discussion that the gonads, *via* the sex steroids, may profoundly affect the rate of catabolism of the adrenal steroids and hence the circulating levels of these hormones. It is perhaps less clear that this may also indirectly control the activity of the adrenal cortex itself. The work of Yates *et al.* (1958) has shown the remarkable parallelism between the adrenal weight and hepatic activity between the two sexes in both rats and hamsters, i.e. the sex with the larger adrenal has a higher hepatic inactivation rate for corticosteroids. The dependence of the two organs on the sexual cycle is equally striking: hepatic activity and adrenal weight increase by 23 and 20% respectively during oestrus in the rat. After partial hepatectomy, adrenal weight decreased in relation to that of sham operated controls (Urquhart *et al.*, 1959), and in rats implanted with Walker 256 tumours the hypertrophy of the adrenal cortex during tumour development is related to the increase in total hepatic reductase activity (Goodlad and Clark, 1961). Similar parallelism between hepatic activity and adrenal weight has been observed after ingestion of sodium deficient diets (Dailey et al., 1960; Eisenstein and Strack, 1961) and after treatment with anabolic steroids (Schriefers et al., 1965).

That the liver affects the adrenal and not *vice versa* was demonstrated by the administration of ACTH or cortisone acetate (Urquhart et al., 1959; Hagen and Troop, 1960). This simulation of a hyperadrenalcortical state did not lead to any increase in total hepatic reductase activity with corticosterone. The liver is thus seen as a vital link in the modulation of adrenal activity by the gonads. Presumably increased hepatic activity leads to decreased plasma corticosteroid levels which stimulates pituitary release of ACTH which in turn increases adrenal weight and activity. It is a real possibility that a release of testosterone during neonatal or possibly fetal life may "programme" the levels of hepatic activity and thus indirectly determine the adrenal activity during adult life (see Price et al., 1975).

F.   CORTICOSTEROID BINDING PROTEIN

In plasma, corticosteroids are bound to a corticosteroid binding globulin (CBG) and it is generally accepted that only that fraction which is unbound is biologically active. Since a high proportion of the steroid is present at any one time in the protein bound form, any variation in the plasma levels of CBG may have pronounced effects on the biological activity of the hormone. In man, CBG levels have been shown to rise during pregnancy and after oestrogen administration (Slaunwhite and Sandberg, 1959; Sandberg and Slaunwhite, 1959; Mills et al., 1960; Sandberg et al., 1967; Seal and Doe, 1967). In rats, progesterone and adrenalectomy increased CBG in both sexes; castration or oestrogen treatment increased CBG levels in male rats whereas testosterone decreased it in females. Lactation and a variety of pathological conditions resulted in a similar decrease (Gala and Westphal, 1965a, b; De Moor et al., 1963). The pituitary appears to be essential for these effects which are abolished by hypophysectomy, although ACTH does not affect CBG levels (Gala and Westphal, 1966). The increased binding of cortisol was also found in mild diabetics (Nelson et al., 1963; Nelson, 1968) and it is possible that the diabetogenic effect of oral contraceptives may be related to these interactions (Wynn and Door, 1966).

Seal and Doe (1967) have carried out an extensive investigation of species differences of the response of CBG to pregnancy. Their results (Table II) show that the rise in CBG in pregnancy observed in man

TABLE II

Effect of pregnancy on corticosteroid-binding globulin concentration in mammals (Seal and Doe, 1967).

Species	Placental type	Cortisol binding Normal	Gravid	G/N
CHIROPTERA				
Pteropus giganteus	Epitheliochorial	105	90	0.9
PRIMATES				
Homo sapiens (20)	Hemochorial	22	55	2.5
Saimiri sciureus (4)	Hemochorial	5	80	16.0
Cercopithecus aethiops sabeus (2)	Hemochorial	28	72	2.5
Nycticebus coucang (2)	Epitheliochorial	26	15	0.6
RODENTIA				
Rattus norvegicus (6)	Hemochorial	100	170	1.7
Mus musculus (12)	Hemochorial	45	190	4.2
Cavia porcellus (12)	Hemochorial	20	600	30.0
LAGOMORPHA				
Oryctolagus cuniculus (3)	Hemochorial	20	140	7.0
CARNIVORA				
Felis domesticus (3)	Endotheliochorial	5	4	0.8
Canis familiaris (6)	Endotheliochorial	6	6	1.0
Procyon lotor (3)	Hemochorial	30	120	4.0
Mephitis mephitis (2)	Endotheliochorial	12	10	0.8
Mustela vison (2)	Endotheliochorial	12	12	1.0
ARTIODACTYLA				
Bos taurus (3)	Syndesmochorial	5	3	0.6
Ovis aries (3)	Syndesmochorial	5	5	1.0
Sus scrofa (3)	Epitheliochorial	4	4	1.0
PERISSODACTYLA				
Equus caballus (3)	Epitheliochorial	10	11	1.1
EDENTATA				
Dasypus novemcinctus (2)	Hemochorial	2.8	4.2	1.5

and the rat does not occur in all species of placental mammals. The observed species distribution of the elevation also indicates that neither is it restricted to any single major phyletic lineage within the eutherian mammals. A correlation was found, however, with the placental vascular relationships in pregnancy. In species where at some stage of pregnancy the maternal blood directly bathes the fetal trophoblast (hemochorial) a rise in CBG and plasma corticosteroids during pregnancy was shown. In species in which the maternal blood is separated from the fetal tissue by one or more maternal layers, no rise in CBG was found. Even within an order, those species known to differ in their type of placentation, such as the slow loris, *Nycticebus coucang* (epitheliochorial), in primates and the racoon, *Procyon lotor* (hemochorial), in the carnivora gave results consistent with the type of placentation rather than that prevalent in the remainder of the order. In the primates and in rats, guinea pigs, cats and dogs the same effects of CBG levels were found after oestrogen administration as occur during pregnancy.

Cortisol is displaced from CBG by progesterone which is bound more strongly to this protein. Thus progesterone may control corticosteroid activity by displacing cortisol from the protein. This may be of especial importance in pregnancy where significant concentrations of progesterone are present (Seal and Doe, 1967).

5. Effects of adrenal hormones on gonadal function

The best evidence for the involvement of the adrenal cortex either directly or indirectly in ovarian function stems from studies in the rat, in which the adrenal cortex appears to control the timing of ovulation. In rats showing a regular 4-day cycle, the concentration of progesterone in peripheral plasma rises slowly between 09.30 and 13.30 h on the day of pro-oestrus and then rapidly during (and after) the "critical period" for luteinizing hormone (LH) release, to reach a maximum at 21.00 h (Feder *et al.*, 1971). The slow increase in progesterone is paralleled by an increase in corticosterone, but the two curves diverge after the "critical period". Dexamethasone treatment at pro-oestrus or dioestrus inhibited ovulation, but progesterone or ACTH administration during the morning of pro-oestrus overcame the blockade. In adrenalectomized animals, the predictability of the cycles was reduced, although most animals continue the cycle at least up to 15 days post operation. The timing of the critical period may be delayed, and its duration extended. The conclusion reached by Feder *et al.* (1971) was that the slow rise in progesterone concentrations in peripheral plasma which preceded LH release was of adrenal origin, and facilitated the release of LH. The effect is only observed at pro-oestrus, at other times, metoestrus for example, progesterone or ACTH administration blocks ovulation.

Similar conclusions have been reached by other authors. In four-day cycling rats, lordotic behaviour is shown after the onset of the critical period for LH release: in five-day rats the mean time of onset is earlier, and shows more variance. Nequin and Schwartz (1971) found that adrenalectomy delayed lordosis in both types of cycle, and combined adrenalectomy and ovariectomy abolished lordosis altogether in significant numbers of animals of both cycle lengths. Adrenal steroids are also involved in sexual receptivity in rhesus monkeys (Everitt and Herbert, 1969, 1971). In rats, stressful operations, such as sham ovariectomy, advance the timing of lordosis. Combined ovariectomy and adrenalectomy, followed by progesterone administration, elicited lordosis in five-day but not four-day rats. When circulating LH was measured in adrenalectomized animals it was found that the LH surge was completely abolished during pro-oestrus (Lawton, 1972), although ovulation still occurred, and pituitary LH was high at oestrus. In suckled rats, in which ovariectomy causes an increase in pituitary LH, adrenalectomy causes its decrease, while corticosteroid treatment to some extent restores it again (Landers and Wagner, 1974). The specificity of progesterone in eliciting the release of LH was shown by Brown-Grant (1974). Ovariectomized animals were treated with oestradiol benzoate or testosterone propionate, and 72 hours later treated with progesterone. This resulted 5 hours later in a marked rise in LH. 20α-hydroxy-4-pregnen-3-one and 5α-pregnane-3,20-dione also significantly enhanced LH secretion. The likelihood that progesterone was indeed the adrenal product which is responsible was shown by Holzbauer *et al.* (1969) and Holzbauer and Godden (1974) since both adrenal tissue and adrenal vein blood show significant increases in pro-oestrus. It should be recalled however, that other adrenal products, notably corticosterone and aldosterone show similar changes (see Section 4, C).

The adrenals are also implicated in the onset of puberty in the rat. Thus puberty is delayed by adrenalectomy (which reduces circulating progesterone, Meijs-Roelofs *et al.*, 1975) at 18 or 25 days of age by up to 7 days, but is unaffected at 35 days (Gorski and Lawton, 1973): the delay caused by earlier adrenalectomy is superimposed on seasonal changes (Ramaley and Bunn, 1972). It is possible that similar processes may occur in man. Plasma levels of androstenedione and oestrone show a significant increment in females earlier than in males, consistent with the earlier onset of puberty in females, and in both sexes androstenedione is the best index of maturation which corresponds with bone age and Tanner's (1962) age of sexual maturation, whereas testosterone does not (Collu and Ducharme, 1975), suggesting that the adrenal cortex rather than the gonads trigger sexual maturation.

The adrenal cortex may also be implicated in sexual maturation in some species of teleost fish. Idler *et al.* (1971) have shown that both peripheral and testicular plasma concentrations of 11-oxotestosterone and 11β-hydroxytestosterone increase during sexual maturation of the Atlantic salmon *Salmo salar* and Simpson and Wright (1977) have found a similar increase in peripheral 11-oxotestosterone concentration with sexual maturation in the rainbow trout *Salmo gairdneri*. The finding (Kime, 1978) that the livers of rainbow trout (*Salmo gairdneri*), pike (*Esox lucius*) and perch (*Perca fluviatilis*) can convert cortisol to androstenetrione and 11β-hydroxyandrostenedione in significant yields *in vitro* suggested the possibility that these compounds may act as precursors of testicular androgens in these species. The isolation of [^{3}H]11-oxotestosterone from the aquarium water 24 h after injection of an immature male rainbow trout with [^{3}H]cortisol confirmed that adrenal cortisol may act as a precursor of 11-oxotestosterone in this species (Kime, unpublished observations). The massive increase in cortisol production during the spawning migration of some species of teleost such as the salmon (Schmidt and Idler, 1962; Robertson *et al.*, 1961) may thus be one of the factors which control the timing of sexual development. In this connection, the findings of Fagerlund and Donaldson (1969) are of particular interest. These authors showed that in the Pacific salmon, *Oncorhynchus nerka*, adrenal corticosteroid secretion was stimulated by gonadal androgens. It is thus possible that in certain situations there may exist an adrenal–hepatic–gonadal–adrenal positive feedback loop which once initiated either by action of gonadotrophins on the testis or by stress and hence ACTH on the adrenal may lead to increased sexual development and the hyperadrenocorticoism observed during the anadromous spawning migration. Further studies on the relation of age and sex to the hepatic catabolism of cortisol in teleosts would be of considerable interest.

Adrenal steroids have also been implicated in the gonadal development of female fish. In the minnow, *Phoxinus. phoxinus*, it has been suggested that stress, mediated through the pituitary-adrenal axis is responsible for the post-spawning regression of the ovaries (Scott, 1964). In the mullet, *Mugil capito*, higher concentrations of 11-oxotestosterone were found in the ovaries of fish from freshwater than in those from seawater. Since this species will not reproduce in freshwater it has been suggested that the androgenic effect of 11-oxotestosterone could be responsible for this lack of fecundity (Eckstein and Eylath, 1970). The origin of this 11-oxotestosterone was not defined, but it is possible that some at least could be of interrenal origin, either directly or via hepatic cleavage of the cortisol side chain, although in other species, e.g. *Anguilla anguilla* (Colombo and Colombo Belvedere, 1976) and *Tilapia aurea* (Eckstein, 1970), the ovary is capable of 11β-hydroxylation *in vitro*. The relative contributions of interrenal, liver and ovary to circulating levels of 11-oxygenated androgens in female fishes have not yet been ascertained.

C_{21} steroids play an important role in the induction of spawning and ovulation in the catfish *Heteropneustes fossilis*. None of the gonadal hormones, oestradiol, testosterone or 11-oxotestosterone, were effective in inducing maturation of oocytes from hypophysectomized catfish either *in vivo* or *in vitro*, but the adrenal steroids 11-deoxycortisol, 11-deoxycorticosterone and 21-deoxycortisol were very effective (Sundararaj and Goswami, 1966; Goswami and Sundararaj, 1971a, 1974). Of the pituitary hormones tested, FSH, TSH, STH, LtH and ACTH only poorly maintained the yolky oocytes and LH produced the best response. HCG was better than PMS and considerably better than LH (Sundararaj and Goswami, 1966; Anand and Sundararaj, 1974). Maturational changes *in vivo* occurred 2 h earlier with deoxycorticosterone acetate than with LH and it was suggested that this time lag represented the duration necessary for LH to act on the interrenal to build up titres of circulating corticosteroids to that required for maturation and ovulation of the oocytes (Goswami and Sundararaj, 1971b). Oocytes did not mature when ovarian pieces were cultured alone, but in the presence of head kidney pieces (which contain interrenal tissue) 9% of the oocytes underwent maturation. Addition of LH to the medium containing ovarian pieces alone evoked maturation of 14% of the oocytes, whereas ACTH was not effective. Stimulation of ovary–head kidney coculture with LH induced maturation in a significantly larger number (50%) of oocytes; porcine ACTH was much less effective. The authors interpreted this to mean that LH, a gonadotrophin, exhibits a much greater adrenocorticotrophic effect than ACTH (Sundararaj and Goswami, 1974). The results could however be explained, at least in part, by difference in structure and specificity between catfish pituitary hormones and the mammalian analogues used in these experiments. Considerable species specificity of pituitary hormones has been reported by several authors. *Rana* LH is 30 to 50 times more potent than ovine LH in stimulating testicular androgen synthesis in *Rana catesbeiana* (Muller, 1977) and in the medaka, *Oryzias latipes*, salmon gonadotrophin had a higher potency than mammalian gonadotrophin in inducing *in vitro* ovulation (Hirose, 1976). The amino acid analysis of a highly purified preparation of carp gonadotrophin showed considerable difference from that of mammalian LH and FSH (Burzawa-Gerard and Fontaine, 1972).

Catfish interrenals produce cortisol and deoxycorticosterone in a

1.4:1.0 ratio. ACTH does not change this ratio, but ovine LH changes it to 1:3. It may be significant that synergism between cortisol and deoxycorticosterone on *in vitro* oocyte maturation can only be demonstrated when the cortisol:deoxycorticosterone ration is 1:2.5 or 1:5 (Sundararaj and Goswami, 1970). A possible role of corticosteroids as a cause of ovulation during the anadromous spawning migration of salmon has been suggested (Sundararaj and Goswami, 1966). The studies of Sundararaj *et al.* have been limited to the catfish and may not be valid for other teleost species. In the medaka, *Oryzias latipes*, cortisol is 4–10 times more effective than progesterone in inducing oocyte maturation and the sensitivity of the oocytes to steroids increases as the time of spawning approaches. Steroid structural requirements for oocyte maturation however appear to be less rigorous in this species than in the catfish since oestradiol and testosterone can induce ovulation as well as cortisol and progesterone (Hirose, 1972). In the rainbow trout, *Salmo gairdneri*, progesterone, its 17α-hydroxy derivative and 17α,20β-dihydroxy-4-pregnen-3-one are effective in inducing *in vitro* oocyte maturation; cortisone, cortisol, deoxycorticosterone acetate, testosterone, oestrone and oestradiol are ineffective (Jalabert *et al.*, 1972; Fostier *et al.*, 1973).

In contrast to the oocytes of the catfish, those of the frog *Rana pipiens* undergo maturation when stimulated even by relatively low concentrations of steroid precursors (pregnenolone, 17α-hydroxy-pregnenolone and dehydroepiandrosterone) and metabolites (pregnanediol), and unlike the catfish, frog oocyte maturation can be induced by androgens (androstenedione and testosterone), progesterone and a wider variety of adrenocortical steroids (Schuetz, 1967).

Although Sundararaj *et al.* consider the adrenal to be the major source of oocyte maturation-inducing corticosteroids, the ovary itself is capable of forming 11-deoxycorticosteroids. 11-Deoxycorticosterone and/or 11-deoxycortisol have been identified after incubation of labelled pregnenolone or progesterone with ovarian tissue preparations of the teleost fishes *Gillichthys mirabilis*, *Microgadus proximus*, *Leptocottus armatus* (Colombo *et al.*, 1973), *Gobius jozo*, *Diplodus annularis*, and *Solea impar* (Colombo and Colombo Belvedere, 1977), the ovary with preovulatory follicles of the lizard *Xantusia vigilis* (Colombo *et al.*, 1974), the corpora lutea of the snake *Storeria dekayi* during pregnancy (Colombo and Yaron, 1976) and the ripe ovary of the newt *Triturus alpestris alpestris* (Colombo *et al.*, 1977). In mammals, an ovarian 21-steroid hydroxylase has been reported in rabbits when raised in darkness from birth (Lucis *et al.*, 1972) and in women with uterine fibroma (Colombo and Pesavento, 1973) and cyclic premen-strual edema (Mahesh and Greenblatt, 1965).

In amphibians, deoxycorticosterone acetate inhibits the incorpora-tion of labelled vitellogenin of *Xenopus laevis* into isolated denuded oocytes of *Rana pipiens* while concomitantly inducing germinal vesicle breakdown and cortical changes (Schuetz *et al.*, 1973; Schuetz, 1967, 1974). 11-Deoxycorticosteroids may both prevent follicular growth and mediate gonadotrophin-induced meiotic maturation of fully grown oocytes (Colombo *et al.*, 1977; Colombo and Colombo Belve-dere, 1975).

We are still far from a full understanding of the role of the adrenal steroids in the reproductive patterns of nonmammalian vertebrates. There are certainly very wide interspecific differences—the role of 21-hydroxylated-11-deoxycorticosteroids which are essential for ovulation in the catfish appear to be of minor importance in some other species. 11-Oxygenated androgens which are important in sexual maturation of male teleosts also appear to play a role in the regulation of fecundity of the female.

One area of ambiguity which arises in these and related studies is that "stress" is frequently observed to have an adverse effect on reproductive performance. Presumably this is adrenal mediated. Paris and Ramaley (1974) showed that in mice subjected to an immobiliza-tion stress, the diurnal rhythm in corticosterone was disturbed, and this is associated with delayed puberty and reduced fertility. In rats, too, the absence of diurnal rhythm in corticosterone concentration in plasma was associated with irregular ovarian cycles (Ramaley, 1975; Nowell, Chapter 6). Possibly this accounts for effects seen in other species, also such as the observations of Liptrap (1970) who found that corticotrophin or corticosteroid administration to sows delays oestrus, shortens it, and delays the peak in oestrogen excretion normally seen at oestrus. It is difficult to see how these effects, apparently at variance with one another, may be due to similar adrenal secretion of steroids all under the control of ACTH, although as noted above, possibly the timing of adrenocortical secretion is critical. It seems more plausible, however, that two different sets of adrenal steroids are involved in these effects, with different mechan-isms for controlling their secretion.

More recently it has been found that LH has a stimulatory effect on the adrenal (Vinson *et al.*, 1975, 1976; see Section 4, C). It would appear that corticosterone itself is affected, although the possibility that progesterone is stimulated has not yet been examined. However, the positive feedback system in which the adrenal is stimulated by LH and in turn facilitates pituitary LH release would seem to be a very unstable one, resulting in an explosive release of both gonadotrophin and steroid until one or the other is exhausted. This may be the mechanism underlying the preovulationary surge in gonadotrophin secretion.

Effects of adrenal steroids on reproduction in males are not extensively documented. The possibility exists that as in the female "stress" impairs reproductive function. Direct evidence for this was found in the work of Bardin and Peterson (1967) who showed that in rats subjected to ether stress, testosterone levels in testicular vein plasma were significantly reduced from $4.2 \pm 0.56\,\mu g$ per 100 ml to 1.4 ± 0.42 (S.E.). They suggest that this may be due to decreased gonadrotrophin secretion. Although the mechanism for this process is still obscure, it would seem to be worth while to investigate it more closely; and it may even have some relevance to studies on the female. Some effects of stress may be extraordinarily long lasting, and animals subjected to light stress prenatally apparently show decreased copulatory success at maturity, and after castration and oestrogen treatment more lordotic behaviour when compared with unstressed controls. Paradoxically however, postnatal stress had no effects, perhaps because of reduced adrenal response to stress in animals up to 10 days (Ward, 1972).

Christian et al. (1965) have suggested that the effects of stress on the gonads may be one of the means of self-regulation of mammalian populations. In a study of the effects of crowding on populations of mice, rats, deer, voles and woodchucks they found that the stress of crowding or of a subordinate position in the social hierarchy caused a significant increase in adrenal weight (see Nowell, Chapter 6). This change was correlated with delayed maturation, lower reproductive function, anoestrus cycles, reduced time in oestrus, delayed implantation and reduced lactation with consequently stunted offspring. Involution of the X-zone in female mice (see Section 6, A) was accelerated suggesting an increase in secretion of androgens. These authors showed that some at least of these effects of social stress in inhibition of reproductive function could be stimulated by ACTH treatment. The absence of corpora lutea in ACTH treated female mice indicated a suppression of LH secretion. Since the inhibitory action of ACTH was apparent even after adrenalectomy in mice, the effect of stress appears to be mediated primarily by the pituitary, although the adrenal may play an important secondary role by increased androgen secretion or by the effects discussed earlier in this section.

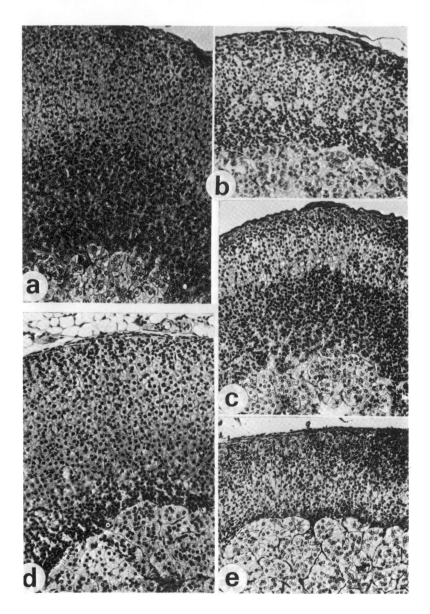

Fig. 4. Experiments to show some pituitary-adrenal relationships in the mouse (from Chester Jones, 1949b). (a) Normal virgin female mouse, age 55 days. The juxtamedullary X zone stands out as a wide deeply stained layer, contrasting with the rest of the cortex. (b) Virgin, hypophysectomized at 40 days of age. Length of time hypophysectomized, 14 days. The zona glomerulosa is not markedly different from that of the control. The degeneration of the zona fasciculata is well advanced. Areas of degeneration are particularly apparent in the region of the inner zona fasciculata and the outer part of the X zone. The X zone has collapsed and is represented by a layer

6.   Adrenal–gonad relationships in the fetus, neonate and in pregnancy

A.   THE X ZONE AND THE FETAL OR TRANSIENT CORTEX

The basic data and literature about these enigmatic zones have been given by Idelman (1978). However because these zones are so intimately bound up with gonadal steroids, gonadotrophins and adrenocorticotrophin, it is worth while to present a summary. This is particularly so as they are very obvious histological entities, yet with no firmly assigned function. In the case of the X zone of the mouse, a well documented species, it may come to occupy up to 50% of the gland in the female (Fig. 4). This depends on the strain and age of the animal, and this account gives a standard pattern but not an inevitable one. In the nulliparous female the X zone persists post-pubertally with gradual declination by "fatty degeneration". Usually during days seven to twelve of first pregnancy, the X zone disappears rapidly, perhaps by the action of ovarian androgens, and a medullary connective tissue capsule is formed. Thus multigravidae do not possess an X zone. In the male, the X zone remains only until puberty when the direct action of testicular androgens cause its collapse by pycnosis of the nuclei and shrinkage of cell cytoplasm. This leads to a settling down on the medulla to form a connective tissue capsule. The appearance of a zona reticularis, of variable size, persistently gives rise to confusion with an X zone.

A major point of interest is that gonadotrophins, probably LH, support the X zone rather than ACTH. This control leads to the continued presence of the X zone in the male castrated before puberty. Moreover, when castrated after puberty, a secondary and prominent X zone appears. This action of the increased secretion of gonadotrophins after removal of the gonads upon the cells of the inner zona fasciculata, known since 1949 (Chester Jones, 1949a, b) may have relevance to recent researches. Also, experimentally, exogenous

---

of pyknotic nuclei lying against the medulla. (c) Virgin, spayed at 43 days, hypophysectomized at 45 days. Length of time hypophysectomized, 14 days. Injections for 12 days of a gonadotrophin primarily LH in nature. The X zone is maintained with the nuclei normal in appearance and the cell cytoplasm staining with eosin. The zona glomerulosa is normal in appearance. The zona fasciculata is degenerating and narrow. (d) Virgin, hypophysectomized for 14 days. Injection of adrenocorticotrophin for 12 days. The zona fasciculata is wide and well maintained. The X zone is collapsed, its nuclei pyknotic, the cell cytoplasm shrunken and non-eosinophilic. The zona glomerulosa has a normal appearance. (e) Virgin, hypophysectomized for 12 days. Human chorionic gonadotrophin for 10 days. The X zone has disappeared and a medullary connective tissue capsule is present. There is a tendency towards hyperemia.

ACTH may elicit adrenal androgens to collapse the mouse X zone (Parkes and Deanesly, 1966).

Another set of animals, the voles, have a juxta-medullary zone comparable to the mouse X zone (Delost, 1953, 1955; Delost and Delost, 1954, 1955; Delost and Dalle, 1970). Here, however, the zone may be resistant to the destructive effect of androgens and persist in both pregnant and lactating females, a similar phenomenon seen in some strains of mice (Shire, 1974, 1976). Nevertheless no firm function has been ascribed to the X zone in these genera and those others noted to possess one (see Idelman, 1978), certainly there is no evidence for steroid production. This might, however, be the interpretation of the findings of Varon *et al.* (1966); they showed that in immature male mice the ratio of corticosterone to cortisol was 1:1 but after sexual maturation or testosterone injections with complete involution of the zone, the ratios rose to 8:1 and 5:1 respectively. However, it is more probable that the changes depend on increased activity of other parts of the adrenal cortex.

The fetal or transient cortex has been noted in primates of the old world but is most fully documented in man. In development, the two parts of the embryonic adrenal apparently do not arise at the same time (Keene and Hewer, 1927; Uotila, 1940; Velican, 1946/47, 1948). The first cortical primordia grow rapidly and reach a considerable size in embryos of 8–9 mm. After the complete separation of gland from the mesothelium on each side, differentiation of the cortex, which is the fetal or transient cortex, continues in embryos of 10–12 mm. After this, the coelomic epithelial cells continued to undergo mitotic division, giving rise to a second proliferation which is destined to form the "permanent" cortex. In the older fetus and to birth the fetal cortex may occupy 85% of the adrenal weight (Deane, 1962). At this stage there is a rim of "permanent" cortex and a large fetal cortex with conspicuous and oesinophilic cells (Fig. 5.). The fetal cortex commences to degenerate just prior to or at birth. At the same time the "permanent" cortex grows, eventually taking on the zonation of the adult gland and thus confirming its nomenclature from fetal to postnatal life.

Despite a superficial resemblance, the human fetal zone and the X zone of the mouse and vole and others are not analogous (Lanman, 1953; Chester Jones, 1957). In all species examined in which a fetal cortex is found—such as *Macaca, Pan,* marmoset *Callithrix, Leontideus* and possibly the armadillo—pregnancy is characterized by excretion of oestriol in the maternal urine. The relationship between these two factors is discussed below.

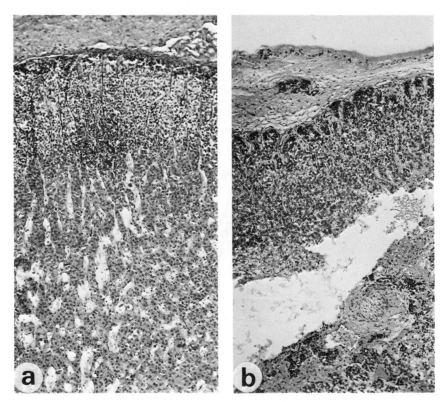

Fig. 5. (a) Part of a section of the adrenal from a still-born baby. To the top is the narrow "permanent" cortex from which, normally, the post-partum cortex develops to give the characteristic appearance. The majority of the adrenal consists of the fetal or transient zone which collapses about birth or soon thereafter (× 85). (b) Part of a section of the adrenal from an anencephalic baby. The transient zone is absent and the cortex is represented only by the narrow "permanent" cortex (× 85). (From Chester Jones, 1955).

In other sub-primate species the role of the fetal adrenal cortex is primarily that of a corticosteroid secretor and apart from its possible role in the initiation of parturition and lactation (see below) is not as intimately involved in C_{19} steroid synthesis as its primate homologue.

Further the occurrence of anencephaly in the human embryo gives rise to speculation about interrelationships (see below). In anencephaly the brain and pituitary are poorly developed to a varying degree. The adrenal and the fetal cortex are only slightly represented or completely absent (Fig. 4b). The permanent cortex seems normal or nearly so. It would seem, too, from Meyer (1912) and Benirscke *et al.*, 1956) that development of the fetal cortex has pursued the usual course until about the fifth month, implying that its degeneration takes place *in utero*.

B.   THE FETO-PLACENTAL UNIT   (General plan in Fig. 6.)

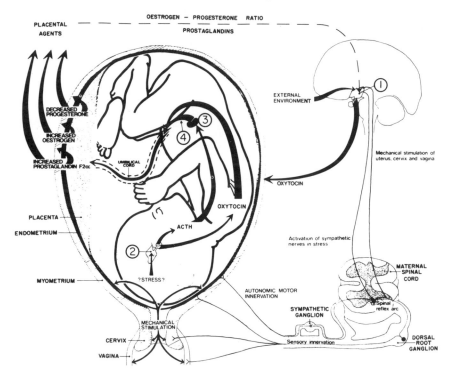

Fig. 6. The feto-placental unit: 1. the maternal pituitary gland; 2. fetal hypothalamus and pituitary; 3. fetal adrenal cortex; 4. increased corticosteroids and oestrogen precursors. (Modified from the Control of Parturition based on data obtained mainly in the sheep and the human. From A. L. R. Findlay, Research in Reproduction, Volume 4, No. 5 (1972), revised 1975. Editor R. G. Edwards).

1.   *Oestrogen biosynthesis*

The endocrine production of steroids can generally be considered as the synthesis of a specific steroid from its precursors by a single gland e.g. oestrogens by the ovary, corticosteroids by the adrenal, and androgens by the testis. As we have seen, this simplistic view is not strictly true and there is considerable interplay in the control and secretion of these glands. In pregnancy, which is marked by a greatly increased steroid production, this interplay between the different endocrine glands reaches its peak.

Although it was once believed that the placenta was the endocrine organ of pregnancy, it soon became clear that steroidogenesis in pregnancy could only be understood by consideration of the fetal endocrine system and the placenta as incomplete parts of a single unit, each possessing only some of the enzymes necessary for steroid biosynthesis (Table III). This concept of the feto-placental unit was

TABLE III

Distribution of enzyme systems in the feto-placental unit.

	Fetus		Placenta
	Adrenal	Liver	
Sulphokinase	+	+	−
Sulphatase	−	−	+
Acetate → cholesterol	+	+	−
Cholesterol → pregnenolone	+	−	+
3β-hydroxysteroid dehydrogenase	−	−	+
Δ^{5-4} isomerase	−	−	+
17-hydroxylase	+		−
17-20 desmolase	+		−
16α-hydroxylase (C_{21})	+	−	−
16α-hydroxylase (C_{18}, C_{19})	−	+	−
Aromatase	−	+	+
17-oxidoreductase			+

put forward by Diczfalusy in 1964 to describe the complete endocrine system. Although the feto-placental unit is capable of biosynthesis of the corticosteroids and gonadal hormones of the adult (see reviews by Diczfalusy, 1966, 1969a, b, 1975; Dell'Aqua et al., 1966; Siiteri and MacDonald, 1966; Liggins, 1972), in our discussion of the adrenal–gonad relations we will concentrate on the role of the fetal adrenal in the biogenesis of oestrogens in pregnancy.

Primate pregnancy is characterized by the production of large amounts of oestriol in the maternal urine and by the large adrenal size composed mainly of a "fetal zone" (see above). The relationship between these two characteristics is shown in the biosynthetic scheme in Fig. 7. Cholesterol formed from acetate in the fetal or maternal system is converted to pregnenolone by either the fetal adrenal (Archer et al., 1971) or the placenta (Diczfalusy, 1969b) but not by the

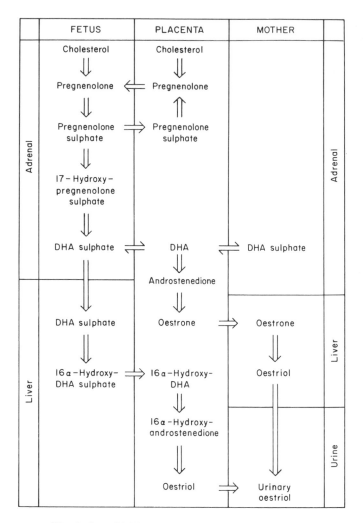

Fig. 7. Steroid biosynthesis in the feto-placental unit.

fetal liver (Telegdy *et al.*, 1972). The placenta is capable of the further conversion of pregnenolone to progesterone whereas little if any such conversion occurs in the human fetus (Diczfalusy, 1969b). Pregnenolone of either fetal or placental origin is converted to the sulphate in the fetal adrenal (Jaffe *et al.*, 1972). The placenta, however, is not capable of sulphurylation of pregnenolone but is able to hydrolyse pregnenolone sulphate to pregnenolone, a reaction which cannot take place in the fetal organism (Diczfalusy, 1969b).

Pregnenolone sulphate is then further converted to DHA sulphate (Jaffe *et al.*, 1972) as the most important precursor of the placental oestrogens. Placental hydrolysis of dehydroepiandrosterone sulphate, followed by conversion to androstenedione and subsequent aromatization leads to oestrone and oestradiol which appear in the maternal urine. The fetal adrenal is not able to carry out the hydrolysis, oxidation or aromatization, although the conversion of androstenedione to oestrone by fetal liver has been demonstrated (Mancuso *et al.*, 1965b, 1968).

Quantitatively, the most important oestrogen in human pregnancy urine is not oestrone or oestradiol but oestriol which accounts for over 90% of the total oestrogen. The formation of this compound may be attributed to the high level of activity of the 16α-hydroxylase enzyme in the fetus, especially the fetal liver. Two pathways for the synthesis of oestriol are possible, a neutral and a phenolic sequence. The neutral pathway is quantitatively the most important sequence and is characterized by 16α-hydroxylation prior to aromatization. Dehydroepiandrosterone sulphate from the fetal adrenals is converted to 16α-hydroxydehydroepiandrosterone sulphate in the fetal liver (Bolte *et al.*, 1966), followed by placental hydrolysis of the sulphate, conversion to 16α-hydroxyandrostenedione, aromatization and reduction at C-17 to give oestriol (Dell'Aqua *et al.*, 1967). A variation of this pathway is the 16α-hydroxylation of pregnenolone sulphate (Jaffe *et al.*, 1972), and conversion to 16α-hydroxydehydroepiandrosterone sulphate in the fetal adrenal (Oakey, 1970; Shahwan *et al.*, 1969). Hydroxylation in the feto-placental unit can be surprisingly specific; the fetal liver readily 16α-hydroxylates dehydroepiandrosterone and oestrone but not pregnenolone (Bolte *et al.*, 1966; Mancuso *et al.*, 1968), whereas the fetal adrenal readily 16α-hydroxylates pregnenolone or progesterone but not dehydroepiandrosterone or oestrone (Shahwan *et al.*, 1969; Jaffe *et al.*, 1972).

A less important sequence for oestriol synthesis is *via* the phenolic pathway. Although fetal liver can convert oestrone to its 16α-derivative, evidence suggests that conversion normally takes place in the maternal liver (Diczfalusy, 1969b; Schwers *et al.*, 1971). The complexity of the enzyme systems involved in the formation of oestriol during pregnancy make it a useful indicator of the health of the fetus (Frandsen, 1969).

Human pregnancy is characterized by the secretion of enormous quantities of dehydroepiandrosterone sulphate (75 mg/day) by the fetal adrenal without at the same time increasing the amounts of potentially harmful androgens or corticosteroids in circulation. This may be a result of the complexity and specificity of the feto-placental enzyme systems which protect the developing fetus from an excess of these steroids (Liggins, 1972). Unlike the adult adrenal, the fetal gland is incapable of converting pregnenolone to progesterone, or dehydroepiandrosterone to androstenedione and relies on the placenta for suitable precursors for androgen and corticosteroid synthesis. Further protection of the fetus from unwanted active hormones is provided by the extremely active sulphurylating enzymes of the fetal organs. Almost all the steroids in the fetal circulation are present as sulphate conjugates, but in contrast the placenta has very active sulphatase systems and is almost devoid of sulphotransferase activity (Warren and Timberlake, 1962). The fetus is thus protected by conjugation from the presence of excessive androgens and oestrogens and yet the placenta may readily obtain its necessary precursors from the fetal circulation by rapid hydrolysis of these esters.

The majority of the information available relates to the function of the feto-placental unit at mid-gestation in the human. Very little information is available on the possible existence of a similar feto-placental relationship in other species. In animals with short gestational periods, the oestrogens and progestins required for the maintenance of pregnancy are supplied by the maternal ovary, the placenta lacking the necessary enzymes for the synthesis of these compounds (Ryan, 1969). Placentae of goats and sows which have a longer gestational period are capable of aromatization but not of progesterone synthesis, while those of most other animals with a long gestation including sheep, cows, horses, humans and macaque monkeys are able to form both oestrogens and progestogens (Ryan, 1969). Clearly it is only in this last class that we should look for similarities to the primate feto-placental pattern. It is, however, only in primates that the adrenal forms a distinct well developed fetal zone which by secretion of dehydroepiandrosterone sulphate provides the source of oestrogen precursors.

2.   Control of the fetal adrenal

There is now considerable evidence that the fetal adrenal, like the adult gland, is stimulated by ACTH. In the rat, high maternal corticosterone levels depressed fetal adrenal weight only from the 18th day of pregnancy whereas maternal adrenalectomy raised fetal adrenal weight only if carried out from the 18th day indicating that fetal adrenocorticotrophic activity begins between the 17th and 18th day of intrauterine development (Milkovic *et al.*, 1973).

In man injection of ACTH into the fetus led to an increase in maternal oestrogen excretion, presumably *via* stimulation of dehydroepiandrosterone sulphate by the fetal adrenal cortex (Strecker *et al.*, 1977). As noted in Section 6, A, above, in anencephalic pregnancy the fetal adrenal develops normally during the first half of gestation after which the fetal zone regresses. Regrowth of the fetal adrenal has been reported following prolonged ACTH treatment of an anencephalic fetus, and atrophy of the normal fetal adrenal occasionally occurs in infants born to mothers treated with high doses of cortisol and its analogues (Lanman, 1962). Urinary oestrogen excretion increased in women pregnant with normal or hydrocephalic fetuses who were given metyrapone, a drug known to stimulate ACTH secretion by inhibition of cortisol synthesis. In contrast, women with anencephalic fetuses showed no change in oestrogen excretion after treatment with this drug (Oakey and Heys, 1970).

The normal early development of the fetal adrenal in anencephaly indicates that during early pregnancy this gland is under the control of another factor. Human chorionic gonadotrophin (hCG) is detectable in the fetus and neonate and it has been supposed to act on the fetal cortex (Rotter, 1949a, b: Chester Jones, 1955). Lauritzen and Lehmann (1967) suggested that hCG may be concerned with the regulation of precursor dehydroepiandrosterone supply for placental oestrogen synthesis. The amount of oestrogen in the placenta may influence production of hCG according to the placental requirement for oestrogen precursor. The existence of a negative feedback control between dehydroepiandrosterone and hCG was demonstrated by the decrease in maternal urinary hCG after adminstration of dehydroepiandrosterone to pregnant women (Lauritzen, 1965). It is perhaps significant that the peak of hCG secretion is reached between the 60th and 80th days of pregnancy and correlates with the beginning of adrenocortical function in the fetus. In late pregnancy, ACTH may take over more and more of the task of stimulating the fetal adrenal gland according to diencephalic impulses, but hCG may continue to act as a regulator of dehydroepiandrosterone available to the placenta. It may also be of some significance that regression of the fetal zone of the adrenal cortex after birth coincides with the elimination of hCG from the neonate (Lauritzen and Lehmann, 1967). In non-pregnant women, administration of large doses of hCG induced a rise in urinary androgen but not of 17-hydroxycorticosteroids and it is possible that this may occur by a mechanism similar to that acting in the fetal adrenal (Pauerstein and Solomon, 1966).

It has also been suggested (Eberlein, 1971) that progesterone secretion by the placenta may exert a control over the development of the fetal adrenal by suppression of the 3β-hydroxysteroid dehydrogenase enzyme and hence cortisol secretion. This in turn would lead to increased ACTH secretion and adrenal hypertrophy. Progesterone may also lead to an increase in fetal ACTH by competing with cortisol for binding sites on the corticosteroid binding globulin. Oakey (1970), reviewing the factors affecting oestrogen and cortisol concentrations in the human fetus, concluded that loss of cortisol from the fetal to the maternal system accounted for the increase in ACTH and hence in androgen production by the fetal adrenal. He suggested that a major factor in facilitating this transport of cortisol to the mother may be the smaller number of binding sites available for cortisol in the fetal than in maternal plasma resulting from competitive binding by progesterone.

Human feto-placental function is distinguished from that of non-primate species by the development of an adrenal fetal zone, and the production of large quantities of hCG, dehydroepiandrosterone sulphate and oestriol. That these all play a role in the maintenance of pregnancy is beyond doubt, but the reason for the evolution of such a system in primates alone remains a mystery.

C. THE ROLE OF THE FETAL ADRENAL IN THE INITIATION OF
    PARTURITION AND LACTATION

1. *Initiation of parturition in sheep*

Whereas most of the work on steroidogenesis in the feto-placental unit has been carried out on man, studies of the initiation of parturition have, mainly for ethical reasons, been centred on sheep. Although many of the factors involved in the initiation of parturition have now been defined (see reviews by Liggins *et al.*, 1972, 1973; Thorburn *et al.*, 1972; Liggins, 1973) the exact nature of the impetus for the initiation of parturition has yet to be clarified.

It is now generally accepted that the impetus for the initiation of parturition comes from the fetus and not from the mother and is mediated by the fetal adrenal cortex. In man fetal anencephaly has long been associated with an extreme prolongation of pregnancy (Rea, 1898; Malplas, 1933) and cases of premature labour without apparent cause have been related to increased fetal adrenal weight

(Anderson *et al.*, 1971). Liggins (1968) has shown that infusion of cortisol into fetal lambs can cause premature delivery after a latent period of 48–72 h, and although dexamethasone was also able to cause premature delivery, corticosterone and deoxycorticosterone were ineffective, indicating that parturition is dependent upon gluco-corticoid rather than mineralocorticoid activity (Liggins, 1969). Fetal plasma corticosteroids have been shown to increase rapidly several days before birth, reaching their maximum concentration at parturi-tion (Bassett and Thorburn, 1969) although maternal plasma corti-costeroid levels were unchanged as parturition approached. Fetal hypophysectomy or adrenalectomy prevents this rise in fetal cortico-steroid concentration and prolongs pregnancy (Liggins *et al.*, 1967; Drost and Holm, 1968). Infusion of ACTH into fetal lambs of more than 88 days gestational age caused parturition on day 4 to day 7 of infusion and at birth the fetal adrenals weighed at least as much as those of normal full term lambs (Liggins, 1968). The increased secretion of corticosteroids by the ACTH stimulated adrenal mani-fests itself not only by the elevated plasma cortisol levels but by the biological effects of cortisol associated with the final stages of gestation e.g. increased pulmonary surfactant and liver glycogen concentration (Liggins, 1968). Corticosteroid binding levels also increase in the fetal lamb in the 20 days prior to parturition; protein bound cortisol increases 13-fold to 39 ng/ml whereas unbound cortisol only increases 10-fold to 2 ng/ml (Liggins *et al.*, 1973).

Changes in progesterone and oestrogen concentrations have also been noted towards the end of the gestational period. Progesterone in maternal peripheral plasma drops during the last few days of pregnancy from 7–11 ng/ml to 0.5–1.0 ng/ml on the day of parturition (Bassett *et al.*, 1969; Thorburn *et al.*, 1972), and a similar fall in progesterone concentration is observed following infusion of either ACTH or dexamethasone into the fetus. In both cases the mean concentration of the hormone is approximately the same as in ewes in labour at term (Liggins, 1969; Liggins *et al.*, 1972; Thorburn *et al.*, 1972). This change in concentration is due to decreased secretion by the placenta rather than to enhanced metabolic clearance (Linzell and Heap, 1968; Mattner and Thorburn, 1971).

In pregnant sheep, total unconjugated oestrogen concentrations in peripheral plasma are low and before day 120 are less than 20 pg/ml. They increase slowly to 25–35 ng/ml five days before term and peak at 75–880 pg/ml 16–24 h before parturition (Challis, 1971; Thorburn *et al.*, 1972). As with progesterone this change in concentration can be stimulated by fetal infusion of dexamethasone (Liggins, 1973). The concentration of prostaglandin $F_{2\alpha}$ is also found to increase in the uterine venous blood of ewes both in premature labour induced by fetal infusion of either dexamethasone or synthetic ACTH (Synacthen, Ciba) and in spontaneous labour (Liggins and Grieves, 1971; Thorburn *et al.*, 1972).

Although artificial manipulation of the levels of progesterone, oestrogen and prostaglandin may affect some of the mechanisms of parturition, only ACTH or cortisol is able to produce premature parturition and to affect the levels of these other compounds in the same manner as observed in spontaneous parturition (Liggins *et al.*, 1973). It thus seems clear that parturition in the sheep is initiated by the fetal hypothalamus *via* increased ACTH and cortisol secretion which causes increased oestrogen and prostaglandin $F_{2\alpha}$ and decreased progesterone secretion by the placenta.

The mechanism by which these changes initiate parturition are less clear. Maternal injection of stilbestrol causes a rise in $PGF_{2\alpha}$ in maternal cotyledons, myometrium and uterine blood to levels similar to those observed during parturition, but this increase is inhibited by progesterone treatment (Thorburn *et al.*, 1972; Liggins, 1973). Maternal administration of either oestrogen or prostaglandin $F_{2\alpha}$ increases the sensitivity of the myometrium to oxytocin, but since $PGF_{2\alpha}$, unlike oestrogen, stimulates uterine smooth muscle *in vitro*, it is likely that oestrogen acts indirectly on the myometrium by increasing prostaglandin synthesis (Liggins, 1973). Progesterone conversely has an inhibitory effect on uterine contractibility in many if not all mammalian species (Csapo, 1969).

Since the release of oxytocin from the posterior pituitary is increased by oestrogen during vaginal distension (Roberts and Share, 1969), reflex stimulation of the release of oxytocin by distension of the cervix and vagina by the descent of the fetus will be enhanced by high pre-partum oestrogen levels, leading to increased contractions of the $PGF_{2\alpha}$ primed myometrium (Fig. 6).

Several links in this chain of events are still obscure; the mechanism by which oestrogen and progesterone levels are affected by cortisol concentrations and the nature of the input to the fetal hypothalamus remain to be clarified. It has been suggested that the latter is related to maturation of the hypothalamic thermoreceptors (Hopkins and Thorburn, 1971; Thorburn *et al.*, 1972). This suggestion was based on the fact that the fetal lamb was hyperthyroid compared to its mother even though the fetal brain was $0.4–0.8°$ higher than the maternal blood. During the last 7–10 days of gestation, fetal thyroxine falls to maternal levels as the fetus becomes aware of its hot and wet

environment, with maturation of the fetal hypothalamic thermoreceptors, and suppresses release of thyrotrophin releasing factor. The
thermal stress would then stimulate corticotrophin releasing factor
which would lead to increased corticosteroids and thence as described
above to parturition. Tal and Sulman (1975) however have recently
shown that rats reared at 37°C are better able to survive the heat
stress if treated with 50 mg/kg/day of dehydroepiandrosterone which
apparently blocks the hypothalamic thermoreceptors. High plasma
dehydroepiandrosterone levels may thus be a factor in the hyperthyr-
oidism of the fetal lamb, and the "maturation" of the fetal thermoreceptors could be simply a reflection of decreased plasma dehydroepiandrosterone concentration resulting from a corticosteroid
stimulated increase in placental aromatase activity as the end of the
gestational period approaches.

2. *Parturition in other species*

The above description of the initiation of parturition in the sheep may
also be valid for other species, but since the endocrine factors in
pregnancy vary so much from species to species (Ryan, 1969) it would
be rather dangerous to overgeneralize.

Increased corticosteroid levels prior to parturition appear to be
common to several other species, e.g. guinea-pig (Gala and Westphal,
1967; Jones, 1974), mice (Barlow *et al.*, 1974), cows (Hoffmann *et al.*,
1973), rats (Gala and Westphal, 1965b; Ota *et al.*, 1974; Martin *et al.*,
1977), goats (Currie *et al.*, 1973; Thorburn *et al.*, 1972) and humans
(Assali *et al.*, 1955; Bayliss *et al.*, 1955). None of these species have
been studied in as much detail as the sheep and values are not
available for the changes in oestrogen, progesterone and prostaglandin concentrations in all of these species. Where such data is available
(bovine: Edqvist *et al.*, 1973; Hoffmann *et al.*, 1973; rat: Martin *et al.*,
1977; rabbit: Nathanielsz and Abel, 1973; horse: Savard, 1961; goat:
Thorburn *et al.*, 1972; pig: Raeside, 1963; Rhesus monkey: Hopper
and Tullner, 1967) the sequence of events appears to follow that of the
sheep and it would seem likely that a similar mechanism of parturition occurs in most mammalian species. Some interspecific variations
in the individual components of the system certainly occur. The
nature of the oestrogen may be oestriol in primates, equilenin in the
mare or oestradiol in sheep and goats. Progesterone may originate in
either the placenta or the corpus luteum and oestrogen may be of
either placental or ovarian origin. Precursors can originate either from
maternal or feto-placental sources. There is at present no explanation

for this interspecific variation, and although the human fetal adrenal cortex certainly plays a role in both pregnancy and parturition there is at present no satisfactory explanation as to why the fetal zone has evolved only in primates.

3. *Lactation*

Adrenal steroid hormones play an important part in the initiation and maintenance of lactation in many species, and the minimal hormonal requirements for lactogenesis in endocrine-deprived animals and in organ culture have been reviewed (Denamur, 1971). Although these requirements vary considerably from species to species, both prolactin and a glucocorticoid are essential for lactation in most of the species studied. Thoman *et al.* (1970a, b) have shown that suckling rats grew at a significantly greater rate when fed by normal rather than adrenalectomized foster mothers. In man it has long been known that adrenalectomy reduces milk secretion and that corticosteroids are important in the development of the mammary gland (Folley, 1956; Lyons *et al.*, 1958; Greenbaum and Darby, 1964; Koch *et al.*, 1968; Thatcher and Tucker, 1970; Denamur, 1971). In both rats and cows, the suckling stimulus increases both ACTH and corticosteroid secretion (Voogt *et al.*, 1969; Wagner, 1969).

As with the initiation of parturition, both progesterone and oestrogen appear to have an effect on the efficiency of lactation. Ovariectomy of rats on the 20th day of pregnancy impaired both parturition and lactation, resulting in prolonged pregnancy and inability to raise the young. Oestrogen therapy immediately after ovariectomy permitted normal parturition and lactation. Prolactin and corticotrophin but not oxytocin improved lactation in 50% of the mothers (Catala and Deis, 1973). Administration of an oestrogen antagonist (MER 25) for the last three weeks of pregnancy in the ewe almost completely inhibited milk secretion despite apparently normal mammary development (Abdul-Karim *et al.*, 1966). In contrast however daily sub-cutaneous injections of oestradiol benzoate from day 3 of lactation decreased litter weight gain and caused regression of the mammary glands. This suppression could be significantly reduced by cortisol acetate treatment but not by prolactin (Mizuno and Sensui, 1973). The increase in oestrogens precedes the increase in prolactin in the pregnant rat, and in this animal oestrogens have also been shown to both stimulate the release of prolactin from the pituitary (see Meites and Nicoll, 1966), and increase the plasma concentration of this hormone (Kwa and Verhofstad, 1967; Kwa *et al.*, 1969; Chen and Meites, 1970; van der Gugten *et al.*, 1970). In the non-pregnant rat plasma prolactin is also higher during pro-oestrus (Niswender *et al.*, 1969) or oestrus (Kwa and Verhofstad, 1967). Progesterone has a general inhibitory effect on lactogenesis (see Denamur, 1971).

Thus the final stages of gestation with the high cortisol and oestrogen and low progesterone levels provide the necessary steroid hormone environment for lactation. Possibly, as in the initiation of parturition, this hormone balance may both stimulate pituitary release of prolactin and sensitize the mammary tissue to this hormone. As discussed above (Section 6, D) the oestrogen and progesterone levels are controlled by the fetal pituitary *via* increased corticosteroid output, but in addition, the dependence of satisfactory lactation on the presence of the maternal adrenal may indicate a more direct effect of corticosteroids on the mammary gland.

Further work will be necessary to define the hormonal inter-relationships in the initiation and maintenance of lactation. The results so far available indicate a close inter-relation between adrenal hormones, gonadal hormones and a number of pituitary hormones of which the most important is prolactin. Considerable interspecific differences are indicated (Denamur, 1971).

References

Aakvaag, A., Vogt, J. H. and Pylling, P. (1970). *Acta endocr., Copnh.* **64,** 103–110.

Abdul-Karim, R. W., Nesbitt, R. E. L. and Prior, J. T. (1966). *Fert. Steril.* **17,** 637–647.

Abraham, G. E. (1974). *J. clin. Endocr. Metab.* **39,** 340–346.

Acevedo, H. F. and Beering, S. C. (1965). *Steroids* **6,** 531–541.

Adams, J. B. (1967). *Biochim. biophys. Acta* **146,** 522–528.

Adams, J. B. and Chulauatnatoz, M. (1967). *Biochim. biophys. Acta* **146,** 509–521.

Adams, J. B. and Poulos, A. (1967). *Biochim. biophys. Acta* **146,** 493–508.

Agate, F. J. (1952). *Ann. N.Y. Acad. Sci.,* **55,** 404–411.

Ahmad, N. and Morse, W. I. (1965). *Can. J. Biochem. Physiol.* **43,** 25–31.

Ahmad, N. and Gower, D. B. (1968). *Biochem. J.* **108,** 23–41.

Alpert, M. (1950). *Endocrinology,* **46,** 166–176.

Anand, T. C., and Sundararaj, B. I. (1974). *Gen. comp. Endocr.* **22,** 154–168.

Anderson, A. B. M., Laurence, K. M., Davies, K., Campbell, H. and Turnbull, A. C. (1971). *J. Obstet, Gynaec. Br. Commonw.* **78,** 481–488.

Anliker, R., Rohr, O. and Marti, M. (1956). *Helv. chim. Acta.* **39,** 1100–1106.

Arai, R., Tajima, H. and Tamaoki, B. (1969). *Gen. comp. Endocr.* **12,** 99–109.

Archer, D. F., Mathur, R. S., Wiqvist, N. and Diczfalusy, E. (1971). *Acta endocr., Copnh.* **66,** 666–678.

Askari, H. A., Monette, G. and Leroux, S. (1970). *Endocrinology* **87,** 1377–1380.

Assali, N. S., Garst, J. B. and Voskian, J. (1955). *J. Lab. clin. Med.* **46,** 385–390.
Axelrod, L. R. and Goldzieher, J. W. (1967). *Acta endocr., Copnh.* **56,** 453–458.
Axelrod, L. R., Goldzieher, J. W. and Ross, S. D. (1965). *Acta endocr., Copnh.* **48,** 392–412.
Axelrod, L. R., Goldzieher, J. W. and Woodhead, D. M. (1969). *J. clin. Endocr. Metab.* **29,** 1481–1488.
Axelrod, L. R., Kraemer, D. C., Burdett, J. and Goldzieher, J. W. (1973). *Acta endocr., Copnh.* **72,** 545–550.
Baggett, B., Engel, L. L., Baleras, L., Lanman, G., Savard, K. and Dorfman R. I. (1959). *Endocrinology* **64,** 600–608.
Baird, D. T. and Guevara, A. (1969). *J. clin. Endocr. Metab.* **29,** 149–156.
Baird, D. T., Horton, R., Longcope, C. and Tait, J. F. (1968). *Perspect. Biol. Med.* **11,** 384–421.
Baird, D. T., Horton, R., Longcope, C. and Tait, J. F. (1969a). *Recent Prog. Horm. Res.* **25,** 611–664.
Baird, D., Uno. A. and Melby, J. C. (1969b). *J. Endocr.* **45,** 135–136.
Baker, D. D. (1937). *Am. J. Anat.* **60,** 231–252.
Baker, D. D. (1938). *J. Morph.* **62,** 3–15.
Bardin, C. W. and Peterson, R. E. (1967). *Endocrinology* **80,** 38–44.
Barlow, J. J. (1964). *J. clin. Endocr. Metab.* **24,** 586–596.
Barlow, S. M., Morrison, P. J. and Sullivan, F. M. (1974). *J. Endocr.* **60,** 473–483.
Barrett, A. M. (1960). *Acta endocr., Copnh. Suppl.* **51,** 421.
Bassett, J. M. and Thorburn, G. D. (1969). *J. Endocr.* **44,** 285–286.
Bassett, J. M., Oxborrow, T. S., Smith, I. D. and Thorburn, G. D. (1969). *J. Endocr.* **45,** 449–457.
Baulieu, E. E. (1960). *J. clin. Endocr. Metab.* **20,** 900–903.
Baulieu, E. E. (1962). *J. clin. Endocr. Metab.* **22,** 501–510.
Baulieu, E. E. (1965). *Rev. Etud. clin. biol.* **10,** 264–270.
Baulieu, E. E. and Dray, F. (1963). *J. clin. Endocr. Metab.* **23,** 1298–1301.
Bayliss, R. I. S., Browne, J. C., Round, B. P. and Steinbeck, A. W. (1955). *Lancet* **i,** 62–64.
Beall, D. (1939). *Nature Lond.* **144,** 76.
Beall, D. (1940). *J. Endocr.* **2,** 81–87.
Beer, J. R., and Meyer, R. K. (1952). *J. Mammal.* **32,** 173.
Begue, R. J., Gustafsson, J. A. and Gustafsson, S. A. (1973). *Eur. J. Biochem.* **40,** 361–366.
Beitins, I. Z., Bayard, F., Kowarski, A. and Migeon, C. J. (1973). *Science N.Y.* **21,** 553–563.
Benirschke, K. and Richart, R. (1964). *Endocrinology* **74,** 382–387.
Benirschke, K., Bloch, E. and Hertig, E. T. (1956). *Endocrinology* **58,** 598–623.
Blaquier, J., Dorfman, R. I. and Forchielli, E. (1967). *Acta endocr., Copnh.* **54,** 208–214.
Blichert-Toft, M., Vejlsted, H., Kehlet, H. and Albrechtsen, R. (1975). *Acta endocr., Copnh.* **78,** 77–85.
Bloch, E., Dorfman, R. I. and Pincus, G. (1954). *Proc. Soc. exp. Biol. Med.* **85,** 106–110.
Bloch, E., Dorfman, R. I. and Pincus, G. (1956). *Archs Biochem. biophys.* **61,** 245–247.
Bloch, E., Cohen, A. I. and Furth, J. (1960). *J. natn. Cancer Inst.* **24,** 97–107.
Bolté, E., Wiqvist, N. and Diczfalusy, E. (1966). *Acta endocr., Copnh.* **52,** 583–597.
Borgstede, H., Henning, H. D. and Zander, J. (1963). *Hoppe-Seyler's Z. physiol. Chem.* **331,** 245–257.

Bosu, W. T. K., Johansson, E. D. B. and Gemzell, C. (1973). *Acta endocr., Copnh.* **74,** 338–347.
Bourne, G. (1949). "The Mammalian Adrenal Gland", Oxford University Press.
Bourne, G. and Zuckermann, S. (1940). *J. Endocr.* **2,** 283–310.
Breuer, H. and Knuppen, R. (1969). *Hoppe Seyler's Z. physiol. Chem.* **350,** 581–590.
Brown-Grant, K. (1974). *J. Endocr.* **62,** 319–332.
Brownie, A. C., Colby, H. D., Gallant, S. and Skelton, F. R. (1970). *Endocrinology* **86,** 1085–1092.
Bryson, M. and Sweat, M. L. (1962). *Archs. Biochem. biophys.* **96,** 1–3.
Bryson, M. J., Young, R. B., Sweat, M. L. and Reynolds, W. A. (1968). *Cancer* **21,** 501–507.
Bulbrook, R. D. and Greenwood, F. C. (1957a). *Br. med. J.* **i,** 662–666.
Bulbrook, R. D. and Greenwood, F. C. (1957b). *Acta endocr., Copnh. Suppl.* **31,** 324–325.
Bullock, L. P. and New, M. (1971). *Endocrinology* **88,** 523–526.
Burger, H. G., Kent, J. R. and Kellie, A. E. (1964). *J. clin. Endocr. Metab.* **24,** 432–441.
Burstein, S. (1968). *Endocrinology* **83,** 485–488.
Burzawa-Gerard, E., and Fontaine, Y. A. (1972). *Gen. comp. Endocr. Suppl.* **3,** 715–728.
Bush, I. E. (1953). *J. Endocr.* **9,** 95–100.
Bush, I. E. and Ferguson, K. A. (1953). *J. Endocr.* **10,** 1–8.
Bush, I. E., Swale J. and Patterson, J. (1956). *Biochem. J.* **62,** 16P–17P.
Cameron, E. H. D., Jones, D., Anderson, A. B. M. and Griffiths, K. (1969). *J. Endocr.* **45,** 215–230.
Carter, S. B. (1954). *J. Endocr.* **13,** 150–160.
Catala, S. and Deis, R. P. (1973). *J. Endocr.* **56,** 219–225.
Catling, P. C. and Vinson, G. P. (1976). *J. Endocr.* **69,** 447–448.
Challis, J. R. G. (1971). *Nature Lond.* **229,** 208.
Chang, E., Mittelman, A. and Dao, T. L. (1963). *J. biol. Chem.* **238,** 913–917.
Cheathum, S. G., Douville, A. W. and Warren, J. G. (1967). *Biochim. biophys. Acta* **137,** 172–178.
Chen, C. L. and Meites, J. (1970). *Endocrinology* **86,** 503–505.
Cheo, K. L. and Loke, K. H. (1968). *Steroids* **11,** 603–608.
Chester Jones, I. (1949a). *Endocrinology* **44,** 427–438.
Chester Jones, I. (1949b). *Endocrinology* **45,** 514–536.
Chester Jones, I. (1952). *Proc. R. Soc. B* **139,** 398–410.
Chester Jones, I. (1955). *Br. med. Bull.* **11,** 156–160.
Chester Jones, I. (1957). "The Adrenal Cortex", Cambridge University Press.
Chester Jones, I. and Roby, C. C. (1954). *J. Endocr.* **10,** 245–250.
Christian, J. J. (1960). *Proc Soc. Exp. Biol. Med.* **104,** 330–332.
Christian, J. J. (1962). *Endocrinology* **71,** 431–447.
Christian, J. J., Lloyd, J. A. and Davis, D. E. (1965). *Rec. Prog. Horm. Res.* **21,** 501–578.
Cohn, G. L. and Mulrow, P. J. (1961). *Proceedings of the Endocrine Society* 43rd Meeting New York, p. 52.
Colas, A. and Heinrichs, L. L. (1965). *Steroids* **5,** 753–764.
Colby, H. D. and Kitay, J. I. (1972). *Endocrinology* **91,** 1523–1527.
Colby, H. D., Witorsch, R. J., Caffrey, J. L. and Kitay, J. I. (1973). *Acta endocr., Copnh.* **74,** 568–575.
Collu, R. and Ducharme, J. R. (1975). *J. steroid. Biochem.* **6,** 869–872.

Colombo, L. and Colombo Belvedere, P. (1975). *Boll. Zool.* **42,** 263–269.
Colombo, L. and Colombo Belvedere, P. (1976). *Gen comp. Endocr.* **28,** 371–385.
Colombo, L and Colombo Belvedere, P. (1977). *Invest. Pesq.* **41,** 147–164.
Colombo, L. and Pesavento, S. (1973). *Gen. comp. Endocr.* **21,** 214.
Colombo, L. and Yaron, Z. (1976). *Gen. comp. Endocr.* **28,** 403–412.
Colombo, L., Bern, H. A., Pieprzyk, J. and Johnson, D. W. (1973). *Gen. comp. Endocr.* **21,** 168–178.
Colombo, L., Yaron, Z., Daniels, E. and Belvedere, P. (1974). *Gen. comp. Endocr.* **24,** 331–337.
Colombo, L., Colombo Belvedere, P., Prando, P., Scaffai, P., and Cisotto, T. (1977). *Gen. comp. Endocr.* **33,** 480–495.
Cortés, J. M., Péron, F. G. and Dorfman, R. I. (1963). *Endocrinology* **73,** 713–720.
Coyne, M. D. and Kitay, J. I. (1969). *Endocrinology* **85,** 1097–1102.
Coyne, M. D. and Kitay, J. I. (1971). *Endocrinology* **89,** 1024–1028.
Crafts, R., Llerana, L. A., Guevara, J. and Lloyd, C. W. (1968). *Steroids* **12,** 151–163.
Critchlow, V., Liebelt, R. A., Bar-Sela, M., Mountcastle, W. and Lipscomb, H. S. (1963). *Am. J. Physiol.* **205,** 807–815.
Csapo, A. I. (1969). *In:* "Progesterone: Its Regulatory Effect on the Myometrium", (G. E. W. Wolstenholme and J. Knight, eds.) Ciba Foundation Study Grp *34* Churchill, London, pp. 13–42.
Currie, W. B., Wong, M. S. F., Cox, R. I. and Thorburn, G. D. (1973). *In* "Endocrine Factors in Labour" (A. Klopper and J. Gardner, eds.) Memoirs of the Society of Endocrinology, **20,** 95–118, Cambridge University Press.
Dailey, R. E., Karickhoff, E. R., Swell, L., Field, H. and Treadwell, C. R. (1960). *Proc. Soc. exp. Biol. Med.* **105,** 326–328.
Dao, T. L. (1953). *Science, N.Y.* **118,** 21–22.
Davies, S. (1937). *Q. Jl. microsc, Sci.* **80,** 81–98.
Dean, F. D., Cole, P. M. and Chester Jones, I. (1959). *J. Endocr.* **18,** iii–iv.
Deane, H. W. (1962). *In:* "Handbuch der experimentellen Pharmakologie" (O. Eichler and A. Farah, eds.) **14** (1), pp. 1–185, Springer Verlag, Berlin.
Deane, H. W. and Greep, R. O. (1946). *Am J. Anat.* **79,** 117–146.
Deane, H. W. and Lyman, C. P. (1954). *Endocrinology* **55,** 300–315.
De Jong, F. H., Baird, D. J. and van der Molen, H. J. (1974). *Acta endocr. Copnh.* **77,** 575–587.
Dell'Acqua, S., Mancuso, S., Benagiano, G., Wiqvist, N. and Diczfalusy, E. (1966). *In* "Proceedings of the Second International Congress on Hormonal Steroids" (L. Martini, F. Fraschini and M. Motta, eds.) pp. 639–645, Excerpta Medica I. C. S. **132,** Amsterdam.
Dell'Acqua, S., Mancuso, S., Eriksson, G., Ruse, J. L., Solomon, S. and Diczfalusy, E. (1967). *Acta endocr., Copnh.* **55,** 401–414.
Delost, P. (1953). *C. r. Séanc. Soc. Biol.* **147,** 1580–1584.
Delost, P. (1955). *J. Physiol., Paris* **47,** 164–167.
Delost, P. (1956). "Les corrélations génito-surrénaliennes chez le campagnol des champs (Microtus arvalis P)" Masson et cie, Paris.
Delost, P. and Dalle, M. (1970). *C. r. Séanc. Soc. Biol., Paris* **164,** 2480–2484.
Delost, P. and Delost, H. (1954). *C. r. Séanc. Soc. Biol., Paris* **148,** 1788–1791.
Delost, P. and Delost, H. (1955). *C. r. Séanc. Soc. Biol., Paris* **149,** 910–914.
De Moor, P., Steeno, O. and Deckx, R. (1963). *Acta endocr., Copnh.* **44,** 107–108.
Denamur, R. (1971). *J. Dairy Res.* **38,** 237–264.
Denef, C. and de Moor, P. (1972). *Endocrinology* **91,** 374–384.

Deshpande, N., Jensen, V., Bulbrook, R. D. and Doouss, T. W. (1967). *Steroids* **9**, 393–404.

Deshpande, N., Jensen, V., Bulbrook, R. D. and Doouss, T. W. (1969). *J. Endocr.* **43**, 135–136.

Deshpande, N., Jensen, V., Carson, P., Bulbrook, R. D. and Doouss, T. W. (1970). *J. Endocr.* **47**, 213–242.

Deshpande, N., Carson, P. and Harley, S. (1971). *J. Endocr.* **50**, 467–484.

Dhom, G., Seebach, H. B. and Stephan, G. (1971). *Z. Zellforsch. mikrosk. Anat.* **116**, 119–135.

Diczfalusy, E. (1964). *Fedn. Proc. Fedn. Am. Socs. exp. Biol.* **23**, 791–798.

Diczfalusy, E. (1966). *In* "Proceedings of the Second International Congress on Hormonal Steroids" (L. Martini, F. Fraschini and M. Motta, eds.) pp. 83–95, Excerpta Media I.C.S. **132**, Amsterdam.

Diczfalusy, E. (1969a). *Acta endocr., Copnh.* **61**, 649–664.

Diczfalusy, E. (1969b). *In* "The Foeto-placental Unit" (A. Pecile and C. Finzi, eds.) pp. 65–109, Excerpta Medica I.C.S. **183**.

Diczfalusy, E. (1975). *In* "Reproductive Endrocrinology" (R. Vokaer and G. De Bock, eds.). pp. 3–19. Pergamon Press, Oxford.

Diczfalusy, E., Notter, G., Edsmyr, R. and Westman, A. (1959). *J. clin. Endocr. Metab.* **19**, 1230–1244.

Dixon, W. R., Phillips, J. G. and Kase, N. (1965). *Steroids* **6**, 81–87.

Doerr, P. and Pirke, K. M. (1975). *Acta endocr., Copnh.* **78**, 531–538.

Döhler, K. D. (1978). *Trends in NeuroScience*, **1(5)**, 138–140.

Donaldson, E. M. and Fagerland, U. H. M. (1969). *J. Fish Res. Bd. Can.* **26**, 1789–1799.

Donaldson, E. M. and Fagerland, U. H. M. (1970). *J. Fish. Res. Bd. Can.* **27**, 2287–2296.

Doous, T. W., Skinner, S. J. M. and Couch, R. A. F. (1975) *J. Endocr.* **66**, 1–12.

Dorfman, R. I. and Ungar, F. (1965). "Metabolism of Steroid Hormones" Academic Press, New York, London and San Francisco.

Dorfman, R. I., Sharma, D. C., Southren, A. L. and Gabrilove, J. L. (1965). *Cancer Res.* **28**, 1125–1128.

Dorfman, R. I., Sharma, D. C., Menon, K. M. J. and Forchielli, E. (1966). *In* "Proceedings of the Second International Congress on Hormonal Steroids" (L. Martini, F. Fraschini and L. Motta, eds.) pp. 391–396 Excerpta Medica **132**, Amsterdam.

Drost, M. and Holm, L. W. (1968). *J. Endocr.* **40**, 293–296.

Easterling, N. E., Simmer, H. H., Dignam, W. J., Frankland, M. V. and Naftolin, F. (1966). *Steroids* **8**, 157–178.

Eberlein W. R. (1971). *In* "The Human Adrenal Cortex" (N. P. Christy, ed.). pp. 317–327, Harper and Row, New York.

Eckstein, B. (1970). *Gen. comp. Endocr.* **14**, 303–312.

Eckstein, B., and Eylath, U. (1970). *Gen. comp. Endocr.* **14**, 396–403.

Eckstein, B. and Klein, A. (1971). *J. steroid Biochem.* **2**, 349–353.

Edqvist, L–E, Ekman, L., Gustafsson, B. and Johansson, E. D. B. (1973). *Acta endocr., Copnh.* **72**, 81–88.

Eik-Nes, K. B. and Hall, P. E. (1962). *Proc. Soc. exp. Biol. Med.* **111**, 280–282.

Eisenstein, A. B. and Strack, I. (1961). *Endocrinology* **68**, 121–124.

Engel, L. L. (1962). *In* "The Human Adrenal Cortex" (H. R. Currie, T. Symington and J. K. Grant), p. 89. Livingstone, Edinburgh.

Engel, L. L. and Dimolene, A. (1963). *J. Endocr.* **26,** 233–240.
Everitt, B. J. and Herbert, J. (1969). *Nature, Lond.* **222,** 1065–1066.
Everitt, B. J. and Herbert, J. (1971). *J. Endocr.* **51,** 575–588.
Ewald, W., Werbin, H. and Chaikoff, I. L. (1964). *Steroids* **4,** 759–776.
Fagerland, U. H. M. and Donaldson, E. M. (1969). *Gen. comp. Endocr.* **12,** 438–448.
Fazekas, A. T. A., Homoki, J. and Teller, W. M. (1974). *J. Endocr.* **61,** 273–276.
Feder, H. H. and Ruf, K. B. (1969). *Endrocrinology* **84,** 171–174.
Feder, H. H., Brown-Grant, K. and Corker, C. S. (1971). *J. Endocr.* **50,** 29–39.
Folley, S. J. (1956). "The Physiology and Biochemistry of Lactation" Oliver and
   Boyd, Edinburgh and London.
Fonzo, D. and Nelson, D. A. (1970). *Acta endocr., Copnh.* **64,** 59–64.
Forchielli, E. and Dorfman, R. I. (1956). *J. biol, Chem.* **223,** 443–448.
Forchielli, E., Brown-Grant, K. and Dorfman, R. I. (1958). *Proc. Soc. Exp. Biol. Med.*
   99, 594–596.
Foster, M. A. (1934). *Am. J. Anat.* **54,** 487–506.
Fostier, A., Jalabert, B., and Terqui, M. (1973). *C. r. hebd. Seanc. Acad, Sci., Paris.*
   277D, 421–424.
Fragachan, F., Nowaczynski, W., Bertranau, E., Kalina, M., Genest, J., Robinson, P.
   and Mion, C. (1969). *Endocrinology* **84,** 98–103.
Frandsen, V. A. (1969). *In* "The Foeto-placental Unit" (A. Pecile and C. Finzi, eds.)
   pp. 369–371, Excerpta Medica I.C.S. **183,** Amsterdam.
Froland, A., Frandsen, V. A. and Johnsen, S. G. (1973). *Acta endocr., Copnh.* **72,**
   182–190.
Gala, R. R. and Westphal, U. (1965a). *Endocrinology* **77,** 841–851.
Gala, R. R. and Westphal, U. (1965b). *Endocrinology* **76,** 1079–1088.
Gala, R. R. and Westphal, U. (1966). *Endocrinology* **78,** 277–285.
Gala, R. R. and Westphal, U. (1967). *Acta endocr., Copnh.* **55,** 47–61.
Gaskin, J. H. and Kitay, J. I. (1970). *Endocrinology* **87,** 779–786.
Gaskin, J. H. and Kitay, J. I. (1971). *Endocrinology* **89,** 1047–1053.
Gemzell, C. A. (1952). *Acta endocr., Copnh.* **11,** 221–228.
Ghraf, R., Hoff, H. G., Lax, E. R. and Schriefers, H. (1975). *J. Endocr.* **67,** 317–326.
Givens, J. R., Anderson, R. N., Ragland, J. B., Wiser, W. L. and Umstot, E. S.
   (1975). *J. clin. Endocr. Metab.* **40,** 988–1000.
Givner, M. L. and Rochefort, J. G. (1972). *Endocrinology* **90,** 1238–1244.
Glenister, D. W. and Yates, F. I. (1961). *Endocrinology* **68,** 747–758.
Goldman, A. S., Gustafsson, J. A. and Stenberg, A. (1974). *Acta endocr., Copnh.* **76,**
   719–728.
Goldzieher, J. W. and Boyd, F. G. (1967). *Acta endocr., Copnh.* **54,** 51–62.
Goldzieher, J. W., Axelrod, L. R. and Weissbein, A. S. (1968). *Acta endocr., Copnh.*
   Suppl. **58,** 5–34.
Goncharov, N. P., Wehrberger, K. and Schubert, K. (1969). *Endocrinology* **55,**
   105–108.
Goncharov, N. P., Wehrberger, K., Schubert, K. and Sheutsova, Z. (1971). *Patol.*
   Fiziol. eksp. Terap. **15,** 31–37.
Goodlad, G. A. J. and Clark, C. M. (1961). *Br. J. Cancer* **15,** 833–837.
Gorski, M. E. and Lawton, J. E. (1973). *Endocrinology* **93,** 1232–1234.
Gorwill, R. H., Snyder, D. L., Lindholm, V. B. and Jaffe, R. B. (1970). *Gen. comp.*
   Endocr. **16,** 21–29.
Goswami, S. V., and Sundararaj, B. I. (1971a). *J. exp. Zool.* **178,** 467–478.
Goswami, S. V., and Sundararaj, B. I. (1971b). *J. exp. Zool.* **178,** 457–466.

Goswami, S. V. and Sundararaj, B. I. (1974). *Gen. comp. Endocr.* **23,** 282–285.

Gower, D. B. (1963). *J. Endocr.* **26,** 173–174.

Gower, D. B. and Ahmad, N. (1966). *Biochem. J.* **100,** 67–68.

Gower, D. B. and Ahmad, N. (1967). *Biochem. J.* **104,** 550–556.

Greenbaum, A. L. and Darby, F. J. (1964). *Biochem. J.* **91,** 307–317.

Greep, R. O., and Chester Jones, I. (1950). *Rec. Prog. Horm. Res.* **5,** 197–254.

Griffiths, K. and Giles, C. A. (1965). *J. Endocr.* **33,** 333–334.

Griffiths, K., Cunningham, D. and Cameron, E. H. D. (1968). *J. Endocr.* **40,** 49–58.

Gustafsson, J. Å. and Stenberg, Å. (1974). *J. biol. Chem.* **249,** 719–723.

Hacik, T. (1968). *Bratisl. lék. Listy,* **20,** 179–187.

Hacik, T. (1969). *Endocr. exp.,* **3,** 147–154.

Hagen, A. A. and Troop, R. C. (1960). *Endocrinology* **67,** 194–203.

Hampl, R. and Starka, V. (1967). *Endocr., exp.,* **1,** 5–13.

Hatai, S. (1913). *Am. J. Anat.,* **15,** 87–119.

Heap, R. B., Holzbauer, M. and Newport, H. M. (1966). *J. Endocr.* **36,** 159–176.

Hechter, O., Macchi, I. A., Korman, H., Frank, E. D. and Frank, H. A. (1955). *Am. J. Physiol.* **182,** 29–34.

Hewitt, W. F., and van Liere, E. J. (1941). *Endocrinology* **28,** 62–64.

Hinsull, S. M. and Crocker, A. D. (1970). *J. Endocr.* **48,** lxxix–lxxx.

Hirai, M., Nakao, T. and Pincus, G. (1968). *Jikeikai med. J.* **15,** 228–248.

Hirose, K. (1972). *Bull. Jap. Soc. scient. Fish.* **38,** 457–461.

Hirose, K. (1976). *J. Fish. Res. Bd Canada* **33,** 989–994.

Hiroshiga, T., and Wada-Okada, S. (1973). *Neuroendocrinology* **12,** 316–319.

Hiroshiga, T., Abe, K., Wada, S. and Kaneko, M. (1973). *Neuroendocrinology* **11,** 306–320.

Hirschmann, H., Decourcy, C., Levy, R. P. and Miller, K. L. (1960). *J. biol. Chem.* **235,** PC 48–49.

Hishida, T. and Kawamoto, N. (1970). *J. exp. Zool.* **173,** 279–283.

Hoffmann, B., Schams, D., Gimenz, T., Ender, M. L., Herrmann, C. and Karg, H. (1973). *Acta endocr., Copnh.* **73,** 385–395.

Hoffmann, W., Knuppen, R. and Breuer, H. (1970). *Hoppe-Seylers Z. physiol. Chem.* **351,** 387–396.

Hofmann, F. G. and Christy, N. P. (1961). *Biochim. biophys. Acta* **54,** 354–356.

Holmes, W. N. (1955). *Anat. Rec.* **122,** 271–293.

Holmes (1966). Personal communication in Parkes and Deanesly.

Holzbauer, M. and Godden, U. (1974). *J. steroid Biochem.* **5,** 109–111.

Holzbauer, M. and Newport, H. M. (1967). *J. Physiol., Lond.* **193,** 131–140.

Holzbauer, M. and Newport, H. M. (1968). *Nature, Lond.* **217,** 967–968.

Holzbauer, M. and Newport, H. M. (1969). *J. Physiol., Lond.* **200,** 821–848.

Holzbauer, M., Newport, H. M., Birmingham, M. K. and Traikov, H. (1969). *Nature, Lond.* **221,** 572–573.

Hopkins, P. S. and Thorburn, G. D. (1971). *Proc. 4th Asian and Oceania Congr. Endocr., Auckland, N. Z.,* Abstr. No. 170.

Hopper, B. R. and Tullner, W. W. (1967). *Steroids* **9,** 517–527.

Horton, R. (1967). Testosterone, Proc. Workshop Conference (J. Tamm, ed.) Georg Thieme Verlag, Stuttgart, 220–225.

Horton, R. and Tait, J. F. (1966). *J. clin. Invest.* **45,** 301–313.

Howes, J. R., Hentges, J. F., and Warwick, A. C. (1960). *Proc. Soc. exp. Biol. Med.* **104,** 322–324.

Hudson, R. W. and Killinger, D. W. (1972). *J. clin. Endocr. Metab.* **34,** 215–224.

Hudson, R. W., Schachter, H. and Killinger, D. W. (1974). *Endocrinology* **95,** 38–47.

Huisn't Veld, L. G. and Dingemanse, E. (1948). *Acta brev. neerl. Physiol.* **16,** 9–15.

Ibayashi, H. and Yamaji, T. (1968). *Folia endocr. jap.* **44,** 858–884.

Ichii, S., Kobayashi, S. and Matsuba, M. (1965). *Steroids* **5,** 123–130.

Idelman, S. (1978). *In* "General, Comparative and Clinical Endocrinology of the Adrenal Cortex" (I. Chester Jones and I. W. Henderson, eds.) Vol. 2, pp. 1–199, Academic Press, London and New York.

Idler, D. R., Horne, D. A. and Sangalang, G. B. (1971). *Gen. comp. Endocr.* **16,** 257–267.

Idler, D. R., Truscott, B., and Stewart, H. C. (1969). *Proc. 3rd Int. Congr. Endocrinol., Mexico 1968.* (C. Gaul, ed.). *Excerpta Med. I. C. S.* **184,** pp 724–729, Amsterdam.

Jackson, C. M. (1913). *Am. J. Anat.* **15,** 1–68.

Jacobs, H. S., Abraham, G. E., Glasser, E. J., Hopper, K. and Kondon, J. J. (1972). *J. Endocr.* **53,** xxxvi–xxxvii.

Jaffe, R. B., Perez-Palacios, G. and Diczfalusy, E. (1972). *J. clin. Endocr. Metab.* **35,** 646–654.

Jalabert, B., Breton, B., and Bry, C. (1972). *C. r. hebd. Séanc. Acad. Sci., Paris.* **275D,** 1139–1142.

Jensen, V., Carson, P. and Deshpande, N. (1971). *J. Endocr.* **50,** 177–178.

Jensen, V., Carson, P. and Deshpande, N. (1972). *J. Endocr.* **55,** 321–331.

Johnsen, S. G., Christiansen, P., Frandsen, V. A., Froland, A. and Nielsen, J. (1971). *Acta endocr., Copnh.* **66,** 587–605.

Jones, C. T. (1974). *Endocrinology* **95,** 1129–1133.

Jones, T. and Griffiths, K. (1968). *J. Endocr.* **42,** 559–565.

Jones, T., Forrest, A. P. M. and Griffiths, K. (1971). *J. Endocr.* **50,** 535–536.

Jones, T., Cameron, E. H. D., Griffiths, K. and Forrest, A. P. M. (1970). Abstr. Third Int. Congr. Hormonal Steroids, Hamburg (V. H. T. James, ed.) p. 106. *Excerpta Medica I.C.S.* **210,** Amsterdam.

Kase, N. and Kowal, J. (1962). *J. clin. Endocr. Metab.* **22,** 925–928.

Kawai, A. and Yates, F. E. (1966). *Endocrinology* **79,** 1040–1046.

Keene, M. F. L. and Hewer, E. E. (1927). *J. Anat.,* **61,** 302–324.

Keller, M., Hauser, A. and Walser, A. (1958). *J. clin. Endocr. Metab.* **18,** 1384–1398.

Keyes, P. H. (1949). *Endocrinology* **44,** 274–277.

Kime, D. E. (1978). *Gen. comp. Endocr.* **35,** 322–328.

Kirschner, M. A., Lipsett, M. B. and Collins, D. R. (1965). *J. clin. Invest.* **44,** 657–665.

Kirschner, M. A. and Bardin, C. W. (1972). *Metabolism* **21,** 667–668.

Kirschner, M. A. and Knorr, D. W. R. (1972). *Acta endocr., Copnh.* **70,** 342–350.

Kitay, J. I. (1961a). *Nature, Lond.* **192,** 358–359.

Kitay, J. I. (1961b). *Endocrinology* **68,** 818–824.

Kitay, J. I. (1963a). *Proc. Soc. exp. Biol. Med.* **112,** 679–683.

Kitay, J. I. (1963b). *Endocrinology* **73,** 253–260.

Kitay, J. I. (1963c). *Endocrinology* **72,** 947–954.

Kitay, J. I. (1963d). *Acta endocr. Copnh.* **43,** 601–608.

Kitay, J. I. (1965a). *Endocrinology* **77,** 1048–1052.

Kitay, J. I. (1965b). *Proc. Soc. exp. Biol. Med.* **120,** 193–196.

Kitay, J. I. (1968). *In* "Functions of the Adrenal Cortex" (K. W. McKerns, ed.). Vol. 2, pp. 775–811. North Holland, Amsterdam.

Kitay, J. I., Coyne, M. D., Newson, W. and Nelson, R. (1965). *Endocrinology* **77,** 902–908.

Kitay, J. I., Coyne, M. D., Nelson, R. and Newsom, W. (1966). *Endocrinology* **78,** 1061–1066.

Kitay, J. I., Coyne, M. D., and Swygert, N. H. (1970). *Endocrinology* **87**, 1257–1265.
Kitay, J. I., Coyne, M. D., Swygert, M. H. and Gaines, K. E. (1971a). *Endocrinology* **89**, 565–570.
Kitay, J. I., Coyne, M. D. and Swygert, N. H. (1971b). *Endocrinology* **89**, 432–438.
Kley, H. K., Nieschlag, E. and Kruskemper, H. L. (1975). *Acta endocr., Copnh.* **79**, 95–101.
Kniewald, Z., Danisi, M. and Martini, L. (1971). *Acta endocr., Copnh.* **68**, 614–624.
Knuppen, R. and Breuer, H. (1962). *Biochim. biophys. Acta* **58**, 147–148.
Knuppen, R. and Breuer, H. (1964). *Hoppe-Seyler's Z. physiol. Chem.* **337**, 159–160.
Knuppen, R., Haupt, O. and Breuer, H. (1964a). *Steroids* **3**, 123–128.
Knuppen, R., Behm, M. and Breuer, H. (1964b). *Hoppe-Seyler's Z. physiol. Chem.* **337**, 145–149.
Knuppen, R., Haupt, M. and Breuer, H. (1965). *J. Endocr.* **33**, 529–530.
Knuppen, R., Haupt, O. and Breuer, H. (1967). *Biochem. J.* **105**, 971–978.
Koch, B., Mielhe-Voloss, C. and Stutinsky, F. (1968). *C. r. hebd. Seanc. Acad. Sci. Press* **264**, 1183–1186.
Koerner, D. R. and Hellman, L. (1964). *Endocrinology* **75**, 592–601.
Koref, O., Stezer, K. and Feher, T. (1971). *Acta endocr., Copnh.* **66**, 727–736.
Korus, W., Schriefers, H., Breuer, H. and Bayer, S. M. (1959). *Acta endocr., Copnh.* **66**, 727–736.
Kowal, T., Forchielli, E. and Dorfman, R. I. (1964a). *Steroids* **3**, 531–549.
Kowal, T., Forchielli, E. and Dorfman, R. I. (1964b). *Steroids* **4**, 77–100.
Kwa, H. G. and Verhofstad, F. (1967). *J. Endocr.* **39**, 455–456.
Kwa, H. G., van der Gugten, A. A. and Verhofstad, F. (1969). *Eur. J. Cancer* **5**, 571–579.
Labhsetwar, A. P. (1972). *J. Endocr.* **52**, 399–400.
Landers, E. and Wagner, W. C. (1974). *Acta endocr., Copnh.* **77**, 498–501.
Lanman, J. T. (1953). *Medicine, Baltimore* **32**, 389–430.
Lanman, J. (1962). *Trans. New Engl. obstet, gynec. Soc.* **16**, 123–130.
Lantos, C. P., Birmingham, M. K. and Traikov, H. (1966). *Acta physiol. latinoam* **16**, 278–281.
Lantos, C. P., Birmingham, M. K. and Traikov, H. (1967). *Acta physiol. latinoam* **17**, 42–54.
Lauritzen, C. (1965). *Acta endocr., Copnh. Suppl.* **100**, 112–113.
Lauritzen, C. and Lehmann, W. D. (1967). *J. Endocr.* **39**, 173–182.
Lawrence, J. R. and Griffiths, K. (1966). *Biochem. J.* **99**, 27–28c.
Lawton, I. E. (1972). *Endocrinology* **90**, 575–579.
Lee, P. A., Kowarski, A., Migeon, C. J. and Blizzard, R. M. (1975). *J. clin. Endocr. Metab.* **40**, 664–669.
Levy, H., Hood, B., Cha, C. H. and Carlo, J. J. (1965). *Steroids* **5**, 677–686.
Liggins, G. C. (1968). *J. Endocr.* **42**, 323–329.
Liggins, G. C. (1969). *J. Endocr.* **43**, 515–523.
Liggins, G. C. (1972). *In* "Human Reproductive Physiology" (R. P. Shearman ed.), pp. 138–197, Blackwell, Oxford.
Liggins, G. C. (1973). *In* "Endocrine Factors in Labour" (A. Klopper and J. Gardner, eds.), pp. 119–140. Memoirs of the Society for Endocrinology, **20**, Cambridge University Press.
Liggins, G. C. and Grieves, S. A. (1971). *Nature, Lond.* **232**, 629–631.
Liggins, G. C., Kennedy, P. C. and Holm, L. W. (1967). *Am. J. Obstet. Gynec.* **98**, 1080–1086.

Liggins, G. C., Grieves, S. A., Kendall, J. Z., and Knox, B. S. (1972). *J. Reprod. Fert. Suppl.* **16,** 85–103.

Liggins, G. C., Fairclough, R. G., Grieves, S. A., Kendall, J. Z. and Knox, B. S. (1973). *Recent Prog. Horm. Res.* **29,** 111–150.

Linet, O. and Lomen, P. (1971). *Acta Endocr., Copnh.* **68,** 303–310.

Ling, A. M. and Loke, K. H. (1966). *Steroids* **8,** 765–775.

Linzell, J. L. and Heap, R. B. (1968). *J. Endocr.* **41,** 433–438.

Liptrap, R. M. (1970). *J. Endocr.* **47,** 197–205.

Lobban, M. C. (1952). *J. Physiol. Lond.* **118,** 565–574.

Loke, K. H. and Gau, L. Y. (1968). *Steroids* **11,** 863–875.

Lombardo, M. E., McMorris, C. and Hudson, P. B. (1959). *Endocrinology* **65,** 426–432.

Lucis, O. J. and Lucis, R. (1970). *Cancer Res.* **30,** 702–708.

Lucis, O. J., Lucis, R., and Kerenyi, N. A. (1972). *Gen. comp. Endocr.* **18,** 605.

Lyons, W. R., Li, C. H. and Johnson, R. E. (1958). *Recent Prog. Horm. Res.* **14,** 219–254.

MacDonald, P. C., Rombart, R. P. and Siiteri, P. K. (1967). *J. clin. Endocr. Metab.* **27,** 1103–1111.

McKeever, S. (1963). *Am. J. Anat.* **113,** 153–167.

MacKinnon, P. C. B. (1978). *Trends in NeuroSciences,* **1(5),** 136–138.

McPhail, M. K. and Read, H. C. (1942). *Anat. Rec.* **84,** 75–89.

Mahesh, V. B., and Greenblatt, R. B. (1965). *In* "Proceedings 47th Meeting of the Endocrinological Society", p. 25.

Mahesh, V. B. and Herrmann, W. (1963). *Steroids* **1,** 51–61.

Malendowicz, L. K. (1974a). *Cell Tiss. Res.* **151,** 537–547.

Malendowicz, L. K. (1974b). *Cell Tiss. Res.* **151,** 525–536.

Malplas, P. (1933). *J. Obstet. Gynaec Br. Emp.* **40,** 1046–1053.

Mancuso, S., Mancuso, F. P., Tillinger, K. G. and Diczfalusy, E. (1965a). *Acta endocr., Copnh.* **49,** 248–261.

Mancuso, S., Dell'Acqua, S., Eriksson, G., Wiqvist, N. and Diczfalusy, E. (1965b). *Steroids* **5,** 183–197.

Mancuso, S., Benagiano, G., Dell'Acqua, S., Shapiro, M., Wiqvist, N. and Diczfalusy, E. (1968). *Acta endocr., Copnh.* **57,** 208–227.

Marks, B. H., Alpert, M. and Kruger, F. A. (1958). *Endocrinology* **63,** 75–81.

Martin, C. E., Cake, M. H., Hartmann, P. E. and Cook, I. F. (1977). *Acta endocr., Copnh.* **84,** 167–176.

Mason, H. L. and Kepler, E. J. (1945). *J. biol. Chem.* **161,** 235–257.

Mattner, P. E. and Thorburn, G. D. (1971). *J. Reprod. Fert.* **24,** 140–141.

Mäusle, E. (1971). *Z. Zellforsch. mikrosk. Anat.* **116,** 136–150.

Maynard, P. V. and Cameron, E. H. D. (1972). *Biochem. J.* **126,** 99–106.

Mehdi, A. Z. and Sandor, T. (1971). *Steroids* **17,** 143–154.

Mehdi, A. Z. and Sandor, T. (1972). *Can. J. Biochem. Physiol.* **50,** 443–446.

Meijs-Roelofs, H. M. A., Vilenbroek, J. T. J., De Jong, F. H. and Welschen, R. (1973). *J. Endocr.* **59,** 295–304.

Meijs-Roelofs, H. M. A., De Greef, W. J. and Vilenbroek, J. T. J. (1975). *J. Endocr.* **64,** 329–336.

Meites, J. and Nicoll, C. S. (1966). *A. Rev. Physiol.* **28,** 57–88.

Meyer, A. S. (1955). *Biochim. biophys. Acta* **17,** 441–442.

Meyer, A. S., Hayano, M., Lindberg, M. C., Gut, M. and Rodgers, O. G. (1955). *Acta endocr., Copnh.* **18,** 148–168.

Meyer, R. (1912). *Virchows Arch. path. Anat.* **210,** 158–164.
Mietkiewski, K., Malendowicz, L., Trosanowicz, R. and Lukaszyk, A. (1969). *Endokrinologie,* **54,** 206–214.
Milewich, L. and Axelrod, L. R. (1972a). *J. Endocr.* **54,** 515–516.
Milewich, L. and Axelrod, L. R. (1972b). *Endocrinology* **91,** 1120–1125.
Milkovic, S., Milkovic, K. and Paunovic, J. (1973). *Endocrinology* **92,** 380–384.
Mills, I. H., Schedl, H. P., Chen, P. S. and Bartter, F. C. (1960). *J. clin. Endocr. Metab.* **20,** 515–528.
Milner, A. J. and Mills, I. H. (1970a). *J. Endocr.* **47,** 369–378.
Milner, A. J. and Mills, I. H. (1970b). *J. Endocr.* **47,** 379–384.
Milner, A. J. and Mills, I. H. (1970c). *J. Endocr.* **48,** 379–387.
Mizuno, H. and Sensui, N. (1973). *Endocr. jap.* **20,** 167–174.
Moser, H. G. and Bernirschke, K. (1962). *Anat. Rec.* **143,** 47–53.
Muller, C. H. (1977). *Gen. comp. Endocr.* **33,** 109–121.
Munro, N. A., Orr, J. C. and Engel, L. C. (1969). *J. Endocr.* **43,** 599–608.
Nathanielsz, P. W. and Abel, M. (1973). *J. Endocr.* **57,** 47–54.
Neher, R. R. and Wettstein, A. (1960a). *Helv. chim. Acta* **43,** 623–628.
Neher, R. R. and Wettstein, A. (1960b). *Helv. chim. Acta* **43,** 1171–1191.
Neher, R. R. and Wettstein, A. (1960c). *Acta endocr. Copnh.* **35,** 1–7.
Nelson, D. H. (1968). *In* "Functions of the Adrenal Cortex" (K. W. McKerns, ed.). Vol. II, pp. 813–827. Appleton Century Crofts, New York.
Nelson, D. H., Tanney, II., Mestman, G., Gieschen, V. W. and Wilson, L. D. (1963). *J. clin. Endocr. Metab.* **23,** 261–265.
Nequin, L. G. and Schwartz, N. B. (1971). *Endocrinology* **88,** 325–331.
Nequin, L. G., Alvarez, J. A. and Campbell, C. S. (1975). *Endocrinology* **97,** 718–724.
Neville, A. M. (1971). *In* "The Human Adrenal Gland and its Relation to Breast Cancer" 1st Tenovus Workshop, Cardiff (K. Griffiths and E. H. D. Cameron, eds.). pp. 28–35. Alpha Omega Alpha, Cardiff.
Neville, A. M. and Webb, J. L. (1965). *Steroids* **6,** 421–426.
Neville, A. M., Anderson, J. M., McCormick, M. H. and Webb, J. L. (1968a). *J. Endocr.* **41,** 541–554.
Neville, A. M., Orr, J. C. and Engel, L. L. (1968b). *Biochem. J.* **107,** 20P.
Neville, A. M., Orr, J. C., Trofimow, N. D. and Engel, L. L. (1969a). *Steroids* **14,** 97–117.
Neville, A. M., Webb, J. L. and Symington, T. (1969b). *Steroids* **13,** 821–833.
Nickerson, P. A. and Morteni, A. (1971). *Am. J. Path.* **64,** 31–44.
Nieschlag, E., Loriaux, D. L., Ruder, H. J., Zucker, I. R., Kirschner, M. A. and Lipsett, M. B. (1973). *J. Endocr.* **57,** 123–134.
Niswender, G. D., Chen, C. L. Midgley, A. R., Meites, J. and Ellis, S. (1969). *Proc. Soc. exp. Biol. Med.* **130,** 793–797.
Nowell, N. W. and Chester Jones, I. (1957). *Acta endocr., Copnh.* **26,** 273–285.
Oakey, R. E. (1970). *Vitams. Horm.* **28,** 1–36.
Oakey, R. E. and Heys, R. F. (1969). *J. Endocr.* **45,** xxiii–xxiv.
Oakey, R. E. and Heys, R. F. (1970). *Acta endocr., Copnh.* **65,** 502–508.
Oertel, G. W. and Eik-Nes, K. (1962). *Endocrinology* **70,** 39–42.
Oertel, G. W., Kaiser, E. and Zimmerman, W. (1963a). *Hoppe-Seyler's Z. physiol. Chem.* **331,** 77–84.
Oertel, G. W., Kaiser, E. and Zimmerman, W. (1963b). *Clinica. chim. Acta* **8,** 154–156.
Oertel, G. W., Groot, K. and Wenzel, D. (1965). *Acta endocr., Copnh.* **49,** 533–540.
Ogle, T. F. and Kitay, J. I. (1979). *Endocrinology,* **104,** 40–44.

Ota, K., Ota, T. and Yokoyama, A. (1974). *J. Endocr.* **61,** 21–28.
Palacious, G. P., Perez, A. E. and Jaffe, R. B. (1968). *J. clin. Endocr. Metab.* **28,** 19–25.
Paris, A. L. and Ramaley, J. A. (1974). *Neuroendocrinology* **15,** 126–136.
Parkes, A. S. (1945). *Physiol. Rev.* **25,** 203–254.
Parkes, A. S. and Deanesly, R. (1966). *In* "Marshalls Physiology of Reproduction" (A. S. Parkes, ed.). Vol. III. pp. 1064–1111. Longmans. 3rd edition, London.
Pauerstein, C. J. and Solomon, D. (1966). *Obstet. Gynec. N.Y.* **28,** 692–699.
Peczenik, O. (1944). *Proc. R. Soc. Edinb. B.* **62,** 59–65.
Perklev, T. (1971). *Acta endocr., Copnh.* **68,** 737–748.
Perklev, T. and Groning, Y. (1969). *Acta endocr., Copnh.* **61,** 449–460.
Phillips, J. G., and Poolsanguan, W. (1978). *J. Endocr.* **77,** 283–291.
Pincus, G. and Romanoff, E. B. (1955). *In* "The Human Adrenal Cortex" (G. E. W. Wolstenholme and M. P. Cameron, eds.). pp. 97–111, Ciba Foundn Colloq. Endocr. **8,** Churchill, London.
Planel, H., Guilhem, A., and Soleilhavoup, J. P. (1960). *C. r. Séanc. Soc Biol.* **154,** 1097–1100.
Plantin, L. O., Diczfalusy, E., and Birke, G. (1957). *Nature, Lond.* **179,** 421.
Poolsanguan, W. (1975). Ph.D. Thesis, University of Hull.
Price, D., Zaaijer, J. J. P., Ortiz, E. and Brinkmann, A. O. (1975). *Am. Zool.,* **15,** (Suppl. 1). 173–195.
Purrott, R. J. and Sage, M. (1969). *Pflügers Arch. ges. Physiol.* **309,** 107–114.
Raeside, J. I. (1963). *J. Reprod. Fert.* **6,** 427–431.
Ramaley, J. A. (1973). *Steroids* **22,** 597–608.
Ramaley, J. A. (1975). *J. Endocr.* **66,** 421–426.
Ramaley, J. A. and Bunn, E. L. (1972). *Endocrinology* **91,** 611–613.
Ramirez, V. D., Moore, D. and McCann, S. M. (1965). *Proc. Soc. exp. Biol. Med.* **118,** 169–173.
Randall, L. O. and Graubard, M. (1940). *Am. J. Physiol.* **131,** 291–295.
Rao, L. G. S. and Taylor, W. (1965). *Biochem. J.* **96,** 172–180.
Raps, D., Barthe, P. L. and Desaulles, P. A. (1971). *Experientia,* **27,** 339–340.
Rea, C. (1898). *J. Am. med. Ass.* **30,** 1166–1167.
Reichstein, T. (1936a). *Helv. chim. Acta* **19,** 29–63.
Reichstein, T. (1963b). *Helv. chim. Acta* **19,** 223–225.
Rennels, E. G. and Singer, E. L. (1970). *Tex. Rep. Biol. Med.* **28,** 303–316.
Revol, A., Mallein, R. and Guinet, P. (1960). *Annls. Biol. clin.* **18,** 565–570.
Richardson, M. C. and Schulster, D. (1972). *J. Endocr.* **55,** 127–139.
Riddle, O. (1923). *Am. J. Physiol.* **66,** 322–339.
Rivarola, M. A. Saez, J. M., Meyer, M. E., Jenkins, M. E. and Migeon, C. J. (1966). *J. clin. Endocr. Metab.* **26,** 1028–1218.
Roaf, R. (1935). *J. Anat.* **70,** 126–135.
Roberts, J. S. and Share, L. (1969). *Endocrinology* **84,** 1076–1081.
Roberts, K. D., van Der Wiele, R. L. and Lieberman, S. (1961). *J. clin. Endocr. Metab.* **21,** 1522–1533.
Robertson, O. H., Krupp, M. A., Favour, C. B., Hane, S. and Thomas, S. F. (1961). *Endocrinology* **68,** 733–746.
Rodier, W. I. and Kitay, J. I. (1974). *Proc. Soc. exp. Biol. med.* **146,** 376–380.
Romanelli, R., Biancalana, D. and Neher, R. (1960). *Acta endocr., Copnh. Suppl.* **51,** 383–384.
Romanoff, E. B., Hudson, P. and Pincus, G. (1953). *J. clin. Endocr. Metab.* **13,** 1546–1547.

Rose, L. I., Williams, G. H., Emerson, K. and Villee, D. B. (1969). *J. clin. Endocr. Metab.* **29,** 1526–1532.

Rosenfeld, R. S., Hellman, L., Roffwarg, H., Wetzmuner, E. R., Fukushima, D. K. and Gallagher, T. F. (1971). *J. clin. Endocr. Metab.* **33,** 87–92.

Rosner, J. M., Charreau, E., Houssay, A. B., Epper, C. (1966). *Endocrinology* **79,** 681–686.

Ross, M. H. (1962). *In* "The Human Adrenal Cortex" (A. R. Currie, T. Symington and J. K. Grant, eds.). pp. 558. Livingstone, Edinburgh.

Rotter, W. (1949a). *Virchows Arch. path. Anat.,* **316,** 590–618.

Rotter, W. (1949b). *Z. Zellforsch.,* **34,** 547–561.

Roy, S. and Mahesh, V. B. (1964). *Endocrinology* **74,** 187–192.

Rubens, R. and Stitch, S. R. (1973). *J. Endocr.* **58,** x.

Rubin, B. L. and Strecker, H. J. (1957). *J. biol. Chem.* **227,** 917–927.

Rubin, B. L. and Strecker, H. J. (1961). *Endocrinology* **69,** 257–267.

Ruhmann-Wennhold, A. and Nelson, D. H. (1970). *Proc. Soc. exp. Biol. Med.* **133,** 493–496.

Ruhmann-Wennhold, A., Johnson, L. R. and Nelson, D. H. (1970). *Biochim. biophys. Acta* **223,** 206–209.

Ryan, K. J. (1969). *In* "The Foeto-placental Unit" (A. Pecile and C. Finzi, eds.). pp. 120–131. Excerpta Medica I.C.S. No. 183, Amsterdam.

Saez, J. M., Loras, B., Morera, A. M., Bertrand, J. (1970). *J. steroid Biochem.* **1,** 355–367.

Saez, J. M., Loras, B., Morera, A. M. and Bertrand, J. (1971). *J. clin. Endocr. Metab.* **32,** 462–469.

Saez, J. M., Morera, A. M., Dazord, A. and Bertrand, J. (1972). *J. Endocr.* **55,** 41–49.

Sakhatskaya, T. S. (1971). *Problemy̆ Endokr. Gormonoter* **17,** 80–82.

Salhanick, H. A. and Berliner, D. L. (1957). *J. biol. Chem.* **227,** 583–590.

Sandberg, A. A. and Slaunwhite, W. R. (1959). *J. clin. Invest.* **38,** 1290–1297.

Sandberg, A. A., Rosenthal, H. E. and Slaunwhite, W. R. (1967). *In* "Proceedings of the Second International Congress on Hormonal steroids" (L. Martini, F. Fraschini and M. Motta, eds.). pp. 707–716. Excerpta Medica I.C.S. **132,** Amsterdam.

Sandor, T., Fazekas, A. G. and Robinson, B. H. (1976). *In:* "General, Comparative and Clinical Endocrinology of the Adrenal Cortex" (I. Chester Jones and I. W. Henderson, eds.). Vol. 1, Chapter 2, pp. 25–125. Academic Press, London, New York and San Francisco.

Savard, K. (1961). *Endocrinology* **68,** 411–416.

Sawyer, C. H. (1975). *Neuroendocrinology,* **17,** 97–124.

Schmidt, P. J. and Idler, D. R. (1962). *Gen. comp. Endocr.* **2,** 204–214.

Schreck, C. B., Flickinger, S. A. and Hopwood, M. L. (1972). *Proc. Soc. exp. Biol. Med.* **140,** 1009–1011.

Schriefers, H. (1967). *Vitams. Horm.* **25,** 271–314.

Schriefers, H., Pittel, M. and Pohl, F. (1962). *Acta endocr., Copnh.* **40,** 140–150.

Schriefers, H., Scharlau, G. and Pohl, F. (1965). *Acta endocr., Copnh.* **48,** 263–271.

Schuetz, A. W. (1967). *Proc. Soc. exp. Biol. Med.* **124,** 1307–1310.

Schuetz, A. W. (1974). *Biol. Reprod.* **10,** 150–178.

Schuetz, A. W., Wallace, R. A., and Dumont, J. (1973). *Cell. Biol. Ann. Meet.* (abstract).

Schwers, J., Vancrombreucq, T., Govaerts, M., Eriksson, G. and Diczfalusy, E. (1971). *Acta endocr., Copnh.* **66,** 637–647.

Scott, D. B. C. (1964). *Ph.D. Thesis,* University of Glasgow.

Seal, U. S. and Doe, R. P. (1967). *In* "Proceedings of the Second International Congress on Hormonal Steroids" (L. Martini, F. Fraschini and M. Motta, eds.). pp. 697–706. Excerpta Medica I.C.S. **132,** Amsterdam.

Selye, H. (1941). *Proc. Soc. exp. Biol. Med.* **46,** 142–146.

Shahwan, M. M., Oakey, R. E. and Stitch, S. R. (1967). *Biochem. J.* **103,** 15P.

Shahwan, M. M., Oakey, R. E. and Stitch, S. R. (1968). *J. Endocr.* **40,** 29–35.

Shahwan, M. M., Oakey, R. E. and Stitch, S. R. (1969). *Acta endocr., Copnh.* **60,** 491–500.

Sharma, D. C., Raheja, M. C., Dorfman, R. I. and Gabrilove, J. L. (1965a). *J. biol. Chem.* **240,** 1045–1053.

Sharma, D. C., Raheja, M. C. Dorfman, R. I. and Gabrilove, J. L. (1965b). *Acta endocr., Copnh.* **50,** 439–451.

Sharma, D. C. and Gabrilove, J. L. (1969). *Comp. Biochem. Physiol.* **31,** 379–390.

Shire, J. G. M. (1974). *J. Endocr.,* **62,** 173–207.

Shire, J. G. M. (1976). *J. Endocr.,* **71,** 445–446.

Short, R. V. (1957). *Ciba Colloq. Endocr.* **11,** 362–378 (Discussion).

Siiteri, P. K. and MacDonald, P. C. (1963). *Steroids* **2,** 713–730.

Siiteri, P. K. and MacDonald, P. C. (1966). *In* "Proceedings of the Second International Congress on Hormonal Steroids" (L. Martini, F. Fraschini and M. Motta, eds.). pp. 726–732. Excerpta Medica I.C.S. **132,** Amsterdam.

Simmer, H. H., Dignam, W. J., Easterling, W. E., Frankland, M. V. and Naftolin, F. (1966). *Steroids* **8,** 179–193.

Simmons, J. E., DiClementi, D. and Maxted, G. (1973). *J. exp. Zool.* **186,** 123–126.

Simpson, T. H. and Wright, R. S. (1970). *J. Endocr.* **46,** 261–268.

Simpson, T. H. and Wright, R. S. (1977). *Steroids* **29,** 383–398.

Singer, S. S. and Sylvester, S. (1976). *Endocrinology* **99,** 1346–1352.

Skowron, S. and Zajaczek, S. (1947). *C. r. Séanc. Soc. Biol.* **141,** 1105–1107.

Slaunwhite, W. R. and Sandberg, A. A. (1959). *J. clin. Invest.* **38,** 384–391.

Smith, C. A. (1959). "The Physiology of the Newborn Infant" 3rd edition, Thomas, Springfield, Ill.

Sneddon, A. and Marrian, F. (1963). *Biochem. J.* **86,** 385–388.

Snyder, J. G. and Wyman, L. C. (1951). *Am. J. Physiol.* **167,** 328–332.

Spector, W. S. (1956). "Handbook of Biological Data" Saunders, Philadelphia.

Stabler, T. A. and Ungar, F. (1970). *Endocrinology* **86,** 1049–1058.

Starka, L. (1965). *Naturwissenschaften.* **52,** 499.

Starkey, W. F. and Schmidt, E. C. H. (1938). *Endocrinology* **23,** 339–344.

Stilling, H. (1898). *Arch. mikrosk. Anast. Entw. Mech.* **52,** 176–195.

Stonesifer, G. L., Lowe, R. H., Cameron, J. L. and Ganis, F. M. (1973). *Ann. Surg.* **178,** 563–564.

Strecker, J. R., Lauritzen, C. and Gossler, W. (1977). *Acta endocr., Copnh. Suppl.* **208,** 88–89.

Sulcova, J. and Starka, L. (1963). *Experientia* **19,** 632–633.

Sundararaj, B. I. and Goswami, S. V. (1966). *J. exp. Zool.* **161,** 287–296.

Sundararaj, B. I. and Goswami, S. V. (1970). *Excerpta Med. Found. Int. Congr. Ser.* **219.** *Proc. Int. Congr. Horm. Steroids 3rd. Hamburg September 7–12,* pp. 966–975.

Sundararaj, B. I. and Goswami, S. V. (1974). *Gen. comp. Endocr.* **23,** 276–281.

Sundersford, J. A. and Aakvaag, A. (1972). *Acta endocr., Copnh.* **71,** 519–529.

Surina, M. N. (1970). *Problemy. Endokr. Gormonoter.* **16,** 90–96.

Swinyard, C. A. (1940). *Anat. Rec.* **76,** 69–79.

Tal, E. and Sulman, F. G. (1975). *J. Endocr.* **67,** 99–103.
Tanner, J. M. (1962). "Growth at Adolescence" 2nd edition. Blackwell, Oxford p. 28.
Telegdy, G., Robin, M. and Diczfalusy, E. (1972). *J. steroid Biochem.* **3,** 693–697.
Thatcher, W. W. and Tucker, H. A. (1970). *Endocrinology* **86,** 237–240.
Thoman, E. B., Conner, R. L. and Levine, S. (1970a). *J. comp. Physiol.* **70,** 364–369.
Thoman, E. B., Sproul, M., Seeler, B. and Levine, S. (1970b). *J. Endocr.* **46,** 297–303.
Thomas, J. P. (1968). *J. clin. Endocr. Metab.* **28,** 1781–1783.
Thorburn, G. D., Nicol, D. H., Bassett, J. M., Shutt, D. A. and Cox, R. I. (1972). *J. Reprod. Fert. Suppl.* **16,** 61–84.
Troop, R. C. (1959). *Endocrinology* **64,** 671–675.
Tsyrlina, E., Chemama, R., Ennuyer, A. and Jayle, M. (1970). *Path. Biol., Paris* **18,** 669–672.
Uotila, U. U. (1940). *Anat. Rec.,* **76,** 183–203.
Urquhart, J., Yates, F. E. and Herbst, A. L. (1959). *Endocrinology* **64,** 816–830.
Van der Gugten, A. A., Sala, M. and Kwa, H. G. (1970). *Acta endocr., Copnh.* **64,** 265–272.
Varon, H. H., Touchstone, J. C. and Christian, J. J. (1966). *Acta endocr., Copnh.* **51,** 488–496.
Velican, C. (1946/47). *Arch. Anat. micr.,* **36,** 316–333.
Velican, C. (1948). *Arch. Anat. micr.* **37,** 73–81.
Vihko, R. (1966). *Acta endocr., Copnh. Suppl.* **109,** 5–67.
Villee, D. B., Rotner, H., Kliman, B., Briefer, C. and Federman, D. D. (1962). *J. clin. Endocr. Metab.* **22,** 726–734.
Villee, D. B., Rotner, H., Kliman, B., Briefer, C. and Federman, D. D. (1967). *J. clin. Endocr. Metab.* **27,** 1112–1122.
Vinson, G. P. (1974a). *Gen comp. Endocr.* **22,** 268–276.
Vinson, G. P. (1974b). *Gen. comp. Endocr.* **22,** 398–399.
Vinson, G. P. and Chester Jones, I. (1963). *J. Endocr.* **26,** 407–414.
Vinson, G. P. and Chester Jones, I. (1964). *J. Endocr.* **29,** 185–191.
Vinson, G. P. and Phillips, J. G. (1976). Unpublished observations.
Vinson, G. P. and Renfree, M. B. (1975). *Gen. comp. Endocr.* **27,** 214–222.
Vinson, G. P., Phillips, J. G., Chester Jones, I. and Tsang, W. N. (1971). *J. Endocr.* **49,** 131–140.
Vinson, G. P., Bell, J. B. G. and Whitehouse, B. J. (1975). *J. Endocr.* **67,** 13P–14P.
Vinson, G. P., Bell, J. B. G. and Whitehouse, B. J. (1976). *J. steroid Biochem.* **7,** 407–411.
Von Euw, J. and Reichstein, T. (1941). *Helv. chim. Acta* **24,** 879–889.
Voogt, J. L., Sar, M. and Meites, J. (1969). *Am. J. Physiol.* **216,** 655–658.
Wagner, W. C. (1969). *J. Am. vet. med. Ass.* **154,** 1395.
Wang, D. Y., Bulbrook, R. D., Sneddon, A. and Hamilton, T. (1967). *J. Endocr.* **38,** 307–318.
Wang, D. Y., Bulbrook, R. D., Thomas, B. S. and Friedman, M. (1968). *J. Endocr.* **42,** 567–588.
Ward, I. L. (1972). *Science N.Y.* **175,** 82–84.
Ward, M. G. and Engel, L. L. (1966). *J. biol. Chem.* **241,** 3417–3453.
Ward, P. J. and Grant, J. K. (1963). *J. Endocr.* **26,** 139–147.
Warren, J. C. and Timberlake, C. E. (1962). *J. clin. Endocr. Metab.* **22,** 1148–1151.
Wassermann, G. F. and Eik-Nes, K. B. (1969). *Acta endocr., Copnh.* **61,** 33–47.

Weinheimer, B., Oertel, G. W., Leppler, W., Blaise, H. and Bette, L. (1966). *In* "Androgens" (A. Vermeulen, ed.). pp. 36–41. Excerpta Medica I.C.S. **101,** Amsterdam.

Weliky, I. and Engel, L. L. (1963). *J. biol. Chem.* **238,** 1302–1307.

Weiss, M. (1975). *Comp. Biochem. Physiol.* **50,** 211–213.

Weiss, M. and Richards, P. G. (1970). *J. Endocr.* **48,** 145–146.

Weisz, J. and Gunsalus, P. (1973). *Endocrinology* **93,** 1057–1065.

West, C. D., Damast, B. and Pearson, O. H. (1958). *J. clin. Invest.* **37,** 341–349.

Wettstein, A. and Anner, G. (1954). *Experientia,* **10,** 397–416.

Whitehouse, B. J. and Vinson, G. P. (1968). *Steroids* **11,** 245–264.

Wieland, R. G., Levy, R. P., Katz, D. and Hirschmann, H. (1963a). *Biochim. biophys. Acta* **78,** 566–568.

Wieland, R. G., Decourcy, C. and Hirschmann, H. (1963b). *Steroids* **2,** 61–70.

Wieland, R. G., Decourcy, C., Levy, R. P., Zala, A. P. and Hirschmann, H. (1965). *J. clin. Invest.* **44,** 159–168.

Wilroy, Jr., R. S. Camacho, A. M., Trouy, R. L. and Hagen, A. A. (1968). *Endocrinology* **83,** 56–60.

Witorsch, R. J. and Kitay, J. I. (1972). *Endocrinology* **90,** 1374–1379.

Wynn, V. and Door, J. W. (1966). *Lancet* **2,** 715–719.

Yates, F. E., Herbst, A. L. and Urquart, J. (1958). *Endocrinology* **63,** 887–902.

Yates, J. and Deshpande, N. (1974). *J. Endocr.* **60,** 27–35.

Yuen, B. H., Kelch, R. P. and Jaffe, R. B. (1974). *Acta endocr., Copnh.* **76,** 117–126.

Zalesky, M. (1934). *Anat. Rec.* **60,** 291–316.

Zieger, G., Lux, B. and Kubatsch, B. (1974). *Acta endocr., Copnh.* **75,** 550–560.

Zizine, L. (1970). *C. r. Séanc. Soc. Biol.* **164,** 2427–2429.

5. Pituitary–Adrenal System Hormones and Adaptive Behaviour

Béla Bohus and David de Wied

Rudolf Magnus Institute for Pharmacology Medical Faculty, University of Utrecht, Vondellaan 6, Utrecht, The Netherlands.

1. Introduction

The concept that the pituitary–adrenal system subserves physiological adaptation of the organism enjoys a history of almost forty years. The stress theory, which is the concept of the non-specificity of the pituitary–adrenal response to a variety of stimuli, was derived by Selye (1950) from observations that an increase in adrenal activity could be elicited by "nocuous" and threatening stimuli such as trauma, haemorrhage, cold, heat, exercise, infections or drugs.

Although Selye (1950) commented that ". . . even mere emotional stress" such as immobilization activates the pituitary–adrenal system, physical and chemical agents have long been regarded as the most important stressors. Observations on animals including man from the early nineteen-fifties resulted in the recognition that the activation of the pituitary–adrenal cortex could be elicited by psychological stimuli as well. Almost 200 publications, reviewed by Mason (1968), indicated that emotional stressors like anxiety, fear and rage are potent activators of the pituitary–adrenal system. Besides stimuli with an aversive character it became apparent that those signalling hope or disappointment could also activate the system. Levine et al. (1972) showed that changes in expectancy during well-established behaviour such as a shift in reinforcement or the withdrawal of reward—are followed by pituitary–adrenal activation.

In parallel with the development of psychoneuroendocrine research concerned with psychological influences on the pituitary–adrenal system, the brain as a target organ of its hormones received more and more attention. As early as 1935, Liddel et al. reported that an extract of the adrenal cortex influences experimental neurosis in sheep. Anderson (1941) showed that hypophysectomy was followed by an inhibition of conditioned salivation and the flexor reflex in the dog. In 1942, evidence of the action of corticosteroids on the central nervous system (CNS) was obtained by reports of Engel and Margolin (1942) and of Thorn and associates (Hoffman et al., 1942) on the altered electroencephalogram (EEG) in Addison's disease and on psychological changes which were sometimes found in that condition. In 1950, Klein and Livingston reported a dramatic improvement in the EEG of epileptic patients during treatment with ACTH. Around 1952, ACTH and cortisone stirred therapeutic hopes and renewed interest in the central influence of pituitary–adrenal hormones. In 1953, Mirsky et al. reported the first extensive experiments on the effect of ACTH on conditioned responses in monkeys and rats. These authors suggested that "adrenocortical hypersecretion may influence the organism . . . either to decrease the effectiveness of an anxiety-producing situation or to eliminate a poorly integrated defence against the anxiety produced by the persistence of a traumatic memory". This conclusion was derived from experiments in which monkeys were trained to press a bar in order to obtain food reward. Bar-pressing was also followed by a tone. Then the monkeys were subjected to fear-conditioning in another situation where the same tone served as the conditioned stimulus. Reinstatement of the original bar-pressing situation led to a rapid and efficient bar-pressing in animals which received ACTH treatment during fear conditioning. Control animals displayed strong "startle" reactions in response to the tone and their bar-pressing behaviour appeared to be impaired. In another experiment they found that ACTH treatment during extinction of an operant avoidance response resulted in a rapid decrease in avoidance performance.

These observations of Mirsky *et al.* (1953) seemed to initiate a whole new area for investigation but unfortunately only a few studies have ensued. Woodbury's (1954) observations on ACTH and corticosteroid effects on brain excitability in the rat, Applezweig's and his associates' (1955, 1959) reports on avoidance behaviour of hypophysectomized rats and studies by Lissák *et al.* (1957) on the effect of ACTH on passive avoidance behaviour of dogs in classical Pavlovian situations may mainly be considered as "reminders" of Mirsky's original studies.

In this paper we attempt to outline the developments of researches on hormones of the pituitary–adrenal system and adaptive behaviour. Such a survey cannot give a detailed description of the experiments although important experimental variables in the effect of these hormones on behaviour will be emphasized. Those who are interested in the details of the problems are referred to the original publications and a number of excellent symposium proceedings, monographs and reviews (de Wied and Weijnen, 1970; Di Giusto *et al.*, 1971; Brain, 1972a; Endröczi, 1972; Levine, 1972; Davidson and Levine, 1972; Smith, 1973; Zimmermann *et al.*, 1973; Gispen *et al.*, 1975a).

2. Behavioural effects of pituitary–adrenal ablation

A. ADRENALECTOMY

Removal of the adrenals does not substantially affect acquisition of conditioned avoidance responses in rats maintained with sodium chloride in their drinking water or by the administration of mineralocorticosteroids such as deoxycorticosterone (Fuller *et al.*, 1956; Moyer, 1958; Moyer and Moshein, 1963; Bohus and Endröczi, 1965; de Wied, 1967; van Delft, 1970; Weiss *et al.*, 1970). A more or less normal salt balance seems necessary to secure physical fitness of the animals in order to execute a conditioned behavioural response. Bohus and Lissák (1968) showed that adrenalectomized rats without maintenance therapy have difficulty in acquiring a one-way active avoidance response (jump on a bench) when the time allotted to execute the conditioned response is 10 sec. Such animals need more time than sham-operated controls to perform the response. Adrenalectomized rats in salt balance may even perform better, as has been reported by Paul and Havlena (1962) in rewarded and by Beatty *et al.* (1970) in avoidance situations. Paul and Havlena (1962) found that adrenalectomized rats maintained on water with NaCl made less errors during the early phases of maze learning than did their controls. They suggested that adrenalectomized animals had less irrelevant fear-motivated responses which interfered with food-seeking behaviour. Beatty *et al.* (1970) reported that adrenalectomy attenuates the deleterious effects of high intensity shock punishment during shuttle-box avoidance acquisition. Since the absence of corticosteroids in adrenalectomized rats results in an increased synthesis and release of ACTH (Hodges and Vernikos–Danellis, 1962) it seemed probable that the observed effects in this later experiment were due to increased ACTH levels rather than to the absence of corticosteroids. Dexamethasone or corticosterone in doses which suppress ACTH release in adrenalectomized rats (Hodges and Jones, 1964) marginally facilitated acquisition of a pole-jumping avoidance response in adrenalectomized rats (van Delft, 1970).

A number of attempts have been made to relate the relationship between performance of an incompletely learned avoidance response and the interval between initial and subsequent training trials to the function of the pituitary–adrenal system. Avoidance performance during relearning of an incompletely acquired shuttle-box avoidance response is the function of the interval between original training and the subsequent relearning session and shows an inverted U-shaped relationship. Relearning performance is high at short (few minutes) and long (several hours) intersession intervals but poor at intermediate intervals (Kamin-effect; Kamin, 1963). Brush (1962) showed that one hour after the original training the avoidance performance is very poor. Subsequently, Brush and Levine (1966) found that plasma corticosterone level after the original learning is correlated with the performance during relearning in the shuttlebox. When relearning is rapid, markedly elevated plasma corticosterone levels were found, while suppressed relearning was correlated with low plasma corticosterone levels. Levine and Brush (1967) further reported that the Kamin-effect is prevented by administration of ACTH or cortisol but not by corticosterone prior to relearning. The observations of Klein (1972) who showed that water stress, lateral anterior hypothalamic stimulation or direct implantation of ACTH in the lateral anterior hypothalamus abolish the Kamin-effect seemed to corroborate the view that the intermediate performance deficit is related to the function of the neuroendocrine system. Subsequent studies demonstrated that removal of the adrenals or the pituitary gland does not affect the Kamin-effect (Marquis and Suboski, 1969; Suboski et al., 1970; Barrett et al., 1971; Klein and Kopish, 1975). Since the performance deficit at intermediate intervals seems to be related to a hypo- rather than a hyperfunction of the pituitary–adrenal system, these studies, although meant to disprove the significance of the neuroendocrine system, substantiate the theory that temporary blockade of pituitary–adrenal system functions may be the underlying cause of the Kamin-effect.

Adrenalectomy interferes with the maintenance of avoidance behaviour in the absence of shock punishment. Adrenalectomized rats are resistant to extinction in active (de Wied, 1967; Bohus et al., 1968; Weiss et al., 1970; Silva, 1974) and passive avoidance behaviour (Bohus, 1974a). Retention of passive avoidance responses appeared to be facilitated in rats adrenalectomized prior to the learning trial (Weiss et al., 1970; Silva, 1973). Because the removal of the adrenals results in an approximately ten-fold increase in circulating ACTH (Hodges and Vernikos–Danellis, 1962; Dallman et al., 1974), it has been suggested that the effects of adrenalectomy on conditioned behaviour are due to the presence of supraphysiological amounts of endogenous ACTH rather than the absence of corticosteroids. Treatment with corticosterone which normalizes the level of circulating ACTH in adrenalectomized rats also similarly affects active avoidance extinction (Weiss et al., 1970).

Adrenalectomy does not influence open field exploratory activity and its rate of habituation (Paul and Havlena, 1962; Davis and Zolovick, 1972; Tamásy et al., 1973), but increases defecation in the open field (Moyer, 1958; Joffe et al., 1972a). Aggressive behaviour is markedly affected by the removal of adrenals. Adrenalectomized, isolated, mice develop aggressive behaviour but they require a longer period of isolation than their controls (Sigg et al., 1966). Attenuation of isolation-induced aggression in adrenalectomized mice was reported by Brain et al. (1971), Harding and Leshner (1972) and by Leshner et al. (1973). Attentuation of aggression is independent of the gonadal system. Leshner et al. (1973) found that aggressive behaviour of adrenalectomized, gonadectomized, isolated mice is not affected by testosterone given as replacement therapy. Furthermore, adrenalectomy in adulthood, even with testosterone administration, reduces aggressiveness of neonatally androgenized, isolated female mice (Leshner and Johnson, 1974). Burge and Edwards (1971) failed to observe any difference in the aggressiveness of isolated, adrenalectomized mice. Comparisons between the various studies are however difficult because of the differences in the measurement of aggression and the use of different strains of mice. Levine and Treiman (1964) reported a differential plasma corticosterone response to stress in different strains of mice. In addition, the behavioural response and effectiveness of hormone treatment in mice is also strain-dependent (Levine and Levin, 1970).

Observations on adrenalectomized rats indicate that an intact adrenal function is not essential in acquiring conditioned behaviour. Under certain conditions adrenalectomy may improve acquisition behaviour. Extinction of conditioned avoidance behaviour in rats or development of isolation-induced aggressive behaviour in mice, however, is markedly reduced in adrenalectomized animals. Since the removal of the adrenals is followed by an increase in pituitary ACTH release, it is difficult to conclude whether the behavioural changes are caused by the absence of corticosteroids or by increased endogenous ACTH levels. The development of supraphysiological release of ACTH in adrenalectomized rats requires a lag period of 1–2 days (Buckingham and Hodges, 1974). Thus, by shortening the interval between the removal of the adrenals and the extinction procedure, the influence of ACTH and/or corticosteroids on extinction can be studied (Bohus, 1974a). Extinction of a one-trial learning passive avoidance response by forced exposure to the shock compartment appeared to be prevented by the prior removal of the adrenals for one hour. Administration of a relatively small dose of corticosterone immediately after adrenalectomy normalized extinction behaviour of adrenalectomized rats. Adrenalectomy immediately after exposure to the shock compartment failed to affect extinction behaviour, i.e. extinction of the passive avoidance response occurred in the same way as in sham-operated control animals. These observations indicate that an intact pituitary–adrenal system function during forced extinction is necessary to extinguish passive avoidance behaviour. Plasma ACTH levels one hour after adrenalectomy are similar to those after sham-adrenalectomy, while corticosterone disappears from the blood (Buckingham and Hodges, 1974). Accordingly, the absence of cortico-sterone rather than the presence of augmented ACTH levels may be the cause of the impaired extinction behaviour of adrenalectomized rats.

B. HYPOPHYSECTOMY

Serious impairment of the acquisition of conditioned avoidance behaviour follows the removal of the pituitary gland. Applezweig and

Baudry (1955) and Applezweig and Moeller (1959) performed experiments on a few rats and found that hypophysectomy reduces the ability to acquire a conditioned shuttle-box avoidance response. Bélanger (1958) using a longer conditioned–unconditioned stimulus interval could not confirm this observation. Subsequent experiments on adenohypophysectomized (de Wied, 1964) or totally hypophysectomized rats (de Wied, 1968; Bohus *et al.*, 1973), however, clearly demonstrated that such animals are substantially inferior in acquiring conditioned avoidance behaviour than their sham-operated controls. Hypophysectomy also appeared to attenuate passive avoidance behaviour (Anderson *et al.*, 1968; Weiss *et al.*, 1970). Impaired one-trial passive avoidance behaviour has been reported by Lissák and Bohus (1972) in the rat when weak or moderate shock punishment was used. Hypophysectomized rats, however, exhibit passive avoidance behaviour almost indistinguishable from that of controls after very high shock intensity punishment. They suggested that fear-motivation rather than learning is impaired in the absence of the pituitary.

Observations in tests other than fear-motivation are scarce and less conclusive. Stone and King (1954) reported that hypophysectomy at an age of 40 days does not substantially affect learning in a relatively simple maze. Hypophysectomized rats were found to be equal or only slightly inferior to their controls of the same age. In a subsequent study, Stone and Obias (1955) showed that rats hypophysectomized at the ages of 15, 30 and 35 days were significantly inferior to matched controls in a 13-choice swimming maze but then only during the second half of the trial series. Rats hypophysectomized at an age of 35 days were later tested on a five-unit light discrimination problem with food reward. Discrimination behaviour of the operated rats did not differ from that of sham-operated controls. Accordingly, these studies did not indicate that the learning ability of hypophysectomized rats is seriously impaired. Slightly inferior performance in the maze might also be due to motivational disturbances as suggested by Stone and Obias (1955).

Since hypophysectomy results in multiple metabolic deficiencies and consequently in physical weakness, it is reasonable to suppose that these factors may be the cause of the deficient avoidance behaviour of hypophysectomized rats. Indeed, substitution therapy of adenohypophysectomized rats with thyroxine, cortisone and testosterone normalized impaired sensory and/or motor function as studied in a straight runway under continuous shock punishment, and improved avoidance acquisition in a shuttle-box (de Wied, 1964; 1971). Due to

its anabolic effect, testosterone alone also resulted in an improvement
of avoidance behaviour but failed to normalize completely the behavioural deficiency of hypophysectomized rats. Growth hormone
treatment improved body growth and appeared to facilitate avoidance
acquisition (de Wied, 1969). A special diet which insured good health
of the hypophysectomized rats resulted in a similar improvement in
conditioned avoidance behaviour (Harris, 1973). These studies therefore suggest that the metabolic disturbances in hypophysectomized
rats may play an important role in impaired performance in active
avoidance situations. That the physical condition is not the primary
cause of the behavioural deficit of the hypophysectomized rats is
suggested by studies on passive avoidance behaviour. This behaviour
is not associated with an increase but a reduction in motor behaviour.
Accordingly, physical weakness and consequent impaired motor
behaviour would result in superior rather than inferior passive
avoidance behaviour. The latter, however, was observed in
hypophysectomized rats (Weiss et $al.$, 1970; Lissák and Bohus,
1972).

Deficiency in the pituitary–adrenal system has been considered as
the underlying cause of the behavioural impairment of hypophysectomized rats. Treatment with adrenal maintenance doses of ACTH
restored avoidance acquisition behaviour of adenohypophysectomized (de Wied, 1964) and hypophysectomized rats (de Wied, 1969;
Weiss et $al.$, 1970) almost to a normal level. These observations
reinforced the notion of Applezweig and Baudry (1955) who, but only
in two hypophysectomized rats, showed some improvement of avoidance acquisition after ACTH treatment. Although the administration
of ACTH restored adrenal function of hypophysectomized rats, the
absence of corticosteroids cannot account for their behavioural deficiency. As has been described, adrenalectomy does not impair
avoidance behaviour. Furthermore, treatment of hypophysectomized
rats with the potent synthetic glucocorticosteroid dexamethasone
failed to improve their avoidance acquisition (de Wied, 1971).
Accordingly, the influence of ACTH on avoidance behaviour of
hypophysectomized rats seemed to be due to an extra-adrenal effect of
the hormone, an effect presumably located in the central nervous
system. Experiments with peptides structurally related to ACTH but
practically devoid of corticotrophic activity substantiated this hypothesis. Treatment of hypophysectomized rats with the ACTH fragments
$ACTH_{1-10}$ or $ACTH_{4-10}$ or with α-MSH which shares the amino
acid sequence 1–13 with ACTH, restored the acquisition of conditioned avoidance behaviour in hypophysectomized rats (de Wied,

1969). The behavioural effect of ACTH fragments could not be due to a metabolic influence of these peptides. $ACTH_{4-10}$ failed to affect body growth, adrenal and testis weights, blood glucose, circulating insulin and free fatty acid levels in the plasma in amounts which normalize avoidance acquisition of hypophysectomized rats. Motor and/or sensory capacities of hypophysectomized rats as studied in a straight runway under continuous shock punishment were only partially restored by $ACTH_{4-10}$ (de Wied, 1969). Deficiences of hypophysectomized rats in conditioned avoidance behaviour may be linked to some extent with metabolic dysfunctions and physical weakness. The studies with metabolically rather inactive peptides suggest that the absence of ACTH and/or related pituitary peptides may be the primary cause of the behavioural impairment in hypophysectomized rats.

The experiments described above exclusively used a long-acting Zn-phosphate preparation of ACTH and ACTH fragments given every other day during the training period. This means that due to the slow absorption of the peptide from a subcutaneous depot, ACTH or fragments were continuously supplied and therefore expected to exert a tonic influence on the central nervous system. Further observations subsequently showed that the presence of the peptides in the organisms, during the daily avoidance sessions only, is a sufficient condition to normalize avoidance behaviour of hypophysectomized rats. Administration of $ACTH_{4-10}$ as a short-acting preparation, dissolved in saline, prior to each daily session, resulted in a normalization of shuttlebox avoidance acquisition in hypophysectomized rats (Bohus *et al.*, 1973). Termination of the treatment after the 7th day of training when $ACTH_{4-10}$-treated rats displayed an average of 80% conditioned avoidance performance led to a progressive decrease in avoidance behaviour despite continued punishment if the rat failed to respond to the conditioned stimulus. Accordingly, the behavioural effect of ACTH-like peptides is of a short term nature which does not extend beyond the actual presence of the peptide in the body. Improvement of avoidance behaviour of hypophysectomized rats by $ACTH_{4-10}$ is probably due to a temporary restoration of fear motivation.

Hypophysectomized rats maintained on a supplementary therapy of thyroxine, cortisone and testosterone are able to acquire conditioned avoidance behaviour but the maintenance of the learned response is impaired. Rapid extinction of the conditioned behaviour took place in these rats (de Wied, 1967). Accordingly, the absence of pituitary peptides, such as ACTH, interferes with the maintenance of learned behaviour. This was confirmed by Weiss *et al.* (1970) who showed that impaired maintenance of avoidance behaviour of hypophysectomized rats can be restored by ACTH treatment.

Behavioural changes of hypophysectomized rats in non-conditioned situations have also been described. Adenohypophysectomized rats are inferior in running speed during escape from shock in a straight runway. ACTH treatment normalizes running speed in a dose-dependent manner (de Wied, 1964) but $ACTH_{4-10}$ failed to restore this behaviour completely in totally hypophysectomized rats (de Wied, 1969). The behavioural response to electric footshock (flinch, jerk, jump and run) appeared even at lower shock intensities in hypophysectomized rats than in sham-operated controls. $ACTH_{1-10}$ failed to modify the threshold in hypophysectomized rats (Gispen *et al.*, 1970). This indicates that the avoidance acquisition deficit is not due to an impaired shock perception and that the normalization of avoidance behaviour by ACTH-like peptides is not mediated by an altered pain perception. Gibbs *et al.* (1973), on the other hand, failed to find differences between hypophysectomized rats and weight-matched controls in a shock responsiveness test. Hypophysectomy led to an increase in exploratory behaviour in solitary situations but failed to affect other behavioural categories such as maintenance, play or fear (Gispen *et al.*, 1973). This indicates that hypophysectomized rats do not possess a disoriented behaviour which might cause the learning deficit. Phillips and Shapiro (1973) reported that lateral hypothalamic or ventral tegmentic self-stimulation is not affected by hypophysectomy. They suggested that the pituitary does not subserve brain stimulation reward.

In conclusion, from the observations on the behavioural consequences of the removal of the anterior pituitary or the whole gland, it is obvious that the absence of pituitary hormone(s) leads to impairment in acquisition and maintenance of fear-motivated behaviour. Although metabolic deficiencies which follow hypophysectomy may also be involved, the absence of ACTH and/or ACTH-like pituitary peptides seems to be of primary importance. On the basis of these observations, de Wied (1969) suggested that the pituitary may manufacture peptides with neurogenic activities (neuropeptides) which may normally operate in the formation and maintenance of conditioned and other adaptive behavioural responses. These neuropeptides may be derived from pituitary hormones (Greven and de Wied, 1973) which may act as prohormones for these entities.

3. Pituitary–adrenal system hormone effects on the behaviour of intact
    animals

A. EFFECTS OF ACTH AND ACTH FRAGMENTS

Mirsky *et al.* (1953) reported that administration of ACTH during the
acquisition of a wheel-turning escape response in the rat diminished
the number of responses. They postulated that ACTH diminishes the
effectiveness of an anxiety-producing stimulus. Subsequent experiments were, however, at variance with this suggestion. Murphy and
Miller (1955) did not observe differences in the acquisition of a
shuttle-box active avoidance response in ACTH-treated and control
rats but the former group appeared to be more resistant to extinction.
Similarly, ACTH treatment failed to affect acquisition behaviour of
posterior lobectomized rats but again the treatment during the
acquisition period appeared to inhibit rapid extinction of the conditioned avoidance response which follows the removal of the posterior lobe of the pituitary (de Wied, 1965). The rate of extinction of
an avoidance response is a function of acquisition. The more effective
the acquisition, the more resistant the response to extinction. This
indicates that ACTH could have affected acquisition. Probably due to
the rapid acquisition of the response an effect may not be detectable.
Indeed, further observations demonstrated that ACTH facilitates
acquisition of a fear-motivated response when the tendency to respond is low. Facilitation of conditioned avoidance reflex activity by
ACTH treatment was shown by Bohus and Endröczi (1965) during
the early phase of conditioning when the avoidance tendency was low.
Such facilitation did not occur, however, if the rats had not scored at
least one avoidance response. This indicated that a certain level of
conditioning has to be developed before ACTH becomes effective. A
reduction of intertrial goal-directed activity was also observed in
ACTH-treated rats. Since intertrial activity is related to the level of
fear motivation (Brush, 1962), it seemed that ACTH had a negative
effect on motivation *per se*. Observations on adrenalectomized rats
maintained with deoxycorticosterone acetate, however, revealed normal avoidance behaviour, but no reduction in goal-directed intertrial
activity which normally occurs in the later phases of conditioning
(Bohus and Endröczi, 1965). Thus, the effect of ACTH on intertrial
activity is probably mediated through the adrenal cortex.

ACTH-affected avoidance behaviour by activation of the adrenal
cortex seemed to be the most obvious postulate from a classical

endocrinological point of view. Observations on adrenalectomized and hypophysectomized rats, however, indicated that the absence of ACTH rather than that of the corticosteroids is deleterious for avoidance behaviour. That ACTH is capable of influencing behaviour, without involvement of the adrenal cortex, was first demonstrated by Miller and Ogawa (1962). They showed that ACTH treatment during the acquisition period in adrenalectomized rats increased resistance to extinction in the same way as previously shown for intact rats (Murphy and Miller, 1955). Moreover, Bohus *et al.* (1968) also repeated their former experiments and showed that ACTH facilitates avoidance performance in the early phase of conditioning in adrenalectomized rats. Intertrial goal-directed activity was not affected by ACTH in the absence of the adrenal cortex. Therefore, it appeared that facilitation of avoidance performance by ACTH is of extra-adrenal nature but reduction of intertrial activity, which may be a measure of generalized fear, requires an intact adrenal cortex.

Subsequent observations at the end of the 1960's reinforced the idea that ACTH facilitates avoidance acquisition and a number of variables determining these behavioural influences of ACTH had also been explored. Beatty *et al.* (1970) showed that ACTH facilitates shuttle-box avoidance acquisition except at high shock intensity punishment when the tendency to respond is low. Facilitation of a Y-maze conditioned avoidance response by ACTH was demonstrated by Ley and Corson (1971). They found that the effect of the treatment depends upon the intensity of the shock and the activity of the rats. ACTH facilitated avoidance behaviour in hypoactive rats at a high shock intensity and in hyperactive rats at a low shock intensity. The number of errors made by the rats in a brightness discrimination paradigm appeared to be higher in hypoactive rats but lower in hyperactive rats treated with ACTH. These authors suggested that the sex of the rats and the diurnal rhythm are additionally important variables which determine the behavioural effects of ACTH (Ley and Corson, 1972). ACTH appeared to be effective in facilitating performance during massed trial avoidance acquisition in the shuttle-box in male rats, but only at the trough of the diurnal adrenal rhythm (Pagano and Loveley, 1972).

Extinction of active avoidance responses appeared to be much more sensitive to treatment with ACTH than acquisition behaviour. Although Murphy and Miller (1955) found that administration of the peptide during the acquisition period resulted in an increased resistance to extinction, more pronounced effects of ACTH on both shuttle-box two-way (de Wied, 1967) and platform jumping one-way avoidance extinction (Bohus *et al.*, 1968) were observed when the peptide was administered throughout the extinction period. Delay of extinction as observed after ACTH administration depended upon the dose of the peptide. Long term administration of ACTH obviously results in hypercorticism. The effect of ACTH on extinction behaviour is, however, not mediated through the adrenal cortex. The extra-adrenal behavioural effect of ACTH was suggested by observations in which ACTH also delayed extinction of avoidance behaviour in adrenalectomized rats (Miller and Ogawa, 1962; Bohus *et al.*, 1968).

That the influence of ACTH is of an extra-adrenal nature and that the behaviourally active moiety of the whole ACTH molecule is located in a small peptide fragment was shown by a large number of observations starting about 1964. Although there were already a few indications that several biological activities of the ACTH molecule— such as that adrenocorticotrophic, melanocyte stimulating, lipolytic effects do not require the whole peptide chain (Hofman *et al.*, 1960; Engel, 1961; Lebovitz, 1967)—observations that fragments of an otherwise biologically active peptide molecule bear neurotropic activity opened a new research trend in neuroendocrine and neurobiological research. It was first observed that synthetic α-MSH and purified β-MSH or even smaller ACTH fragments (ACTHβ_{1-10}) were as active as ACTH in delaying extinction of a pole jumping and shuttle-box conditioned avoidance response (de Wied, 1966; Bohus and de Wied, 1966). ACTH$_{11-24}$ was ineffective (de Wied, 1966). Subsequent experiments then demonstrated that the fragment ACTH$_{4-10}$ possesses full behavioural activity that leads to a delay of extinction of active avoidance behaviour (Greven and de Wied, 1967). Therefore, this heptapeptide was regarded as containing the minimal requirements for the behavioural effect of ACTH-like peptides (de Wied, 1969). Further structure–activity studies, however, revealed that the sequence ACTH$_{4-7}$ (H-Met-Glu-His-Phe-OH) contains the essential elements for the behavioural effect of ACTH analogues. Using a pole jumping avoidance test, it was shown that shortening the peptide ACTH$_{1-10}$ from the amino end up to ACTH$_{4-10}$ did not influence the behavioural activity as measured on the delay of extinction. Shortening of the peptide ACTH$_{4-10}$ from the carboxyl end gave no marked changes in the behavioural effect until the amino acid residue at position 7 was removed (Greven and de Wied, 1973). These studies also indicated that the peptide core essential for the behavioural activity is different from that required for MSH-activity.

Hofman *et al.* (1960) demonstrated that tryptophan in position 9 is essential for MSH-activity because $ACTH_{1-8}$ has lost this biological effect. MSH-activity seems to require the sequence $ACTH_{6-9}$ (Otsuka and Inouye, 1964). A key role which the amino acid residue phenylalanine at position 7 plays in the mediation of the behavioural effect of ACTH-like peptides was first demonstrated using an analogue of $ACTH_{1-10}$ in which phenylalanine in position 7 was replaced by its D-enantiomer. In contrast to $ACTH_{1-10}$ "all-L" which delays the extinction of a shuttle-box conditioned avoidance response, [D-Phe^7] $ACTH_{1-10}$ administration led to a facilitation of extinction (Bohus and de Wied, 1966). Subsequent observations then revealed that the heptapeptide [D-Phe^7] $ACTH_{4-10}$ and the tetrapeptide [D-Phe^7] $ACTH_{4-7}$ are as active as the decapeptide [D-Phe^7] $ACTH_{1-10}$ D-isomer (Greven and de Wied, 1967; de Wied *et al.*, 1975a). As in intact rats, administration of [D-Phe^7] $ACTH_{1-10}$ led to rapid extinction of a conditioned shuttle-box avoidance response in rats hypophysectomized prior to the extinction training (Bohus and de Wied, 1966). Thus the behavioural effect of the D-analogue is not due to a competitive inhibition of the effect of ACTH or ACTH-like peptides of pituitary origin. Replacement of other amino acids in the sequence [Lys^8]-$ACTH_{4-9}$ by D-enantiomers does not induce facilitation of extinction. Instead, these peptides delay extinction of the avoidance response as do "all-L" peptides derived from ACTH. Such substitutions potentiate the inhibitory effect of the peptide on extinction of a pole jumping avoidance response (Greven and de Wied, 1973; de Wied *et al.*, 1975a). This potentiation may be due to an enhanced resistance to proteolytic breakdown of the D-isomer peptides (Witter *et al.*, 1975). Substitution of certain amino acids of the active sequence may potentiate the behavioural effect, presumably due to increasing the resistance against breakdown by proteolytic enzymes. Greven and de Wied (1973) reported that the peptide sequence [Met^4 (O)-D-Lys^8-Phe^9] $ACTH_{4-9}$ is a thousandfold more potent than $ACTH_{4-10}$ and subcutaneous administration of this peptide in nanogram quantities delays extinction of a pole-jumping avoidance response (Table I).

That phenylalanine in position 7 is of major importance in the behavioural action of ACTH is also suggested by observations showing that the tripeptide $ACTH_{7-9}$ (de Wied *et al.*, 1975a) and dogfish α- and β-MSH (van Wimersma Greidanus *et al.*, 1975) which contain this tripeptide (Lowry and Chadwick, 1970; Bennett *et al.*, 1974) inhibit avoidance extinction. The same "message" as in $ACTH_{4-10}$ may be conveyed by the sequence 7–9. This becomes

TABLE I

Behaviourally active amino acid sequence 4–10 of ACTH which is contained in the molecules α- and β-MSH and β-LPH (porcine)

ACTH (C-terminal portion)

39	38	37	36	35	34	33	32	31	30	29	28	27	26	25	24	23	22	21	20	19
PHE	-GLU	-LEU	-PRO	-PHE	-ALA	-GLU	-ALA	-LEU	-ASP	-ASP	-GLU	-ALA	-GLY	-ASP	-PRO	-TYR	-VAL	-LYS	-VAL	-PRO

18	17	16	15
ARG	ARG	LYS	LYS

α-MSH

1	2	3	[4	5	6	7	8	9	10]	11	12	13	14	15
SER	-TYR	-SER	-MET	-GLU	-HIS	-PHE	-ARG	-TRY	-GLY	-LYS	-PRO	-VAL	-GLY	-LYS

β-MSH

1	2	3	4	5	6	7	8	9	10	[11	12	13	14	15	16	17]	18	19	20	21	22
ALA	-GLU	-LYS	-ASP	-GLY	-PRO	-TYR	-ARG			-MET	-GLU	-HIS	-PHE	-ARG	-TRY	-GLY	-SER	-PRO	-PRO	-LYS	-ASP

β-LPH

1	2	3	4	43	44	45	46	[47	48	49	50	51	52	53]	54	55	91
GLU	-LEU	-GLY	-THR	-GLY	-PRO	-TYR	-LYS	-MET	-GLU	-HIS	-PHE	-ARG	-TRY	-GLY	-SER	-PRO	GLU

The boxed region (positions 4–10 of ACTH): MET-GLU-HIS-PHE-ARG-TRY-GLY is common to all the molecules.

expressed only after chain elongation. In fact, $ACTH_{7-16}$ is as potent as $ACTH_{4-7}$ in delaying extinction of the pole jumping avoidance response (de Wied et al., 1975a). Since both sequences share the residue phenylalanine[7], this amino acid may be considered as a key for the behavioural effect. Final conclusions regarding the topochemical requirements have to await data on in vitro interaction between peptides and their putative receptors.

The observations on conditioned avoidance behaviour of rats clearly indicate the influence of ACTH and related peptides on both acquisition and extinction processes. The behaviour of rats during active avoidance acquisition is accompanied by a high level of arousal and pituitary-adrenal system activation, and the tendency to respond is rather high. These factors may account for the relatively insensitive state of the organism for exogenously supplied ACTH or ACTH-like peptides. Extinction behaviour, however, is often affected by these peptides. The behaviour of animals during extinction of conditioned avoidance behaviour when shock punishment is omitted is more labile. This allows more "space" for modulating effects on behaviour. Avoidance behaviour during extinction is determined by the retrieval of the acquired fear-motivation and by active inhibitory processes which develop in the absence of reinforcement. ACTH and related peptides maintain fear-motivated responses presumably due to preserving the motivational value of environmental stimuli. It is worth mentioning that the influence of ACTH and related peptides is of a short term nature and does not extend beyond the actual presence of exogenously administered peptide in the organism. Discontinuation of the administration of ACTH during extinction sessions leads to rapid extinction of the conditioned avoidance response (Bohus et al., 1968). Administration of a single dose of $ACTH_{1-10}$ after the first extinction session delays extinction of the pole-jumping avoidance response during subsequent extinction sessions for 4–6 h (van Wimersma Griedanus and de Wied, 1971). Facilitation of passive avoidance learning and retention by ACTH and related peptides has also been demonstrated. Levine and Jones (1965) reported that administration of ACTH enhances the acquisition of passive avoidance behaviour. The passive avoidance was evoked by inhibition of a previously stabilized bar-press response for water following the presentation of a punishing electric shock. The treatment was more effective when given during both the acquisition and the retention period than when given during acquisition only. In a subsequent experiment Guth et al. (1971a) analysed the effect of ACTH treatment in a three-stage thirst-versus-fear conflict situation. Facilitation of passive avoidance behaviour was observed in ACTH-treated rats when the peptide was given at any one stage—approach training, avoidance training or retention test, respectively. Rats receiving the peptide during both avoidance training and retention testing did not show facilitated passive avoidance behaviour. Pappas and Gray (1971) found that ACTH increases the latency to resume 200 licks for water in thirsty rats after shock punishment within a single session. Lissák and Bohus (1972) demonstrated that administration of $ACTH_{1-24}$ as a long acting preparation prior to the single learning trial increased passive avoidance latencies in a situation where the innate dark preference of the rat was punished by electric footshock. The magnitude of the effect of the peptide appeared to depend upon the intensity of the punishment: the lower the shock intensity, the more effective the treatment. Schneider et al. (1974) found that ACTH improves conditioned suppression of a drinking response when the peptide is given before training but not when administered before the retention test, or if the rats were conditioned in the morning at the trough of the circadian cycle of pituitary–adrenal activity. Facilitated passive avoidance behaviour in a one-trial learning "step-out" situation was observed by Gray (1975) in rats receiving ACTH both during training and testing. Decreased passive avoidance was seen, on the other hand, in rats which received ACTH during testing only. He concluded that the experiment demonstrated a state-dependent learning under the influence of ACTH.

ACTH-like peptides given prior to the retention test of a one-trial learning "step-through" situation facilitate passive avoidance behaviour. As in active avoidance situations, fragments of ACTH such as $ACTH_{1-10}$ $ACTH_{4-10}$ or $ACTH_{4-7}$ increase passive avoidance latencies in rats (Greven and de Wied, 1973). This effect is of a short term nature. Facilitation of passive avoidance behaviour is seen during the test session 1 h after a single injection of ACTH-like peptides. Avoidance latency, however, 24 h after the injection, does not differ significantly from that of saline treated control rats. Interestingly, administration of $[D-Phe^7] ACTH_{4-10}$, which in active avoidance situations acts opposite to "all-L" peptides, facilitates passive avoidance behaviour (Greven and de Wied, 1973). The behavioural effect of this peptide, however, is of a considerably longer duration. The mechanism through which $[D-Phe^7] ACTH_{4-10}$ and the "all-L" peptides exert their influence on passive avoidance behaviour is probably different. Both specific response suppression and generalized emotional arousal may lead to passive avoidance behaviour. Rigter and Popping (1976) also found the same effects of

$ACTH_{4-10}$ and D-Phe[7] $ACTH_{4-10}$ on passive avoidance behaviour of the rat in a situation in which the adverse stimulus was not a footshock. Both peptides delayed extinction of a conditioned taste aversion response induced by a single administration of sugar dissolved in water plus LiCl which made the animal sick.

Passive avoidance behaviour is an aversively motivated response which always involves conflict between approach and avoidance or avoidance and avoidance tendencies. The aversive stimulus at the learning trial leads to an inhibition of the previously acquired approach response motivated by thirst or hunger. Inhibition of the explorative tendency or of a genetically determined dark preference response in the rat (step-out or step-through situations) is also used to induce passive avoidance behaviour. The advantage of a passive avoidance task is that the aversive motivation level may be well controlled and that the performance measures are more representative of learning differences, since differences in motor capacity or general activity play a relatively minor role. Therefore, ACTH-induced facilitation of both active and passive avoidance behaviour suggests a common mechanism(s) which is involved in both these behaviours, probably fear motivation. Although most studies report facilitation of passive avoidance behaviour by ACTH and related peptides, there is no agreement on the most effective, if any, treatment schedules. First, the whole ACTH molecule was used in most of the studies and frequently the peptide was given as a long-acting preparation. Therefore, an influence mediated through the adrenal cortex might have interacted with the intrinsic behavioural effect of the peptide. Secondly, ACTH was sometimes given before the passive avoidance learning period, i.e. during approach training. Therefore, ACTH might have influenced approach behaviour which then affected the balance between approach and avoidance tendencies.

The influence of ACTH and related peptides on rewarded behaviour has been demonstrated in a number of studies. Guth *et al.* (1971b) reported that ACTH increases the rate of a bar-press response for water only when extraneous noise and the motivation level were stringently controlled. In other studies, ACTH failed to influence continuously rewarded approach behaviour but it modified drug or partial reinforcement induced behavioural changes. Leonard (1969) showed that ACTH administration partially antagonizes the deleterious effect of sodium barbitone on running time for a food reward in a multiple T-maze but failed to influence the effect of the drug on the errors made in the maze. Gray *et al.* (1971) have studied the influence of ACTH on the acquisition and extinction of a partially reinforced runway response where food was used as reward. Partial reinforcement results in an increased running speed during acquisition and in delayed extinction as compared with continuous reinforcement. A relatively high dose of ACTH administered during acquisition appeared to block the effect of partial reinforcement. Partially reinforced rats receiving ACTH behaved as continuously reinforced rats both during acquisition and extinction. ACTH had no effect on the performance of continuously reinforced rats. That this behavioural effect of ACTH is of extra-adrenal nature has been demonstrated by Garrud et al. (1977). Administration of $ACTH_{4-10}$ also blocks the effect of partial reinforcement. The observations of Gray et al. (1971) were contrary to expectation. Gray (1967) proposed that frustrative non-reward, like partial reinforcement and punishment, acts on behaviour by way of a common physiological mechanism. Since ACTH enhances the effect of punishment, enhancement of the effect would have been expected. Thus, ACTH and related peptides might not act on the same physiological mechanism under fear-frustration conditions unless the peptides modify motivational mechanisms (facilitation at a low and attenuation at a high level) or subverse reinforcement. Recent knowledge on the behavioural mechanisms involved in peptide effects favour the motivational hypothesis. However, it is questionable whether later findings of Gray and Garrud (1977) are in accord. They studied bar-pressing behaviour of the rats with high and low reward rate intrusion periods under the influence of $ACTH_{4-10}$. High rates of reward intrusion increase whilst low rates decrease bar pressing. $ACTH_{4-10}$ attenuates both effects. These findings suggest that $ACTH_{4-10}$ may attenuate behavioural adjustment to both favourable and unfavourable changes. In contrast, Isaacson et al. (1976) suggested that $ACTH_{4-10}$ treatment results in a better use of information without a general effect on learning per se. They find that $ACTH_{4-10}$ improves correct performance in the first trial in a four-table choice situation for water reward in the rat.

ACTH and related peptides also markedly influence extinction performance in approach situations. Gray (1971) showed that extinction of a straight runway response motivated by hunger is delayed by treatment with ACTH during extinction. In subsequent experiments, Garrud et al. (1974) found that the influence of treatment with $ACTH_{4-10}$ is similar to that of ACTH. Moreover, they reported that [D-Phe^7] $ACTH_{4-10}$ facilitates extinction of the runway response, which effect of the peptide is similar to that observed in active avoidance paradigms. Bohus et al. (1975) showed that $ACTH_{4-10}$ administration before each extinction session results in a delay of extinction of a sexually-motivated runway response in male rats which were allowed to copulate during acquisition training in the goal-compartment with a highly receptive female. $ACTH_{4-10}$, however, failed to affect running behaviour during extinction in those rats that were not allowed contact with the receptive female during the training period.

All these observations indicate that the influence of ACTH-like peptides on behaviour is not restricted to fear-motivated reactions. Acquisition and extinction behaviour is equally affected by ACTH whether the rewarded response is motivated by thirst, hunger or sex. That reward is an essential requisite for the behavioural effects of peptides in approach situations is suggested by experiments of Bohus *et al.* (1975), since extinction behaviour of non-rewarded males, although motivated to seek contact with a receptive female, remained unaffected by $ACTH_{4-10}$. Therefore, ACTH and related peptides do not seem to affect motivation *per se*. Some of the variables which determine their effect on avoidance behaviour, such as the motivation level, seem to play a role in approach behaviour as well. Extinction of the approach response again appeared to be sensitive to modification by ACTH-like peptides in addition to behavioural paradigms such as a partial reinforcement schedule where uncertainty is introduced. The "stressful" character of stimuli associated with approach behaviour has only recently been recognized and the aversive quality may be as strong as in avoidance behaviour. Observations by Levine *et al.* (1972) demonstrate that any departure from expected schedules such as withdrawal of reward (extinction) or a reinforcement shift causes the release of ACTH. This means that uncertainty may be regarded as a stressful stimulus. The common character of "stressfulness" may then account for the similarities in the effect of ACTH-like peptides in approach and avoidance situations and presuppose common mechanisms involved in their behavioural actions.

Since both α- and β-MSH share amino acid sequences with ACTH (1–13 in α-MSH and 5–14 in β-MSH), it is not surprising that the effects of MSH on behaviour are similar to those of ACTH. α-MSH appeared to restore impaired avoidance extinction of posterior lobectomized rats (de Wied, 1965). Furthermore, delay of extinction of conditioned active avoidance responses (de Wied, 1966; de Wied and Bohus, 1966) and facilitated retention of a passive avoidance response were found in α- and β-MSH-treated intact rats (Greven and de Wied, 1973). Kastin and his coworkers (1973a, 1975), interested in the biological effect of MSH in mammals, performed a number of behavioural experiments and obtained in principle similar results to those found in animals treated with ACTH fragments. Thus, α- and β-MSH: facilitate passive avoidance behaviour (Sandman et al., 1971a; Dempsey et al., 1972); facilitate shuttle-box acquisition at low shock intensity punishment (Stratton and Kastin, 1974); delay extinction of a T-maze response running for food (Sandman et al., 1969); and increase resistance to extinction of a lever-press response under a fixed ratio reinforcement schedule (Kastin et al., 1974). In addition, α-MSH treated rats reverse faster in a brightness discrimination test than controls (Sandman et al., 1972). α-MSH facilitates visual discrimination learning in albino but not in pigmented rats. Reversal of this response is, however, facilitated by α-MSH in both strains of animals (Sandman et al., 1973). Kastin et al. (1973, 1975) suggested that α-MSH exerts an extrapigmentary effect on behaviour in animals and man and that the mechanism of action is probably facilitation of (visual) attention. However, it is questionable whether α-MSH is a natural hormone in man, a species without an intermediate lobe. In addition, neither α- nor β-MSH has been detected as such in man, although β-MSH may be generated from pituitary LPH (Lowry and Scott, 1975).

In the analysis of the behavioural effects of pituitary peptides, interest has also been focused upon the relation between memory processes and pituitary function. de Wied and Bohus (1966) suggested that posterior pituitary principles, presumably vasopressin, influence long term memory processes as compared with the short term effect of α-MSH on avoidance behaviour—probably due to enhancement of trial-to-trial memory. Extensive work on vasopressin and its analogues substantiated the hypothesis on the involvement of the hypothalamo-neurohypophysial system in long term memory processes (de Wied et al., 1975b). Experiments in relation to the effect of ACTH on memory processes mainly made use of the influence of the peptide on amnesia induced by protein synthesis inhibitors, anoxia or electroconvulsive shock (ECS). Flexner and Flexner (1971) reported that, in mice, amnesia for a Y-maze avoidance response induced by intracerebral injection of puromycin is prevented by ACTH given as a gel preparation for up to 3 days before, or within 16 h after, learning. Puromycin was given 1 day after the single acquisition session or after ACTH injection. Retention was tested 1–2 weeks later. Subsequent experiments in which highly purified ACTH was used, failed to replicate these findings (Lande et al., 1972), though ACTH gel was injected only prior to acquisition. The authors suggested that the anti-amnesic effect found by Flexner and Flexner (1971) might have been due to vasopressin impurity of their ACTH.

Quinton (cited by Rigter *et al.*, 1975) administered ACTH prior to the learning or to the retention test and in the latter case found it attenuated puromycin-induced amnesia. Rigter *et al.* (1974) then showed that $ACTH_{4-10}$ alleviates amnesia for a one-trial learning passive avoidance response induced by post-learning application of CO_2 in the rat. The peptide is effective when given before the retention test but treatment prior to the learning trial did not abolish amnesia. Rigter and van Riezen (1975) also showed that alleviation of amnesia by $ACTH_{4-10}$ does not depend upon the amnesic agent and the behavioural task. $ACTH_{4-10}$ administered before the retention test attenuated ECS-induced amnesia of a one-trial learning thirst-motivated response of the rat. Furthermore, the anti-amnesic effect of $ACTH_{4-10}$ was demonstrated even 2 weeks after the amnesic treatment (Rigter *et al.*, 1975). These authors suggested that $ACTH_{4-10}$ promotes retrieval of memory which is impaired by amnesic treatment. Recovery from ECS-induced amnesia was reported by Keyes (1974) in the rat by administration of ACTH 4 h after the learning and ECS treatment. He suggested that the peptide reactivates the internal physiological state present during learning and therefore promotes the retrieval of memory. Gold and Buskirk (1976a), on the other hand, suggested that post-training ACTH administration influences memory storage processes. They observed that a low dose of ACTH given immediately after the learning of an inhibitory avoidance task in rats enhances the retention of the response. A tenfold higher dose of the peptide resulted in an attenuation of the retention.

These observations suggest that ACTH and related peptides influence memory processes. The majority of the observations favour the hypothesis that they affect retrieval rather than storage of memory. The retrieval hypothesis seems to fit with the observations that ACTH-like peptides facilitate retention of passive avoidance responses and delay extinction of conditioned responses. The absence of an effect of peptide treatment on amnesia before the training does not exclude an influence on storage processes. Observations with post-learning administration of the peptide represent a better approach to study memory processes because the behavioural effect of the treatment after training cannot easily be attributed to influences on motivation, performance pain, and other factors unrelated to memory (McGaugh, 1961). It is possible that the anti-amnesic effect of post-learning ACTH treatment is mediated through the adrenal cortex in contrast to the retrieval effect of an extra-adrenal nature. Barondes and Cohen (1968) reported that cycloheximide-induced amnesia is reduced by administration of a mixture of cortisol and corticosterone 3 h after the learning session. Nakajima (1975a) showed that cortisol or corticosterone given immediately and not more than 3 h after the learning trial antagonizes cycloheximide-induced amnesia. Cortisol-succinate had the same effect after bilateral injection into the hippocampus within 5 min of the learning trial (Cottrell, 1975). Adrenalectomy prior to learning protects memory against the amnesia induced by puromycin or cycloheximide and direct mediation by the adrenal cortex is uncertain (Flexner and Flexner, 1970; Nakajima, 1975a).

Species differences may complicate the picture presented by effects of ACTH on conditioned response performance. Endröczi and Lissák (1962) showed that ACTH administration in the cat facilitates extinction of a conditioned approach response. The treatment also appeared to suppress intertrial goal-directed activity. In the rat, ACTH-induced suppression of intertrial activity is mediated through the adrenal cortex (Bohus et al., 1968), and it is possible that the behavioural effects in the cat may also be adrenal-dependent. High doses of ACTH, given intravenously, appeared to block conditioned avoidance and escape behaviour, conditioned EEG arousal and vasomotor reaction in the rabbit (Korányi et al., 1965/66; Korányi and Endröczi, 1967). Cortisol failed to affect these responses. Since the same authors demonstrated that ACTH inhibits a polysynaptic spinal reflex arc, it is probable that the peptide exerts its inhibitory action at a lower (spinal) level in this species.

ACTH and related peptides also influence "non-learned" behavioural patterns such as sexual and aggressive ones. These peptides may, in addition, induce behavioural reactions if injected into the cerebral ventricles. Ferrari (1958) reported that intraventricular administration of ACTH or MSH induces stretching and yawning in several mammalian species. These behavioural reactions were later called the stretching–yawning syndrome (SYS). This appears approximately 30 min after intraventricular injection, lasts for several hours and is episodic in character. Peptide fragments like $ACTH_{4-10}$ or α- and β-MSH, β-MSH_{8-22} and β-LPH also induce SYS after intraventricular injection both in the rabbit and the rat (Ferrari et al., 1963; Bertolini et al. 1975). $ACTH_{5-10}$ appeared to be much less active (Ferrari et al., 1963). It was suggested that the peptide entity $ACTH_{5-10}$ is the active core inducing SYS (Bertolini et al., 1975). A synthetic $ACTH_{1-25}$ analogue, in which methionine in position 4 was replaced by leucine (Doepfner, 1966), appeared to be highly active (Bertolini et al., 1968). Gessa et al. (1967) found that the onset of SYS is more rapid when $ACTH_{1-24}$ is injected into the 3rd ventricle or into the hypothalamus. They also showed that SYS is accompanied by cerebral cortical arousal as indicated by electroencephalographic studies in the cat. Izumi *et al.* (1973) reported that Zn given intraventricularly induces behavioural changes which in some respects are similar to those found after ACTH analogues. The behavioural effect of Zn occurs much more quickly after injection and has a different quality. It can be distinguished from ACTH-induced SYS. Although the Zn content of β-MSH was high, that of ACTH$_{1-24}$ was low. Baldwin *et al.* (1974) showed that intraventricular administration of ACTH$_{4-10}$ in the female rabbit is less effective in inducing SYS than ACTH$_{1-24}$. Further, they found that prior saline administration in the ventricle and intracardial blood collection attenuates the effect of ACTH$_{1-24}$ and completely abolishes that of ACTH$_{4-10}$.

Ferrari *et al.* (1963) and Izumi *et al.* (1973) reported that SYS is preceded by increased grooming activity in rodents. The excessive grooming induced by intraventricularly administered ACTH and related peptides in the rat has recently been analysed by Gispen *et al.* (1975b). Intraventricular ACTH$_{1-24}$ induces a large scale of elements of maintenance behaviour repertoire such as vibrating, washing, grooming, scratching, licking paw, licking tail, etc. This behaviour was observed for one hour after peptide administration and did not depend upon intact pituitary, adrenals and gonads and the sex of the rat. The authors showed that fragments such as ACTH$_{1-16}$, α- and β-MSH are as potent as ACTH$_{1-24}$ while ACTH$_{1-10}$ or ACTH$_{4-10}$ were inactive. Gispen *et al.* (1975b) suggested that although the presence of the ACTH$_{5-10}$ sequence is of importance, C-terminal elongation of the peptide is necessary for the expression of the effect. It is of interest that the D-phenylalanine[7] enantiomers of both ACTH$_{1-10}$ and ACTH$_{4-10}$ also induce excessive grooming (Gispen *et al.*, 1975b). These fragments lack the C-terminal elongation, as do their "all-L" congeners which are ineffective. It may be that [D-Phe[7]] ACTH fragments are topochemically better equipped to activate the receptor than similar "all-L" fragments without C-terminal elongation.

Bertolini *et al.* (1968, 1969) reported that intraventricular administration of ACTH$_{1-24}$ or α-MSH, besides the induction of SYS, elicits certain elements of sexual behaviour such as erection, ejaculation, sexual posturing and genital licking in the male cat and rabbit even in the absence of a receptive female partner. Castration prevents these behavioural effects of ACTH while testosterone replacement has a restorative effect. In the squirrel monkey, McLean (1973) found that implantation of ACTH$_{1-24}$ or α-MSH in solid form in the septopreoptic region resulted in episodes of full penile erection, stretching, yawning and scratching of the body. Haun and Haltmeyer (1975) using male rabbits confirmed the behaviour observations of Bertolini *et al.* (1968, 1969) but they also showed that intraventricular saline injection and heart puncture for blood collection before intraventricular $ACTH_{1-24}$ administration prevented the induction of sexual behaviour elements. Intraventricular $ACTH_{1-24}$, however, invariably resulted in an increase in plasma LH and testosterone levels. Plasma corticosterone concentration appeared to be higher but only in those animals which failed to show an elevation of the corticosterone level 90 min after saline injection and cardiac puncture. It is of importance to note that male rabbits during sexual excitation induced by intraventricularly administered ACTH failed to copulate with receptive female partners (Bertolini *et al.*, 1968, 1969).

In contrast to these observations, Korányi *et al.* (1965/66) reported that intravenous administration of ACTH suppressed sexual behaviour of the male rabbit in a novel situation but not in their home-cage. Cortisol alone was not effective but, like testosterone pretreatment, prevented ACTH-induced behavioural suppression. In the rat, Soulairac *et al.* (1953) found that systemic ACTH treatment increased the number of intromissions and ejaculations within a fixed observation interval. Bertolini *et al.* (1975) reported that intraventricular administration of $ACTH_{1-24}$ shortens ejaculation latency in sexually experienced male rats. On the other hand, Bohus *et al.* (1975) observed that repeated systemic doses of $ACTH_{4-10}$ increased intromission and ejaculation latencies, whereas ejaculation was followed by a longer post-ejaculatory-interval in castrated male rats treated with a threshold dose of testosterone. Replacement therapy with a larger dose of testosterone prevented the effect of $ACTH_{4-10}$ and the peptide also failed to affect copulatory behaviour in castrated rats without supplementary therapy. The same authors also showed that $ACTH_{4-10}$ increased resistance to extinction of a sexually motivated conditioned runway response in the male rat but failed to affect the urge to seek contact with a receptive female when copulation was prevented with the incentive female.

Systemic administration of $ACTH_{1-24}$ to ovariectomized oestrogen-primed female rats induces lordosis behaviour through increasing progesterone output from the adrenals (Feder and Ruf, 1969). Meyerson and Bohus (1976) reported that, unlike $ACTH_{1-24}$, systemic administration of $ACTH_{4-10}$ failed to replace progesterone for induction of lordosis in oestrogen-primed, ovariectomized female rats. This indicates that the peptide fragment does not stimulate adrenal progesterone output in contrast to the full ACTH molecule. The ACTH fragment also failed to modify oestrogen + progesterone-induced lordosis behaviour. Meyerson and Bohus (1976), furthermore, found that $ACTH_{4-10}$ enhances adaptation to the testing environment, but it does not influence sexual motivation in ovariectomized, oestrogen-treated female rats.

Intraventricular administration of $ACTH_{1-24}$ induces a posture resembling lordosis in the female rabbit. Plasma LH and progesterone levels are elevated by this treatment and ovulation is observed (Baldwin et al., 1974; Haun and Haltmeyer, 1975; Sawyer et al., 1975). In female rabbits, plasma corticosterone levels were always increased after intraventricular ACTH injection (Haun and Haltmeyer, 1975). $ACTH_{4-10}$ on the other hand, when given intraventricularly, failed to induce sexual posture and LH release regularly while ovulation did not occur (Baldwin et al., 1974; Sawyer et al., 1975). It seems, therefore, that ACTH-induced lordosis and LH release and ovulation are represented in a molecule larger than the $ACTH_{4-10}$ fragment. Baldwin et al. (1974) reported additionally that intraventricular injection of LHRH (LH releasing hormone), although stimulating LH release and inducing ovulation, fails to evoke a lordosis response in the absence of an active male partner. This indicates that lordosis-like behaviour as induced by intraventricular $ACTH_{1-24}$ is not mediated through LHRH and subsequent LH-release.

These observations suggest that ACTH and related peptides may influence both male and female sexual behaviour. However, it is not yet possible to depict a uniform view on the nature and significance of these effects. The route of administration seems to be of importance because stretching, yawning and excessive grooming and sexual excitation cannot be induced by systemic administration of ACTH (Bertolini et al., 1969). This is different from the influence of ACTH and related peptides on conditioned avoidance and approach behaviour. ACTH-like peptides whether given systemically or intracerebrally affect avoidance behaviour in the same way. It might be that the difference is due to the peptide moiety which induces sexual excitation and which modifies conditioned behaviour. The observation that short peptide fragments such as $ACTH_{4-10}$ which markedly affect avoidance behaviour fail to induce sexual excitation and excessive grooming (Baldwin et al., 1974; Gispen et al., 1975b) favours this suggestion. It is known that $ACTH_{1-24}$ is rapidly metabolized in the body into small fragments (Lowry and McMartin, 1974) and that only small quantities of such fragments enter the brain (Verhoef and Witter, 1976). Furthermore, relatively high amounts of ACTH given intraventricularly are necessary to induce sexual excitation and SYS. Effective amounts are above those required to modify conditioned behaviour. It might be, therefore, that neither the required peptide moiety nor insufficient amounts reach the brain after systemic administration. If ACTH fragments are physiologically involved in these processes, one might postulate that these reach the active sites in the brain through release from the pituitary directly into the CSF.

Another interpretation may arise from a behavioural cause. Male and female postures resembling sexual activity after intraventricular ACTH administration were observed in solitary situations. Bertolini et al. (1969) noted that "sexually excited" male rabbits did not copulate with female partners. This indicates that intraventricular ACTH induces a "pseudo sexual excitation" rather than influencing the normal sexual repertoire of the animal. In experiments in which systemic administration of ACTH and related peptides was used to study sexual behavior in a social context, changes in copulatory behavioural patterns could have been due to the interaction of the peptide with environmental factors rather than with sexual motivation per se.

Agonistic behaviour includes aggressive, submissive and defensive responses in competitive situations (Scott and Fredericson, 1951). Social hierarchy and conflict therein influences pituitary–adrenal function (Brain, 1972a). ACTH has been shown to affect certain forms of agonistic behaviour. Brain et al. (1971) reported that ACTH treatment reduces isolation-induced aggressiveness in male mice. Although $ACTH_{4-10}$ treatment failed to affect isolation-induced aggression (Brain, 1972b), the behavioural effect of ACTH appeared to be of an extra-adrenal nature. Leshner et al. (1973) found that ACTH is effective in adrenalectomized rats maintained on a tracer dose of corticosterone and the effect of the peptide is independent of the testes. Poole and Brain (1974) showed that ACTH increases the "attackability" of intact and adrenalectomized mice. Leshner et al. (1975) reported that ACTH treatment facilitates "avoidance of attack" in male mice.

ACTH and ACTH fragments do not seem to influence spontaneous behaviour of rats in a novel environment (Moyer, 1966; Bohus and de Wied, 1966; Weijnen and Slangen, 1970). Similarly α-MSH failed to influence locomotor activity, and food and water intake in the rat (Kastin et al., 1973b). A tendency towards increased activity was found by Nockton et al. (1972) in an open-field situation with rats receiving α-MSH and footshock prior to the test. α-MSH appeared to facilitate locomotor activity of rats bearing lesions in the septal area

(Brown *et al.*, 1974). Increased locomotor activity was found by
Sakamoto (1966) in mice given β-MSH but synthetic α-MSH and
ACTH appeared to be ineffective. In the rabbit, drowsiness, hyperpnoea and hypotension have been observed after β-MSH or ACTH
administration (Dyster–Aas and Krakau, 1965) while in the rat
β-MSH induces drowsiness, hyperpnoea, hypertension and piloerec-
tion (Sakamoto, 1966). These observations indicate that the general
effect of the peptides related to ACTH depends upon both their
structure and the species studied.

In conclusion, much evidence collected during the last 10 years has
established the profound effect of ACTH and related peptides on
various forms of adaptive behaviour. The most prominent feature of the
behavioural effect of ACTH being that it is not mediated through the
adrenal cortex and that it does not require the entire molecule. A new
class of peptide hormones has been discovered consisting of small
fragments of the ACTH molecule. The behavioural activity of these
"neuropeptides" (de Wied, 1969) which originate in the pituitary
gland and for which ACTH, MSH and LPH may serve as prohormones is independent of such classical biological activities as those on
the adrenal cortex, on melanocytes and on fat mobilization.

B.   EFFECTS OF CORTICOSTEROIDS

Early behavioural experiments with ACTH suggested, in accord with
current endocrinological views, that corticosteroids are the primary
candidates for the behavioural effects of ACTH. However, ACTH
fragments devoid of corticotrophic effects were as active as the whole
molecule. In addition, many of the behavioural effects of corticosteroids appeared to be opposite to those of ACTH.

Overall effects of corticosteroids on acquisition of conditioned
avoidance behaviour were minimal. Long term administration of
corticosteroids such as cortisone in a oneway active avoidance test
(Bohus and Lissák, 1968) or dexamethasone in a shuttle-box avoidance procedure (Conner and Levine, 1969; Beatty *et al.*, 1970) failed
to influence acquisition of conditioned avoidance behaviour in the rat.
Levine and Brush (1967) however reported that ACTH or cortisol
treatment normalizes the poor performance during relearning of an
incompletely acquired avoidance behaviour. As discussed earlier, the
avoidance performance during relearning of an incompletely acquired
shuttle-box avoidance response is a function of the interval between
original training and the subsequent relearning session and shows an
inverted U-shaped relationship (Kamin, 1963). Brush and Levine

(1966) found that the plasma corticosterone level after the original learning is correlated with the performance during relearning in the shuttle-box. Suppressed relearning was correlated with low plasma corticosterone levels. Therefore, the observations of Levine and Brush (1967) indicated that reinstatement of pituitary–adrenal activity by ACTH or cortisol treatment prevents the suppression of performance. Dexamethasone, on the other hand, failed to affect avoidance performance subsequent to incompletely learned shuttle-box avoidance behaviour (Kasper–Pandi et al., 1970). Stolk (1972) suggested that an intact pituitary–adrenal function is necessary for the maintenance of poor avoidance performance. He showed that the avoidance behaviour of poorly performing female rats in a shuttle-box deteriorates further after adrenalectomy. Corticosterone treatment restored the avoidance behaviour towards that of the original poor performance level. In sham-operated rats corticosterone substantially improved avoidance performance. Other observations suggested that corticosteroids may stabilize avoidance behaviour. Wertheim et al. (1967) found that dexamethasone but also ACTH treatment leads to fewer responses and shocks received and a higher frequency of long duration interresponse pauses in a free-operant (Sidman) avoidance schedule. These changes in performance may be characterized as response suppression which results in better timing behaviour. A similar effect of corticosteroids has been observed by Bohus and Lissák (1968) in a conditioned one-way active avoidance situation. They found that cortisone treatment did not affect conditioned response performance per sé but led to a decrease in intertrial responses. This might also be interpreted as a more efficient timing of avoidance behaviour. Later Endröczi (1972) showed that administration of corticosterone did not affect conditioned avoidance performance in a shuttle box during the early and the stabilized phases of the training (respectively, low and high response performance). However, the corticosteroid markedly suppressed avoidance performance at the middle stage when the avoidance level was around 50%. In this respect it is worth noting that ACTH treatment facilitated avoidance performance of intact and adrenalectomized rats at a similar stage in a one-way avoidance procedure (Bohus and Endröczi, 1965; Bohus et al., 1968). Extinction behaviour appeared to be a more sensitive behavioural substrate of corticosteroids. Corticosteroids such as corticosterone, dexamethasone and cortisone appeared to facilitate extinction of conditioned active avoidance responses in a dose-dependent manner (de Wied, 1967; Bohus and Lissák, 1968). The mineralocorticosteroid aldosterone was less active than the glucocorticosteroids (de Wied, 1967).

Impairment of passive avoidance behaviour in the rat by corticosteroids has been demonstrated in a number of conflict situations. Bohus *et al.* (1970a) showed that administration of cortisone 3 h before a single learning trial leads to a suppression of immediate passive avoidance in a light–dark conflict situation. Treatment immediately before learning resulted in suppression of delayed passive avoidance behaviour in the rat which was studied 24 or 48 h after the single learning trial. The effect of cortisone appeared to be a function of the intensity of the punishing footshock and the dose of cortisone. The higher the intensity, the more cortisone was needed to suppress passive avoidance behaviour. The same authors demonstrated that suppression of a water-rewarded discriminative conditioned approach response by punishing footshock is inhibited by dexamethasone administered prior to passive avoidance learning. Dexamethasone-treated rats showed less suppression of non-punished approach response, and passive avoidance of the punished response was extinguished more rapidly than in the controls.

Thus, corticosteroid treatment besides facilitating extinction of passive avoidance behaviour may enhance discrimination between punished and non-punished responses. More recent observations by Garrud (1975) corroborate this view. He employed conditioned suppression of a bar-pressing response for reward as the measure of avoidance behaviour and used punished and non-punished conditioned stimuli. Corticosterone administration failed to influence conditioned suppression of the bar-pressing response in his experiments. However, corticosteroid-treated rats increased their base-line response rate which was further increased during the presentation of the non-punished conditioned stimulus.

Furthermore, it was observed that passive avoidance behaviour in a thirst-versus-fear conflict situation in the rat is suppressed by cortisone, cortisol and corticosterone but not by deoxycorticosterone (Bohus, 1971, 1973). Corticosterone, the main physiological glucocorticosteroid of the rat adrenal cortex, had the greatest effect. It was suggested that facilitation of active avoidance extinction and suppression of passive avoidance behaviour by corticosteroids may be the result of fear reduction. Pappas and Gray (1971) reported that dexamethasone, when given before learning or before the retention test, reduced the latency in resumption of drinking in a fear-versus-thirst conflict situation. Dexamethasone given before both learning and retention appeared to be ineffective. They concluded that the presence or absence of dexamethasone serves as an additional cue which is associated with the punishing shock. Removal of this cue in

adrenal funct
tions of the yc
1966; Denenb
1957, 1958; L
cortical respor
Levine and I
reduced depe
during infanc
increased or c
adulthood. Ba
and adrenoco
responds to in
this steroid, a
nization in a w
to subsequent
ment of this h
ences are mec
Denenberg an
(1973) in a cri
involvement of
effects. The c
evidence of a c
cal reactivity
infancy does n
Grota (1972) a
dexamethason
blocking the cc
does not preve
reactivity in w
   There are
maternal pitu
behaviour of
adrenal system
ment and fun
(Milkovic et al.
male offspring
better in an ay
tumour in the
behaviour of t
affected by ma
influence of pi
mized and de

rats receiving dexamethasone only prior to learning or removal of the non-steroid state at learning by administering dexamethasone prior to the retention test decreases fear. In contrast, Endröczi and Nyakas (1971, 1972) and Endröczi (1972) reported that corticosterone or dexamethasone administration to rats adrenalectomized 2–3 weeks earlier facilitated, instead of inhibiting, passive avoidance learning in a thirst-versus-fear conflict situation. These steroids also appeared to suppress the exploratory behaviour of thirsty rats in a multiple maze and to increase the latency to drinking. These authors concluded that facilitation of passive avoidance is unrelated to an increase in fear but rather due to inhibition of the integration of the goal-directed motor response, that is of the approach tendency. In the intact thirsty rat, none of the corticosteroids affected exploratory behaviour, water intake or running for water in a discriminative conditioned situation (Bohus et al., 1970a; Bohus, 1971, 1975a). Passive avoidance behaviour represents an inhibition by punishment of a previously established approach response whether this response is motivated by hunger, thirst or innate dark preference. Accordingly, increased passive avoidance latencies may be the result of either decreased approach tendency or facilitated fear motivation. Conversely, decreased passive avoidance may be the result of fear suppression or increased approach motivation. Thus, corticosteroids seem to affect different motivational systems in the presence or the absence of the adrenals which may result in seemingly conflicting observations.

   There are relatively few reports on the influence of corticosteroids on rewarded behaviour. Levine (1968) found that both dexamethasone and ACTH treatment improves DRL (differential reinforcement of low rate response) performance of the rat which again indicates better timing behaviour of the treated rats. The influence of corticosterone treatment on the acquisition of a space discriminative conditioned approach response for water reward in the rat was studied by Bohus (1973). It was found that suppression of the acquisition of this response by corticosterone depends upon the deprivation level and consequently on the motivation of the rat. Marked retardation of the acquisition was found in corticosterone-treated rats kept on a 12 h deprivation schedule while the corticosteroid failed to influence behaviour in rats with 23 h water deprivation. Hennessy et al. (1973) failed to find an effect of corticosterone, cortisol or ACTH treatment on the acquisition of a runway response for food reward in rats deprived for 23 h but corticosteroid-treated rats ran faster in the early phase of extinction training under continued treatment. They performed their experiments at the end of the dark phase of the daily

light-
rever
watei
steroi
new
reinfi
impri

En
the a
was 1
(1962
respo
cortic
or ap
peptii

Ne
the ci
1968)
conte
ward.
and r
assess
admii
ral ef
reduc
found
1968)
respo:
pressi
treate
DRL
hyper
(How
and E
et al.,
Increi
McKi
tions :
the c
devel(

A r
early

(conditioned fear) appeared to increase open field activity in the offspring of adrenalectomized mothers but it was decreased in dexamethasone-treated mothers. Stress effects on adult conditioned shuttle-box avoidance behaviour are affected by maternal treatment with dexamethasone but only in the female offspring. Stressing the mothers led to a decrease in avoidance performance of the offspring while dexamethasone treatment appeared to block these stress effects. On the other hand, manipulation of the maternal pituitary–adrenal system by hypophysectomy (Obias, 1957), adrenalectomy (Havlena and Werboff, 1963) or dexamethasone in the drinking water during gestation (Grota, 1972) failed to affect the adult behaviour of offspring. Hence the role of pituitary–adrenal system hormones in prenatal stress effects on adult behaviour cannot be easily determined from presently available observations.

Administration of corticosteroids in adulthood seems to influence a number of behavioural responses other than conditioned behaviour. In contrast to ACTH and adrenalectomy, corticosteroid administration facilitates aggressive behaviour. Kostowski (1967) reported that corticosterone facilitates shock-induced aggression in mice. Kostowski et al. (1970) found that cortisol facilitates murine aggressiveness of rats and isolation-induced aggression in mice. Deoxycorticosterone decreased murine aggression but failed to affect isolation-induced aggression. Banerjee (1971a) showed that cortisol facilitates isolation-induced aggression in mice which had no aggressive experience but failed to influence the behaviour of experienced mice. Brain et al. (1971) reported that dexamethasone administration leads to facilitation of aggression which follows isolation of mice. Dexamethasone also appeared to be effective in restoring aggressiveness of adrenalectomized, isolated mice (Leshner et al., 1973). Facilitation of maternal aggression by cortisol occurs in the rat (Endröczi et al., 1958). Thus, several forms of aggressive behaviour are under the control of pituitary–adrenal system hormones in an opposite way. However, it is not clear whether ACTH or corticosteroids determine the agonistic behavioural response. Brain (1975) has recently argued that isolation is less stressful at least for mice, which develop isolation-induced aggression, than group housing or crowding. Since adrenocortical activity in animals in the lower range of the dominant–subordinate hierarchy seems to be higher (Louch and Higgenbotham, 1967; Varon et al., 1966; Leshner and Candland, 1972), it might be that high levels of ACTH rather than high levels of corticosteroids influence the appearance of agonistic behaviour, i.e. a low level of aggressiveness and a higher level of fear. Fear and aggressiveness often appear to be inversely related (Brain and Nowell, 1969, 1970). Since ACTH facilitates fear-motivated behaviour but reduces aggression, it follows that it may act agonistically on the behaviour of subordinate rats. Leshner *et al.* (1975) found that ACTH treated mice showed a high tendency to avoid attack. Therefore, corticosteroids may play little if any role in aggressive behaviour except as a result of suppression of ACTH release.

Data on the behaviour of rats exposed to a novel environment and treated with corticosteroids are sometimes conflicting. No changes in the open-field behaviour of water-deprived rats were found by Bohus *et al.* (1970a) after administration of a single dose of dexamethasone, previously shown to affect passive avoidance behaviour. Miyabo *et al.* (1972) found that repeated administration of cortisol-free alcohol or cortisol-21-sulphate in a relatively large dose facilitated exploratory behaviour of rats in an open-field system when treatment was started soon after weaning, though it was ineffective when started at an "adult" (46 days) age. The authors suggested that cortisol affects behaviour through direct action on the central nervous system rather than *via* the intermediary metabolism or suppression of ACTH release, since cortisol-21-sulphate has no classical glucocorticosteroid activity. Endröczi (1972), on the other hand, observed that both dexamethasone and corticosterone suppress exploratory activity of water deprived, adrenalectomized rats in a multiple maze. It is not clear whether the absence of the adrenals, the age of the animals, or the different behavioural situation is the variable which might have influenced the behavioural response to corticosteroid administration.

Adrenal corticosteroids affect intracranial self-stimulation behaviour of the rat. Slusher (1965) reported that cortisol and dexamethasone increase the self-stimulation rate in the lateral hypothalamus medial forebrain bundle (MFB) region. Adrenalectomy resulted in similar changes. Endröczi (1972) showed that cortisol administration suppresses self-stimulation rate in the midline septal area. This suppression could be reinstated by increasing the intensity of the stimulation current. Corticosterone and cortisol, however, failed to influence self-stimulation in the MFB region if stimulation resulted in a high response rate in control periods. Dexamethasone, on the other hand, suppressed self-stimulation rate both in rostral septal and MFB regions. It seems, therefore, that localization in the brain, the corticosteroids used and the rate of stimulation are variables which determine the effect of corticosteroids on self-stimulation.

Pituitary–adrenal influences on spontaneous running activity of the rat have long been known. In 1936, Richter observed that adrenalectomy attenuates running wheel activity which could be normalized by treatment with adrenocortical extract. Pedersen–Bjergaard and Tonnesen (1954) found that a high dose of orally administered cortisol increased running activity. Dexamethasone given in the drinking water appeared to facilitate activity but only during the nocturnal phase (Kendall, 1970). Peak activity appeared 5–7 days after the onset of treatment and activity remained above the control level for 1–2 weeks after its cessation. Suppression of the pituitary–adrenal function by implantation of cortisol acetate in the median eminence of the hypothalamus led to a reduction in activity, which was restored by dexamethasone administration in the drinking water. Beatty et al. (1971) have reported similar findings. They suggested that influence is related to the effect of dexamethasone in decreasing the body weight since food deprivation and consequent body weight loss also increases running activity. Amelioration of the effect of adrenalectomy or adrenal demedullation on running activity by corticosterone treatment has been reported by Leshner (1971). This treatment, however, reduced the activity of intact rats but increased the activity of intact rats deprived of food. Taken together, it can be concluded that corticosteroids play an important role in the regulation of spontaneous running activity which has been suggested to serve a regulatory function in the metabolic economy of the organism (Brobeck, 1945). The differences in the effect of dexamethasone and corticosterone in intact rats may be related to the glucocorticosteroid activity of these two steroids or to their varying effect on body weight.

There are at least three important questions which deserve detailed discussion concerning the behavioural effects of corticosteroid. First, corticosteroids may have an intrinsic effect on brain mechanisms or an indirect effect, through blockade of ACTH release, which removes the opposite behavioural changes of ACTH and related peptides. Administration of corticosteroids, particularly when repeated, is known to block stress-induced ACTH release. The effect of glucocorticosteroids on extinction of a shuttle-box avoidance response, however, does not correlate with the rate of inhibition of ACTH release (de Wied, 1967). Attenuation of passive avoidance learning by a single injection of various corticosteroids, also, is not closely correlated with the suppression of ACTH release except when a high dose of cortisol or 6-dehydro-16-methylene-cortisol is used (Bohus, 1973). In addition, both corticosterone and dexamethasone facilitate avoidance extinction in hypophysectomized rats maintained on a substitution therapy (de Wied, 1967). Thus, the influence of these corticosteroids on behaviour is independent of ACTH and an intrinsic property of the steroid molecule in the central nervous system. Experiments with implantation of cortisol directly into the brain support this view. Bohus (1968) found that cortisol implants in the median eminence, which effectively inhibit the release of ACTH, had only a modest effect on the rate of extinction of an active avoidance response in the rat. The rate of extinction was correlated with the rate of suppression of ACTH release. Implantation of cortisol in the mesencephalic reticular formation, which hardly reduces ACTH release, markedly facilitates extinction. Similarly, no correlation was found between attenuation of passive avoidance behaviour and suppression of ACTH release in rats bearing corticosteroid implants in a number of brain regions (Bohus, 1973).

A second problem of importance is the relation between behaviour and other biological activities of corticosteroids and whether species specific glucocorticosteroids have a preferential behavioural action. de Wied (1967) reported that aldosterone had only little effect on extinction of avoidance behaviour as compared with glucocorticoids in the rat. Similarly, deoxycorticosterone whether given systemically or implanted in the brain failed to affect the passive avoidance behaviour of the rat (Bohus, 1973). Thus, these studies seemed to indicate that the behavioural effect of corticosteroids is primarily bound to glucocorticosteroid properties. Gray (1976) found that the influence of prestimulation such as footshock, airblast, or handling on subsequent operant avoidance behaviour is dependent upon miner-alocorticosteroids. Facilitation of avoidance performance by these prestimulations is blocked by adrenalectomy but not by hypophysectomy. Administration of aldosterone or deoxycorticoster-one but not of corticosterone normalizes the prestimulation effect on subsequent behaviour. van Wimersma Greidanus (1970) reported that progesterone, 19-norprogesterone and pregnenolone are as po-tent as corticosterone in facilitating extinction of a pole-jumping avoidance response. Hydroxidione, testosterone, and oestradiol appeared to be ineffective.

Selye (1942) originally observed that certain hormonal steroids possess hypnotic effects. Gyermek et al. (1967) found that progester-one metabolites, which might be responsible for sedative effects during pregnancy (Figdor et al., 1957; P'An and Laubach, 1964), impair operant avoidance behaviour in the rat. However, progester-one and pregnenolone appeared to be effective only in relatively high doses. Banerjee (1971b) found that progesterone impairs acquisition and performance of a pole climbing avoidance response in female rats only. Electrophysiological studies also suggest a central depressant effect of progesterone (Kawakami and Sawyer, 1959; Arai *et al.*, 1967; Heuser, 1967; Gyermek *et al.*, 1967). However, van Wimersma Greidanus *et al.* (1973) failed to find changes in open-field activity of male rats following progesterone administration in doses which already facilitate avoidance extinction. Hydroxydione, on the other hand, which has no effect on conditioned behaviour, suppresses exploratory activity. Thus, the behavioural effects of pregnene steroids are not due to hypnotic properties. Taken together, there is no agreement on the specificity of the steroid structure and the behavioural action. The assumption that the behavioural paradigm determines whether glucocorticosteroids, mineralocorticosteroids or other pregnene-type steroids are involved, may sound strange but no other alternative is provided by the available data to explain the contradictory results. The observations cited above are mainly derived from studies on fear-motivated behaviour. Avoidance responses in different situations may be controlled by different brain mechanisms with different anatomical localization. Therefore, it might be that some brain structures may differentiate between steroid structures while others do not. That this is not a pure speculation can be derived from studies on the localization of corticosteroid effects on behaviour and on the uptake of steroids in the central nervous system. These issues will be discussed later.

The third problem is the use of dexamethasone in behavioural experiments. This synthetic steroid is known not only as an effective inhibitor of ACTH release but also as a potent glucocorticosteroid with slight mineralocorticosteroid activity. A large number of studies used dexamethasone to suppress ACTH release in order to demonstrate the behavioural effect of "chemical hypophysectomy". Most of these studies failed to substantiate the hypothesis that suppression of ACTH release leads to behavioural changes similar to those found after hypophysectomy (e.g. Conner and Levine, 1969; Beatty *et al.*, 1970; Kasper–Pandi *et al.*, 1970, etc.). Levine and Levin (1970) reported the only evidence suggesting that dexamethasone-induced blockade of ACTH release affects behaviour. They showed that A/JAX strain mice take less shock but have higher plasma corticosterone levels than the DBA/2 strain in a thirst-versus-fear conflict situation. Dexamethasone given prior to acquisition impairs passive avoidance behaviour in A/JAX mice but not in the DBA/2 strain. On the other hand, de Wied (1967) showed that dexamethasone is effective in facilitating extinction of a shuttle box avoidance response in hypophysectomized rats in the same way as in intact animals. Furthermore, de Wied (1971) found that dexamethasone reduces the already inferior shuttle box avoidance acquisition in hypophysecto-mized rats. Similar behavioural effects of ACTH and dexamethasone were found by Wertheim *et al.* (1967) in a Sidman–avoidance situation and by Levine (1968) in rats on a DRL schedule. They suggested that the behavioural effect of ACTH is mediated through the adrenal cortex. Accordingly, an intrinsic glucocorticosteroid rather than ACTH release suppressive activity may be the origin of the behavioural effect of dexamethasone. Therefore, the use of dexamethasone as a tool to block ACTH release in behavioural experiments is not justified. The fact that the uptake and distribution of this synthetic steroid in the brain is different from that of corticosterone in the rat (de Kloet *et al.*, 1975) suggests that observa-tions with dexamethasone may be misleading as far as physiological mechanisms are concerned.

In conclusion, corticosteroids profoundly affect various forms of adaptive behaviour. The behavioural influence of corticosteroids is mostly suppressive which in certain situations may support the effectiveness of the performance. It further seems that the behavioural effects of corticosteroids are due to an intrinsic activity in the brain and not primarily mediated through the suppression of pituitary ACTH release.

4. The site(s) and mechanism(s) of the behavioural action of pituitary–adrenal system hormones

A. BEHAVIOURAL EVIDENCE FOR THE LOCI OF THE EFFECT OF ACTH, ACTH FRAGMENTS AND CORTICOSTEROIDS IN THE BRAIN

The site(s) of action of pituitary–adrenal system hormones in the central nervous system has been explored by attempting to block their behavioural effect by destruction of various brain regions or by implanting ACTH-related peptides or corticosteroids directly in the brain. The importance of the thalamic parafascicular region in mediating the behavioural effect of ACTH-like peptides was sug-gested by Bohus and de Wied (1967a). Bilateral destruction of this area in the rat did not materially affect active avoidance learning but resulted in facilitation of extinction (Bohus and de Wied, 1967b). Unlike its action in intact rats, α-MSH failed to affect the extinction behaviour of rats bearing lesions in the parafascicular nuclei. Subse-quent experiments demonstrated that parafascicular lesions also block the behavioural effect of $ACTH_{4-10}$ (van Wimersma Greidanus *et al.*, 1974). More direct evidence of the involvement of the posterior thalamic area was obtained by implanting peptides in this region. In a similar way to peptides given by systemic injection, $ACTH_{1-10}$ implantation led to a delay of extinction of an active avoidance response, while [D-Phe^7] $ACTH_{1-10}$ facilitated extinction (van Wimersma Greidanus and de Wied, 1971). The parafascicular nuclei and the centrum medianum as parts of the non-specific thalamic nuclei play an important role in the maintenance of behaviour. Lesions in this area impair the retention of avoidance behaviour (Cardo, 1965; Cardo and Valade, 1965; Bohus and de Wied, 1967b; Delacour, 1969) and of instrumental avoidance and approach responses (Delacour et al., 1966; Alexinsky and Houcine, 1973) and cause retention deficit in visual learning (Thompson et al., 1970). Electrical stimulation of the centrum medianum-parafascicular complex improves avoidance performance (Cardo, 1967). Delacour (1970, 1971) suggested that this complex is involved in the interaction between defensive motivation and the integration of sensory inputs.

Although the implantation studies of van Wimersma Greidanus and de Wied (1971) failed to demonstrate effective sites in the brain other than the parafascicular area, recent evidence suggests that the limbic forebrain may also be involved in the mediation of the behavioural effect of ACTH-like peptides. van Wimersma Greidanus and de Wied (1976) showed that bilateral destruction of the anterodorsal hippocampus prevents the effect of $ACTH_{4-10}$ on extinction of an active avoidance response. In addition, systemic administration of ACTH normalizes deficient shuttle-box acquisition behaviour of rats bearing lesions in the amygdaloid nuclei (Bush et al., 1973). Thus, it seems that the primary locus of action of ACTH-like peptides is in the posterior thalamic area but the appearance of the behavioural effects requires an intact limbic–midbrain system. It remains to be determined whether the limbic forebrain serves as an executive mechanism of peptide effects or whether additional sites of action exist especially in the septal–hippocampal complex. The limbic–midbrain system is significantly involved as the site of the behavioural action of corticosteroids, which seem to have multiple sites of action therein (Bohus, 1975a). Implantation of cortisol in minute amounts in the mesencephalic reticular formation, in the CA_1 and CA_2 layers of the dorsal hippocampus, the medial septal nuclei, the anterior hypothalamic area, the amygdaloid complex, or the posterior thalamic area appeared to mimic the behavioural effect of systemically administered corticosteroids. Implants located in other brain areas such as the medial, dorsal and posterior hypothalamus, anterior or lateral thalamus, the cerebral cortex and the ventral hippocampus appeared to be ineffective (Bohus, 1968, 1970a). Implantation of corticosterone or dexamethasone and of progesterone in the thalamic parafascicular area also leads to facilitation of avoidance extinction (van Wimersma Greidanus and de Wied, 1969; van Wimersma Greidanus *et al.*, 1973). Endröczi (1972) showed that cortisol implantation in the preoptic region, the antero–lateral hypothalamus, and in the septal nuclei facilitates active avoidance extinction in adrenalectomized rats. On the basis of implantation studies Bohus (1970a) suggested that as corticosteroids have multiple sites of action in the brain, their influence may have a dual character, i.e. enhancement of forebrain inhibition and suppression of ascending reticular activation, both of which promote extinction of active avoidance responses. Subsequent implantation experiments using a fear-versus-thirst conflict situation as the behavioural paradigm, however, indicated that the influence of corticosteroids on the limbic forebrain is suppressive rather than facilitative. Implantation of various corticosteroids such as cortisone, cortisol and corticosterone in the dorsal hippocampus, medial septum, anterior hypothalamus and the midline thalamus suppresses passive avoidance behaviour similarly to corticosteroids given systemically. The effects on retention of passive avoidance behaviour immediately after the learning trial or 24 h thereafter, however, appeared to depend on the site of implantation. This indicated that corticosteroids may selectively influence brain mechanisms which are involved in immediate or delayed retrieval of learned information. Deoxycorticosterone and 16-methyl-6-dehydro-cortisol, which is a potent inhibitor of ACTH release without substantial intrinsic glucocorticosteroid activity, failed to influence passive avoidance behaviour when implanted in the same structures (Bohus, 1971, 1973, 1975a). The importance of the hippocampus as a corticosteroid sensing structure in the mediation of the behavioural effect of corticosteroids, and the need for a stringent control of the intracerebral implantation procedure as suggested by septal implantation studies has been extensively discussed by Bohus (1975a). Endröczi (1972) has also emphasized the role of the septal-hippocampal complex in the central nervous effect of corticosteroids. It has been shown that impaired passive avoidance behaviour and increased exploratory activity in rats with septal lesions or surgical hippocampectomy remain unaltered after corticosteroid administration. In the intact rat, both behaviours are profoundly affected by corticosteroids.

The opposite behavioural effect of ACTH and corticosteroids on extinction of active avoidance behaviour is most apparent in the parafascicular thalamic area. Implantation of cortisone in the pos-

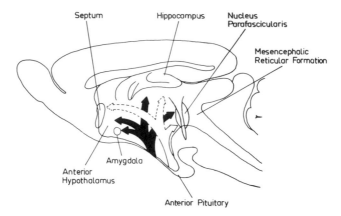

Fig. 1. The sites of behavioural effects of ACTH and related peptides (empty arrows) and of corticosteroids (black arrows) in the limbic structures. The hormones probably reach their site of action *via* cerebrospinal transport.

terior thalamus prevents the effect of systemically administered ACTH. ACTH treatment, however, counteracts the effects of cortisone implants in forebrain areas, i.e. prevents rapid extinction of the avoidance response (Bohus, 1970b). Studies with ablation of different brain areas or with intracerebral implantation of ACTH-like peptides and corticosteroids have demonstrated the central nervous localization of the behavioural effects of pituitary–adrenal system hormones. The observations generally agree that the limbic–midbrain system plays an important role in the mediation of the behavioural effect of these hormones and that this system contains site(s) sensitive to ACTH-like peptides or corticosteroids (Fig. 1).

B.   TRANSPORT AND UPTAKE OF HORMONES IN THE BRAIN AND THEIR
     MODE OF ACTION AT THE CELLULAR LEVEL

The knowledge on the exact site of action of hormones in the brain cannot be complete until the transport to, and the uptake of hormones by specific receptors, which translate the hormone-carried information, is exactly known. Although research is in progress concerning the transport and uptake of ACTH and related peptides in the brain, little is known. On the other hand, uptake and receptor mechanisms for corticosteroids in the brain have been extensively investigated and the trend of recent research is to relate the biochemical findings to their functional activity.

The route(s) through which pituitary ACTH-like peptides reach the brain is not known; one route might be the re-uptake of peptides from the general circulation. Peripheral administration of these peptides mimics this transport route. Pelletier et al. (1975) using I^{125}-labelled α-MSH showed that, after intravenous administration of the peptide, strong labelling can be found in the choroid plexus, ependymal cells and the meninges, indicating that MSH and/or metabolites penetrate into the cerebrospinal fluid. High labelling was found around the cerebral arteries as well. Concentration of radioactivity was further observed in the corpus striatum and the reticular nuclei of the posterior thalamus. This latter localization seems to reinforce the notion that the primary site of behavioural action of ACTH-like peptides is in the posterior thalamus. The physiological route of transport however is most probably not the general circulation but rather the cerebrospinal fluid. Peptides injected into the cerebral ventricles are effective on conditioned behaviour in much lower quantities than when given peripherally. Furthermore, certain behavioural effects of peptides such as stretching, yawning, and extensive grooming can only be induced by intraventricular administration. Bioassayable MSH-activity has been found in the cerebrospinal fluid of monkeys and man by Rudman et al. (1973). Allen et al. (1974) reported that the CSF of the human being contains slightly more immunoreactive ACTH than the plasma. Since they found that the blood–CSF barrier is relatively impermeable to ACTH in man and cat, they suggested that the ACTH present in CSF may enter by a mechanism which bypasses the blood–CSF barrier. Such a bypass mechanism might be a direct leakage of ACTH from the pituitary surface into the adjacent basilar cisterns or a retrograde transport of the peptide through the stalk and basal hypothalamus to the third ventricle. They also suggested that at least part of the immunoreactivity in the CSF might be due to ACTH fragments. Direct evidence for the presence of these fragments in the CSF is not yet available. The fact that less than 1% of a peripherally administered $ACTH_{4-9}$ analogue with high metabolic stability is taken up in the brain (Verhoef and Witter, 1976), indicates that the blood–brain barrier, although rather impermeable to small ACTH fragments, takes sufficient amounts of peptide to elicit behavioural effects. Not more than 1 nanogram of this particular stabile peptide (H-Met(O)–Glu–His–Phe–D-Lys–Phe–(OH)) is needed to affect conditioned avoidance behaviour.

Although structural activity studies on the behavioural effects of ACTH-like peptides indicate the sequential requirements for peptide–receptor interaction (de Wied et al., 1975a), isolation of specific receptors for ACTH-like peptides in the brain has not been successful. It is of interest, however, that ACTH-like peptides have an appreciable affinity for stereo-specific opiate binding sites in synaptosomal plasma membranes of the rat brain. Terenius (1975) showed that both $ACTH_{1-28}$ and $ACTH_{4-10}$ have affinity for binding *in vitro*. Subsequent studies (Terenius *et al.*, 1975) pointed to an active site around $ACTH_{4-10}$ but a second affinity site might be present in a sequence adjacent to the 4–10 core. Recent observations suggest the presence of an endogenous ligand of morphine of a peptide nature in the CSF and in brain tissue (Terenius and Wahlström, 1975; Simantov *et al.*, 1976). These peptides appear to be present in the sequence 61–91 (C-peptide, Bradbury *et al.*, 1976), 61–76 (α-endorphin, Guillemin *et al.*, 1976) and 61–65 (Met- or Leu-enkephalin, Hughes *et al.*, 1975) of β-LPH. Interestingly, $ACTH_{4-10}$ represents the amino acid sequence 47–53 of the β-LPH molecule. It is believed that such peptides are derived from pituitary γ-LPH. These may be generated in the CNS by brain enzymes. These studies support the hypothesis that pituitary peptides act as prohormones for biologically active brain peptides in the same way as $ACTH_{4-10}$ is derived from ACTH or LPH to affect adaptive behavioural responses.

Uptake of corticosteroids in the brain and the characteristics of their receptors have been extensively delineated by McEwen *et al.* (1975) and by de Kloet and McEwen (1976). McEwen *et al.* (1969, 1970) and Knizley (1972) showed that labelled corticosterone is preferentially taken up by the hippocampus and the septum of adrenalectomized rats following its systemic injection. Although the radioactivity is distributed more or less uniformly along the length of the hippocampus, autoradiographic observations showed an intense concentration of corticosterone in the pyramidal neuronal zones of the CA_1 and CA_2 regions. Subcellular fractionation studies revealed that corticosterone is taken up by the cell nuclei and this uptake has a saturation point. Uptake in the soluble cytosol fraction, on the other hand, is not to saturation (McEwen and Wallach, 1973). From a behavioural point of view, it is of interest that dexamethasone does not have a regional preference for uptake as does corticosterone (de Kloet *et al.*, 1975). Progesterone is also taken up by hippocampal cells but is found only in the cytosol fraction. It is suggested that this steroid may compete with corticosterone in such a way that progesterone blocks the cytosol binding sites and thus the transfer of corticosterone to the nucleus (McEwen and Wallach, 1973). Deoxycorticosterone uptake in the brain seems to be highest in areas corresponding to the reticular formation, in a form of an anaesthetic metabolite

(Kraulis *et al.*, 1975). Autoradiographic studies also showed the existence of glucocorticoid-containing neurons in the hippocampus, septum and amygdala and neocortex (Stumpf and Sar, 1971; Gerlach and McEwen, 1972). These observations reinforce the notion that the limbic–midbrain system, especially the septum and the hippocampus, may be of primary importance in the mediation of the behavioural effects of corticosteroids. Furthermore, the specificity of the receptors and the central function of corticosterone in binding to these receptors may be an explanation for its preferential behavioural effect in which the hippocampus may play a central role.

This review does not include in its scope a discussion of the biochemical basis of the effect of pituitary–adrenal system hormones on brain and behaviour in detail but it is helpful to summarize the studies on the "cellular site(s)" of action.

The neurochemical effect of ACTH and ACTH fragments on macromolecular events in the brain has recently been reviewed by Schotman *et al.* (1976). Gispen *et al.* (1971) showed that hypophysectomy leads to a marked decrease in RNA content and to large polysomes in the brain stem and in the frontal cortex of the rat. Improvement of avoidance learning in hypophysectomized rats after the administration of $ACTH_{1-10}$ is associated with an increase in the polysome content of the brain stem. Peptide treatment without avoidance training does not restore the polysome pattern of hypophysectomized rats. Hypophysectomy decreases the incorporation of radioactive labelled leucine into rapidly turning over proteins in the brain stem, and $ACTH_{1-10}$ restores the incorporation towards normal. The stimulated incorporation of leucine by $ACTH_{1-10}$ is most marked in proteins of soluble and membrane origin. [D-Phe^7] $ACTH_{1-10}$ which has an effect opposite to that of [L-Phe^7] $ACTH_{1-10}$ on avoidance behaviour, further decreases the already impaired incorporation of radioactive leucine of hypophysectomized rats (Schotman *et al.*, 1972; Reith *et al.*, 1974, 1975). These observations suggest that a disturbance in protein synthesis in the brain stem may be responsible for the deficient behaviour of hypophysectomized rats and that these processes depend on ACTH and related peptides. Similar effects of ACTH fragments on brain protein synthesis have been reported by Reading and Dewar (1971), Rudman *et al.* (1974), Lloyd (1974) and by Dunn *et al.* (1976) in intact animals. Jakoubek *et al.* (1970, 1971, 1972) reported that the administration of a crude ACTH preparation results in inhibition of uridine incorporation into RNA in intact mice. Recently, Schotman *et al.* (1976) have shown in the rat that this effect of ACTH is mediated by the adrenal cortex. ACTH in adrenalectomized rats has the reverse effect and stimulates uridine incorporation.

These biochemical effects of ACTH fragments may not represent the first neurochemical event but a further step in the train of events which ultimately leads to a functional response of the nerve cells. According to the "second messenger" concept (Sutherland, 1972), polypeptide hormones interact with a receptor on the surface of the plasma membrane of the effector cell which results in an increased production of cyclic adenosyl monophosphate (cAMP). The cyclic nucleotide (second messenger) mediates information of peptide-cell membrane binding and subsequently triggers a number of biochemical events in the effector cell. That ACTH interacts with effector cells of the adrenal cortex in this way has been shown by Hofman et al. (1970) and Sayers et al. (1974). Observations of Wiegant and Gispen (1975) suggest that ACTH fragments affect brain metabolism in a similar way.

In contrast to ACTH fragments which exert their effect on the cell surface, glucocorticosteroids act intracellularly at the genomic level and certain enzymes in the brain are regulated via hormonal effects on the genome (McEwen et al. 1975; de Kloet and McEwen, 1976; De Vellis and Kukes, 1973).

Pituitary–adrenal system hormones may influence neurotransmitter systems in the brain (Versteeg and Wurtman, 1976). Increased noradrenaline (NA) turnover in several brain areas has been found after ACTH administration or adrenalectomy (Javoy et al., 1968; Fuxe et al., 1970, 1973). Corticosterone and dexamethasone normalize the increased NA turnover in adrenalectomized rats (Hökfelt and Fuxe, 1972). Conversely, hypophysectomy decreases NA turnover (Fuxe et al., 1970; Versteeg et al., 1972). Although $ACTH_{4-10}$ treatment increases NA turnover in intact rats (Versteeg, 1973; Leonard, 1974), it has no effect in adrenalectomized or hypophysectomized rats (Versteeg and Wurtman, 1975). Effects of pituitary–adrenal system hormones on central dopamine and serotonin systems have also been reported (Azmitia et al., 1970; Azmitia and McEwen, 1974; Versteeg et al., 1972; Friedman et al., 1973; Fuxe et al., 1973). Fuxe et al. (1973) postulated a correlation between central NA turnover and avoidance behaviour of the rat. It is, however, rather premature to assign a single transmitter system to the control of various forms of adaptive behaviour.

In conclusion, the behavioural effects of pituitary–adrenal system hormones seem to be localized in the limbic–midbrain system. The significance of this system in various forms of adaptive behaviour has for a long time been of interest in neurobiological research and it has been considered to play a key role in the organization of such processes as emotion, motivation and memory (see Papez, 1937; McLean, 1970; Isaacson, 1974). Putative receptors, at least for corticosteroids, have been found within this system and other biochemical studies suggest that either at the membrane or the genome level, ACTH and corticosteroids may have a modulatory function.

C. ELECTROPHYSIOLOGICAL EVIDENCE FOR THE CENTRAL NERVOUS ACTION OF PITUITARY–ADRENAL SYSTEM HORMONES

Early clinical studies on the effect of pituitary–adrenal system hormones on electrical activity of the brain stimulated experimental investigations in this respect. Torda and Wolff (1952) showed that a single administration of ACTH induces an increase in the electrical activity of the brain including an increase in EEG voltage, occasional spiking, paroxysmal runs of low frequency and high voltage waves, and lowers the convulsion threshold for pentamethylene tetrazol. ACTH appeared to be active in intact, hypophysectomized and adrenalectomized rats. ACTH-induced hyperexcitability of the brain does not, therefore, depend on the presence of the adrenal cortex. Woodbury (1952) studied the influence of ACTH on the electroconvulsive threshold (EST) in the rat. In contrast to Torda and Wolff (1952), his measure of convulsive activity was the appearance of tonic-clonic seizures which develop at much higher electroshock intensity than the electrocortical spiking. ACTH alone slightly increased EST and prevented the substantial increase in EST as a result of deoxycorticosterone treatment. The peptide, however, failed to affect cortisone-induced decrease in EST. Wasserman et al. (1965), on the other hand, reported that ACTH causes a reduction in the threshold for minimal clonic electroshock seizures in young rats and this effect appeared to be independent of the adrenal cortex. Accordingly, ACTH may alter brain excitability but the physiological significance of this effect is not clear.

Subsequent electrophysiological observations with ACTH and related peptides showed that β-MSH but not ACTH facilitates the ventral root response to dorsal root stimulation in the cat with low spinal transection (Krivoy and Guillemin, 1961). Nicolov (1967) found that ACTH increases the electrical activity of the spinal cord of the dog. Cortisol induced similar changes indicating that the effect of ACTH was mediated through the adrenal cortex. Korányi and

Endröczi (1967) however found that ACTH inhibits the polysynaptic flexor reflex in the rabbit spinal cord. More recent observations by Krivoy *et al.* (1974) showed again a facilitatory effect of ACTH and related peptides on the spinal cord. They found that $ACTH_{1-24}$ and β-MSH antagonize the depressive effect of morphine on mono- and polysynaptic reflexes in the cat spinal cord.

Studies on peripheral nerve and skeletal muscle indicate an action of ACTH and ACTH fragments on the neuromuscular junction. ACTH administration increases muscle action potential amplitude and contraction height and delays fatigue in normal, adrenalectomized and hypophysectomized rats (Strand *et al.*, 1973/74). Subsequent experiments of Strand and Cayer (1975) showed that ACTH fragments like $ACTH_{4-10}$ and α- and β-MSH facilitate muscle and nerve action potentials in hypophysectomized rats. The authors suggested that ACTH and ACTH-fragments act as modulators of peripheral excitatory systems, capable of raising a lowered response to normal levels rather than increasing normal parameters.

The studies on the spinal cord and the peripheral nerve indicated the neurotropic activity of ACTH and related peptides independent of the adrenal cortex. These peptides also influence brain electrical activity at the supraspinal level. Although the number of studies with ACTH fragments is increasing, most data have been obtained with whole ACTH, so that changes may have been due to a genuine neurotropic effect of the peptide or mediated through the adrenal cortex. Dyster-Aas and Krakau (1965) found slow wave activity and the absence of cortical activation after arousing stimuli in rabbits given α-MSH. Suppression of conditioned EEG arousal after ACTH administration was also observed by Korányi *et al.* (1965/66) in the rabbit. ACTH in the same species decreases electrical activity of the hippocampus and the reticular formation. Cortisol seems to have opposite effects on the same structures as judged from the EEG analysis of unanesthetized rabbits (Kawakami *et al.*, 1966). Korányi and Endröczi (1967) reported that ACTH administration in the rabbit is followed by a burst of spindles with a relatively short latency in the septal area and in the somatomotor cortex. One may recall that ACTH treatment in this species, in contrast to the rat, suppresses conditioned avoidance behaviour. In the rat, Sandman *et al.* (1971b) found high voltage, slow activity in the cortex, resembling limbic activity, after the administration of α-MSH. A similar effect of α-MSH on EEG activity was found by Denman *et al.* (1972) in the frog. Sawyer *et al.* (1968) showed that ACTH increases multiple unit activity (MUA) with a relatively short latency in the basal hypothalamus and dorsolateral thalamus and was followed by a depression of long latency. The latter change was absent in adrenalectomized rats but could be mimicked by dexamethasone administration. Studies by Steiner *et al.* (1969) using direct iontophoretic application of ACTH in the hypothalamus and the midbrain showed an increase in single unit activity after peptide administration while dexamethasone led to a decrease in activity. Increased hypothalamic single unit activity after systemic ACTH administration in the rat was observed by van Delft and Kitay (1972). Pfaff *et al.* (1971) found that ACTH administration in intact and hypophysectomized freely-moving or urethane anesthetized rats increased while corticosterone decreased unit activity in the dorsal hippocampus. Phillips and Dafny (1971) found cells in the CA_3 layer of the rat hippocampus which decreased their activity after cortisol administration while other cells increased their activity. Michal (1974) observed, on the other hand, that MUA activity in the rat hippocampus is inhibited by local dexamethasone application.

Gray (1974) reported that $ACTH_{4-10}$ increases the threshold of septal driving of hippocampal theta activity in the frequency of 7.7 Hz in the rat, while corticosterone selectively lowers the threshold to evoke theta rhythm with a frequency of about 7 Hz. These findings seem to indicate that ACTH and corticosteroids affect the excitability of either the medial septal "pacemaker" cells (Stumpf, 1965) or hippocampal theta generating cells in an opposite way, but only at selective frequencies. In an earlier experiment, Gray (1972) showed that septal driving of hippocampal theta activity with a frequency of 7.7 Hz decreases the resistance to extinction of a rewarded runway response. Corticosterone affects this behaviour in a similar way (Garrud *et al.*, 1974). In the rat, a hippocampal frequency of 7 Hz generally accompanies the performance of fixed action patterns, such as eating, drinking, and grooming (Routtenberg, 1968). The frequency around 7.7 Hz was related to novelty or frustrative nonreward (Gray and Ball, 1970), while frequencies of 8.5 to 10 Hz accompanied the initiation of the performance of learned patterns of behaviour (Elazar and Adey, 1967; Gray and Ball, 1970; Vanderwolf, 1969). Although corticosterone-influences and "extinction" frequencies are slightly different, the effect of corticosterone on the septal–hippocampal complex may be of functional significance in corticosteroid-induced enhancement of behavioural extinction. A shift in the dominant frequency of hippocampal rhythmic slow activity during prestimulus "facing" periods of a conditioned food reinforcement operant response was observed by Urban *et al.* (1974) in the dog after $ACTH_{4-10}$ administration. This peptide led to a shift in the dominant frequency in the direction of lower frequencies without affecting the stabilized behaviour. It is of interest that $ACTH_{4-10}$ induces a shift in the dominant frequency of hippocampal and posterior thalamic rhythmic slow activity (RSA or theta activity) from 7.0 to 7.5 Hz and increases the appearance of 7.5 to 9.0 Hz components when RSA is evoked by electrical stimulation of the reticular formation of the freely moving rat (Urban and de Wied, 1975). Since similar frequency shifts are produced by increasing the strength of stimulation it was suggested that the ACTH fragment increases the state of arousal in limbic-midbrain structures. Activation of the reticular activating system which induces cortical arousal is accompanied by theta activity in the hippocampus (Green and Arduini, 1954), and the frequency of the theta rhythm upon the intensity of the stimulation of the midbrain reticular formation (Klemm, 1972). These observations and the fact that higher theta frequencies are associated with the initiation of performance may support the hypothesis that augmentation of the state of arousal by $ACTH_{4-10}$ in the limbic–midbrain system increases the probability of a given behavioural performance (de Wied et al., 1975b).

A facilitatory influence of $ACTH_{1-24}$ administered intraventricularly in the rabbit on brain electrical activity has been shown by Baldwin et al. (1974). They found an increased MUA in the region of the lateral preoptic-diagonal band of Broca and the periventricular preoptic area. Facilitation of a central nervous vigilance regulating system was suggested from studies with ACTH fragments by Wolt-huis and de Wied (1976). They found that $ACTH_{4-10}$ without affecting the latencies and amplitudes of visually evoked primary potentials, diminished the amplitude of the late components at different light intensities. An inhibitory influence of ACTH on brain electrical activity was reported by Korányi et al. (1971a). They found that ACTH decreases MUA in the mesencephalic reticular formation, midline thalamic nuclei and in the hypothalamus during attentive behaviour and paradoxical sleep and in response to sensory stimulation both in intact and adrenalectomized cats. They suggested that both ACTH and corticosteroids decrease brain excitability and in this way might influence habituation processes. In a further study Korányi et al. (1971b) found that MUA activity during slow wave sleep is decreased by cortisol in the mesencephalic reticular formation, the medial forebrain bundle, the medial preoptic area, the centromedian thalamic nuclei and the fornix. ACTH administration resulted in a decrease of spontaneous MUA and the response to reticular stimulation in the medial preoptic area while increased

MUA was observed in the fornix and the medial forebrain bundle in response to reticular stimulation.

Feldman *et al.* (1961) reported that ACTH and also cortisol and adrenocortical extract increases the amplitude of negative potentials in the multisynaptic system of neurons extending through the central part of the midbrain and the diencephalon. They suggested that the site of action of corticosteroids is in the mesencephalic reticular formation. Slusher *et al.* (1966) recorded unit activity in the diencephalon and the midbrain of curarized cats and showed that the activity of the cells increased after cortisol injection but this acceleration was followed by a decrease in unit activity. Intracerebral injection of cortisol was followed by an acceleration of midbrain units but the appearance of the effect occurred much later than after systemic injection. Intrahypothalamic local administration of cortisol caused acceleration of unit activity in the same cells in the midbrain and deceleration in other cells. This indicates that cortisol may influence different cell populations and that the response pattern may be biphasic—acceleration followed by deceleration. Endröczi *et al.* (1968) reported a similar biphasic response pattern after systemic administration of cortisone in curarized rats. The amplitude of evoked potentials in the brain stem reticular formation and in the medial hypothalamus after sciatic nerve stimulation was first increased by cortisone and followed by attenuated potentials 30–40 min later. Martin *et al.* (1975) found that cortisol increases the occurrence of spontaneous RSA in the rabbit hippocampus but RSA evoked by sensory stimulation appeared to be unaffected. McGowan and Timiras (1975), on the other hand, showed that corticosterone increases the amplitude of evoked potentials in the dorsal hippocampus and in the thalamus of the rat after somatosensory and visual stimulation.

Profound influences of corticosteroids on brain excitability were found by Woodbury *et al.* (1957). Cortisol and cortisone markedly decreased EST in the rat, while deoxycorticosterone and deoxycortisol increased the threshold. It is of interest that corticosterone had no substantial effect on this parameter of brain excitability. The threshold to evoke propagated seizure activity by stimulation of the dorsal hippocampus appeared to be lowered by cortisone in the cat (Lissák and Endröczi, 1962) and by cortisol in the rat (Conforti and Feldman, 1975). No changes were observed in the threshold in other structures such as the amygdala or the reticular formation. Feldman (1966) found that intraventricular administration of cortisol in the cat leads to convulsions. Convulsions were observed in rats, cats and monkeys even following systemic administration of deoxycortisol

(Heuser and Eidelberg, 1961; Heuser *et al.*, 1965). Endröczi (1969) showed that local injection of cortisol, cortisone and deoxycortisol into the rat hippocampus markedly decreased the threshold to evoke propagated seizures. Corticosterone had only slight effects while deoxycorticosterone increased the threshold. Feldman (1971) reported that local infusion of cortisol into the hippocampus or into the posterior hypothalamus or the midbrain reticular formation initiated epileptic discharge. The significance of the increased susceptibility for convulsions for the behavioural action of corticosteroids is not clear. It is known that electroconvulsive shock and stimulation of the hippocampus either electrically or chemically which is followed by epileptiform discharges results in amnesia and performance deficits (see Nakajima, 1975b). Thus, corticosteroid-induced epileptiform discharges may explain at least some of the suppressive effects of adrenocortical hormones on behaviour. There are, however, some findings which disagree with this idea. First, corticosterone in the rat does not substantially affect the threshold of evocation to propagated seizures while it markedly influences conditioned behaviour. Secondly, corticosterone, even if it is administered directly in the layer of the apical dentrites of the pyramidal cells in CA_1 and CA_2 regions of the dorsal hippocampus, does not evoke spontaneous discharges (Endröczi, 1969). It is more likely that the convulsive phenomena are of a pharmacological nature and stimulate the occurrence of epilepsy which has often been noted in patients on steroid therapy (e.g. Streifler and Feldman, 1953; Glaser, 1953).

Adrenalectomy decreases EST and the susceptibility to audiogenic seizures (Woodbury, 1958; Woodbury and Vernadakis, 1967). On the other hand, Feldman and Robinson (1968) found an increased CNS excitability after adrenalectomy. Chambers *et al.* (1963) reported that transmission is retarded, the latency is extended and the peak of multisynaptic components of cortical potentials evoked by reticular formation stimulation is delayed in adrenalectomized cats receiving deoxycorticosterone. Cortisone restored the response partially while dexamethasone appeared to be more effective in this respect. Impairment of frequency filtering capacity of the non-specific thalamic nuclei and the absence of electroencephalographic after-reaction after mesencephalic reticular formation stimulation was observed in adrenalectomized rats by Korányi and Endröczi (1970). Systemic administration of cortisol but not of aldosterone normalized these alterations indicating that the functional deficits were due to the absence of glucocorticosteroids rather than to that of mineralocorticosteroids.

Adrenalectomy abolishes the circadian rhythm in paradoxical sleep patterns in the rat. It is re-established by cortisol administration though now out of phase (Johnson and Sawyer, 1971). Hypophysectomy, on the other hand, results in a diminution of the total sleeping time in the rat. Paradoxical sleep is more reduced than slow wave sleep (Valatx et al., 1975). ACTH treatment progressively increased the duration of paradoxical sleep in hypophysectomized rats. These authors suggested that hypophysectomy alters the regulation or triggering of sleep rather than basic sleep mechanisms. Although direct evidence is not yet available, the behavioural deficits of hypophysectomized rats may be related to alteration of the sleep cycle. Lucero (1970), Leconte et al. (1973), Smith et al. (1972) and others demonstrated a close correlation between paradoxical sleep duration, learning and memory in the rat.

Electrophysiological observations indicate that ACTH and related peptides and corticosteroids, as on avoidance behaviour, affect brain electrical activity in an opposite way. Most of the data suggest that ACTH and related peptides facilitate, whilst corticosteroids suppress, electrical activity of the limbic–midbrain system. This supports the hypothesis that the behavioural effect of these hormones is mediated through this system. This conclusion is obviously oversimplified. A great number of contradictory observations exist, probably as a result of differences in the various species, in the forms of anaesthesia used, and in the use of species–specific or aspecific corticosteroids. Most of the electrophysiological studies were aimed at exploring the feedback function of the pituitary–adrenal system hormones in the brain, though, obviously, more is required to determine the precise electrophysiological effects.

D. PITUITARY–ADRENAL SYSTEM HORMONES AND PSYCHOPHYSIOLOGICAL CORRELATES OF ADAPTIVE BEHAVIOUR

Autonomic responses have long been used as an index of emotional behaviour. Cardiovascular changes, particularly in heart rate, have been regarded as a measure of psychological processes underlying behaviour such as motivation, attention, and arousal. Data on autonomic correlates of emotional behaviour as affected by pituitary–adrenal system hormones are relatively scarce. Bohus (1975b) showed that $ACTH_{4-10}$ which facilitates the retention of a passive avoidance response of the rat alters heart rate during this behaviour. Mean heart rate of control rats decreases relative to pre-learning rate, and the degree of the bradycardic response to fear as a result of shock punishment depends upon its intensity during the single learning trial

(Bohus, 1974b). In contrast, facilitated passive avoidance in rats treated with $ACTH_{4-10}$ is accompanied by increased mean heart rate relative to the pre-learning values. According to the unidimensional activation theory, increased heart rate indicates an increased arousal. Bohus (1977) have further analysed the electrocardiographic changes during the entire passive avoidance period and found that $ACTH_{4-10}$ treatment results in a change in the distribution pattern of the interbeat (R–R) intervals rather than causing a shift in the dominant R–R interval length which is typical for this emotional behaviour. $ACTH_{4-10}$ increased appearance of short R–R intervals. In contrast, the distribution histogram in control rats showed a stepwise decline in the direction of longer R–R intervals. This may indicate that sympathetic influences on the heart rate, which are normally minimal, are facilitated by $ACTH_{4-10}$. Contemporary psychophysiological research maintains that heart rate does not provide a simple, unidimensional index of brain processes underlying behaviour such as motivational or affective states. Moreover, it seems that the vagally mediated heart rate in a mildly stressful behavioural paradigm is related to attentional and expectancy processes while sympathetic influences are evoked by more intense stress in which the organism is actively engaged in the preparation or execution of activities that will cope with the stress (Obrist et al., 1974; Pribram and McGuiness, 1975). Accordingly, increased sympathetic activity in $ACTH_{4-10}$-treated rats may reflect a facilitated arousal state which increases the probability of a given behavioural performance in emotional situations. Observations in a classically conditioned emotional situation corroborate this view. $ACTH_{4-10}$-treated rats showed accelerated heart rates during extinction of a classically conditioned fear response. Parallel to this generalized emotional activation (arousal), extinction of the conditioned cardiac response which appeared to be a heart rate deceleration was delayed in the peptide treated rats (Bohus, 1973). That is to say, facilitated arousal increases the probability that a learned autonomic response will occur. This observation also indicates that ACTH-like peptides not only affect instrumental but also learned autonomic responses. As with the effect of corticosteroids on extinction behaviour, corticosterone treatment facilitates extinction of a classically conditioned cardiac response and attenuates the generalized emotional activation (Bohus, 1973). Thus, corticosterone attenuates arousal and decreases the probability that learned autonomic responses occur.

Species differences may, however, complicate the picture. Korányi and Endröczi (1967) reported that ACTH administration inhibits conditioned vasomotor responses in the rabbit. Furthermore, the conditioning paradigm in psychophysiological studies may determine the autonomic response pattern. In experiments in which fear conditioning was preceded by an intensive habituation training, and the conditioned cardiac response was rather weak, cortisone-treated rats showed an increased heart rate during extinction (Bohus *et al.*, 1970b). Psychophysiological observations in man receiving ACTH-related peptides were also reported. Endröczi *et al.* (1970) reported the recovery of a stimulus-specific EEG arousal pattern in habituated subjects after the administration of $ACTH_{1-10}$. Increased somatosensory evoked responses were found in subjects receiving α-MSH (Kastin *et al.*, 1971). $ACTH_{4-10}$ appeared to increase the power output of the 7–12 Hz and higher than 12 Hz frequence bands of occipital EEG and a recovery of alpha-blocking response occurred in habituated persons during attention, memory and anxiety tasks. $ACTH_{4-10}$ has a slight effect on contingent negative variation but not on any measures of autonomic activity (Miller *et al.*, 1974).

E.   BEHAVIOURAL EFFECTS OF ACTH-LIKE PEPTIDES AND
     CORTICOSTEROIDS

Recent advances on the nature of the behavioural effects of ACTH-like peptides and corticosteroids on the pituitary and brain processes has necessitated a re-evaluation of some of our interpretations outlined above.

Most of the earlier data favoured a hypothesis that retrieval of the memory is improved by peptides. However, Gold and Buskirk (1976a) suggested that storage of memory is also affected by ACTH since administration of low doses of the peptides after the learning trials facilitated retention of a passive avoidance response while a high dose impaired it. The facilitatory effect depends upon the interval between the learning trial and the treatment. Delayed injection (2 h) had no effect on later retention (Gold and Buskirk, 1976b). Correction of the deficit of a discriminative active avoidance response in hypophysectomized rats by post-training administration of ACTH also appeared to be time-dependent (Gold *et al.*, 1977). Corticosterone administration failed to mimic the effects of ACTH (Gold and Buskirk, 1976b) so that an extra-adrenal effect of the peptide is likely, particularly since later retention of an avoidance response in mice is facilitated by post-learning injection of $ACTH_{4-10}$ in a time-dependent manner (Flood *et al.*, 1976). Post-trial administration of the 7–D–Phe analogue of $ACTH_{4-10}$, which appeared to affect avoidance behaviour oppositely to that of $ACTH_{4-10}$, impairs later retention, potentiates anisomycin-induced amnesia and opposes the memory-facilitating effect of $ACTH_{4-10}$. On the other hand, van Wimersma Greidanus (1977) failed to observe an effect of $ACTH_{4-10}$ on memory storage in the rat even when a high dose of the peptide was administered after the learning of a passive avoidance response. Thus ACTH and related peptides may influence memory storage (consolidation) processes; the variables that determine the effectiveness of the peptide or its physiological significance are unclear.

In recent years a rather unconventional technique has been applied to aversive stimuli, and named taste aversion, flavour toxicosis, bait shyness (Revusky and Garcia, 1970; Garcia et al., 1974). In the technique a preferred, mostly sweetened substance is presented to the animal and then followed by the injection of toxic substances or X-rays to make the animal sick. On re-exposure to the preferred substances the animal avoids consumption of the substance for a considerable time. The technique shares certain properties with the conventional shock-induced passive avoidance—e.g. single exposure to the aversive stimulus—but the delay of more than 30 min between the preferred substance and toxic treatment may reflect a peculiar form of learning. Levine and his associates (Kendler et al., 1976; Hennessey et al., 1976; Levine et al., 1977) investigated the effects of pituitary–adrenal system hormones on conditioned taste aversion in the rat using LiCl injection as the aversive stimulus. Dexamethasone phosphate, administered prior to the LiCl injection to suppress pituitary–adrenal activation by the toxic drug, attenuated taste aversion. Conversely, ACTH treatment during recovery prolonged extinction of taste aversion. $ACTH_{4-10}$ had a similar but less pronounced effect, a finding in accordance with the observations of Rigter and Popping (1976). Administration of ACTH prior to the learning trial tended to attenuate the aversion, probably as a result of the elevated endogenous corticosterone levels. The effect of dexamethasone is most probably due to an intrinsic behavioural effect, independent of the suppression of pituitary ACTH release. Moreover intracerebral implantation of cortisol which blocks the increase in plasma corticosterone fails to alter the recovery of taste aversion (Levine et al., 1977). These observations using this unconventional learning model are thus consonant with data obtained with conventional passive avoidance techniques, and suggest a primarily retrieval effect of ACTH and related peptides.

Hormonal sensitivity of the developing nervous system (organizational effects) has been extensively studied in relation to gonadal steroids and thyroid hormones. Developmental effects of corticosteroids have been discussed earlier in this Chapter (cf Section III, B). Recently, a number of observations suggest that ACTH related peptides affect the developing brain and thereby later behaviour. In the rat, ACTH injections from postnatal days 3 to 5 resulted in a facilitated passive avoidance behaviour at 22 to 24 days of age (Nyakas, 1973). ACTH and a number of analogues injected subcutaneously as long-acting zinc-phosphate preparations into rats at an age of 3 days accelerate eye-opening. $ACTH_{1-39}$, $ACTH_{1-24}$, $ACTH_{1-18}$, and $ACTH_{1-16}$ were effective. Smaller N-terminal fragments failed to affect eye-opening (Van der Helm–Hylkema and de Wied, 1976). Beckwith et al. (1977) injected α-MSH intraperitoneally into rat pups from two through seven days of age and found that it affected the efficiency of response performance on a differential reinforcement of low rate responding (DRL) operant schedule at a juvenile age. The treatment also improved performance in learning a discrimination problem, the reversal of the problem and an extra-dimensional shift at adult age of the males (but not of the females). In shuttle-box avoidance behaviour tests performed at 90 days of age rats treated either neonatally or as adults displayed similar facilitated acquisition and extinction patterns. The latter finding disagrees with findings in other laboratories. The authors argue that the strict acquisition criterion applied in their study contrasts with that of others in which enhancement of retention during extinction followed "incomplete" learning. Closer inspection of Beckwith et al.'s (1977) data, however, indicates that their rats showed a rather poor performance during the acquisition criterion. Champney et al. (1976) injected a potentiated ACTH analogue (8–D–Lys–9–Phe) $ACTH_{4-9}$ into the cerebral ventricles 23 h after birth and studied adult behaviour. Passive avoidance behaviour and the learning of a discrimination problem were unaffected. However, both neonatal and adult treatments facilitated reversal learning and visual orientation in both sexes. These effects were taken to indicate that the peptide enhanced attention.

In summary, these observations clearly indicate developmental effects of ACTH-like peptides but the critical period for these effects has yet to be determined. Van der Helm–Hylkema and de Wied (1976) showed that administration of ACTH and its analogues on the 12th day of age does not affect eye-opening. It seems that "organizational" and "activational" effects of the peptides are of a similar character.

Significant observations have recently been published on the uptake of ACTH fragments by the brain. Greenberg *et al.* (1976) determined the brain uptake index for α-MSH after rapid intracarotid arterial injection. The peptide readily penetrated and was uniformly distributed within the brain. Kastin *et al.* (1976a) found that radioactivity associated with ^{3}H-α-MSH labelled at tyrosine, position 2, after intracarotid injection, is mostly localized in the occipital cortex, cerebellum and in the pons-medulla region. The lowest activity was found in the pons-medulla when ^{3}H-tyrosine was injected into the carotid artery. Uptake of a modified ACTH analogue (7–^{3}H–Phe– D–D–Lys ACTH$_{4-9}$) by isolated brain nuclei was studied by Verhoef *et al.* (1977a) after intracerebroventricular administration of the peptide. The highest uptake of the peptide as such was found in the dorsal and fimbrial septal nuclei. Subsequently the uptake in the septum was shown to be higher in hypophysectomized than in intact rats. The specificity of the uptake is indicated by the fact that treatment of hypophysectomized rats with ACTH$_{1-24}$ or ACTH$_{4-10}$ decreases the uptake of the labelled analogue. The uptake remains unchanged after the administration of (D–Phe^7) ACTH$_{4-10}$, β- LPH$_{61-76}$ or ACTH$_{11-24}$ (Verhoef *et al.*, 1977b).

The physiological route for the transport of ACTH and related peptides into the brain is most probably along the pituitary stalk and basal hypothalamus. Oliver *et al.* (1977) studied the transport of several pituitary hormones in the pituitary stalk vasculature by determining hormone concentrations at the proximal end of the stalk. They suggest that, among other pituitary hormones, ACTH and α-MSH are transported retrogradely through certain vascular channels along the pituitary stalk toward the hypothalamus. Appearance of (^{3}H–Phe^7–D–Lys^8) ACTH$_{4-9}$ in the brain after intrapituitary and intracellular injection of the peptide was studied by Mezey *et al.* (1978). The evidence suggests transport of the peptide from the pituitary via the stalk to the hypothalamus. Transport to other brain areas may occur via the cerebrospinal fluid.

Neuropeptides related to ACTH may, however, be produced by the brain itself. Earlier observations (Guillemin *et al.*, 1962; Schally *et al.*, 1962; Rudman *et al.*, 1973), suggesting the presence of biologically active and immunoreactive ACTH- and MSH-like material in the brain, have been confirmed (Oliver *et al.*, 1976; Swaab, 1976; Krieger *et al.*, 1977a, b; Loh and Gainer, 1977). Large amounts of bioassayable, immunoreactive ACTH were found in the median eminence and the basal hypothalamus and smaller quantities in the hippocampus, cerebral cortex, cerebellum, thalamus, preoptic region, amygdala and septum (Krieger *et al.*, 1977a, b). Immunoreactive MSH-like and both immuno- and bioactive ACTH-like materials are present in the brain long after hypophysectomy (Oliver *et al.*, 1976; Swaab, 1976; Vaudry *et al.*, 1976; Krieger *et al.*, 1977a, b). Accordingly, peptides related to ACTH in the brain may not be of pituitary origin. The functional significance of these "brain-borne" peptides is however not clear. As noted earlier, removal of the pituitary gland results in impaired avoidance behaviour, an effect corrected by peripheral administration of small amounts of ACTH or related peptides (de Wied, 1964; Bohus *et al.*, 1973; Gold *et al.*, 1977). Thus, it may well be that ACTH- or MSH-like materials are present in the brain in behaviourally inactive forms and the generation of active principles is impaired after the removal of the pituitary gland.

During the last few years, significant advances have been made with regard to the opiate-like peptides of pituitary and brain origin. The C-terminus of β-LPH (β-LPH$_{61-91}$) and its fragments (endorphins, enkephalins) have profound opiate-like activities (see Snyder and Simantov, 1977). Potent antinociceptive activity, especially of β-endorphin (β-LPH$_{61-91}$), was described by several groups of investigators. Intracerebroventricular or intracysternal administration of β-endorphin produces catalepsy which can be prevented by opiate antagonists (Bloom *et al.*, 1976; Jacquet and Marks, 1976; Chretien *et al.*, 1977; Loh and Li, 1977; Guillemin *et al.*, 1977). Met-enkephalin (β-LPH$_{61-65}$) α- and γ-endorphin (β-LPH$_{61-76}$ and $_{61-77}$) failed to produce catalepsy (Guillemin *et al.*, 1977). All these peptides however caused acute episodes of wet shake behaviour after intraventricular injection; the response can also be prevented by opiate antagonists (Bloom *et al.*, 1976; Guillemin *et al.*, 1977). Subanalgesic but still high doses of β-endorphin disrupted male sexual behaviour (Meyerson and Terenius, 1977) and lever pressing behaviour for food reward (Lichtblau *et al.*, 1977). These effects of the peptide could also be reversed by opiate antagonists.

As noted earlier, ACTH and its analogues have low affinity for rat brain opiate receptors *in vitro* (Terenius, 1975; Terenius *et al.*, 1975). Like morphine (Gispen and Wiegant, 1976), ACTH and a number of its analogues induce excessive grooming in the rat after intracerebroventricular administration (Gispen *et al.*, 1975). ACTH-induced excessive grooming can be attenuated by opiate antagonists (Gispen and Wiegant, 1976). C-terminus fragments of β-LPH administered intraventricularly also induce grooming. β-endorphin appears to be the most potent but α-endorphin and β-LPH$_{61-69}$ possess some activity (Gispen *et al.*, 1976). Thus, when ACTH-like peptides in μg amounts are administered centrally, a certain behavioural response resembling that of opiate-like peptides is induced. An action on brain opiate receptors is apparent. Profound behavioural effects of endorphins and enkephalins, dissociated from the opiate-like effects and independent of opiate receptors, have also been observed. Compared with $ACTH_{4-10}$, Met-enkephalin administered subcutaneously in µg quantities, appeared to be as potent, while β-LPH_{61-69} and α-endorphin were 10 times more potent in delaying extinction of a pole-jumping avoidance response (de Wied, 1977). Rapid extinction of a pole-jumping avoidance response as caused by the specific opiate antagonist naltrexone, could be prevented by α-endorphin, Met-enkephalin and ACTH analogues (de Wied et al., 1978). β-LPH, Met-enkephalin and Leu-enkephalin administered peripherally in low doses appeared to alleviate CO_2-induced amnesia in the rat. The opiate antagonist naloxone did not block the antiamnesic effects of these peptides (Rigter et al., 1977a, b). Met-enkephalin administered intraperitoneally improved maze performance by rats. $(D$-$Phe^2)$-Met-enkephalin, a compound almost devoid of opiate activity, had a comparable effect (Kastin et al., 1976b). The amino acid sequence $ACTH_{4-10}$, which was designated as the behaviourally active entity of ACTH-like peptides (de Wied, 1969), is shared by ACTH, α- and β-MSH and by β-LPH. It represents the sequence β-LPH_{47-53} in the N-terminus of β-LPH. It is not yet known whether ACTH or β-LPH is the prohormone from which $ACTH_{4-10}$ is generated. Several observations suggest a close relationship between ACTH, β-LPH and endorphins. A common precursor to ACTH, β-LPH and endorphins ("very big" ACTH) has been suggested (Orth and Nicholson, 1977; Mains et al., 1977). Immunological studies showed that β-LPH, endorphins and ACTH occur in the same pituitary cells (Moriarty, 1973; Phifer et al., 1974; Bloom et al., 1977) and within the same secretory granules (Moriarty, 1973). Furthermore, stimuli that alter ACTH release from the pituitary alter β-LPH release in a parallel manner (Abe et al., 1969; Gilkes et al., 1975; Chretien et al., 1977). $ACTH_{4-8}$ (β-LPH_{47-51}) may be generated from β-LPH by tryptic digestion (Graf, 1976). Structure–activity studies showed that the sequence $ACTH_{4-7}$ contains the essential elements for the behavioural effect of ACTH analogues (Greven and de Wied, 1973). The same "message" as in $ACTH_{4-10}$, however, may also be conveyed by the sequence $ACTH_{7-9}$ which may become expressed by chain elongation (de Wied et al., 1975). $ACTH_{7-16}$ appeared to be as active as $ACTH_{4-10}$. An analogue of $ACTH_{7-16}$ is one million-fold more potent than $ACTH_{7-16}$. A doublet of basic lysine residues seems to be essential for this great behavioural potency at exactly the same distance from $ACTH_{7-9}$ as is the case in natural ACTH (Greven and de Wied, 1977). Since only ACTH fulfils this structural requirement, it may be that different behaviourally active fragments originate from ACTH and β-LPH.

Neurochemical responses, possibly underlying the mechanisms of behavioural action of ACTH-like peptides, have been discussed by Dunn and Gispen (1977). Much evidence suggests that ACTH and related peptides affect cerebral protein synthesis, RNA synthesis, protein phosphorylation, cyclic nucleotide metabolism and the turnover of dopamine, noradrenaline and serotonin (see also p. 310 of this chapter). There is however no evidence that these responses are primary. The authors postulate that cell surface receptors may occur only in certain brain cells and that more than one type exists. The latter suggestion is based upon the diversity of behavioural and biochemical effects of ACTH, ACTH-fragments and analogues with D-amino acid isomers. The authors suggest that receptor interaction stimulates the synthesis of cyclic AMP and/or GMP and probably prostaglandins. The cyclic nucleotides would then act via protein kinases to activate or inhibit cellular chemical processes. An action on cyclic AMP may be involved in an interaction with transmitter release. Clearly there are many potential mechanisms but identification of specific receptors in localized brain areas is essential before the cellular bases of behavioural effects of neuropeptides are understood.

Observations related to the behavioural action of corticosteroids have aimed at elucidating the neurochemical and cellular bases for these effects. Kovacs et al. (1976, 1977) investigated the influence of corticosterone on active and passive avoidance behaviour in relation to brain serotonin content in the rat. At low and high doses corticosterone affects the extinction of an active avoidance response and the passive avoidance behaviour in the opposite manner. The behavioural effects were correlated with opposite changes in serotonin contents, primarily in the hypothalamus and mesencephalon. Low doses of corticosterone (100–500 μg) facilitated avoidance extinction and passive avoidance behaviour and increased serotonin levels. Conversely, high doses of the steroid (1.0 mg/100 g) resulted in resistance to extinction and impaired passive avoidance behaviour with decreased serotonin levels. These observations and experiments with drugs interacting with the serotoninergic system suggested possible involvement of the central serotoninergic system in the mediation of the behavioural actions of corticosterone.

The role of putative hippocampal corticosterone receptors in the behavioural action of steroids has been studied by Bohus and de Kloet

al. (1977) studied the influence of synthetic α-MSH and β-MSH$_{1-22}$ on mental functions in men: α-MSH improved, and β-MSH$_{1-22}$ reduced, verbal memory performance without an effect on visual memory. The peptides failed to influence contingent negative variation (CNV), which is a sensitive measure of the stimulant and depressant effects of psychotropic drugs.

These data clearly indicate behavioural effects of ACTH-like peptides in normal man. The postulated psychological processes behind the peptide effects in men (attention, arousal, vigilance, motivation, short-term memory, etc.) are more or less the same as in animal experiments. Negative findings have been reported in man. In animal research the importance of variables such as motivational level, task complexity, insensitivity of measure, etc. as the cause of ineffectiveness of the peptide are generally recognized. In contrast, these variables, not well studied in human research, are often oversimplified by some. It may also be that the young, healthy volunteers participating perform almost optimally in the performance tasks. This leaves a rather narrow space to influence mental performance by the peptides which have none of the "side" effects of most psychostimulant drugs. One might, therefore, expect more obvious mental effects of ACTH-like peptides in patients where psychological functions are impaired.

Indeed, Endroczi (1972) reported that ACTH$_{1-10}$ relieved symptoms of depression in depressive patients. In a more extensive study Sandman *et al.* (1976) found that ACTH$_{4-10}$ improves attention of stimulus processing in mentally retarded adult men. Ferris *et al.* (1976) found that ACTH$_{4-10}$ affects cognitive functions in impaired geriatric patients. The effect of the peptide on various cognitive functions (improvement or impairment) depended upon the severity of cognitive alterations and the dose of the peptides. Effects of ACTH$_{4-10}$ on electroconvulsive treatment(ECT)-induced memory dysfunctions in psychiatric patients have been studied by Small *et al.* (1977). Although there were some positive effects of the peptide after a single ECT, no effects were found between seizures after five or six ECTs. Accordingly, if ACTH$_{4-10}$ does exert anti-amnesic effects in man, then they are rather subtle.

The number of clinical observations is obviously insufficient to draw any conclusions. It remains an inviting prospect to explore further the potential applicability of these peptides in the mentally ill and to determine whether decreased or increased production of ACTH-like or other neuropeptides contribute to mental dysfunctions in young, adult and elderly people.

The developments as reviewed here may add to a better under-standing of in what way ACTH and related peptides and corticoster-oids affect brain processes contrariwise and indeed the physiological significance of these actions. We suggest that the pituitary gland and the pituitary–adrenal axis represent two separate systems affecting brain functions. Neuropeptides originating from prohormones like ACTH and β-LPH are transported to the brain and act on different receptors. These peptides may be released from the pituitary as a result of enzymatic fragmentation of the parent molecule. A possibil-ity is that such fragments are formed by brain enzymes eventually from brain-borne parent molecules. It remains to be seen whether ACTH- and β-LPH related peptides serve different physiological functions in the brain. How far these peptides are involved in pathological processes is unknown, but it is possible that these components exert a number of CNS-effects.

The pituitary–adrenal system affects brain function via corticoster-oids. The effects may be permissive and delayed due to receptor occupation and generation of genomic processes. Non-genomic effects of steroids may rapidly modulate brain processes by interacting with the neurotransmitter systems either by activating enzymes or chang-ing the permeability and excitability of nerve cells and neurons.

5.   Concluding remarks

Pituitary–adrenal system hormones affect the formation and mainten-ance of various learned (conditioned) and innate behavioural pat-terns. Although definition of adaptive behaviour is difficult without implying an anthropomorphic view, one may assume that behaviou-ral responses to changes in the environment which serve the survival of the individual or of the species may be specified as adaptive behaviour. Learning, memory but also extinction and forgetting are considered as the highest form of behavioural adaptation. Stimuli evoking fear, anxiety, hope or disappointment or signals associated with reward (food, water or sexual contact) play an important role in behavioural adaptation. Since ACTH shares its behavioural effects with α- and β-MSH and smaller fragments of these peptides and glucocorticosteroids but also precursors of the adrenal steroid synth-esis such as pregnenolone and progesterone influence avoidance behaviour one may question whether or not it is worthwhile to discuss the behavioural function of the pituitary–adrenal system *per se*. The fact that many of the behavioural effects of ACTH and related peptides and of corticosteroids are opposite raises further doubts on this question (Table II).

Several attempts have been made to relate the reactivity of the pituitary–adrenal system to behavioural performance. A negative correlation was found between the pretraining level of plasma corticosterone and the total number of conditioned avoidance responses in a one-way avoidance situation in the rat (van Delft, 1970). Similarly, low plasma corticosterone levels as measured after the last conditioning session were associated with superior shuttle-box avoidance performance in the rat (Endröczi, 1972). Mason et al. (1968) reported that the lowest 17-hydroxy-corticosteroid excretion during conditioning in monkeys was associated with the highest lever-pressing rate in a Sidman avoidance situation. In contrast, rats which showed superior performance in a one-way conditioned avoidance test had the highest corticosterone secretion rate under pentobarbital anaesthesia (Bohus et al., 1963). Wertheim et al. (1969) found a positive correlation between pituitary–adrenal reactivity to ether stress and subsequent avoidance performance in a free-operant situation. They also reported that higher resting plasma corticosterone levels were correlated with a greater proficiency of the avoidance performance. Lovely et al. (1972) found that individually housed rats had higher basal corticosterone levels and their avoidance behaviour was also superior to rats housed in groups. A positive correlation between the pituitary–adrenal response to ether stress and conditioned avoidance performance as a function of age in rats was suggested by Johnston et al. (1974). Passive avoidance behaviour appeared to be positively correlated with pituitary–adrenal activity in rats (Endröczi et al., 1957; Endröczi, 1972) and in dogs (Lissák et al., 1957).

These reports do not allow a definite conclusion regarding the relation between behavioural performance and the activity of the pituitary–adrenal system. The contradictory data may be the result of differences in the various behavioural paradigms. The most obvious differences however arise because some authors measured the plasma samples during behaviour while others determined pituitary–adrenal reactivity to stress in a situation unrelated to the behaviour. In addition, the plasma concentration of corticosteroids does not reflect the concentration of pituitary–adrenal system hormones in the brain. It is more likely that plasma corticosterone levels mirror the magnitude of processes which are associated with the behaviour rather than the rate at which it is modulated by the pituitary–adrenal system. Ursin et al. (1975) found a progressive decrease in plasma corticosterone levels in the course of shuttle-box avoidance conditioning in the

TABLE II

Effect of pituitary–adrenal system hormones on avoidance and approach behaviour

Pituitary–adrenal manipulation	Treatment	Avoidance behaviour[a]		Approach behaviour[a]	
		Learning	Retention	Learning	Retention
Adrenalectomy	None	Enhanced (?)	Enhanced	Enhanced (?)	Enhanced
	Corticosterone	Normal	Normal	—	—
	Dexamethasone	Enhanced (?)	—	—	—
	Deoxycorticosterone	Normal	Enhanced	—	—
Hypophysectomy	None	Deficient	N.S.	Deficient (?)	—
	"Suppl. therapy"	Normal	Attenuated	—	—
	ACTH	Normal	Normal	—	—
	MSH	Normal	—	—	—
	ACTH$_{4-10}$	Normal	Enhanced	—	—
	D-Phe^7 ACTH$_{4-10}$	Attenuated	Attenuated	—	—
	Dexamethasone	Attenuated	—	—	—
None	ACTH	Enhanced (?)	Enhanced	Enhanced (?)	Enhanced
	MSH	Enhanced (?)	Enhanced	Normal	Enhanced
	ACTH$_{4-10}$	—	Enhanced	—	Enhanced
	D-Phe^7 ACTH$_{4-10}$	Attenuated (?)	Attenuated	Attenuated (?)	Attenuated
	Corticosterone	Attenuated (?)	Attenuated	Attenuated (?)	Attenuated
	Cortisol	Attenuated (?)	Attenuated	—	—
	Cortisone	Attenuated (?)	Attenuated	—	—
	Dexamethasone	Normal	Attenuated	Normal	Attenuated
	Deoxycorticosterone	—	Normal	Normal	—
	Aldosterone	—	Normal	—	—
	Pregnenolone	—	Attenuated	—	—
	Progesterone	Attenuated (?)	Attenuated	—	—

[a] (?) The effect strongly depends on experimental (mainly motivational) variables.—No data are available

rat. According to these authors this decline comprised three stages in avoidance learning. In the beginning, the rats received a high number of aversive stimuli which then became predictable due to classical conditioning. In a final stage, the avoidance performance was high and the rat coped efficiently. They further showed that changes in plasma corticosterone levels may reflect the cause of avoidance behaviour deficits in rats with limbic system lesions. Auerbach and Carlton (1971) found that electroshock-induced amnesia for a learned passive avoidance response is associated with an absence of the rise in plasma corticosterone levels. Rigter (1975), using CO_2 as the amnesic agent, corroborated this finding and additionally showed that the plasma corticosterone response is normalized when CO_2-induced amnesia is prevented by the administration of $ACTH_{4-10}$. Accordingly, plasma corticosterone levels reflect attenuated fear due to amnesic treatment while reinstatement of fear is followed by pituitary–adrenal activation. Nevertheless, the pituitary–adrenal system is very active in many behavioural situations and therefore one may expect that alterations in the levels of these hormones modulate ongoing behaviour associated with various forms of stress. There are at least two alternative ways through which physiological modulation of brain functions may be realized. The opposite effects of ACTH and related peptides and of corticosteroids on behaviour may be the result of the time lag between the release of ACTH and the subsequent increase in plasma corticosterone level. ACTH-like peptides are released immediately after the onset of psychological stimulus and probably reach their site of action via the cerebral ventricles. By increasing the state of arousal of limbic–midbrain structures, the probability increases that a given behavioural response occurs. The function of corticosteroids would then be to normalize the increased excitability of these structures. This explanation presupposes a primary function of pituitary peptides in the modulation of behaviour. The second alternative would be that the hormonal sensitivity of limbic–midbrain mechanisms is primarily determined by ongoing nervous processes rather than by the availability of either of these hormones. This hypothesis presupposes a primary role of central nervous processes underlying behaviour and a secondary, unspecific, role of pituitary–adrenal system hormones. A number of arguments can be found for and against these alternatives. For example, recent data indicate that corticosteroids may well be made available more quickly and in increased quantities than previously thought. Psychological stresses can rapidly increase plasma corticosterone levels (Bassett and Cairncross, 1975; Bohus, 1975c). On the other hand, there is a lag time between exogenous administration of peptides related to ACTH and corticosteroids and the onset of their behavioural action. This lag time seems to be independent of the route of administration, be it systemic or intraventricular.

Several hypotheses have been developed to explain the psychological substrate of the behavioural effects of pituitary–adrenal system hormones. Some of these hypotheses were obviously premature since they were based upon a single observation (state dependency, cue value of hormones, etc.). Other hypotheses, such as reinstatement of motivational cues (Levine and Brush, 1967), memory (de Wied and Bohus, 1966), excitability and fear (Weiss *et al.*, 1970), and facilitation of (visual) attention (Kastin *et al.*, 1975), are better specified but still need further investigation. A more detailed behavioural analysis of the influence of pituitary–adrenal system hormones may lead to a re-evaluation and extension of some of the hypotheses such as adaptation versus increased motivational value of environmental stimuli (de Wied *et al.*, 1968, 1975b), internal inhibition versus suppression of exploratory activity (Lissak and Endroczi, 1964; Endroczi, 1972), fear suppression versus elimination of non-relevant behavioural responses (Bohus and Lissak, 1968; Bohus, 1975a).

References

Abe, K., Nicholson, W. E., Liddle, G. W., Orth, D. N. and Island, D. P. (1969). *J. clin. Invest.* **48**, 1580–1585.

Ader, R. (1970). *Physiol. Behav.* **5**, 837–839.

Ader, R. and Grota, L J. (1973). *In* "Drug Effects on Neuroendocrine Regulation" (E. Zimmermann, W. W. Gispen, B. H. Marks and D. de Wied, eds.) "Progress in Brain Research" Vol. 39, pp. 395–405. Elsevier, Amsterdam.

Alexinsky, T. and Houcine, O. (1973). *Brain Res.* **55**, 149–158.

Allen, J. P., Kendall, J. W., McGilvra, R. and Vancura, C. (1974). *J. clin. Endocr. Metab.* **38**, 586–593.

Anderson, O. D. (1941). *In* "The Genetic and Endocrine Basis for Differences in Form and Behaviour" (C. R. Stockard, ed.). *Am. Anat. Memoirs* **19**, 647–753.

Anderson, D. C., Winn, W. and Tam, T. (1968). *J. comp. physiol. Psychol.* **66**, 497–499.

Applezweig, M. H. and Baudry, F. D. (1955). *Psychol Rep.* **1**, 417–420.

Applezweig, M. H. and Moeller, G. (1959). *Acta Psychol.* **15**, 602.

Arai, Y., Hiroi, M., Mitra, J. and Gorski, R. A. (1967). *Neuroendocrinology* **2**, 275–282.

Ashton, H., Millman, J. E., Telford, R., Thompson, J. W., Davies, T. F., Hall, R., Shuster, S., Thody, A. J., Coy, D. H. and Kastin, S. J. (1977). *Psychopharmacology* **55**, 165–172.

Auerbach, P. and Carlton, P. L. (1971). *Science, N.Y.* **173**, 1148–1149.

Azmitia, E. C. and McEwen, B. S. (1974). *Brain Res.* **78**, 291–302.

Azmitia, E. C., Jr. and McEwen, B. S. (1976). *J. Neurochem.* **27**, 773–778.

Azmitia, E. C., Algeri, S. and Costa, E. (1970). *Science*, N.Y. **169**, 201–203.
Balázs, R. and Cotterrell, M. (1972). *Nature*, Lond. **236**, 348–350.
Baldwin, D. M., Haun, C. K. and Sawyer, C. H. (1974). *Brain Res.* **80**, 291–301.
Banerjee, U. (1971a). *Communs Behav. Biol.* **6**, 163–170.
Banerjee, U. (1971b). *Neuroendocrinology* **7**, 278–290.
Barondes, S. H. and Cohen, H. D. (1968). *Proc. natn. Acad. Sci. U.S.A.* **61**, 923–929.
Barrett, R. J., Leith, N. J. and Ray, O. S. (1971). *Physiol. Behav.* **7**, 663–665.
Bassett, J. R. and Cairncross, K. D. (1975). *Pharmac. Biochem. Behav.* **3**, 139–142.
Beatty, P. A., Beatty, W. W., Bowman, R. E. and Gilchrist, J. C. (1970). *Physiol. Behav.* **5**, 939–944.
Beatty, W. W., Scouten, C. W. and Beatty, P. A. (1971). *Physiol. Behav.* **7**, 869–871.
Beckwith, B. E., Sandman, C. A., Hothersall, D. and Kastin, A. J. (1977). *Physiol. Behav.* **18**, 63–71.
Bélanger, D. (1958). *Can. J. Psychol.* **12**, 171–178.
Bennett, H. P. J., Lowry, P. J., McMartin, C. and Scott, A. P. (1974). *Biochem. J.* **141**, 439–444.
Bertolini, A., Gessa, G. L., Vergoni, W. and Ferrari, W. (1968). *Life Sciences*, **7**, 1203–1206.
Bertolini, A., Vergoni, W., Gessa, G. L. and Ferrari, W. (1969). *Nature, Lond.* **221**, 667–669.
Bertolini, A., Gessa, G. L. and Ferrari, W. (1975). *In* "Sexual Behaviour: Pharmacology and Biochemistry" (M. Sandler and G. L. Gessa, eds.) pp. 247–257. Raven Press, New York.
Bloom, F., Segal, D., Ling, N. and Guillemin, R. (1976). *Science*, N.Y. **194**, 630–632.
Bloom, F., Battenberg, E., Rossier, J., Ling, N., Leppaluoto, J., Vargo, T. M. and Guillemin, R. (1977). *Life Sciences* **20**, 43–48.
Bohus, B. (1968). *Neuroendocrinology* **3**, 355–365.
Bohus, B. (1970a). *In* "Pituitary, Adrenal and the Brain" (D. de Wied and J. A. W. M. Weijnen, eds.) "Progress in Brain Research" Vol. 32, pp. 171–184. Elsevier, Amsterdam.
Bohus, B. (1970b). *Acta physiol. hung.* **38**, 217–223.
Bohus, B. (1971). *In* "Hormonal Steroids" Proceedings Third Int. Congress Hormonal Steroids (V.H.T. James and L. Martini, eds.) Excerpta Medica *I.C.S.* No. 219, pp. 752–758. Amsterdam.
Bohus, B. (1973). *In* "Drug Effects on Neuroendocrine Regulation" (E. Zimmermann, W. H. Gispen, B. H. Marks and D. de Wied, eds.) "Progress in Brain Research" Vol. 39, pp. 407–420. Elsevier, Amsterdam.
Bohus, B. (1974a). *Brain Res.* **66**, 366–367.
Bohus, B. (1974b). *Biotelemetry* **1**, 193–210.
Bohus, B. (1975a). *In* "The Hippocampus" (R. L. Isaacson and K. H. Pribram, eds.) Vol. 1, pp. 313–353. Plenum Publ. Co., New York.
Bohus, B. (1975b) *In* "Hormones, Homeostasis and the Brain" (W. H. Gispen, Tj. B. van Wimersma Greidanus, B. Bohus and D. de Wied, eds.) "Progress in Brain Research" Vol. 42, pp. 275–283. Elsevier, Amsterdam.
Bohus, B. (1975c) *In* "Les Endocrine et la Milieu" (H. P. Klotz, ed.) "Problèmes Actuels d'Endocrinologie et de Nutrition" Série No. 19, pp. 55–62. Expansion Scientifique Française, Paris.
Bohus, B. (1977). *In* "Hypertension and Brain Mechanisms" (W. de Jong, A. P. Provoost and A. P. Shapiro, eds.) "Progress in Brain Research" Vol. 47, pp. 227–288. Elsevier, Amsterdam.

Bohus, B. and de Wied, D. (1966). *Science, N.Y.* **153**, 318–320.
Bonus, B. and de Wied, D. (1967a). *Physiol. Behav.* **2**, 221–223.
Bohus, B. and de Wied, D. (1967b). *J. comp. physiol. Psychol.* **64**, 26–30.
Bohus, B. and Endröczi, E. (1965). *Acta physiol. hung.* **26**, 183–189.
Bohus, B. and de Kloet, E. R. (1977). *J. Endocr.* **72**, 64P–65P.
Bohus, B. and Lissák, K. (1968). *Int. J. Neuropharmacol.* **7**, 301–306.
Bohus, B., Endröczi, E. and Lissák, K. (1963). *Acta physiol. hung.* **24**, 79–83.
Bohus, B., Nyakas, Cs. and Endröczi, E. (1968). *Int. J. Neuropharmacol.* **7**, 307–314.
Bohus, B., Grubits, J., Kovacs, G. and Lissák, K. (1970a). *Acta physiol. hung.* **38**, 381–391.
Bohus, B., Grubits, J. and Lissák, K. (1970b). *Acta physiol. hung.* **37**, 265–272.
Bohus, B., Gispen, W. H. and de Wied, D. (1973). *Neuroendocrinology* **11**, 137–143.
Bohus, B., Hendrickx, H. H. L., van Kolfschoten, A. A. and Krediet, T. G. (1975). *In* "Sexual Behaviour: Pharmacology and Biochemistry" (M. Sandler and G. L. Gessa, eds.) pp. 269–275. Raven Press, New York.
Bradbury, A. F., Smyth, D. G., Snell, C. R., Birdsall, N. J. M. and Hulme, E. C. (1976). *Nature, Lond.* **260**, 793–795.
Brain, P. F. (1972a). *Behav. Biol.* **7**, 453–477.
Brain, P. F. (1972b). *Neuroendocrinology* **10**, 371–376.
Brain, P. F. (1975). *Life Sciences*, **16**, 187–200.
Brain, P. F. and Nowell, N. W. (1969). *Physiol. Behav.* **4**, 945–947.
Brain, P. F. and Nowell, N. W. (1970). *Physiol. Behav.* **5**, 259–261.
Brain, P. F., Nowell, N. W. and Wouters, A. (1971). *Physiol. Behav.* **6**, 27–29.
Brobeck, J. R. (1945). *Am. J. Physiol.* **143**, 1–5.
Brown, G. M., Uhlir, I. V., Seggie, J., Schally, A. V. and Kastin, A. J. (1974). *Endocrinology* **94**, 583–587.
Brush, F. R. (1962). *J. comp. physiol. Psychol.* **55**, 888–892.
Brush, F. R. and Levine, S. (1966). *Physiol. Behav.* **1**, 309–311.
Buckingham, J. C. and Hodges, J. R. (1974). *J. Endocr.* **63**, 213–222.
Burge, K. G. and Edwards, D. A. (1971). *Physiol. Behav.* **7**, 885–888.
Bush, D. F., Lovely, R. H. and Pagano, R. R. (1973). *J. comp. physiol. Psychol.* **83**, 168–172.
Cardo, B. (1965). *Psychol. Franc.* **10**, 344–351.
Cardo, B. (1967). *Physiol. Behav.* **2**, 245–248.
Cardo, B. and Valade, F. (1965) *C. r. hebd. Seanc. Acad. Sci., Paris,* **261**, 1399–1402.
Chambers, W. F., Freedman, S. L. and Sawyer, C. H. (1963). *Expl Neurol.* **8**, 458–469.
Champney, R. F., Sahley, T. L. and Sandman, C. A. (1976). *Pharmac. Biochem. Behav.* **5**, suppl. 1, 3–9.
Chrétien, M., Seidah, N. G., Benjannet, S., Dragon, N., Routhier, R., Motomatsu, T., Crine, P. and Lis, M. (1977). *Ann. N.Y. Acad. Sci.* **297**, 84–105.
Conforti, N. and Feldman, S. (1975). *J. neurol. Sci.* **26**, 29–38.
Conner, R. L. and Levine, S. (1969). *Horm. Behav.* **1**, 73–83.
Cotterrell, M., Balazs, R. and Johnson, A. L. (1972). *J. Neurochem.* **19**, 2151–2167.
Cottrell, G. A. (1975). M. A. Thesis. Dalhousie University.
Dallman, M. F., Demanincor, D. and Shinsako, J. (1974). *Endocrinology* **95**, 65–73.
Davidson, J. M. and Levine, S. (1972). *A. Rev. Physiol.* **34**, 375–408.
Davis, M. and Zolovick, A. J. (1972). *Physiol. Behav.* **8**, 579–584.
Delacour, J. (1969). *Physiol. Behav.* **4**, 969–974.
Delacour, J. (1970). *In* "Pituitary, Adrenal and the Brain" (D. de Wied and J. A. W.

M. Weijnen, eds.) "Progress in Brain Research" Vol. 32, pp. 158–170. Elsevier, Amsterdam.

Delacour, J. (1971). *Neuropsychologia* **9,** 157–174.

Delacour, J., Albe–Fessard, D. and Libouban, S. (1966). *Neuropsychologia* **4,** 101–112.

van Delft, A. M. L. (1970). Ph.D. Thesis, University of Utrecht.

van Delft, S. M. L. and Kitay, J. i. (1972). *Neuroendocrinology* **9,** 188–196.

Dempsey, G. L., Kastin, A. J. and Schally, A. V. (1972). *Horm. Behav.* **3,** 333–337.

Denenberg, V. H. (1964). *Psychol. Rev.* **71,** 335–351.

Denenberg, V. H. and Zarrow, M. X. (1971). *In* "Early Childhood: The Development of Self-Regulatory Mechanisms" (D. H. Walcher and D. L. Peters, eds.) pp. 39–64. Academic Press, New York, London and San Francisco.

Denenberg, V. H., Brumaghim, J. T., Haltmeyer, G. C. and Zarrow, M. X. (1967). *Endocrinology* **81,** 1047–1052.

Denman, P. M., Miller, L. H., Sandman, C. A., Schally, A. V. and Kastin, A. J. (1972). *J. comp. physiol. Psychol.* **80,** 59–65.

De Vellis, J. and Kukes, G. (1973). *Tex. Rep. Biol. Med.* **31,** 271–293.

Diez, J. A., Sze, P. Y. and Ginsburg, B. E. (1976). *Brain Res.* **104,** 396–400.

Di Giusto, E. L., Cairncross, K. and King, M. G. (1971). *Psychol. Bull.* **75,** 432–444.

Doepfner, W. (1966). *Experientia* **22,** 527–528.

Dornbush, R. L. and Nikolovski, O. (1976). *Pharmacol. Biochem. Behav.* **5,** suppl. 1, 69–72.

Dunn, A. J. and Gispen, W. H. (1977). *Biobehav. Rev.* **1,** 15–23.

Dunn, A. J., Iuvone, P. M. and Rees, H. D. (1976). *Pharmacol. Biochem. Behav.* **5,** 139–145.

Dyster–Aas, H. K. and Krakau, C. E. T. (1965). *Acta endocr. Copnh.* **48,** 409–419.

Elazar, Z. and Adey, W. R. (1967). *Electroenceph. clin. Neurophysiol.* **23,** 225–240.

Endröczi, E. (1969). *In* "Results in Neurophysiology, Neuroendocrinology, Neuropharmacology and Behaviour" (K. Lissák, ed.) "Recent Developments of Neurobiology in Hungary" Vol. 2, pp. 27–46. Akadémiai Kiadó, Budapest.

Endröczi, E. (1972). "Limbic System, Learning and Pituitary–Adrenal Function". Akadémiai Kiadó, Budapest.

Endröczi, E. and Lissák, K. (1962). *Acta physiol. hung.* **21,** 257–263.

Endröczi, E. and Nyakas, Cs. (1971). *Acta physiol. hung.* **39,** 351–360.

Endröczi, E. and Nyakas, Cs. (1972). *Acta physiol. hung.* **41,** 55–61.

Endröczi, E., Telegdy, G. and Lissák, K. (1957). *Acta physiol. hung.* **11,** 393–398.

Endröczi, E., Lissák, K. and Telegdy, G. (1958). *Acta physiol. hung.* **14,** 353–357.

Endröczi, E., Lissák, K., Korányi, L. and Nyakas, Cs. (1968) *Acta physiol. hung.* **33,** 375–382.

Endröczi, E., Lissák, K., Fekete, T. and de Wied, D. (1970). *In* "Pituitary, Adrenal and the Brain" (D. de Wied and J. A. W. M. Weijnen, eds.) "Progress in Brain Research" Vol. 32, pp. 254–261. Elsevier, Amsterdam.

Engel, F. L. (1961). *Vitam. and Horm.* **19,** 189–202.

Engel, G. L. and Margolin, S. G. (1942). *Arch. intern. Med.* **70,** 236–259.

Eysenck, H. J. (1975). *In* "Emotions. Their parameter and measurement" (L. Levi, ed.) pp. 439–467. Raven Press, New York.

Feder, H. H. and Ruf, K. B. (1969). *Endocrinology* **69,** 171–174.

Feldman, S. (1966). *Epilepsia* **7,** 271–282.

Feldman, S. (1971). *Epilepsia* **12,** 249–262.

Feldman, S. and Robinson, S. (1968). *J. Neurol. Sci.* **6,** 1–8.

Feldman, S. and Sarne, Y. (1970). *Brain Res.* **23,** 67–75.

Feldman, S., Todt, J. C. and Porter, R. W. (1961). *Neurology* **11**, 109–115.

Ferrari, W. (1958). *Nature, Lond.* **181**, 925–926.

Ferrari, W., Gessa, G. L. and Vargiu, L. (1963). *Ann. N.Y. Acad. Sci.* **104**, 330–343.

Ferris, S. H., Sathananthan, G., Gershon, S., Clark, C. and Moshinsky, J. (1976). *Pharmacol. Biochem. Behav.* **5**, suppl. 1, 73–78.

Figdor, S. K., Kodet, M. J., Bloom, B. M., Agnello, E. J., P'An, S. I. and Laubach, G. D. (1957). *J. Pharmac. exp. Ther.* **119**, 299–309.

Flexner, J. B. and Flexner, L. B. (1970). *Proc. natn. Acad. Sci. U.S.A.* **66**, 48–52.

Flexner, J. B. and Flexner, L. B. (1971). *Proc. natn. Acad. Sci. U.S.A.* **68**, 2519–2521.

Flood, J. F., Jarvik, M. E., Bennett, E. L. and Orme, A. E. (1976). *Pharmacol. Biochem. Behav.* **5**, suppl. 1, 41–51.

Friedman, E., Friedman, J. and Gershon, S. (1973). *Science, N.Y.* **182**, 831–832.

Fuller, J. L., Chambers, R. M. and Fuller, R. P. (1956). *Psychosom. Med.* **29**, 323–328.

Fuxe, K., Corrodi, H., Hökfelt, T. and Jonsson, G. (1970). *In* "Pituitary, Adrenal and the Brain" (D. de Wied and J. A. W. M. Weijnen, eds.) "Progress in Brain Research" Vol. 32, pp. 42–56. Elsevier, Amsterdam.

Fuxe, K., Hökfelt, T., Jonsson, G., Levine, S., Lidbrink, P. and Löfström, A. (1973). *In* "Brain–Pituitary–Adrenal Interrelationships" (A. Brodish and E. S. Redgate, eds.) pp. 239–269. Karger, Basel.

Gaillard, A. W. K. and Sanders, A. F. (1975). *Psychopharmacologia* **42**, 201–208.

Garcia, J., Hankins, W. G. and Rusiniak, K. W. (1974). *Science, N.Y.* **185**, 824–831.

Garrud, P., Gray, J. A. and de Wied, D. (1974). *Physiol. Behav.* **12**, 109–119.

Garrud, P., Gray, J. A., Rickwood, L. and Coen, C. (1977). *Physiol. Behav.* **18**, 813–818.

Gerlach, J. L. and McEwen, B. S. (1972). *Science, N.Y.* **175**, 1133–1136.

Gessa, G. L., Pisano,, M., Vargiu, L., Crabai, F. and Ferrari, W. (1967). *Revue can. Biol.* **26**, 229–236.

Gibbs, J., Sechzer, J. A., Smith, G. P., Conners, R. and Weiss, J. M. (1973). *J. comp. physiol. Psychol.* **82**, 165–169.

Gilkes, J. J. H., Bloomfield, G. A., Scott, A. P., Lowry, P. J., Ratcliffe, J. G., Landon, J. and Rees, L H. (1975). *J. clin. Endocr. Metab.* **40**, 450–457.

Gispen, W. H. and Wiegant, V. M. (1976). *Neurosci. Letters* **2**, 159–164.

Gispen, W. H., van Wimersma Greidanus, Tj.B. and de Wied, D. (1970). *Physiol. Behav.* **5**, 143–146.

Gispen, W. H., de Wied, D., Schotman, P. and Jansz, H. S. (1971). *Brain Res.* **31**, 341–351.

Gispen, W. H., van der Poel, A. M. and van Wimersma Greidanus, Tj.B. (1973). *Physiol. Behav.* **10**, 345–350.

Gispen, W. H., van Wimersma Greidanus, Tj.B., Bohus, B. and de Wied, D. (1975a). "Hormones. Homeostasis and the Brain" "Progress in Brain Research" Vol. 42. Elsevier, Amsterdam.

Gispen, W. H., Wiegant, V. M., Greven, H. M. and de Wied, D. (1975b). *Life Sciences*, **17**, 645–652.

Gispen, W. H., Wiegant, V. M., Bradbury, A. F., Hulme, E. C., Smyth, D. G., Snell, C. R. and de Wied, D. (1976). *Nature, Lond.* **264**, 794–795.

Glaser, G. H. (1953). *Epilepsia* **2**, 7–14.

Gold, P. E. and Van Buskirk, R. (1976a). *Behav. Biol.* **16**, 387–400.

Gold, P. E. and Van Buskirk, R. (1976b). *Horm. Behav.* **7**, 509–517.

Gold, P. E., Rose, R. P., Spanis, C. W. and Hankins, L. L. (1977). *Horm. Behav.* **8**, 363–371.

Gráf, L. (1976). *Acta Biochim. Biophys. hung.* **11**, 267–277.
Gray, J. A. (1967). *Adv. Sci.* **23**, 595–605.
Gray, J. A. (1971). *Nature, Lond.* **229**, 52–53.
Gray, J. A. (1972). *Physiol. Behav.* **8**, 481–490.
Gray, J. A. (1974). Presented at Roussel UCLAF Round Table Discussion: "Brain Oligopeptides", Paris.
Gray, J. A. and Ball, G. G. (1970). *Science, N.Y.* **168**, 1246–1248.
Gray, J. A. and Garrud, P. (1977). *In* "Neuropeptide Influences on the Brain and Behaviour" (L. H. Miller, C. A. Sandman and A. J. Kastin, eds.) pp. 201–212. Raven Press, New York.
Gray, J. A., Mayes, A. R. and Wilson, M. (1971). *Neuropharmacology* **10**, 223–230.
Gray, P. (1975). *J. comp. physiol. Psychol.* **88**, 281–284.
Gray, P. (1976). *J. comp. physiol. Psychol.* **90**, 1–17.
Green, J. D. and Arduini, A. (1954). *J. Neurophysiol.* **17**, 533–557.
Greenberg, R., Whalley, C. E., Jourdikian, F., Mendelson, I. S., Walter, R., Nikolics, K., Coy, D. H., Schally, A. V. and Kastin, A. J. (1976). *Pharmacol. Biochem. Behav.* **5**, suppl. 1, 151–158.
Greven, H. M. and de Wied, D. (1967). *Eur. J. Pharmacol.* **2**, 14–16.
Greven, H. M. and de Wied, D. (1977). *In* "Frontiers of Hormone Research" (Tj.B. van Wimersma Greidanus, ed.) Vol. 4, pp. 140–152. Karger, Basel.
Greven, H. M. and de Wied, D. (1973). *In* "Drug Effects on Neuroendocrine Regulation" (E. Zimmermann, W. H. Gispen, B. H. Marks and D. de Wied, Eds.) "Progress in Brain Research" Vol. 39, pp. 429–442. Elsevier, Amsterdam.
Grota, L. J. (1972). Paper presented at the Meeting of Eastern Psychol. Ass., Boston.
Grota, L. J. and Ader, R. (1972). *Psychon. Sci.* **28**, 10–12.
Guillemin, R., Schally, A. V., Lipscomb, H. S., Andersen, R. N. and Long, J. M. (1962). *Endocrinology* **70**, 471–477.
Guillemin, R., Ling, N. and Burgus, R. (1976) *C. r. hebd. Seanc. Acad. Sci., Paris,* **282**, 1–5.
Guillemin, R., Ling, N., Lazarus, L., Burgus, R., Minick, S., Bloom. F., Nicoll, R., Siggins, G. and Segal, D. (1977). *Ann. N.Y. Acad. Sci.* **297**, 131–156.
Guth, S., Seward, J. P. and Levine, S. (1971a). *Horm. Behav.* **2** 127–138.
Guth, S., Levine, S. and Seward, J. P. (1971b). *Physiol. Behav.* **7**, 196–200.
Gyermek, L., Genther, G. and Fleming, N. (1967). *Int. J. Neuropharmacol.* **6**, 191–198.
Haltmeyer, G. C., Denenberg, V. H., Thatcher, J. and Zarrow, M. X. (1966). *Nature, Lond* **212**, 1371–1373.
Haltmeyer, G. C., Denenberg, V. H. and Zarrow, M. X. (1967). *Physiol. Behav.* **2**, 61–63.
Harding, C. F. and Leshner, A. I. (1972). *Physiol. Behav.* **8**, 437–440.
Harris, R. K. (1973). *J. comp. physiol. Psychol.* **82**, 254–260.
Haun, C. K. and Haltmeyer, G. C. (1975). *Neuroendocrinology* **19**, 201–213.
Havlena, J. and Werboff, J. (1963). *Psychol. Rep.* **12**, 348–350.
van der Helm–Hylkema, H. and de Wied, D. (1976). *Life Sciences,* **18**, 1099–1104.
Hennessy, J. W., Cohen, M. E. and Rosen, A. J. (1973). *Physiol. Behav.* **11**, 767–770.
Hennessy, J. W., Smotherman, W. P. and Levine, S. (1976). *Behav. Biol.* **16**, 413–424.
Heuser, G. (1967). *Anesthesiology* **28**, 173–183.
Heuser, G. and Eidelberg, E. (1961). *Endocrinology* **69**, 915–924.
Heuser, G., Ling, G. M. and Buchwald, N. A. (1965). *Archs Neurol. Psychiat Chicago,* **13**, 195–203.

Hodges, J. R. and Jones, M. T. (1964). *J. Physiol., Lond.* **173,** 190–200.

Hodges, J. R. and Vernikos. Danellis, J. (1962). *Acta endocr. Copnh.* **39,** 79–86.

Hoffman, W. C., Lewis, R. A. and Thorn, G. W. (1942). *Bull. Johns Hopkins Hosp.* **70,** 335–361.

Hofman, K., Thompson, T. A., Woolner, M. E., Spühler, G., Yajima, H., Cipera, J. D. and Schwartz, E. T. (1960). *J. Am. chem. Soc.* **82,** 3721–3726.

Hofman, K., Andreatta, R., Bohn, H. and Moroder, L. (1970). *J. med. Chem.* **13,** 339–345.

Hökfelt, T. and Fuxe, K. (1972). *In* "Brain–Endocrine Interaction. Median Eminence: Structure and Function" (K. M. Knigge, D. E. Scott and A. Weindl, eds.) pp. 181–223. Karger, Basel.

Howard, E. (1965). *J. Neurochem.* **12,** 181–191.

Howard, E. (1968). *Expl. Neurol.* **22,** 191–208.

Howard, E. (1973). *J. comp. physiol. Psychol.* **85,** 211–220.

Howard, E. and Granoff, D. M. (1968). *Expl. Neurol.* **22,** 661–673.

Hughes, J., Smith, T. W., Kosterlitz, H. W., Fothergill, L. A., Morgan, B. A. and Morris, H. R. (1975). *Nature, Lond.* **258,** 557–579.

Isaacson, R. L. (1974). "The Limbic System". Plenum Press, New York.

Isaacson, R. L., Dunn, A. J., Rees, H. D. and Waldock, B. (1976) *Physiol. Psychol.* **4,** 159–162.

Izumi, K., Donaldson, J. and Barbeau, A. (1973). *Life Sciences* **12,** 203–210.

Jacquet, Y. F. and Marks, N. (1976). *Science, N.Y.* **194,** 632–635.

Jakoubek, B., Semiginovsky, B., Kraus, M. and Erdossova, R. (1970). *Life Sciences* **9,** 1169–1179.

Jakoubek, B. Semiginovsky, B. and Dedicova, A. (1971). *Brain Res.* **25,** 133–141.

Jakoubek, B., Buresova, M., Hajek, I., Etrychova, J., Pavlik, A. and Dedicova, A. (1972). *Brain Res.* **43,** 417–428.

Javoy, F., Glowinski, J. and Kordon, C. (1968). *Europ. J. Pharm* **4,** 103–104.

Joffe, J. M., Mulick, J. A. and Rawson, R. A. (1972a). *Horm. Behav.* **3,** 87–96.

Joffe, J. M., Milkovic, K. and Levine, S. (1972b). *Physiol. Behav.* **8,** 425–430.

Johnson, J. H. and Sawyer, C. H. (1971). *Endocrinology* **89,** 507–512.

Johnston, R. E., Miya, T. S. and Paolino, R. M. (1974). *Physiol. Behav.* **12,** 305–308.

Kamin, L. J. (1963). *J. comp. physiol. Psychol.* **56,** 713–718.

Kasper–Pandi, P., Hansing, R. and Usher, D. R. (1970). *Physiol. Behav.* **5,** 361–363.

Kastin, A. J., Miller, L. H., Gonzalez–Barcema, D., Hawley, W. D., Dyster–Aas, H. K., Schally, A. V., Parra, M. L. V. and Velasco, M. (1971). *Physiol. Behav.* **7,** 893–896.

Kastin, A. J., Miller, L. M., Nockton, R., Sandman, C. A., Schally, A. V. and Stratton, L. O. (1973a). *In* "Drug Effects on Neuroendocrine Regulation" (E. Zimmermann, W. H. Gispen, B. H. Marks and D. de Wied, eds.) "Progress in Brain Research" Vol. 39, pp. 461–470. Elsevier, Amsterdam.

Kastin, A. J., Miller, M. C., Ferrell, L. and Schally, A. V. (1973b). *Physiol. Behav.* **10,** 399–401.

Kastin, A. J., Dempsey, G. L., Leblanc, B., Dyster–Aas, K. and Schally, A. V. (1974). *Horm. Behav.* **5,** 135–139.

Kastin, A. J., Sandman, C. A., Stratton, L. O., Schally, A. V. and Miller, L. H. (1975). *In* "Hormones, Homeostasis and the Brain" (W. H. Gispen, Tj. B. van

Wimersma Greidanus, B. Bohus and D. de Wied, eds.). "Progress in Brain Research" Vol. 42, pp. 143–150. Elsevier, Amsterdam.
Kastin, A. J., Nissen, C., Nikolics, K., Medzihradszky, K., Coy, D. H., Teplan, I. and Schally, A. V. (1976a). *Brain Res. Bull.* **1,** 19–26.
Kastin, A. J., Scollan, E. L., King, M. E., Schally, A. V. and Coy, D. H. (1976b). *Pharmacol. Biochem. Behav.* **5,** 691–695.
Kawakami, M. and Sawyer, C. H. (1959). *Endocrinology* **65,** 631–643.
Kawakami, M., Koshino, T. and Hattori, Y. (1966). *Jap. J. Physiol.* **16,** 551–569.
Kendall, J. W. (1970). *Horm. Behav.* **1,** 327–336.
Kendler, K., Hennessy, J. W., Smotherman, W. P. and Levine, S. (1976). *Behav. Biol.* **17,** 225–229.
Keyes, J. B. (1974). *Physiol. Psychol.* **2,** 307–309.
Klein, S. B. (1972). *J. comp. physiol. Psychol.* **79,** 341–359.
Klein, S. B. and Kopish, R. M. (1975). *Behav. Biol.* **13,** 377–381.
Klein, R. and Livingston, S. (1950). *J. Pediat.* **37,** 733–746.
Klemm, W. R. (1972). *Brain Res.* **41,** 331–344.
de Kloet, R. and McEwen, B. S. (1976). *In* "Molecular and Functional Neurobiology" (W. H. Gispen, ed.) pp. 258–307. Elsevier, Amsterdam.
de Kloet, R., Wallach, G. and McEwen, B. S. (1975). *Endocrinology* **96,** 598–609.
Knizley, H., Jr. (1972). *J. Neurochem.* **19,** 2737–2745.
Korányi, L. and Endröczi, E. (1967). *Neuroendocrinology* **2,** 65–75.
Korányi, L. and Endröczi, E. (1970). *In* "Pituitary, Adrenal and the Brain" (D. de Wied and J. A. W. M. Weijnen, eds.) "Progress in Brain Research" Vol. 32, pp. 120–130. Elsevier, Amsterdam.
Korányi, L., Endröczi, E. and Tárnok, F. (1965/66). *Neuroendocrinology* **1,** 144–157.
Korányi, L., Beyer, C. and Guzman–Flores, C. (1971a). *Physiol. Behav.* **7,** 321–329.
Korányi, L., Beyer, C. and Guzmán–Flores, C. (1971b). *Physiol. Behav.* **7,** 331–335.
Kostowski, W. (1967). *Diss. Pharmac. Pharmacol.* **19,** 619–623.
Kostowski, W., Rewerski, W. and Piechodcki, T. (1970). *Neuroendocrinology* **6,** 311–318.
Kovács, G. L., Telegdy, G. and Lissák, K. (1976). *Psychoneuroendocrinology* **1,** 219–230.
Kovács, G. L., Telegdy, G. and Lissák, K. (1977). *Horm. Behav.* **8,** 155–165.
Kraulis, I., Foldes, G., Traikov, H., Dubrovsky, B. and Birmingham, M. K. (1975). *Brain Res.* **88,** 1–14.
Krieger, D. T., Liotta, A. and Brownstein, M. J. (1977a). *Proc. natn. Acad. Sci. U.S.A.* **74,** 648–652.
Krieger, D. T., Liotta, A. and Brownstein, M. J. (1977b). *Brain Res.* **128,** 575–579.
Krivoy, W. A. and Guillemin, R. (1961). *Endocrinology* **69,** 170–175.
Krivoy, W. A., Kroeger, D., Taylor, A. N. and Zimmermann, E. (1974). *Europ. J. Pharmacol.* **27,** 339–345.
Lande, S., Flexner, J. B. and Flexner, L. B. (1972). *Proc. natn. Acad, Sci. U.S.A.* **69,** 558–560.
Lebovitz, H. E. (1967). *In* "Introduction to Clinical Neuroendocrinology" (E. Bajusz, ed.) pp. 243–253. Williams and Wilkins Co., Baltimore.
Leconte, P., Hennevin, E. et Bloch, V. (1973). *Brain Res.* **49,** 367–379.
Leonard, B. E. (1969). *Int. J. Neuropharmacol.* **8,** 427–435.
Leonard, B. (1974). *Arch. int. Pharmacodyn.* **207,** 242–253.
Leshner, A. I. (1971). *Physiol. Behav.* **6,** 551–558.
Leshner, A. I. and Candland, D. K. (1972). *Physiol. Behav.* **8,** 441–445.
Leshner, A. I. and Johnson, A. E. (1974). *Physiol. Behav.* **13,** 703–705.

Leshner, A. I., Walker, W. A., Johnson, A. E., Kelling, J. S., Kreisler, S. J. and Svare, B. B. (1973). *Physiol. Behav.* **11**, 705–711.
Leshner, A. I., Moyer, J. A. and Walker, W. A. (1975). *Physiol. Behav.* **15**, 689–693.
Levine, S. (1957). *J. comp. physiol. Psychol.* **50**, 609–612.
Levine, S. (1958). *J. comp. physiol. Psychol.* **51**, 230–233.
Levine, S. (1962). *Science N.Y.*, **135**, 795–796.
Levine, S. (1968). *In* "Nebraska Symposium on Motivation" (W. J. Arnold, ed.), pp. 85–101. University of Nebraska Press, Lincoln, Nebraska.
Levine, S. (1970). *In* "Pituitary, Adrenal and the Brain" (D. de Wied and J. A. W. M. Weijnen, eds.) "Progress in Brain Research" Vol. 32, pp. 79–85. Elsevier, Amsterdam.
Levine, S. (1972): "Hormones and Behaviour" Academic Press, New York, London and San Francisco.
Levine, S. and Brush, F. R. (1967). *Physiol. Behav.* **2**, 385–388.
Levine, S. and Jones, L. E. (1965). *J. comp. physiol. Psychol.* **59**, 357–360.
Levine, S. and Levin, R. (1970). *Horm. Behav.* **1**, 105–110.
Levine, S. and Mullins, R. J., Jr. (1966). *Science, N.Y.* **152**, 1585–1592.
Levine, S. and Treiman, D. M. (1964). *Endocrinology* **75**, 142–144.
Levine, S., Chevalier, J. A. and Korchin, S. J. (1956). *J. Personality* **24**, 475–493.
Levine, S., Haltmeyer, G. C., Karas, G. G. and Denenberg, V. H. (1967). *Physiol. Behav.* **2**, 55–59.
Levine, S., Goldman, L. and Coover, G. D. (1972). *In* "Physiology, Emotion and Psychosomatic Illness". CIBA Foundation Symposium 8 (new series) pp. 281–291. Elsevier/Excerpta Medica/North Holland, Amsterdam.
Levine, S., Smotherman, W. P. and Hennessy, J. W. (1977). *In* "Neuropeptide Influences on the Brain and Behaviour". (L. H. Miller, C. A. Sandman and A. J. Kastin, eds.). pp. 163–177. Raven Press, New York.
Ley, K. F. and Corson, J. A. (1971). *Experientia* **27**, 958–959.
Ley, K. F. and Corson, J. A. (1972). *Int. J. Psychobiol.* **2**, 265–271.
Lichtblau, L., Fossom, L. H. and Sparber, S. B. (1977). *Life Sciences* **21**, 927–932.
Liddel, H. S., Anderson, O., Kotyuka, O. and Hartman, F. A. (1935). *Arch. Neurol. Psychiat. (Chicago)* **34**, 973–993.
Lissák, K. and Bohus, B. (1972). *Int. J. Psychobiol.* **2**, 103–115.
Lissák, K. and Endröczi, E. (1962). *In* "Physiologie de l'Hippocampe". Colloq. Int. CNRS No. 107, pp. 463–473 (J. Cadilhac, ed.). Editions CNRS, Paris.
Lissák, K. and Endröczi, E. (1964). *In* "Major Problems in Neuroendocrinology". (E. Bajusz and G. Jasmin, eds.), pp. 1–16. Karger, Basel.
Lissák, K., Endröczi, E. und Medgyesi, P. (1957). *Pflügers Arch, ges. Physiol.* **265**, 117–124.
Lloyd, G. (1974). Ph.D. Thesis, University of Edinburgh.
Loh, H. H. and Li, C. H. (1977). *Ann. N.Y. Acad. Sci.* **297**, 115–128.
Loh, Y. P. and Gainer, H. (1977). *Brain Res.* **130**, 169–175.
Louch, C. D. and Higginbotham, M. (1967). *Gen. comp. Endocr.* **8**, 441–444.
Lovely, R. H., Pagano, R. R. and Paolino, R. M. (1972). *J. comp. physiol. Psychol.* **81**, 331–335.
Lowry, P. J. and Chadwick, A. (1970). *Biochem. J.* **118**, 713–718.
Lowry, P. J. and McMartin, C. (1974). *Biochem. J.* **138**, 87–95.
Lowry, P. J. and Scott, A. P. (1975). *Gen. comp. Endocr.* **26**, 16–23.
Lucero, M. (1970). *Brain Res.* **20**, 319–322.

Mains, R. E., Eipper, B. A. and Ling, N. (1977). *Proc. natn. Acad. Sci. U.S.A.* **74,** 3014–3018.
Marquis, H. A. and Suboski, M. D. (1969). *Proc. 77th Ann. Conv. Am. Psychol. Ass.* **4,** 207–208.
Martin, S. M., Moberg, G. P. and Horowitz, J. M. (1975). *Brain Res.* **93,** 535–542.
Mason, J. W. (1968). *Psychosom. Med.* **30,** 576–607.
Mason, J. W., Brady, J. V. and Tolliver, G. A. (1968). *Psychosom. Med.* **30,** 608–630.
McEwen, B. S. and Wallach, G. (1973). *Brain Res.* **57,** 373–386.
McEwen, B. S., Weiss, J. M. and Schwartz, L. S. (1969). *Brain Res.* **16,** 227–241.
McEwen, B. S., Weiss, J. M. and Schwartz, L. S. (1970). *Brain Res.* **17,** 471–482.
McEwen, B. S., Gerlach, J. L. and Micco, D. J. (1975). *In* "The Hippocampus" (R. L. Isaacson and K. H. Pribram, eds.) Vol. 1, pp. 285–322. Plenum Press, New York.
McEwen, B. S., Krey, L. C. and Luine, V. N. (1978). *In* "The Hypothalamus" (S. Reichlin, R. J. Baldessarini and J. B. Martin, eds.) pp. 255–268. Raven Press, New York.
McGaugh, J. L. (1961). *Psychol. Rep.* **8,** 99–104.
McGowan–Sass, B. K. and Timiras, P. S. (1975). *In* "The Hippocampus" (R. L. Isaacson and K. H. Pribram, eds.) Vol. 1, pp. 355–374. Plenum Press, New York.
McLean, P. D. (1970). *In* "The Neurosciences, Second Study Program" (F. O. Schmitt, ed.) pp. 336–349. Rockefeller University Press, New York.
McLean, P. D. (1973). *In* "Hormones and Brain Function" (K. Lissák, ed.), pp. 379–389. Plenum Press, New York.
Mezey, E., Palkovits, M., de Kloet, E. R., Verhoef, J. and de Wied, D. (1978). *Life Sciences* **22,** 831–838.
Meyerson, B. J. and Bohus, B. (1976). *Pharmacol. Biochem. Behav.* **5,** 539–545.
Meyerson, B. J. and Terenius, L. (1977). *Europ. J. Pharmacol.* **42,** 191–192.
Michal, E. K. (1974). *Brain Res.* **65,** 180–183.
Milković, S., Milković, K., Sencar, I. and Paunović, J. (1970). *In* "Pituitary, Adrenal and the Brain" (D. de Wied and J. A. W. M. Weijnen, eds.) "Progress in Brain Research" Vol. 32, pp. 71–85. Elsevier, Amsterdam.
Miller, R. E. and Ogawa, N. (1962). *J. comp. physiol. Psychol.* **55,** 211–213.
Miller, L. H., Kastin, A. J., Sandman, C. A., Fink, M. and Van Veen, W. J. (1974). *Pharmacol. Biochem. Behav.* **2,** 663–668.
Miller, L. H., Harris, L. C., van Riezen, H. and Kastin, A. J. (1976). *Pharmacol. Biochem. Behav.* **5,** suppl. 1, 17–21.
Miller, L. H., Fischer, S. C., Groves, G. A., Rudrauff, M. E. and Kastin, A. J. (1977). *Pharmacol. Biochem. Behav.* **7,** 417–419.
Mirsky, I. A., Miller, R. and Stein, M. (1953). *Psychosom. Med.* **15,** 574–584.
Miyabo, S., Hisada, T., Ueno, K., Kishida, S. and Kitanaka, I. (1972). *Horm. Behav.* **3,** 227–236.
Moriarty, G. C. (1973). *J. Histochem. Cytochem.* **21,** 855–894.
Moyer, K. E. (1958). *J. genet. Psychol.* **92,** 11–16.
Moyer, K. E. (1966). *J. genet. Psychol.* **108,** 297–302.
Moyer, K. E. and Moshein, P. (1963). *J. comp. physiol. Psychol.* **56,** 163–166.
Murphy, J. V. and Miller, R. E. (1955). *J. comp. physiol. Psychol.* **48,** 47–49.
Nakajima, S. (1975a). *J. comp. physiol. Psychol.* **88,** 378–385.
Nakajima, S. (1975b). *In* "The Hippocampus" (R. L. Isaacson and K. H. Pribram, eds.) Vol. 1, pp. 393–413. Plenum Press, New York.
Neckers, L. and Sze, P. Y. (1975). *Brain Res.* **93,** 123–132.

Nicolov, N. (1967). *Folia med. (Plovdiv)* **9**, 249–255.
Nockton, R., Kastin, A. J., Elder, S. T. and Schally, A. V. (1972). *Horm. Behav.* **3**, 339–344.
Nyakas, Cs. (1973). *In* "Hormones and Brain Function" (K. Lissák, ed.) pp. 83–89. Plenum Press, New York.
Nyakas, Cs. and Endröczi, E. (1972). *Acta physiol. hung.* **42**, 231–241.
Obias, M. D. (1957). *J. comp. physiol. Psychol.* **50**, 120–124.
Obrist, P. A., Lawler, J. E. and Gaebelein, C. J. (1974). *In* "Limbic and Autonomic Nervous Systems Research" (L. V. DiCara, ed.) pp. 311–334. Plenum Press, New York.
Oliver, C., Eskay, R. L. and Porter, J. C. (1976). Abstract No. 594, Vth Int. Congress of Endocrinology, July 18–24, Hamburg.
Oliver, C., Mical, R. S. and Porter, J. C. (1977). *Endocrinology* **101**, 598–604.
Olton, D. S., Johnson, C. T. and Howard, E. (1974). *Devl. Psychobiol.* **8**, 55–61.
Orth, D. N. and Nicholson, W. E. (1977). *Ann. N.Y. Acad. Sci.* **297**, 27–46.
Otsuka, H. and Inouye, K. (1964). *Bull. Chem. Soc. Jap.* **37**, 1465–1471.
Pagano, R. R. and Lovely, R. H. (1972). *Physiol. Behav.* **8**, 721–723.
P'An, S. I. and Laubach, G. D. (1964). *In* "Methods in Hormone Research" (R. Dorfman, ed.) Vol. 20, pp. 415–475. Academic Press, New York, London and San Francisco.
Papez, J. W. (1937). *Arch. Neurol. Psychiat. Chicago*, **38**, 725–744.
Pappas, B. A. and Gray, P. (1971). *Physiol. Behav.* **6**, 127–130.
Paul, C. and Havlena, J. (1962). *J. Psychosom. Res.* **6**, 153–156.
Pedersen–Bjergaard, K. and Tonnesen, M. (1954). *Acta endocr.* **17**, 329–337.
Pelletier, G., Labrie, F., Kastin, A. J. and Schally, A. V. (1975). *Pharmacol. Biochem. Behav.* **3**, 671–674.
Pfaff, D. W., Silva, M. T. A. and Weiss, J. M. (1971). *Science, N.Y.* **172**, 394–395.
Phifer, R. F., Orth, D. N. and Spicer, S. S. (1974). *J. clin. Endocr. Metab.* **39**, 684–692.
Phillips, M. I. and Dafny, N. (1971). *Brain Res.* **25**, 651–655.
Phillips, A. G. and Shapiro, M. (1973). *Physiol. Behav.* **10**, 351–353.
Poole, A. E. and Brain, P. (1974). *In* "Integrative Hypothalamic Activity" (D. F. Swaab and J. P. Schadé, eds.) "Progress in Brain Research" Vol. 41, pp. 465–472. Elsevier, Amsterdam.
Pribram, K. H. and McGuiness, D. (1975). *Psychol. Rev.* **82**, 116–149.
Reading, H. W. and Dewar, A. J. (1971). Abstract 3rd Meeting ISN, Budapest, p. 199.
Rees, F. D., Stumpf, W. E. and Sar, M. (1975). *In* "Anatomical Neuroendocrinology" (W. E. Stumpf and L. D. Grant, eds.) pp. 262–269. Karger, Basel.
Reith, M. E. A., Schotman, P. and Gispen, W. H. (1974). *Brain Res.* **81**, 571–575.
Reith, M. E. A., Schotman, P. and Gispen, W. H. (1975). *Neurobiology*, **5**, 355–368.
Revusky, S. and Garcia, J. (1970). *In* "The Psychology of Learning and Motivation: Advances in Research and Theory" (G. H. Bower, ed.) Vol. 4, pp. 1–84. Academic Press, New York, London and San Francisco.
Richter, C. P. (1936). *Endocrinology* **20**, 657–666.
Rigter, H. (1975). *Behav. Biol.* **15**, 207–211.
Rigter, H. and van Riezen, H. (1975). *Physiol. Behav.* **14**, 563–566.
Rigter, H. and Popping, A. (1976). *Psychopharmacologia* **46**, 255–261.
Rigter, H., van Riezen, H. and de Wied, D. (1974). *Physiol. Behav.* **13**, 381–388.
Rigter, H., Elbertse, R. and van Riezen, H. (1975). *In* "Hormones, Homeostasis and

the Brain" (W. H. Gispen, Tj.B. van Wimersma Greidanus, B. Bohus and D. de Wied, eds.) Vol. 42, pp. 163–171. Elsevier, Amsterdam.

Rigter, H., Shuster, S. and Thody, A. J. (1977a). *J. Pharm. Pharmacol.* **29,** 110–111.

Rigter, H., Greven, H. and van Riezen, H. (1977b). *Neuropharmacology* **16,** 545–547.

Routtenberg, A. (1968). *Physiol. Behav.* **3,** 533–535.

Rudman, D., Del Rio, A. E., Hollins, B. M., Houser, D. H., Keeling, M. E., Sutin, J., Scott, J. W., Sears, R. A. and Rosenberg, M. Z. (1973). *Endocrinology* **92,** 372–379.

Rudman, D., Scott, J. W., Del Rio, A. E., Houser, D. H. and Sheen, S. (1974). *Am. J. Physiol.* **226,** 687–692.

Ruf, K. and Steiner, F. A. (1967). *Science, N.Y.* **156,** 667–668.

Sakamoto, A. (1966). *Nature, Lond.* **211,** 1370–1371.

Sandman, C. A., Kastin, A. J. and Schally, A. V. (1969). *Experientia* **25,** 1001–1002.

Sandman, C. A., Kastin, A. J. and Schally, A. V. (1971a). *Physiol. Behav.* **6,** 45–48.

Sandman, C. A., Denman, P. M., Miller, L. H. and Knott, J. R. (1971b). *J. comp. physiol. Psychol.* **76,** 103–109.

Sandman, C. A., Miller, L. H., Kastin, A. J. and Schally, A. V. (1972). *J. comp. physiol. Psychol.* **80,** 54–58.

Sandman, C. A., Alexander, W. D. and Kastin, A. J. (1973). *Physiol. Behav.* **11,** 613–617.

Sandman, C. A., George, J. M., Nolan, J. D., van Riezen, H. and Kastin, A. J. (1975). *Physiol. Behav.* **15,** 427–431.

Sandman, C. A., George, J., Walker, B. B. and Nolan, J. D. (1976). *Pharmacol. Biochem. Behav.* **5,** suppl. 1, 23–28.

Sandman, C. A., George, J., McCanne, T. R., Nolan, J. D., Kaswan, J. and Kastin, A. J. (1977). *J. clin. Endocrin. Metab.* **44,** 884–891.

Sawyer, C. H., Kawakami, M., Meyerson, B., Whitmoyer, D. I. and Lilley, J. J. (1968). *Brain Res.* **10,** 213–226.

Sawyer, C. H., Baldwin, D. M. and Haun, C. K. (1975). *In* "Sexual Behaviour: Pharmacology and Biochemistry" (M. Sandler and G. L. Gesaa, eds.) pp. 259–268. Raven Press, New York.

Sayers, G., Beall, R. J. and Seelig, S. (1974). *In* "Biochemistry of Hormones" (H. V. Rickenburg, ed.) pp. 25–60. Butterworth Univ. Park Press, London.

Schally, A. V., Lipscomb, H. S., Long, J. H., Dear, W. E. and Guillemin, R. (1962). *Endocrinology* **70,** 478–480.

Schapiro, S. (1968). *Gen. comp. Endocr.* **10,** 214–228.

Schapiro, S. and Norman, R. (1967). *Science, N.Y.* **155,** 1279–1281.

Schapiro, S., Salas, M. and Vukovich, K. (1970). *Science, N.Y.* **168,** 147–151.

Schneider, A. M., Weinberg, J. and Weissberg, R. (1974). *Physiol. Behav.* **13,** 633–636.

Schotman, P., Gispen, W. H., Jansz, H. S. and de Wied, D. (1972). *Brain Res.* **46,** 349–362.

Schotman, P., Reith, M. E. A., van Wimersma Greidanus, Tj. B., Gispen, W. H. and de Wied, D. (1976). *In* "Molecular and Functional Neurobiology" (W. H. Gispen, ed.) pp. 310–344. Elsevier, Amsterdam.

Scott, J. P. and Frederickson, E. (1951). *Physiol. Zoöl.* **24,** 273–309.

Selye, H. (1942). *Endocrinology* **30,** 437–458.

Selye, H. (1950). "Stress. The Physiology and Pathology of Exposure to Stress". Acta Medical Publication, Montreal.

Sigg, E. B., Day, C. and Colombo, C. (1966). *Endocrinology* **78,** 679–684.

Silva, M. T. A. (1973). *Behav. Biol.* **9,** 553–562.

Silva, M. T. A. (1974). *Physiol. Psychol.* **2,** 171–174.

Simantov, R., Kuhar, M. J., Pasternak, G. W. and Snyder, S. H. (1976). *Brain Res.* **106,** 189–197.

Slusher, M. A. (1965). *Proc. Soc. exp. Biol. Med.* **120,** 617–620.

Slusher, M. A., Hyde, J. E. and Laufer, M. (1966). *J. Neurophysiol.* **29,** 157–169.

Small, J. G., Small, I. F., Milstein, V. and Dian, D. A. (1977). *Acta psychiat. scand.* **55,** 241–250.

Smith, G. P. (1973). *In* "Progress in Physiological Psychology" (E. Stellar and J. M. Sprague, eds.) Vol. 5, pp. 299–343. Academic Press, New York, London and San Francisco.

Smith, C. T., Kitahama, K., Valatx, J.-L. and Jouvet, M. (1972). *C. r. hebd. Seanc., Acad. Sci., Paris,* **275,** 1283–1286.

Smith, D. J., Joffe, J. M. and Heseltine, G. F. D. (1975). *Physiol. Behav.* **15,** 461–469.

Snyder, S. H. and Simantov, R. (1977). *J. Neurochem.* **28,** 13–20.

Soulairac, A., Desclaux, P., Soulairac, M.-L. et Teysseyre, J. (1953). *J. Physiol., Paris* **45,** 527–531.

Steiner, F. A., Ruf, K. and Akert, K. (1969). *Brain Res.* **12,** 74–85.

Stolk, J. M. (1972). *Fedn Proc. Fedn Am. Scocs exp. Biol.,* **31,** 551.

Stone, C. P. and King, F. A. (1954). *J. comp. physiol. Psychol.* **47,** 213–219.

Stone, C. P. and Obias, M. D. (1955). *J. comp. physiol. Psychol.* **48,** 404–411.

Strand, F. L. and Cayer, A. (1975). *In* "Hormones, Homeostasis and the Brain" (W. H. Gispen, Tj. B. van Wimersma Greidanus, B. Bohus and D. de Wied, eds.). "Progress in Brain Research" Vol. 42, pp. 187–194. Elsevier, Amsterdam.

Strand, F. L., Stoboy, H. and Cayer, A. (1973/74). *Neuroendocrinology* **13,** 1–20.

Stratton, L. O. and Kastin, A. J. (1974). *Horm. Behav.* **5,** 149–155.

Streifler, M. and Feldman, S. (1953). *Confin. neurol. (Basel)* **13,** 16–27.

Stumpf, C. (1965). *Int. Rev. Neurobiol.* **8,** 77–138.

Stumpf, W. E. and Sar, M. (1971). *In* "Hormonal Steroids" (V. H. T. James and L. Martini, eds.) pp. 503–507. Excerpta Medica, *I.C.S. 219,* Amsterdam.

Stumpf, W. E. and Sar, M. (1975). *In* "Anatomical Neuroendocrinology" (W. E. Stumpf and L. D. Grant, eds.) pp. 254–261. Karger, Basel.

Suboski, M. D., Marquis, H. A., Black, M. and Platenius, P. (1970). *Physiol. Behav.* **5,** 283–289.

Sutherland, E. W. (1972). *Science, N.Y.* **177,** 401–408.

Swaab, D. F. (1976). Abstract No. 285, Vth Int. Congress of Endocrinology, July 18–24, Hamburg.

Swanson, H. H. and McKeag, A. M. (1969). *Horm. Behav.* **1,** 1–5.

Tamásy, V., Korányi, L., Lissák, K. and Jandala, M. (1973). *Physiol. Behav.* **10,** 995–1000.

Terenius, L. (1975). *J. Pharm. Pharmacol.* **27,** 450–452.

Terenius, L. and Wahlström, A. (1975). *Life Sciences* **16,** 1759–1764.

Terenius, L., Gispen, W. H. and de Wied, D. (1975). *Europ. J. Pharmacol.* **33,** 395–399.

Thompson, R., Truax, T. and Thorne, M. (1970). *Brain Behav. Evol.* **3,** 261–284.

Torda, C. and Wolff, H. G. (1952). *Am. J. Physiol.* **168,** 406–413.

Urban, I. and de Wied, D. (1976). *Expl Brain Res.* **24,** 325–334.

Urban, I., Lopes da Silva, F. H., Storm van Leeuwen, W. and de Wied, D. (1974). *Brain Res.* **69,** 361–365.

Ursin, H., Coover, G. D., Køhler, C., Deryck, M., Sagvolden, T., and Levine, S. (1975). *In* "Hormones, Homeostasis and the Brain" (W. H. Gispen, Tj. B. van

Wimersma Greidanus, B. Bohus and D. de Wied, eds.) "Progress in Brain Research" Vol. 42, pp. 263–274. Elsevier, Amsterdam.

Valatx, J.-L., Chouvet, G. and Jouvet, M. (1975). *In* "Hormones, Homeostasis and the Brain" (W. H. Gispen, Tj. B. van Wimersma Greidabus, B. Bohus and D. de Wied, eds.). "Progress in Brain Research" Vol. 42, pp. 115–120. Elsevier, Amsterdam.

Vanderwolf, C. H. (1969). *Electroenceph. clin. Neurophysiol.* **26**, 407–418.

Varon, H. H., Touchstone, J. C. and Christian, J. J. (1966). *Acta endocr., Copnh* **51**, 488–496.

Vaudry, H., Oliver, C., Vaillant, R. and Kraicer, J. (1976). Abstract No. 667, Vth Int. Congress of Endocrinology, Hamburg, July 18–24.

Verhoef, J. and Witter, A. (1976). *Pharmacol. Biochem. Behav.* **4**, 583–590.

Verhoef, J., Palkovits, M. and Witter, A. (1977a). *Brain Res.* **126**, 89–104.

Verhoef, J., Witter, A. and de Wied, D. (1977b). *Brain Res.* **131**, 117–128.

Versteeg, D. H. G. (1973). *Brain Res.* **49**, 483–485.

Versteeg, D. H. G. and Wurtman, R. J. (1975). *Brain Res.* **93**, 552–557.

Versteeg, D. H. G. and Wurtman, R. J. (1976). *In* "Molecular and Functional Neurobiology" (W. H. Gispen, ed.) pp. 201–234. Elsevier, Amsterdam.

Versteeg, D. H. G., Gispen, W. H., Schotman, P., Witter, A. and de Wied, D. (1972). *Adv. biochem. Psychopharmacol.* **6**, 219–239.

Wasserman, M. J., Belton, N. R. and Millichap, J. G. (1965). *Neurology* **15**, 1136–1141.

Weijnen, J. A. W. M. and Slangen, J. L. (1970). *In* "Pituitary Adrenal and the Brain" (D. de Wied and J. A. W. M. Weijnen, eds.) "Progress in Brain Research" Vol. 32, pp. 221–235. Elsevier, Amsterdam.

Weiss, J. M., McEwen, B. S., Silva, M. T. and Kalkut, M. (1970). *Am. J. Physiol.* **218**, 864–868.

Wertheim, G. A., Conner, R. L. and Levine, S. (1967). *J. exp. anal. Behav.* **10**, 555–563.

Wertheim, G. A., Conner, R. L. and Levine, S. (1969). *Physiol. Behav.* **4**, 41–44.

de Wied, D. (1964). *Am. J. Physiol.* **207**, 255–259.

de Wied, D. (1965). *Int. J. Neuropharmacol.* **4**, 157–167.

de Wied, D. (1966). *Proc. Soc. exp. Biol. Med.* **122**, 28–32.

de Wied, D. (1967). *In* "Hormonal Steroids" (L. Martini, F. Fraschini and M. Motta, eds) Excerpta Medica I.C.S. 132, pp. 945–951, Amsterdam.

de Wied, D. (1968). *In* "Progress in Endocrinology, (C. Gual, ed.) Excerpta Medica I.C.S. 184, pp. 310–316, Amsterdam.

de Wied, D. (1969). *In* "Frontiers in Neuroendocrinology 1969" (W. F. Ganong and L. Martini, eds.) pp. 97–140. Oxford University Press, New York.

de Wied, D. (1971). *In* "Normal and Abnormal Development of Brain and Behaviour" (G. B. Stoelinga and J. J. van der Werff ten Bosch, eds.) pp. 315–322. Leiden University Press, Leiden.

de Wied, D. (1977). *Ann. N.Y. Acad. Sci.* **297**, 263–272.

de Wied, D. and Bohus, B. (1966). *Nature, Lond.* **212**, 1484–1486.

de Wied, D. and Weijnen, J. A. W. M. (1970). "Pituitary, Adrenal and the Brain". Progress in Brain Research Vol. 32. Elsevier, Amsterdam.

de Wied, D., Bohus, B. and Greven, H. M. (1968). *In* "Endocrinology and Human Behaviour" (R. P. Michael, ed.) pp. 188–199, Oxford University Press, Oxford.

de Wied, D. Witter, A. and Greven, H. M. (1975a). *Biochem. Pharmacol.* **24**, 1463–1468.

de Wied, D., Bohus, B., Gispen, W. H., Urban, I. and van Wimersma Greidanus, Tj. B. (1975b). *In* "CNS and Behavioural Pharmacology" (M. Airaksinen, ed.). Proc. Sixth Int. Congress Pharmacology Vol. 3, pp. 1930.

de Wied, D., Bohus, B., van Ree, J. M. and Urban, I. (1978). *J. Pharmacol. exp. Ther.* **204,** 570–580.

Wiegant, V. M. and Gispen, W. H. (1975). *Exp. Brain Res.* **23,** suppl. 219.

van Wimersma Greidanus, Tj. B. (1970). *In* "Pituitary, Adrenal and the Brain" (D. de Wied and J. A. W. M. Weijnen, eds.). "Progress in Brain Research" Vol. 32, pp. 185–191. Elsevier, Amsterdam.

van Wimersma Greidanus, Tj. B. (1977). *In* "Frontiers in Hormone Research" (Tj. B. van Wimersma Greidanus, ed.). Vol. 4, pp. 129–139. Karger, Basel.

van Wimersma Greidanus, Tj. B. and de Wied, D. (1969). *Physiol. Behav.* **4,** 365–370.

van Wimersma Greidanus, Tj. B. and de Wied, D. (1971). *Neuroendocrinology* **7,** 291–301.

van Wimersma Greidanus, Tj. B. and de Wied, D. (1976). *Pharmacol. Biochem. Behav.* **5,** suppl. 1, 29–33.

van Wimersma Greidanus, Tj. B., Wijnen, H., Deurloo, J. and de Wied, D. (1973). *Horm. Behav.* **4,** 19–30.

van Wimersma Greidanus, Tj. B., Bohus, B. and de Wied, D. (1974). *Neuroendocrinology* **14,** 280–288.

van Wimersma Greidanus, Tj. B., Lowry, P. J., Scott, A. P., Rees, L. H. and de Wied, D. (1975). *Horm. Behav.* **6,** 319–327.

Witter, A., Greven, H. M. and de Wied, D. (1975). *J. Pharmacol. exp. Ther.* **193,** 853–860.

Wolthuis, O. L. and de Wied, D. (1976). *Pharmacol. Biochem. Behav.* **4,** 273–278.

Woodbury, D. M. (1952). *J. Pharmacol. exp. Ther.* **105,** 27–36.

Woodbury, D. M. (1954). *Recent Progr. Horm. Res.* **10,** 65–107.

Woodbury, D. M. (1958). *Pharmacol. Rev.* **10,** 275–357.

Woodbury, D. M. and Vernadakis, A. (1967). *In* "Neuroendocrinology" (L. Martini and W. F. Ganong, eds.) Vol. 2. pp. 335–375. Academic Press, New York, London and San Francisco.

Woodbury, D. M., Timiras, P. S. and Vernadakis, A. (1957). *In* "Hormones, Brain Function, and Behaviour" (H. Hoagland, ed.) pp. 27–54. Academic Press, New York, London and San Francisco.

Zimmermann, E., Gispen, W. H., Marks, B. H. and de Wied, D. (1973). "Drug Effects on Neuroendocrine Regulation" Progress in Brain Research Vol. 39. Elsevier, Amsterdam.

6. Adrenocortical Function in Relation to Mammalian Population Densities and Hierarchies

N. W. Nowell

Department of Zoology, University of Hull, Hull HU6 7RX, England

1. Introduction

The concept of homeostasis, first formulated in terms of modern physiology by W. B. Cannon in 1929, is now extensively applied to most aspects of zoology, from the environment down to the cell. This chapter considers some of the interactions between those endocrine and behavioural mechanisms that participate in homeostasis both of whole populations and of the individuals within it.

The integration of an individual's homeostatic mechanisms principally occurs at the level of the hypothalamus. From here, pituitary–adrenocortical and autonomic activity elicit both the metabolic reactions and behavioural responses that alleviate environmental stresses. For example endocrine and behavioural responses to an abiotic factor such as hypothermia are designed, with the help of learned experience, to overcome the stress of cold and to return body temperature to normal; in the same way analogous responses to injury, caused for example by a biotic factor such as an aggressive male conspecific, permit metabolic responses to the injury and prevent further attack. Thus homeostasis of the individual is maintained (Fig. 1).

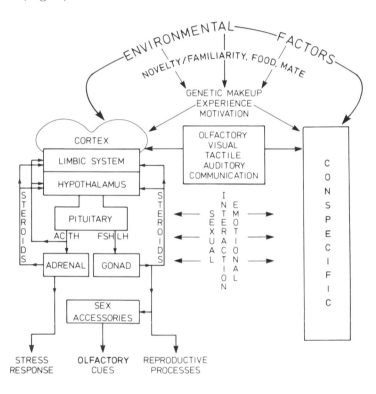

Fig. 1. Scheme representing the integrating mechanisms and inter-relationships between the environment, the endocrine system and behaviour.

In highly dense populations behavioural mechanisms, such as territoriality and hierarchy formation, cannot always maintain population homeostasis, social strife results and the pituitary–adrenal axis is activated. Long-term and excessive activity of this system may however, itself act, by negative means, to control population growth and consequently assist in long-term population homeostasis.

Moreover an additional feature is that gonadal hormones feed back on the hypothalamus and modify sexual behaviour in male and female. There is increasing evidence that pituitary and adrenal hormones facilitate the acquisition and modulation of behavioural responses to changing environmental demands. The causes and effects of pituitary–adrenocortical function in the context of mammalian social behaviours will be considered.

2. Animal populations and social stress

Ecologists are not agreed on the mechanisms whereby animal population numbers are controlled and a general model that accounts for all the variables involved in population control is awaited. The more popular theories differ primarily in their classifications of the variables, especially with regard to the dependence or independence of population density. Secondly the extent to which either of these variables predominates over the other is contentious.

Some ecologists emphasize the harshness of the environment and this in part depends on population density which, in extreme conditions, may destroy a population. Other investigators stress the competition for resources, which may occur at densities below or near to the maximum carrying capacity of the environment.

The balance between population size and the environment obviously involves a highly complex series of interactions between the two and the many excessive increases or decreases in populations seen in the field are indicative of a temporary loss of such balance. It is also important to point out (den Boer, 1968) that both population and environmental heterogeneity reduce the chances of violent fluctuations, particularly when migrations are present. Thus different genotypes will survive particular environmental variables better than others.

A lengthy discussion of the relative importance of different variables is not intended in this chapter, and both environment and intraspecific competition are considered alongside one another. This approach follows the opinion that intraspecific competition is a density-dependent factor which operates, in the face of environmental change, when resources threaten to become, or are, scarce.

Wynne-Edwards (1962) has argued that many species have evolved mechanisms to limit their numbers below the maximum carrying capacity of their environments. He (*op cit*) quotes numerous examples of visual, auditory and olfactory mechanisms of communication whereby animals assess their own density and, by behavioural means, limit excessive population increases. Evidence for limitation by such altruistic means as starvation, emigration and failure to breed in the presence of unexhausted resources is, however, lacking. It would seem more probable that such responses are physiological and behavioural consequences of intraspecific stress.

Activities directly related to population balance, relying upon visual, auditory and olfactory communication, include the maintenance of personal space, the acquisition and retention of territory and the development of hierarchies within groups. These activities thus tend to reduce intraspecific competition. However, as population densities increase, this tendency is reversed and the principal manifestation of such reversal is an increase in intraspecific agonistic behaviour.

Not all mammals have defined territories and when they do they may use them for quite different purposes. Probably the most consistent uses are for reproduction and food acquisition so that territorial activity is most common in the breeding season. The area may have quite distinct boundaries with visual, olfactory and tactile landmarks enabling the owner to develop a familiarity with the area. Such familiarity is ideal for courtship, mating and care of young and donates advantage to the owner over intruders. The tendency to threaten or attack an intruder is known to decrease as an animal moves away from the centre of its territory.

Several studies have suggested that with increased numbers in a population, territorial activity is superseded by the formation of a characteristic hierarchical system. This may be very stable, only broken by changes in the composition of the group and it may be linear in nature as a result of frequent agonistic encounters at different levels from the highest dominant to the lowest subordinate males. Such hierarchies are common in primates, colonial carnivores and ungulates but are not universal; in most rodents one male of a group assumes the dominant role whilst others form a non-linear hierarchy, although extreme subordinates may exist at the bottom.

The agonistic behaviour of territoriality and hierarchy is characteristic of males and is in part androgen-dependent. It frequently involves physical attack, but may only consist of threatening behaviour such as physical display, noise or release of an aversive odour. Whatever the form of such agonism it may be regarded as a potential or an actual threat, and be considered as a physical or, more especially, a psychological stressor to the recipient with previous experience of defeat by the aggressor. Clearly encounters between individuals increase with increasing population density and encounters between two males threaten resources such as food, personal space, territory, mates, as well as personal safety which then trigger agonistic behaviour.

The high levels of social stress in populations of increasing density has focused attention upon the hypothalamo–hypophysial–adrenocortical axis largely because this participates in stress-response mechanisms. Also, many reports of the diminished fecundity of dense populations invoke the hypothalamo–hypophysial gonadal system. Studies in the field, laboratory and clinic support the contention that some form of reciprocal relationship exists between the function of these two endocrine axes, a notion first advanced by Selye in 1950. In addition, studies of population dynamics have suggested that adrenocortical hormones directly affect population numbers by influencing either somatic tissue responses to influence infection or by influencing the central nervous system to modulate avoidance, emotional or sexual behaviours.

A. SOCIAL STRESS

Much information, both field and laboratory, concerns the effects of social interaction upon pituitary–adrenocortical function, particularly in relation to survival and the maintenance of normal reproductive potential; the mammals considered experience varying degrees of stress usually because of increases in their density. Of necessity, a large proportion of this work has employed laboratory animals, or wild animals maintained in the laboratory, and extrapolation from these studies to "natural" environments must be made with caution.

Some enlightening studies on natural populations of woodchucks by Bronson (1963, 1964, 1967) revealed a close correlation between adrenocortical stimulation and the agonistic behaviour which occurred in the breeding season. Very large increases in adrenal weight and depletion of lipid from the zonae fasciculata and reticularis occurred as fighting increased during the breeding season, and again in autumn when there was extensive movement of the young adults. Such increases in adrenocortical activity, associated with the breeding season, typical of many sciurids, are related both to increases in encounters during sexual behaviour and to an overall increase in numbers. Christian (1955a, b; 1959) has calculated that the adrenal weight of laboratory mice increases with the logarithm of population size. Similar relative increases have also been reported in wild rats (Christian and Davis, 1956; Barnett, 1958), wild deer (Welch, 1962), wild rabbits (Myers, 1970) and in the monkey (Mason, 1959). The notorious lemming is no exception in that both male and female lemmings, taken from peak populations, have adrenal glands with *in vitro* secretory capacities some 20 to 60 times greater than those of low-density populations. This reflected maximal stimulation, since no further increase in corticoid production resulted from ACTH injection into such animals (Andrews, 1970; Andrews and Strohbehn, 1971; Table I).

TABLE I

Relative adrenal weights and corticosterone secretory rates of male lemmings, voles and mice under various conditions of population density and rearing (from Andrews, 1970).

Animal type and season	n	Adrenal weight mg (fresh)/100 g body wt. Mean ± S.E.	Corticosterone secretory rate µg/mg dry wt./h Mean ± S.E.
Brown lemming (*Lemmus trimucronatus*)			
Summer low-density	36	26.91 ± 2.10	0.82 ± 0.10
Summer high-density	36	44.02 ± 4.12	15.86 ± 1.53
Winter low-density	24	30.12 ± 2.59	1.56 ± 0.28
Lab reared (pair housed)	18	35.68 ± 2.02	1.08 ± 0.12
Lab reared (group housed)	9	29.03 ± 4.06	4.30 ± 0.41
Collared lemming (*Dicrostonyx groenlandicus*)			
Summer	12	31.26 ± 2.51	4.21 ± 0.34
Winter	8	32.81 ± 2.09	5.08 ± 0.29
Tundra vole (*Microtus oeconomus*)			
Summer	10	34.03 ± 6.71	3.10 ± 0.21
Winter	8	36.29 ± 5.96	4.12 ± 0.27
Red-backed vole (*Clethrionomys rutilus*)			
Summer	8	6.61 ± 1.02	1.98 ± 0.26
Deer mouse (*Peromyscus leucopus*)			
Summer	48	26.90 ± 1.56	7.56 ± 0.43
Winter	24	23.21 ± 0.96	11.02 ± 1.36
Lab-reared (pair housed)	18	20.80 ± 0.88	8.33 ± 1.03

Such increases in pituitary/adrenocortical activity negatively corre-late with social rank (Davis and Christian, 1957) and Bronson and Eleftheriou (1964) have related strain differences in adrenocortical response to crowding to the genetic predisposition of mice to aggres-sive behaviour. In addition to the studies of Bronson and Eleftheriou

(1964), other experiments have shown that increased adrenal size is invariably associated with an increased corticosteroid secretion (Barrett and Stockham, 1963; Varon et al., 1966; Brain and Nowell, 1970). Further Bronson and Eleftheriou (1964, 1965a) showed that defeat produced a blood corticosterone peak within an hour to a level that remained for up to 24 h, suggesting that a prolonged release of ACTH occurred after defeat (Table II).

The subordinate animal in a population undoubtedly suffers the greatest stress (Davis and Christian, 1957; Louch and Higginbotham, 1967) and may either opt out to take no further part in agonistic behaviour or emigrate from the area. The relationship between social rank and adrenocortical activation is most clearly demonstrated experimentally when mice, isolated for some time, are grouped together. The resulting social turmoil is the direct cause of an overall increase in adrenal activity. The differential stress suffered by individuals, however, results in greater variation in such groups compared with relatively unstressed isolates (Welch, 1964; Brain and Nowell, 1969, 1970, 1971a). The establishment of a fairly stable hierarchy is accompanied by a return of adrenocortical activity to levels approaching those of the isolates. High levels of adrenocortical activity may be maintained, however, and Sassenrath (1970) reported an extreme example in grouped rhesus monkeys in which the urinary corticoid levels of subordinate males remained 3 to 10 times greater than those of isolates for the 2 year period whilst they remained in the group.

Laboratory studies, as well as those on natural populations, have delineated some important variables that affect the development and maintenance of social status. Not only must seasonal and daily rhythms be controlled, but genetic, sexual, experiential and environmental factors are also important. For example, a dominant male, either in a strange environment, or when paired and with an unusual amount of fighting (Brain and Nowell, 1970), displays a greater increase in adrenocortical activity than when in familiar surroundings or in a stable group. Moreover, the presence of an unstable hierarchy produces high levels of stress in the dominant animal. For example dominant male squirrel monkeys have higher urinary 17-hydroxycorticosteroids than their subordinates, and Leshner and Candland (1972) have suggested that this reflects the stress of the unusual number of aggressive encounters necessary to maintain social position (Fig. 2).

These findings question the relative effects of physical and psychological factors as mediators of the response. The suggestion that

TABLE II

Mean plasma corticosterone levels (μg/100 ml) following daily exposure to trained fighting mice for either 1 or 4 days (1 fight per day) or kept in isolation and killed on the same time schedule as fighter-exposed mice. Means represent the average of 3 plasma samples, each pooled from 5 mice (Bronson and Eleftheriou, 1965a).

Corticosterone fraction	No. of fights	Time after last fight (h)[a]					
		1/3	1	3	6	12	24
Unbound	1	13.7 (11–18)[b]	10.7 (9–13)	6.0 (4–7)	6.7 (4–9)	2.3 (2–3)	3.7 (0–6)
	4	7.7 (7–8)	8.3 (2–18)	0.0	0.0	0.7 (0–2)	0.0
	None	0.0	0.0	0.0	1.7 (0–4)	0.0	0.7 (0–2)
Bound	1	7.3 (6–9)	6.6 (6–8)	8.3 (7–10)	8.6 (8–9)	7.6 (7–8)	8.3 (7–9)
	4	7.6 (6–9)	9.0 (8–10)	10.0 (9–11)	9.7 (9–10)	8.3 (8–9)	8.3 (6–10)
	None	7.6 (6–9)	7.0 (6–9)	9.3 (8–11)	8.6 (8–9)	8.3 (8–9)	8.6 (7–10)
Total	1	21.0 (17–27)	17.3 (15–19)	14.3 (12–17)	15.3 (13–18)	10.0 (9–11)	12.0 (9–14)
	4	15.3 (14–17)	17.3 (11–28)	10.0 (9–11)	9.7 (9–10)	9.0 (8–11)	8.3 (6–10)
	None	7.6 (6–9)	7.0 (6–9)	9.3 (8–11)	10.3 (8–13)	8.3 (8–9)	9.3 (7–11)

[a] All fights ended at 10:15 a.m.; [b] Range in parentheses

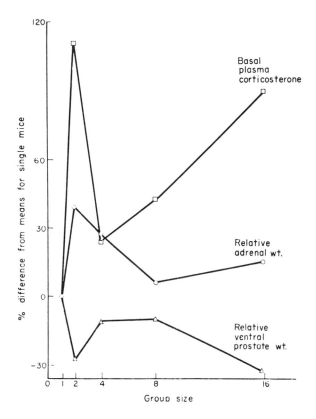

Fig. 2. Relative adrenal weight, basal plasma corticosterone concentrations and relative ventral prostate weight as a function of group size in mice. Values are expressed as percentage differences compared with mice housed singly (Brain and Nowell, 1970).

wounding was the chief stressful factor (Southwick and Bland, 1959) is not a confirmed correlation (Christian, 1959b). The number of contacts between individuals appears to be a more important factor than the density of the group. Moreover the prospect of battle and subsequent injury can be more stressful than the physical damage itself. Thus, Bronson and Eleftheriou (1965b) demonstrated that, in previously defeated mice, a subsequent association with a fighter, without physical contact, induced an adrenocortical response which was as great as that following physical defeat (Table III). In addition the mere spectacle of a fight increases adrenocortical function in the monkey (Mason, 1959), and in the wild rabbit, physical stress has little effect while psychological stress greatly increases the volume of

TABLE III

Plasma corticosterone concentrations in mice in response to physical stress (from
Bronson and Eleftheriou, 1965b)

	Unbound corticosterone—µg/100 ml. plasma		
	Groups		
	1	2	3
Day			
6	9.0	11.1	3.8
7	9.5	11.2	1.2
9	3.5	10.9	4.5
Mean ± S.E.	7.13 ± 1.6	11.1 ± 1.1	3.2 ± 0.9

Group 1: Physical defeat for 6–9 days. Group 2: Defeat for 5 days then put in presence
only of fighter. Group 3: Continuously exposed to presence only of fighter.

the zona fasciculata, at the expense of the zona glomerulosa (Myers,
1967).

Various endocrinological studies have held increases in pituitary–
adrenocortical function, directly or indirectly, to be responsible for
decreases in reproductive potential or survival. These studies strongly
support, for whatever reason, a reciprocal relationship between
ACTH and gonadotrophin secretion (Selye, 1939, 1950), a suggestion
championed by Christian (1950) in relation to population density
control. This subject has been extensively reviewed by Christian
(1971a, b; 1975). Inhibition of maturation, juvenile mortality and
suppression of reproduction are frequently quoted to result from high
population densities and social strife (*Microtus*: Pitelka, 1957, Murray,
1965; cotton rat: Green, 1964; Norway rat: Andrews *et al.*, 1972,
Brooks and Barnes, 1972; snowshoe hare: Meslow and Keith, 1968;
house mouse: Lidicker, 1966). These findings from natural populations have been supported by many laboratory experiments, employing fixed populations without free movement into or out of the
area (Fig. 3). Increased pituitary–adrenal function associated with
morphological, physiological and behavioural manifestations of reduced gonadotrophin activity has been reported in, for example,
grouped male (Christian, 1955a, b; Christian *et al.*, 1965; Bronson,
1967; Brain and Nowell, 1969, 1970, 1971a, b; Lloyd, 1971; Bronson,
1973) and anoestrus female mice (Christian, 1960); grouped female
mice also display intrauterine mortality and impaired lactation, and
consequently undernourished young are characteristic (Christian and
Le Munyan, 1958).

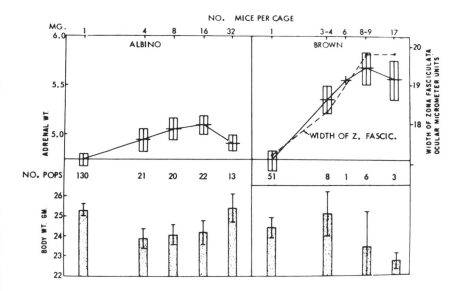

Fig. 3. Mean body and adrenal weights (± S.E.) of male albino and brown house mice caged in groups to show influence of population densities. Adrenal hypertrophy was greater in the more aggressive brown than the docile albino mice (from Christian *et al.*, 1965; see also Christian, 1959a, 1963).

Bliss *et al.* (1972) showed that the acute stress of forced swimming in rats and the chronic stress of mixed grouping in male mice greatly reduced testicular testosterone production without affecting its half-life (Table IV). In grouped male mice, previously isolated, the intense fighting of hierarchy establishment correlated with quite dramatic decreases in plasma levels of FSH and LH (Bronson, 1973, Fig. 4). On days 1–3 after hierarchy formation, the raised plasma cortico-sterone concentrations became normal, but plasma LH remained low until the end of the experimental period at 14 days. Fluctuations in plasma testosterone levels of rhesus monkeys were also shown in the experiments by Rose *et al.* (1972). These levels were increased above baseline in the presence of females and decreased following defeat by other males. The increase in the presence of females could be related to dominant behaviour and/or increased sexual activity (Fig. 5).

In man surgical stress may decrease blood levels of both LH and testosterone (Matsumoto *et al.*, 1970; Carstensen *et al.*, 1973), although Nakashima *et al.* (1975) found decreased testosterone in the presence of increased LH after anaesthesia and major surgery. Seyler

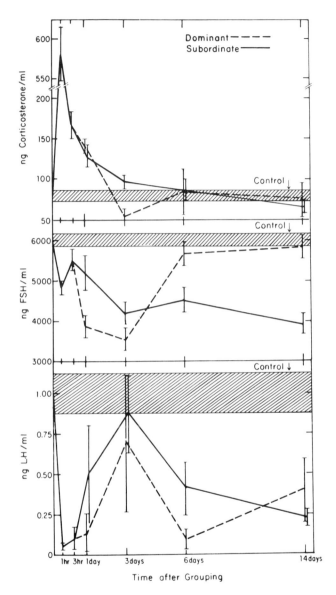

Fig. 4. Mean concentrations (± S.E.) of corticosterone, FSH and LH in plasma of CF–1 male mice housed one or four per cage. Concentrations of FSH and LH are expressed in ng equivalents of NIH–FSH S1. Each point represents a mean of 4–5 samples in the case of dominant (– – – –) males and 10–14 samples from subordinates (———). Hatched areas give the pooled mean ± one SE for all isolated control animals killed throughout the 14 day period (from Bronson, 1973).

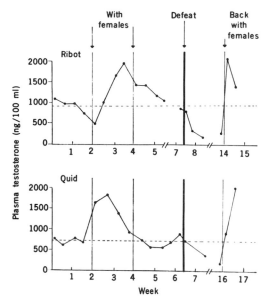

Fig. 5. Plasma testosterone concentrations in two male rhesus monkeys (Ribot and Quid) over a period of 4 months. Values on discontinuous weeks are not connected. After defeat, both males drop below baseline levels, horizontal broken lines (– – – –), determined from weeks 1 and 2 prior to access to the females. Within 2 to 4 days after reintroduction to the females, both animals showed a rise in plasma testosterone equivalent to that initially experienced. (From Rose *et al.*, 1972).

TABLE IV

Effect of swimming and chronic psychosocial stress on plasma and testicular testosterone at termination of stress. All values have been converted to percentages followed by the standard error of the mean. Numbers in parentheses give number of animals in each group (from Bliss *et al.*, 1972).

Procedure	Plasma		Testes	
	Controls	Experimentals	Controls	Experimentals
Rats—Swimming 1 h	100 ± 3% (5)	46 ± 2% (10)	100 ± 9% (5)	34 ± 2% (5)
Rats—Swimming 1 h	100 ± 2% (5)	19 ± 3% (10)		
Mice—Psychosocial stress 3 wks			100 ± 7% (5)	47 ± 4% (6)

and Reichlin (1973) also found an increased serum LH following ether stress. In the male rat, Howland *et al.* (1974) found that ether stress increased LH and FSH whereas surgery decreased LH and that fasting decreased testosterone as well as LH and FSH, while Euker *et al.* (1975) found the acute stress of moving a cage of rats from one room to another followed by serial blood sampling increased serum LH over a 90 minute experimental period. Such conflicting reports suggest that stresses of short duration have different effects to more severe chronic ones. Thus Liptrap and Raeside (1975) found in the boar that injections of rapid-acting porcine ACTH increased plasma testosterone while long-acting ACTH, known to maintain high blood levels of corticosteroids in both boar and rat, actually lowered plasma testosterone (Fig. 6). These two effects depended upon the adrenals and the authors suggested that although short-term stress levels of corticosteroids stimulate testicular androgen production, long-term stress levels have an opposite effect, possibly by decreasing LH output (see above). This physiological sequence might occur in mixed animal groups in which brief stresses enhance reproductive and agonistic behaviour, whilst the longer periods of strife depress reproductive and agonistic behaviour.

In female mice Jarrett (1965) also demonstrated that chronic ACTH decreased the reproductive potential, an effect mediated by the adrenals. C_{11} steroids directly suppress gonadotrophin secretion (Robson and Sharaf, 1952; Jarrett, 1965; Hagino, 1968; Hagino *et al.*, 1969; Liptrap, 1970) and cortisol implants into the basomedial hypothalamus of the immature rat inhibit both ACTH and gonadotrophin secretion (Smith *et al.*, 1971; Tables V and VI). These authors emphasize the high susceptibility of the young animal to these suppressive actions. Gonadotrophin may also be suppressed by the negative feed-back of adrenal androgens (Byrnes and Shipley, 1950; Christian, 1960; Christian *et al.*, 1965), an effect greater in the young animal than in the adult (Ramirez and McCann, 1963, 1964; Smith and Davidson, 1967) and one which would therefore delay maturation. In fact, low doses of adrenal androgens inhibit maturation of house mice (Christian *et al.*, 1965). According to Andrews (1968) the increased quantities of adrenocortical progesterone secreted as a result of stress-induced ACTH-release augment the negative feedback effect of the adrenal androgens.

Although much of the evidence is circumstantial for natural populations, it is probable that susceptibility to infection and disease is a major control of population numbers. This may be regarded as a direct result of increased "glucocorticoid" secretion by the stressed

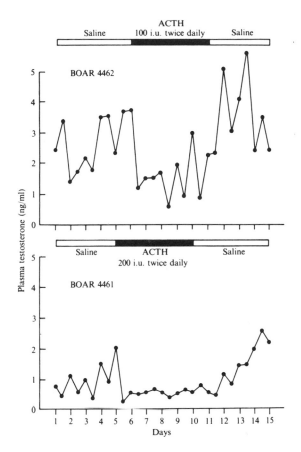

Fig. 6. Effects of twice daily injections of saline (white bars) or ACTH (black bars) on plasma testosterone concentrations in two adult boars (4462 and 4461). Blood samples were collected and injections given at 08.00 and 20.00 h. (From Liptrap and Raeside, 1975).

subordinate, immature and alien members of populations. Christian (1968) found that the ratio cortisol:corticosterone was higher in stressed immature and subordinate mice than in mature and dominant conspecifics and suggested that this renders them more susceptible to disease since cortisol is more likely than corticosterone to increase disease susceptibility.

The "glucocorticoids" are known to depress immune responses, cellular responses to injury and defence against parasitic infections (Kass *et al.*, 1955; Bowen *et al.*, 1957; Hobson, 1960; Oliver, 1962; Campbell, 1963; Chang and Rasmussen, 1965; Sen *et al.*, 1965;

TABLE V

Effects of intracerebral cortisol implants on female reproductive and adrenal function of rats; autopsy 3 weeks following implantation (from Smith et al., 1971).

Treatment	n	Body weight (g)	Vaginal opening (days)	Number of animals showing vaginal oestrous cycles	Mean number of vaginal oestrous cycles	Adrenal weight (mg)	Corticosterone (μ/100 ml plasma) basal	stress
Chol–MBH	9	126 ± 2[a]	37.1 ± 1.5	9	1.9 ± 0.20[a]		19.2 ± 2.5	74.3 ± 4.1
Cortisol–MBH	11	137 ± 3	38.9 ± 2.4	1[b]	0.09 ± 0.3[d]	16.1 ± 2.4[c]	9.5 ± 0.9[b]	27.4 ± 6.9[e]
Adrenalectomy	10	154 ± 3	38.6 ± 1.5	6	1.2 ± 0.42	—	10.2 ± 2.2	14.3 ± 2.5

Chol = cholesterol; MBH = medical basal hypothalamus. [a] Mean ± standard error. [b] Fisher's t exact probability test, P < 0.005 compared with chol; P ≤ 0.005 compared with adrenalectomy. [c] t-test, P < 0.001 compared with cholesterol; P < 0.02 compared with adrenalectomy. [d] t-test, P < 0.001 compared with cholesterol; NS compared with adrenalectomy. [e] t-test, P < 0.001 compared with cholesterol.

TABLE VI

Effects of intracerebral cortisol implants on immature male rats (from Smith et al., 1971).

Treatment	N	Body weight (g)	Prostate (mg)	Adrenal (mg)	Corticosterone (µg/100 ml plasma)	
					basal	stress
Chol–MBH	8	184 ± 9	97.6 ± 7.5	24.0 ± 1.3	9.9 ± 1.1	37.4 ± 5.0
Cortisol–MBH	7	147 ± 4[a]	17.9 ± 3.5[b]	13.4 ± 1.3[c]	4.4 ± 0.8[b]	9.9 ± 4.4[c]
Adrenalectomy	6	163 ± 9	94.0 ± 7.5	—	8.8 ± 0.6	8.8 ± 2.7
Adrenalectomy (partial)	4	194 ± 7	99.3 ± 6.5	13.4 ± 5.2	8.5 ± 3.4	25.0 ± 6.2

[a] t-test $P < 0.01$ compared with cholesterol: partial adrenalectomy, NS compared with adrenalectomy. [b] t-test $P < 0.001$ compared with cholesterol; partial adrenalectomy, adrenalectomy groups. [c] t-test $P < 0.001$ compared with cholesterol.

Frenkel and Lunde, 1966; Noble, 1971). Experimentally, densely grouped animals are more susceptible to infection (Christian and Williamson, 1958; Davis and Read, 1958; Noble, 1962; Vessey, 1964; Plaut *et al.*, 1969) and to the pathogenic effects of some drugs and physical agents (Christian, 1968). Such effects on natural population numbers may or may not be primary (Bull, 1955, 1957, 1964; Mykytowycz, 1962).

The "mineralocorticoids" have, to a very limited extent, been related to mortality within animal populations. Christian *et al.* (1965) suggested that the mass mortalities of sika deer in Japan resulted from prolonged hypersecretion of ACTH in response to high population density, which eventually produced fatal hypokalaemia, itself consequent upon prolonged sodium retention. Other reports of impaired population growths may also invoke mechanisms related to sodium homeostasis. In areas where the soil is deficient in sodium and where populations are dense, animals have zonae glomerulosae that are wider than those of animals from sodium-rich environments. The combination of sodium deficiency and high population density may thus have fatal effects.

B. AGONISTIC BEHAVIOUR

An emotional behaviour of particular interest in the present context is inter-male agonism. Historically the study of agonistic behaviour has evolved separately from those of other emotions and is justly considered separately.

The term "social strife" implies competition for resources with increased contact between individuals and a measure of agonistic behaviour between them; the latter is a major biotic stressor, particularly for the subordinate members of the group. It has been proposed that stressors are environmental factors whose perception does not coincide with a central neural representation of previous environmental experience. This concept concurs with Sokolov's (1960) model for habituation (see below) and can be regarded as a coding of the individual's previous experience (reinforcement) upon which an individual bases subsequent homeostatic responses. In a more general sense, such coding is analogous to the "settings" for blood sugar, temperature and hormonal titres involving the hypothalamic control of homeostasis in response to abiotic change. Thus the greater the difference between the *observed* and the *expected*, the greater is the stress and the greater the "fear" reaction to such novelty. This fear reaction is manifested endocrinologically as increased sympathetic nervous

and pituitary/adrenal activities and behaviourally as a central choice
between attack and avoidance (or fleeing). These choices might well
have evolved selectively to remove undesirable stressors which
threaten the individual's homeostasis. Both Hebb (1946) and Galef
(1970) have argued that these two behaviours are at opposite ends of a
continuum with common causal stimuli. The latter vary in quality but
have a common component, novelty, the intensity of which ranges
from frustration to extreme pain. Competition within groups of
animals can provide a wide range of stimuli and produce either attack
or avoidance between conspecifics. This eventual response depends
upon a number of variables:

(1) Duration and intensity of the novelty—generally, the more
intense or prolonged the stimulus the more intense is the response,
although long duration or high intensity levels can transform an
aggressive response into one of avoidance (fleeing or freezing).

(2) Motivation—high levels of goal-directed behaviour, for example
for food or sex, might override a novelty stimulus which does not
interfere with such behaviour, but might reinforce the response to a
stimulus that frustrates it.

(3) Hormones

  (a) Androgens—aggressive behaviour is generally characteristic of
      the male, castration reduces this behaviour and testosterone, to
      varying degrees, restores it. The presence of testicular
      androgens in the neonate renders CNS elements responsive to
      these hormones in the adult; conversely, neonatal castration
      dramatically diminishes responsiveness.

  (b) Pituitary–adrenal hormones—these will be considered in more
      detail below.

  (c) Catecholamines—adrenal medullary catecholamines, nor-
      adrenaline and adrenaline, are associated with sympathetic
      activity and their turnover is intimately involved in the arousal
      response to noxious stimuli. The ratio adrenaline to nor-
      adrenaline is increased by glucocorticoids which stimulate
      phenylethanolamine N-methyltransferase (PNMT) activity
      (Wurtman and Axelrod, 1966). It has been postulated that
      nor-adrenaline is related more to outwardly directed aggressive
      behaviour and adrenaline to inwardly directed anxious be-
      haviour (Funkenstein, 1955, 1956; Elmadjian et al., 1958;
      Arguelles et al., 1972) so that long-term social stress, suffered by
      subordinate animals, might render them less inclined to be
      aggressive and more to be of an anxious nature.

(4) Experience—experiences in both neonatal and adult life have a

vital modulating influence upon individual responses to novelty, and frequently override the other three variables. Variety of environmental experience, in terms of Sokolov's model, enables more effective and economic response to change. Specific experiences, on the other hand, act as response reinforcers to subsequent similar experiences. For example, for agonistic encounters, victory in an encounter with a conspecific enhances the chances of victory in subsequent encounters whereas defeat tends to lead to further defeat. Repeated contacts with familiar dominant/subordinate conspecifics thereby leads to habituation, loss of novelty and a consequent lowering of the level of agonistic behaviour. Low levels of environmental experience produce less effective, more uneconomic responses to change so that, for example, laboratory animals which have been isolated in individual cages are more sensitive to contact with conspecifics and react more aggressively—particularly after an initial experience of victory.

With few exceptions and given the above variables the following conclusion applies: a dominant male, who has experienced previous victories against conspecifics, when coming into contact with a completely strange intruder on his familiar breeding/ feeding area, if not highly motivated towards other goals, is likely to display very high levels of aggression and to be victorious in encounters with the intruder.

A causal relationship exists between androgens and agonistic behaviour while one associated between pituitary–adrenocortical function and this behaviour is more tenuous. The latter interaction must recognize that aggressive behaviour itself is responsible for increases in pituitary–adrenal function.

Much of the evidence is circumstantial. The high potential aggressiveness of the isolate or of the dominant members of a group has been causally related to the relatively low levels of pituitary–adrenocortical activity in these animals. Conversely, the declination in the mean levels of aggressive behaviour after the establishment of an hierarchy in a newly mixed group of animals has been causally related to the prolonged increase in pituitary–adrenal activity in such groups. Since ACTH and corticosteroids influence fear-motivated and avoidance behaviours as well as the performance of orienting responses and subsequent habituation to novel stimuli, it is reasonable to argue that these hormones, possibly by acting on the limbic system/hypothalamic areas, modulate individual aggressive/submissive responses to threat of or actual attack by a conspecific.

The earliest work in this field (Sigg et al., 1966; Sigg, 1969) emphasized that long-term grouping suppressed aggressive potential,

irrespective of androgen treatment, but failed to demonstrate significant effects of adrenalectomy upon the development of aggressive
potential in isolated male mice. An attenuation in this development
was ascribed to either negative metabolic changes or to deficient
gonadal function indicated by lower prostate weights in the adrenalectomized animals. Later Brain *et al.* (1971a) found decreased aggressive potentials in 13-day adrenalectomized or ACTH-injected mice
which were increased by dexamethasone injections (Table VIII). Such

TABLE VII

Effects of adrenalectomy, dexamethasone and ACTH on isolation-induced aggression
in mature male mice (from Brain *et al.*, 1971a).

Treatment	Composite aggression score of mice
Adrenalectomy	16.5[a]
Sham adrenalectomy	24.5
Dexamethasone injected	33.0[a]
Sham injected	23.0
ACTH injected	19.0[a]
Sham injected	29.0

P < 0.05 compared with experimental controls.

findings support the notion that the pituitary/adrenal system is
directly involved in modulating aggressive behaviour. More recent
investigations on naive isolated mice have produced equivocal results
and a clear picture has yet to emerge.

A recent review by Leshner (1975) puts some of the problems into
perspective, though more emphasis must be placed upon the temporal
and dose relationships of administered hormones to observed behavioural responses. The more indirect effects of pituitary–adrenocortical hormone manipulation such as those on general metabolism,
sexual function and catecholamine metabolism must also be taken
into account. Moreover the intimate relationship between ACTH and
the corticosteroids in both feedback and stress-response often pre-
cludes the definition of the active hormone in an effect. The corticosteroids may act directly by opposing ACTH, or by blocking the
production of ACTH which itself modulates behaviour. Such problems necessitate very careful experimental control.

Corticosterone injections increase aggressiveness in isolated mice
(Kostowski *et al.*, 1970) and several studies have confirmed the data of
Brain *et al.* (1971) showing decreased aggressiveness following adre-

nalectomy. Candland and Leshner (1974) extended this finding and restored aggressive behaviour in isolated adrenalectomized mice by corticosterone injection; a dose of 200 μg daily established an upper threshold (higher doses actually decreased aggression). This work supported previous findings that low levels of ACTH injected for relatively short periods of time increased aggressiveness (Brain and Evans, 1973; Leshner *et al.*, 1973) a response which, according to Leshner *et al.* (*op cit*), reflected the small increases in corticosteroid output which resulted from the treatment.

However, there are many indications that these effects are conse-quent upon the negative feedback effects of corticosterone upon ACTH release. Certainly ACTH has more clearly defined direct effects upon the CNS than the corticosteroids. Leshner *et al.* (1973), for example, showed that ACTH decreased aggressive behaviour in adrenalectomized mice which were kept on a fixed maintenance dose of corticosterone and that the response was dose-dependent. Long term ACTH injection regimes decrease aggressiveness (Brain, 1971; Brain *et al.*, 1971b; Leshner *et al.*, 1973). In the latter study this effect was considered independent of adrenocortical or gonadal actions and supported the earlier finding of Harding and Leshner (1972) that grouping reduced aggressiveness in both intact and adrenalectomized mice. Recent studies (Nock and Leshner, 1976; Moyer and Leshner, 1976) indicate that the corticosteroids are critical in the development of the avoidance responses by the subordinate animals, whereas ACTH is critical in the suppression of aggression.

Both ACTH and the glucocorticoids have been related to fear and learning and many reports suggest that corticosteroids oppose ACTH actions within the CNS. Endroczi (1972a) has proposed that a balance between ACTH and corticosteroids produces a state of non-motivational behaviour and reflects the "tonic" level of pituit-ary–adrenal adaptation. Moreover prolonged stress increased corti-costeroid production more than ACTH in response to a subsequent standard stress (Fig. 7); that is, the balance of ACTH to corticoster-oids can change (Endroczi, 1972a), a finding very germane to the probable mutual antagonism of these hormones. According to Levine (1968a), increased pituitary–adrenocortical activity supports goal-directed motor patterns during exposure to novelty by both activation and inhibition. In avoidance, corticosteroids facilitate basal forebrain inhibitory reactions, which can be prevented by ACTH which also facilitates the acquisition of conditioned avoidance responses by opposing the non-specific tonic effect of corticosteroids.

Corticosteroids facilitate the extinction of conditioned avoidance

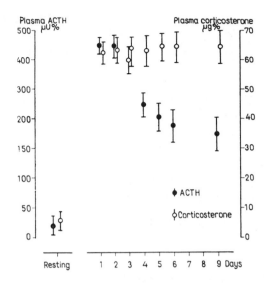

Fig. 7. Plasma ACTH and corticosterone concentrations in rats given 10 electric shocks within one minute each day and killed in the 8th and 20th min thereafter (from Endroczi, 1972a).

responses (de Weid, 1966; Bohus and Lissák, 1968) and the latter authors recorded a high level of intertrial responses following adrenalectomy, concluding that this was a result of a higher level of fear. Similarly, Weiss *et al.* (1970) concluded that ACTH increases generalized fear or anxiety and that corticosteroids restore this excitability to normal. Thus, in a novel situation, hypophysectomized rats showed less, and adrenalectomized ones more, fear than normal. Both Weiss *et al.* (1970) and Kasper-Pandi *et al.* (1970) point out that the most significant effects of pituitary–adrenal hormones occur with mild generalized fear and that these responses involve neural, including autonomic, stimulation. On the other hand, ACTH and adrenalectomy facilitate acquisition of the avoidance of high but not moderate unconditional stimuli intensity in the rat (Beatty *et al.*, 1970). The many contradictory findings in the literature may result from such variables as the post-operative or post-injection time-lapse before behavioural testing, the route and quantity of hormone injected and, for the behavioural tests themselves, the timing, quantity and quality of the aversive stimuli as well as the intensity of any motivation on the part of the animal. However it is likely that the facilitating effect of glucocorticoids upon CNS inhibitory mechanisms suppresses irrelevant responses to aversive stimuli (Kamin, 1963; Brush and

Levine, 1966), an effect producing a more efficient response which is susceptible to disinhibition by ACTH (Levine, 1968a). Thus the reported lower levels of aggressiveness in the presence of high levels of ACTH might well be due to the maintenance of irrelevant responses at the expense of the more specific aggressive response.

If aggressive behaviour towards aversive stimuli is fear-motivated hormonal manipulations may influence experimental results. As noted, isolation of male mice increases their reactivity to a novel environment whilst their pituitary/adrenocortical system is relatively inactive. In squirrel monkeys Candland and Leshner (1974) showed that of a number of isolated males those which had the highest urinary 17-OH-corticosteroid levels were the ones which became dominant after grouping. This, presumably, was due to their greater reactivity to the strangeness of the metabolism cage and to urine collection. As in most grouping studies, however, immediately after the establishment of a dominance order, a reversal was seen and the subordinate members showed higher urinary 17-OH-corticosteroids, reflecting their greater stress experience.

It would seem that the low pituitary–adrenocortical responses of dominant animals in the early stages of new contact with conspecifics reflects a low level of social stress and a fear-motivated behaviour manifested as aggression. It is suggested that in the initial stages of contact such responses are characteristic of most members of a group. However, if the duration of social interaction and its resultant stress is prolonged, continued release of ACTH and/or corticosteroids occurs concomitant with more widespread avoidance behaviour by the more subordinate members of the group. The latter with their lower androgen levels and enlarged adrenal cortex which is highly responsive to ACTH (Stark *et al.*, 1963; Varon *et al.*, 1966; Solem, 1966; Sassenrath, 1970; Endroczi, 1972a, b) produce relatively large quantities of corticosteroids for long periods. These quantities of steroids may suppress fear-motivated responses, cause habituation and possibly, if the individual has experienced regular, prolonged high intensity stresses, a complete opting out of the social hierarchy.

Under these circumstances low levels of ACTH and corticosteroids support aggressive behaviour in the presence of relatively high androgen levels. Experience of victory maintains such a hormonal balance in dominant animals. When this hormonal balance has been reversed by repeated defeat, as in subordinate animals, avoidance or "freezing" behaviour results. Leshner *et al.* (1975) found that mice learn to avoid the attacks of conspecifics which have previously defeated them when levels of ACTH and corticosteroids are high. They did not demonstrate any antagonistic effects between these two hormones as in previous experiments using electric shock avoidance.

The modulation of agonistic responses by high levels of corticosteroids may occur in other ways. The influence of the C_{11} steroids upon PNMT in the adrenal medulla promotes nor-adrenaline synthesis so that anxious rather than aggressive responses are seen in animals subjected to repeated social stress. In this context Koboyashi et al. (1974) suggested that the relatively high ratio of adrenal to gonadal steroids in blood of animals after long-term stress directly reduces monoamine oxidase activity and induces the accumulation of false transmitters so that cholinergic responses dominate adrenergic responses in the CNS. This alteration, they suggest, is responsible for a swing away from narrow well-focused attention, enhancing behavioural responses to short-term, moderate levels of stress; broad-focused attention, enhancing performances that demand perceptual restructuring, characteristic of long-term stress, then predominates.

This swing may account for the different responses of dominant and subordinate males in a group, and supports the data of Eleftheriou and Church (1968) in which repeatedly defeated mice displayed changed nor-adrenaline/serotonin concentrations in limbic/hypothalamic areas indicative of a shift towards tropotrophic dominance. Such a parasympathetic bias would be manifested in a decreased level of central activation, motor activity and metabolic rate and a lowered responsiveness to environmental stimuli. Whether this shift is causally related to the relatively high levels of corticosteroids or, as suggested by Breggin (1965), to the high levels of adrenaline also released in defeat, is unknown.

One outstanding feature common to much of the experimental evidence for a relationship between pituitary–adrenal hormones and agonistic behaviour is their temporal relationships. Thus in the boar Liptrap and Raeside (1975), supporting Koboyashi et al. (1974), found that short-term increases in blood levels of ACTH/corticosterone elevated testosterone levels, whilst long-term increases actually decreased testosterone levels. These effects were concluded to be mediated by the adrenal and clearly, if generally applicable, must be highly significant in suppressing aggressive behaviour in stressed subordinates.

The attenuated pituitary–adrenal response after repeated daily exposure of mice to trained fighters (Bronson and Eleftheriou, 1965a) is another temporal feature, possibly attributable to corticosteroid-mediated habituation in the defeated animal. The effect may also however be related to the shift in hypothalamic ergotrophic/tropo-trophic balance.

Ng *et al.* (1973), in a study of a growing population of BALB/C mice, recorded that four pairs of mice increased to a total of 2,200 in an 18 month period. This population density produced reproductive failure and a number of the population became subordinate and withdrawn displaying a greater adrenal PNMT activity and catecho-lamine turnover. A reciprocal relationship between social stress and reproductive potential and a relationship between long-term social stress and type of social reactivity were thus apparent.

In conclusion pituitary–adrenocortical hormones may modulate aggressive behaviour in a variety of ways: (a) The maintenance of a reasonable level of general metabolism; (b) Direct effects upon the CNS to influence arousal and the acquisition and maintenance of fear-motivated behaviour; (c) The control of adreno-medullary catecholamine metabolism; (d) An effect on the hypothalamic control of the ergotrophic/tropotrophic balance.

3. Emotional behaviour

The study of animal population must consider individual physiolo-gical and behavioural responses to environmental stimuli, particular-ly those presented by their conspecifics. Historically there have been two obvious approaches to the concept of emotion: (a) the physiolo-gical and behavioural approach and (b) the cognitive approach. This general field has attracted physiologists, ethologists and, rarely, psychologists; such various approaches account, in part, for the divergent theories of emotion. Contemporary theories have been influenced by the relative emphasis placed on *peripheral*, as against *central*, correlates of emotion.

The James–Lange theory (Lange and James, 1922) accounted for emotional feelings and experience rather than behaviour and had a "peripheral" emphasis. Afferent and efferent connections between peripheral organs (somatic and visceral) and the cortex were consi-dered to be uninterrupted so that central feelings of emotion are related to the awareness of the responses of peripheral tissues to appropriate stimuli. The Cannon–Bard theory on the other hand, (Bard, 1928, 1939) used "central" factors and gave the thalamus, via the cortex, responsibility for feelings of emotion, and under appropri-ate circumstances cortical inhibition allows the thalamus to evoke emotional behavioural reactions via the hypothalamus. Thus the hypothalamus assumed a principal effector role and Cannon (1932) propounded his Emergency Theory to account for increased sympathetic and adrenomedullary activity in emotional states with the "fight or flight" connotation. Further, Selye's General Adaptation Syndrome accounted for the temporal reaction to non-specific stress, again via the hypothalamus, and involved sympathetic arousal, followed by pituitary–adrenal activation. More recently the elements of the limbic system and their role in modulating the hypothalamic effector mechanism have been delineated (Brady, 1962; Endroczi, 1972a). In particular the hippocampus and the amygdala have been considered. The amygdala clearly regulates emotional behaviour by inhibiting motivations and modulating the satiation of drives such as eating and sex (Goddard, 1964)—thus amygdalectomy can result in loss of discrimination between what is edible and what is not and choice of a mate. The hippocampus is also inhibitory, excluding stimulus patterns from attention, and has efferent control over sensory systems (Douglas, 1967). Habituation and passive avoidance behaviour in the presence of novel stimuli and the extinction of learned responses are therefore influenced.

Qualitative differences between autonomic and adrenal responses to different emotional states are equivocal although more refined approaches, both in the provision of stimuli and in the monitoring of emotional feelings, behaviour and peripheral responses, may resolve the difficulties. Schachter and Singer (1962) introduced a new concept proposing that cognition determines the behavioural response to a specific stimulus to produce a general state of neuroendocrine arousal. Cognitive input after adrenaline or placebo injection was examined in man, and self-rated moods in the presence of "stooges", either euphoric or angry, revealed that the feeling of physiological arousal was produced by adrenaline and this effect was specifically related to the apparent mood of the "stooge".

Historically the study of emotional behaviour and its relationship with the neurophysiological mechanisms were initially concerned with man. The classification of stimuli and their cognition was therefore a relatively easy matter. More recently, however, neurophysiological and biochemical approaches have involved other animals. Unfortunately direct interpretation of cognition by such animals of a stimulus is difficult; comparison with our own cognition and extrapolation of overt behavioural responses are made and assumed to be related to the specific stimuli known in man. Cognition must be related to the quality, intensity and duration of the stimulus, to the state of motivation, to arousal levels and to previous experience. In other words, a wide range of variables exists.

The majority of "emotionality" tests in mammals have employed the open-field experiment. Its obvious methodological limitations and the interpretation and correlation of results have been reviewed (Archer, 1975; Walch and Cummins, 1976) and doubts as to the validity of many concepts have emerged. It is quite obvious that in future there must be careful control of as many variables as possible. Different levels and the time course of factors under investigation should be tested, since the physiological and behavioural mechanisms of "emotional" behaviour do not operate independently.

Manipulation of the pituitary–adrenal system also presents many problems not only because it is to some degree a self-regulating feedback system but also because it affects metabolism in general. In the CNS, where ACTH and the corticosteroids act in opposition to one another, blockage, ablation or injection experiments have their obvious weaknesses. In addition, the close involvement of both the autonomic and adrenal systems in behavioural and physiological responses render design and interpretation of experiments extremely difficult. Previous experience and type of stress (physical or psychological) imposed must be carefully controlled and time studies are essential if valid conclusions on the long-term effects of hormones are to be obtained. In this connection, high susceptibility to psychological or emotional stress in the "emotional" animal, compared with the "non-emotional" one, may be reversed when these animals are presented with physical stresses with smaller elements of emotion. Such conditions may well apply in studies such as those of Candland and Leshner (1974) in which squirrel monkeys which later showed dominance in a group, reacted more to urine collection than those which became subordinate in the group, a state of affairs reversed by subsequent hierarchy formation, aggressive encounters and physical defeat.

A positive relationship exists between emotional reactivity and pituitary–adrenal function in the face of novelty or psychological stress and there are many examples in which pituitary–adrenal function has been correlated with emotional arousal. It is also now well accepted that ACTH and the glucocorticoids directly modulate some fear-motivated behaviours. Thus there is increasing interest in the simultaneous monitoring of emotional behaviour and pituitary–adrenal activity. Indeed it has been suggested that hormones of the pituitary–adrenal system actually mediate in the changes in emotional reactivity to the environment (Levine, 1962a, b, 1970; Denenberg and Zarrow, 1971). Before considering this hypothesis further, a brief consideration of the effects of both infantile and adult experience upon "emotionality" and pituitary–adrenal function is pertinent.

4. Experience

A. NEONATAL EXPERIENCE

The differentiation of the hypothalamo–pituitary–gonadal system and the physiological and behavioural mechanisms controlled by it in the adult can be permanently altered by changes in the blood levels of androgen during critical periods of development; in mammals the period of greatest sensitivity is in early neonatal life (Harris, 1964). With this system in mind, attempts have been made to establish analogous responses within the hypothalamo–pituitary–adrenal system during neonatal life.

The most frequent method of infantile stimulation has been to handle the pups for varying periods of time. This technique certainly imposes more than one form of stimulus upon the pup (see Russell, 1971). The common factor involved in this tactile stress, and the associated imposed changes in both body temperature and maternal behaviour, is thought to be the increased environmental stimulation with all its neuroendocrine consequences, including activation of the hypothalamo–pituitary–adrenal axis. This system is of course similarly affected by such techniques as electric shock, and by the fostering of young to so-called rat "aunts".

In considering neonatal changes with respect to the hypothalamo–pituitary–adrenal system, Zarrow et al. (1968) suggested that there are three prerequisites for the concept that changing corticosteroid levels in the neonate "condition" hypothalamic structures responsible for CRF release:

(i) The neonate must be able to release corticosteroids in amounts sufficient to induce an effect; in young rats, adult levels of corticosteroids have been obtained at day 1 and, although there was a fall off in adrenal responsiveness at day 7, this could be prevented by ACTH administration on day 5 (Phillpott et al., 1969a, b). Structural changes, indicative of permanent increases in activity, occur in the hypothalamus and adrenal cortex after early ACTH treatment (Palkovits and Mitro, 1968a, b).

(ii) Handling, or other early stimulation, must be sufficient to induce a significant adrenocortical response in the neonate; handling of rat pups, when only two days old, significantly increases plasma corticosterone within 30 minutes (Denenberg et al., 1967). Such a procedure has also resulted in an earlier maturation of the stress-response mechanism (Levine, 1968a) and the 24 hour rhythm of adrenal activity has been advanced in the progeny of handled rats (Ader and Deitchman, 1970).

(iii) Available corticosterone must be taken up by hypothalamic neurones; Zarrow et al. (1968) showed that after intraperitoneal injection ^{14}C-4-corticosterone accumulated preferentially in the hypothalamus within 10 minutes.

It has been widely proposed that stresses imposed at these critical and relatively rapid periods of development not only advance the maturation of the hypothalamo–pituitary–adrenal system, but also permanently modify the "setting" of this system. Thus, as with the habituation model of Sokolov (1960), the more varied the infantile experience, the more selective is the response to later stimuli.

The manipulated compared with the non-manipulated animal responds with only a moderate plasma corticosterone rise after a mild stress such as handling (Levine et al., 1967; Ader, 1968; Fig. 8) but with a relatively large rise in response to a more severe stress such as electric shock (Haltmeyer et al., 1967). In the latter experiments a more rapid post-shock decline in corticosterone release occurred in the manipulated than in the non-manipulated animals.

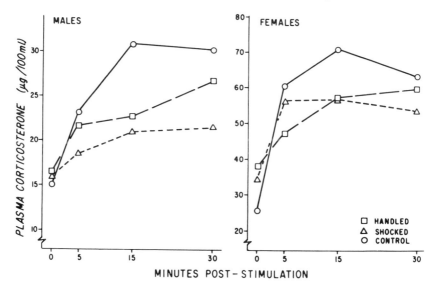

Fig. 8. The plasma corticosterone response in adults after the "reaction-to-handling test" in rats handled, shocked or unmanipulated during the first 20 days of life (From Ader, 1968).

Zarrow et al. (1972) found that, in spite of the fact that handled rats release less corticosterone after the mild stress of open-field testing than non-handled ones, the latter had smaller increases in hypothalamic CRF. The difference between the two groups was suggested to depend upon the rate of release of CRF which is supported more recently by Grota (1976) who found that handled rats do not differ from non-handled ones in either *in vitro* adrenocortical response to a standard ACTH dose, or in the metabolic half-life of corticosterone.

These modifications in pituitary–adrenal function are often paralleled by behavioural changes perhaps also reflecting Sokolov's model, in that animals stimulated as infants frequently show less emotion and more rapid habituation to novelty than controls when adult. Thus Denenberg *et al.* (1962), Levine (1962), Denenberg (1963, 1969), Goldman (1969) and Anderson *et al.* (1972) have all demonstrated such decreased emotionality in open-field tested animals (Fig. 9).

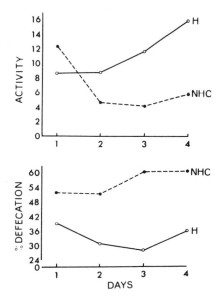

Fig. 9. Open-field activity and defaecation in handled (H) and non-handled (NHC) rats during four days of open-field testing (From Denenberg, 1969).

Increased activity in the open-field (Cowley and Widdowson, 1965; Wells *et al.*, 1969) and improved performance in the "T"-maze (Bernstein, 1952) are other behavioural modifications described. Gray (1971) pointed out that male rats bred for low fearfulness or early-handled rats were superior in active avoidance learning and the early-handled animals were likened to females—the latter are less fearful and display a more intense adrenocortical response to noxious stress than the more fearful male.

Aggressiveness, a form of emotional response to a novel stimulus (e.g. another animal) and a behaviour characteristic of males and not females, is also reduced by infantile stimulation, stresses such as handling (Becker and Gaudet, 1968) and, in the case of mice, rearing with rat aunts (Denenberg et al., 1966; Hudgens et al., 1967, 1968).

B.   ADULT EXPERIENCE

It is well known that the "setting" of the hypothalamo–pituitary–adrenal axis in the adult mammal is not static and varies with experience of stress. In both wild and laboratory environments, animals which have experienced frequent social stresses have enlarged, more active, adrenals which have a greater potential for further stress-responses than those of their less-stressed conspecifics. On the other hand, the repeated presentation of the same stress generally results in habituation and a fall-off in pituitary–adrenal function (Bronson and Eleftheriou, 1965a) as in the subordinate animals after repeated attacks by the same dominant animal. In stable hierarchies, therefore, both social status and relative pituitary–adrenal activity, once established, usually remain unchanged whereas in unstable groups of changing composition increased pituitary–adrenal activity is characteristic.

Behavioural modifications, principally in emotional reactivity, also occur as a result of adult experience. For example, the group-housed animal, with experience of relatively high levels of social stress, is less emotional and fearful than the individually caged one (Thiessen, 1964a, b; Essman, 1966; Morrison and Thatcher, 1969). The latter authors demonstrated that, with increased numbers of rats per cage, adrenal size increased and emotionality, tested in the open field, decreased. This agrees with the hypothesis of Denenberg and Haltmeyer (1967) of a monotonic decrement in emotionality with increasing intensity of early stimulation. In agreement with comments upon infantile rats (Gray, 1971), the experiments of Morrison and Thatcher (1969) and Levitt and Bennett (1972) showed that crowding of adult rats increased learning ability.

Statements about emotional reactivity are supported by studies on aggressiveness. It is generally true that there is a reciprocal relationship between the potentiality for aggressiveness and experience of social interaction. Thus the isolated animal, having experienced a low level of environmental stimulation and low pituitary–adrenal activity, has a high aggressive potential when grouped. Grouping is followed initially by a high level of aggressive behaviour and results in a reversal of the responses shown by the isolate (Brain et al., 1971). It would seem, then, that the aggressiveness of the isolate characterizes the fear-motivated responses of the more emotional animal. This conclusion is further supported by the fact that aggressive mice exhibit far less exploratory behaviour than non-aggressive ones (Valzelli, 1969, 1971), and in addition isolated rats are less active than group-housed ones in both the open field and the Greek cross maze (Hoyenga and Lekan, 1970; Thompson and Lippman, 1972).

Such evidence suggests that loss of emotionality is frequently associated with, though not necessarily causally related to, previous experience of stress and its accompanying pituitary–adrenal response.

Various laboratory experiments have attempted to establish a direct causal relationship. Swanson and McKeag (1969) for example injected cortisol into neonatal rats with the rationale that it would be analogous to the stress of neonatal stimulation. The injected adults did show an increase in open-field activity, a sign of less emotionality. Rosecrans (1970) however concluded that rats of different emotionalities did not differ basically in their pituitary–adrenal function although the highly emotional ones failed to habituate to novelty and reacted to a novel environment (strange cage) with a greater increase in pituitary–adrenal activity than the less emotional ones. Pituitary–adrenal function was taken to reflect and not cause emotional reactivity. Moyer (1958) found that adrenalectomy of the rat increased emotional elimination when tested later in the open-field, a finding confirmed by Joffe et al. (1972), who also observed reduced locomotor activity in the adrenalectomized animals; these two effects suggest a direct relationship between corticosteroids and loss of emotionality.

Ader (1970, 1975) has criticized the idea that pituitary–adrenal hormones directly mediate emotionality and, though not disagreeing with the basic findings on infantile stimulation, he suggests that the behavioural and endocrine changes are independent, although possibly mediated by the same central source. Doubt was cast firstly on whether the neonate can respond to neurogenic stress and secondly, whether there is an increased response to intense stresses in the adult after early stimulation; a lowered response was found in previously handled rats irrespective of the intensity of electric shock stress. Thirdly, Ader also questions the effectiveness of "early" pre-weaning as against post-weaning stimulation and provides evidence that the latter reduced pituitary–adrenal responsiveness to electric shock as effectively as pre-weaning stimulation (Fig. 10).

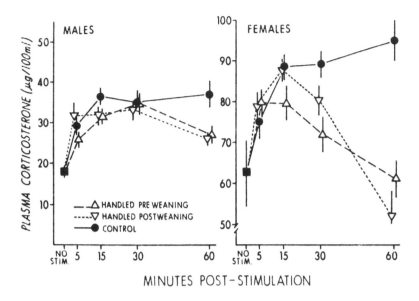

Fig. 10. Adult plasma corticosterone response to electric shock in unmanipulated (control) rats and rats handled daily throughout the pre- and post-weaning period of life. Means ± S.E. (From Ader, 1970).

Stern *et al.* (1973) have also suggested that the relationship between open-field behaviour and pituitary–adrenal function is merely correlational. However, lactating rats with below-normal corticosterone responses to the stress of open-field were found to defaecate more, in subsequent open-field tests, than controls (a sign of their greater emotionality); it was concluded that these animals and others with median eminence implants of cortisol showed unaltered open-field behaviour as a correlate of altered pituitary–adrenal activity.

Many questions remain regarding both the behavioural and the endocrine aspects of these studies. To what extent do tests such as the open-field one measure "emotionality"? How standardized are the tests and how well controlled and standardized are the genetic, behavioural and endocrine backgrounds of the subjects?

Studies on the effects of corticosteroids on specific sites in the CNS have provided considerable insight into possible direct roles of these hormones on behaviour. There are some clear relations between the position of the steroid-sensitive areas involved with ACTH release and those involved with motivational and emotional behaviour. Experimental approaches have included lesions and implantations and interpretation of responses have been hampered because these procedures affect adjacent areas. The several sites of uptake of radioactive corticosteroids include hippocampus, reticular formation, amygdala, septum, hypothalamus and cortex, the hippocampus being predominant.

Cortisol implanted into the mesencephalic reticular formation facilitates extinction of conditioned avoidance responses (Van Wimersma Greidanus, 1970). Moreover the steroids which facilitate the extinction of the pole-jumping avoidance response also fulfilled the criteria for the anaesthetic activity of C-21 steroids. This is an interesting point, probably relevant to the fact that extremely subordinate animals in a group suffer high levels of social stress, opt out and retire from the group. Bohus (1968) made cortisol implants into the rostral mesencephalic reticular formation and into the median eminence and observed a direct inhibitory effect upon both avoidance behaviour and ACTH release. More recently cortisol implanted into the medial thalamus, anterior hypothalamus, rostral septum or amygdala was shown to facilitate extinction of conditioned avoidance response (Bohus, 1970).

Pffaf *et al.* (1971), using telemetric recording of unit activity, demonstrated that corticosteroids have the opposite effect to ACTH

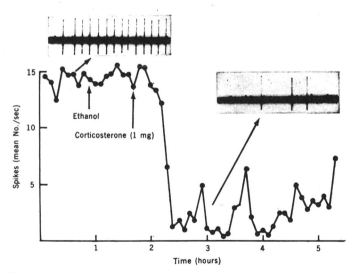

Fig. 11. Single unit activity in the hippocampus of a freely moving hypophysectomized female rat after a single i.p. injection of corticosterone in ethanol. A control ethanol injection had no effect. One-second samples from unit activity before (left) and after (right) injection are shown (from Pfaff *et al.*, 1971).

in the hippocampal region by reducing unit firing, in some cases in the same cells (Fig. 11). This region accumulates the highest concentrations of injected radioactive corticosterone (Pffaf et al., 1971) and McEwen et al. (1971) have suggested that, because its binding is saturated at very low levels of circulating corticosterone, the region regulates the basal and possibly the diurnal blood levels of these hormones.

There is other evidence for a close relationship between corticosteroid availability and hippocampal function (Antelman and Brown, 1972). Lesions in this area resulted in higher levels of circulating ACTH, an increase in fear-motivated behaviour, in the acquisition of avoidance responses and in resistance to extinction. Fear was concluded to have become an inbuilt part of the animal, as a consequence of the chronically elevated ACTH levels. High levels of corticosteroids also existed, obviously as a direct result of the high ACTH levels, but the corticosteroids, in the absence of the hippocampus, were unable to elicit inhibition. The hippocampus plays an essential role in the habituation to novel aversive stimuli which follows a period of fear-motivation and its presence is necessary for corticosteroids to promote this inhibitory effect (Green et al., 1967; Olton and Isaacson, 1968; Rabe and Haddad, 1969).

Assuming that the limbic system, acting via the hypothalamus, significantly modulates both behavioural and endocrine responses to emotional stimuli, major questions remain: do the pituitary–adrenal hormones feed back upon this area, possibly in opposition to one another? If so, do they act non-specifically to (a) change the thresholds of all settings which govern endocrine and behavioural responses or (b) support "permissively" changes in "settings" differentially, as a result of specific experiences and the differential habituation within the limbic system which results from them?

A prerequisite in a consideration of antagonistic effects of ACTH and corticosteroids upon emotional behaviour in general, akin to their effects upon fear-motivated behaviour and habituation, is that the ratio of the two hormones presented to the CNS, with time, varies with behavioural/endocrine experience. This seems to be the case. Thus the enlarged adrenals of subordinate animals are more highly reactive to ACTH compared with less subordinate animals (Varon et al., 1966; Solem, 1966; Sassenrath, 1970). In addition, Endroczi (1972a) has clearly demonstrated a dissociation between blood levels of ACTH and corticosterone in response to repeated daily electric shock in the rat (see page 370). It remains to be seen whether such changes are causally related to changes in emotional reactivity.

5. Sensory systems

The efficiency of sensory processes is clearly relevant to behavioural responses and there are several indications that the adrenocorticosteroids play a significant rôle in modulating at least some of these processes.

Henkin and Daly (1968) pointed out that human hypoadrenocorticism is accompanied by more acute auditory, gustatory and olfactory sensitivities and subnormal perceptual abilities. Henkin (1970) has since confirmed that the glucocorticoids inhibit the detection, and increase the perception of incoming sensory and proprioceptive signals. The detection of some odours increased by a factor of 1000 in patients suffering from adrenocortical insufficiency and this became normal after the administration of glucocorticoids but not mineralocorticoids. Detection and perception are inversely related and follow the pattern of the circadian rhythm of corticosteroid secretion. In the rat the olfactory threshold to pyridine decreases after adrenalectomy (Sakellaris, 1972).

It is tempting to speculate that glucocorticoids modulate behaviour by changing sensory sensitivities. For example, the peripheral olfactory input is separated from the hypothalamus by only two synapses, and in addition there is a relayed pathway through the basal and lateral amygdaloid nuclei. Such behavioural activity as feeding and sexual and aggressive activities might well be influenced, and the lowering of sensory input in the subordinate animal, with its high corticosteroid levels, would be expected to facilitate habituation and reduce emotional reactivity.

6. Locomotor activity

The pituitary–adrenocortical system might affect behaviour by modulating locomotor activity, an essential part of many forms of behaviour. An increase in activity may act positively to enhance a specific behaviour, or negatively by enhancing alternative behaviours.

Although increased sensory inputs are frequently correlated with increases in both general activity and pituitary–adrenal function, evidence for a direct effect of these hormones upon activity is by no means conclusive. Some experiments indicate, however, that corticosteroids act, in opposition to ACTH, to increase activity. Thus dexamethasone stimulates running activity in the male rat (Kendall, 1970) and adrenalectomy has the opposite effect; the latter response was reversed by corticosterone replacement (Leshner, 1971). This may be a "metabolic" effect and not a direct action on the CNS and the latter author suggested that blood levels of free fatty acids may provide a key link between corticosteroids and activity. Whatever the mechanism, if such an antagonism between ACTH and corticosteroids exists, then it could, at least in part, account for some of the reports of decreased aggressive behaviour following ACTH and its reversal by corticosteroids.

7. Sexual behaviour

The concept that the social stress which populations of high density experience limits population growth because of a reciprocal loss of pituitary–gonadal function has already been discussed. Although there is much evidence for a reduced fecundity, information on the behavioural aspect of sexual reproduction in this context is slight. Reduced gonadal steroid production, on a long-term basis, must presumably affect such behaviour adversely.

There is both clinical and experimental evidence that population limitation results from stress during sexual differentiation. Thus Ward (1972) has shown that *in utero* demasculinization can occur if pregnant female rats are exposed to repeated stress. Male progeny of such mothers showed behavioural responses indicative of demasculinization when injected with testosterone and paired with females, and of feminization when injected with ovarian hormones and placed with males. Such effects probably result from competitive inhibition of the testosterone-dependent sexual differentiation of CNS elements, following the release of weak adrenal androgens such as androstenedione provoked by high levels of ACTH.

The adrenogenital syndrome, resulting from an enzyme defect, is characterized by increased androgenic steroids alongside deficient cortisol production. Increased ACTH production results in varying degrees of cortical hyperplasia. The extent of the masculinizing effects of these excess androgens depends, apart from their concentration, upon the age of the individual when the abnormality develops and, in fact, psychological masculinization need not be a permanent effect (Money and Erhardt, 1968).

Studying the more direct effects of adrenal steroids upon female sexual behaviour Davidson et al. (1968) found no effect of adrenalectomy upon oestrogen-induced oestrous behaviour in rats. Also Jacobel and Rogers (1971) failed to record any effects of this operation upon natural oestrous behaviour in the rat. Adrenocortical hormones can however exert a positive effect upon this behaviour. The sequential release of oestrogen, followed by progesterone, is essential for the development of mating behaviour in most strains of rats, and Nequin and Schwartz (1971) in a series of experiments involving ovariectomy and adrenalectomy, only one hour before a behaviour test on the day of pro-oestrus, concluded that adrenal progesterone plays a significant rôle in the timing of mating behaviour.

In addition, female sexual behaviour can be enhanced under conditions of stress and this reaction has been ascribed to the stimulation of adrenal progesterone release by high ACTH titres (Feder and Ruf, 1969; Paris *et al.*, 1971). This conclusion is supported by data showing changes in pituitary–adrenal activity around the oestrous cycle with an increase on the day of pro-oestrus (Raps *et al.*, 1971). Furthermore, Lawton (1972) demonstrated that stress advanced and adrenalectomy abolished the LH surge, an effect attributed to the removal of adrenal progesterone.

Investigations into the sexual receptivity of the rhesus monkey by Everitt and Herbert (1969, 1970) have shown that this behaviour depends upon the direct action of adrenal androgens upon the central nervous system. The suppression of such behaviour, by adrenalectomy or dexamethasone treatment, concomitantly suppressed the mounting behaviour of the male. More recently Everitt *et al.* (1972) showed that receptivity by the adrenalectomized, ovariectomized female could be restored by androstenedione but not by dehydroepiandrosterone. This action of adrenal androgens may only be a characteristic of primates, whilst sub-primates rely upon oestrogens and progestins.

In the male there is very little evidence for a direct effect of adrenal steroids on sexual behaviour. Bloch and Davidson (1968) showed that adrenalectomy has no effect upon sexual behaviour in the male rat.

References

Ader, R. (1968). *J. comp. Physiol. Psychol.* **66**, 264–268.
Ader, R. (1970). *Physiol. Behav.* **5**, 837–839.
Ader, R. (1975). *In*: "Hormonal Correlates of Behaviour" 1. (Ed. B. E. Eleftheriou and R. L. Sprott) pp. 7–33. New York.
Ader, R. and Deitchman, R. (1970). *J. comp. Physiol. Psychol.* **71**, 492–496.
Anderson, C. O., Denenberg, V. H. and Zarrow, M. X. (1972). *Behaviour* **43**, 97–120.
Andrews, R. V. (1968). *Endocrinology* **83**, 1387–1389.
Andrews, R. V. (1970). *Acta endocr. Copnh.* **65**, 639–644.

Andrews, R. V. and Strohbehn, R. (1971). *Comp. Biochem. Physiol.* **38**, 183–201.
Andrews, R. V., Belknap, J., Southard, J., Lorincz, M. and Hess, S. (1972). *Comp. Biochem. Physiol.* **41A**, 149.
Antelman, S. M. and Brown, T. S. (1972). *Physiol. Behav.* **9**, 15–20.
Arguelles, A. E., Martinez, M. A., Hoffman, C., Ortiz, G. A. and Chekherdemian, M. (1972). *Hormones* **3**, 167–174.
Archer, J. (1975). *Behav. Biol.* **14**, 451–479.
Bard, P. (1928). *Am. J. Physiol.* **84**, 490–515.
Bard, P. (1939). *Res. Publs. Ass. for Research of Nervous and Mental Disease* **19**, 190–218.
Barnett, S. A. (1958). *J. psychosom. Res.* **3**, 1–11.
Barrett, A. M. and Stockham, M. A. (1963). *J. endocr.* **26**, 97–105.
Beatty, P. A., Beatty, W. W., Bowman, D. E. and Gilchrist, J. C. (1970). *Physiol. Behav.* **5**, 939–944.
Becker, G. and Gaudet, I. J. (1968). *Psychonom. Sci.* **11**, 115–116.
Bernstein, L. (1952). *Psychol. Bull.* **49**, 38–40.
Bliss, E. L., Frischat, A. and Samuels, L. (1972). *Life Sci.* **11**, 231–238.
Bloch, G. J. and Davidson, J. M. (1968). *Physiol. Behav.* **3**, 461–465.
Bohus, B. (1968). Neuroendocrinology **3**, 355–365.
Bohus, B. (1970). *In*: "Pituitary Adrenal and the Brain". (D. de Wied and J. A. W. H. Weijnen, Eds.) *Prog. Brain Res.* **32**, 181–183.
Bohus, B. and Lissák, K. (1968). *Intl. J. Neuropharmacol.* **7**, 301–396.
Bowen, S. T., Gowen, J. W. and Tauber, O. E. (1957). *Proc. Soc. exp. Biol. Med.* **94**, 476–479.
Brady, J. V. (1962). *In*: "Experimental Foundations of Clinical Psychology" (Ed. A. J. Brachrach) pp. 343–385. Basic Books, New York.
Brain, P. F. (1971). *J. endocr.* **51**, xviii–xix.
Brain, P. F. and Evans, C. M. (1973). *J. endoct.* **57**, xxxix–xl.
Brain, P. F. and Nowell, N. W. (1969). *J. endocr.* **46**, xvi–xvii.
Brain, P. F. and Nowell, N. W. (1970). *Physiol. Behav.* **5**, 907–910.
Brain, P. F. and Nowell, N. W. (1971a). *Gen. comp. Endocr.* **16**, 149–154.
Brain, P. F. and Nowell, N. W. (1971b). *Gen. comp. Endocr.* **16**, 155–159.
Brain, P. F., Nowell, N. W. and Wouters, A. (1971). *Physiol. Behav.* **6**, 27–29.
Breggin, P. R. (1965). *Archs. gen. Psychiat.* **12**, 255–259.
Bronson, F. H. (1963). *Ecology* **44**, 637–643.
Bronson, F. H. (1964). *Anim. Behav*, **12**, 470–478.
Bronson, F. H. (1967). *In*: "Husbandry of Laboratory Animals", (Ed. M. L. Conalty). pp. 513–542. Academic Press, New York and London.
Bronson, F. H. (1973). *Physiol. Behav.* **10**, 947–951.
Bronson, F. H. and Eleftheriou, B. E. (1964). *Gen. comp. Endocr.* **4**, 9–14.
Bronson, F. H. and Eleftheriou, B. E. (1965a). *Proc. Soc. exp. Biol. Med.* **118**, 146–149.
Bronson, F. H. and Eleftheriou, B. E. (1965b). *Science, N.Y.* **147**, 627–628.
Brooks, J. E. and Barnes, A. M. (1972). *Vector News* **19**, 5.
Brush, F. R. and Levine, S. (1966). *Physiol. Behav.* **1**, 309–311.
Bull, P. C. (1955). *Nature (Lond.)* **175**, 218.
Bull, P. C. (1957). *Proc. N. Z. ecol. Soc.* **5**, 11–12.
Bull, P. C. (1964). *N.Z. C.S.I.R.O. Bull.* **158**, 1–145.
Byrnes, W. W. and Shipley, E. G. (1950). *Proc. Soc. exp. Biol. Med.* **74**, 308–310.
Campbell, W. A. (1963). *J. Parasit.* **49**, 628–632.
Candland, D. K. and Leshner, A. I. (1974). *In*: "Limbic and Autonomic Nervous System Research " (Ed. L. V. Di Cara) pp. 137–163. Plenum Press, New York.

Cannon, W. B. (1932). "The Wisdom of the Body." W. B. Norton and Co, New York.
Carstensen, H., Amer, I., Wide, L. and Amer, B. (1973). *J. Steroid Biochem.* **4,** 605–611.
Chang, S. and Rasmussen, A. F. (1965). *Nature (Lond.)* **205,** 623–624.
Christian, J. J. (1950). *J. Mammal.* **31,** 247–259.
Christian, J. J. (1955a). *Am. J. Physiol.* **181,** 477–480.
Christian, J. J. (1955b). *Am. J. Physiol.* **182,** 292–300.
Christian, J. J. (1959a). *In*: "Comparative Endocrinology" (A. Gorbman Ed.) pp. 71–97, Wiley, New York.
Christian, J. J. (1959b). *Proc. Soc. exp. Biol. Med.* **101,** 166–168.
Christian, J. J. (1960). *Proc. Soc. exp. Biol. Med.* **104,** 330–332.
Christian, J. J. (1963). *Military Med.* **128,** 571–603.
Christian, J. J. (1968). *Am. J. Epidemiol.* **87,** 255–264.
Christian, J. J. (1971a). *In*: "The Action of Hormones: Genes to Population" (P. Foa ed.) pp. 471–499. Thomas, Springfield.
Christian, J. J. (1971b). *J. Mammal.* **52,** 556–567.
Christian, J. J. (1975). *In*: "Hormonal Correlates of Behaviour" (B. E. Eleftheriou and R. L. Sprott, eds.) Vol. 1, pp. 205–274. Plenum Press, New York.
Christian, J. J. and Davis, D. E. (1956). *J. Mammal.* **37,** 475–486.
Christian, J. J. and Le Munyan, C. D. (1958). *Endocrinology* **63,** 517–529.
Christian, J. J. and Williamson, H. O. (1958). *Proc. Soc. exp. Biol. Med.* **99,** 385–387.
Christian, J. J., Lloyd, J. A. and Davis, D. E. (1965). *Recent Prog. Horm. Res.* **21,** 501–578.
Cowley, J. J. and Widdowson, E. M. (1965). *Br. J. Nutr.* **19,** 397–406.
Davidson, J. M., Rodgers, C. H., Smith, E. R. and Bloch, G. H. (1968). *Endocrinology* **82,** 193–195.
Davis, D. E. and Christian, J. J. (1957). *Proc. Soc. exp. Biol. Med.* **94,** 728–731.
Davis, D. E. and Read, C. P. (1958). *Proc. Soc. exp. Biol. Med.* **99,** 269–272.
den Boer, P. J. (1968). *Acta biotheor.* **18,** 165–194.
Denenberg, V. H. (1963). *Sci. Am.* **208,** 138–146.
Denenberg, V. H. (1969). *Ann. N.Y. Acad. Sci.* **159,** 852–859.
Denenberg, V. H., Brumaghim, J. T., Haltmeyer, G. C. and Zarrow, M. X. (1967). *Endocrinology* **81,** 1047–1052.
Denenberg, V. H., Hudgens, G. A. and Zarrow, M. X. (1966). *Psychol. Rep.* **18,** 451–456.
Denenberg, V. H. and Haltmeyer, G. C. (1967). *J. comp. Physiol. Psychol.* **63,** 394–396.
Denenberg, V. H., Morton, J. R. C., Kline, N. J. and Grota, L. J. (1962). *Can. J. Psychol.* **16,** 72–76.
Denenberg, V. H. and Zarrow, M. X. (1971). *In*: "Early Childhood: The Development of Self-regulatory Mechanisms". (Ed. D. H. Walcher and D. L. Peters) pp. 39–64. Academic Press, New York and London.
Douglas, R. J. (1967). *Psychol. Bull.* **67,** 416–442.
Eleftheriou, B. E. and Church, R. L. (1968). *Physiol. Behav.* **3,** 977–980.
Elmadjian, F., Hope, J. M. and Lamson, E. T. (1958). *Rec. Prog. Horm. Res.* **14,** 513–553.
Endroczi, E. (1972a). "Limbic System Learning and Pituitary–Adrenal Function" (Akademiai Kiado, Budapest).
Endroczi, E. (1972b). *In*: "Hormones and Behaviour. pp. 173–207. (Ed. S. Levine). Academic Press. New York and London.
Essman, J. (1966). *Anim. Behav.* **14,** 406–409.

Euker, J. S., Meites, J. and Riegle, G. D. (1975). *Endocrinology* **96,** 85–92.
Everitt, B. J. and Herbert, J. (1969). *Nature, Lond.* **222,** 1065–1066.
Everitt, B. J. and Herbert, J. (1970). *J. endocr.* **48,** xxxviii.
Everitt, B. J., Herbert, J. and Hamer, J. D. (1972). *Physiol. Behav.* **8,** 409–415.
Feder, M. and Ruf, K. B. (1969). *Endocrinology* **84,** 172–174.
Frenkel, J. K. and Lunde, M. N. (1966). *J. Infect. Dis.* **116,** 414–424.
Funkenstein, D. H. (1955). *Sci. Am.* **192,** 74–80.
Funkenstein, D. H. (1956). *J. nerv. ment. Dis.* **124,** 58–68.
Galef, B. G. (1970). *J. comp. Physiol. Psychol.* **70,** 370–381.
Goddard, G. V. (1964). *Psychol. Bull.* **62,** 89–109.
Goldman, P. S. (1969). *Ann. N.Y. Acad. Sci.* **159,** 640–650.
Gray, J. A. (1971). *Acta Psychol.* **35,** 29–46.
Green, P. M. (1964). Ph.D. thesis, Oklahoma State Univ. pp. 1–101.
Green, R. H., Beatty, W. W. and Schwartzbaum, J. S. (1967). *J. comp. Physiol. Psychol.*
    64, 444–453.
Grota, L. J. (1976). *Dev. Psychobiol.* **9,** 211–216.
Hagino, N. (1968). *Jap. J. Physiol.* **18,** 350–355.
Hagino, N., Watanabe, M. and Goldzieher, J. (1969). *Endocrinology* **84,** 308–314.
Haltmeyer, G. C., Denenberg, V. H. and Zarrow, M. X. (1967). *Physiol. Behav.* **2,**
    61–63.
Harding, C. F. and Leshner, A. I. (1972). *Physiol. Behav.* **8,** 437–440.
Harris, G. W. (1964). *Endocrinology* **75,** 627–648.
Hebb, D. O. (1946). *Psychol. Dev.* **53,** 259–276.
Henkin, R. I. (1970). *In*: "Pituitary Adrenal and the Brain". (D. de Wied and J. A.
    W. M. Weijnen, Eds.) *Prog. Brain Res.* **32,** 270–294.
Henkin, R. I. and Daly, R. L. (1968). *J. clin. Invest.* **47,** 1269–1280.
Hobson, D. (1960). *Br. J. Exp. Path.* **51,** 251–258.
Howland, B. E., Beaton, D. B. and Jack, M. I. (1974). *Experientia.* **30,** 1223–1225.
Hoyenga, K. T. and Lekan, R. K. (1970). *Psychonom. Sci.* **20,** 56.
Hudgens, G. A., Denenberg, V. H. and Zarrow, M. X. (1967). *J. comp. Physiol.*
    Psychol. **63,** 304–308.
Hudgens, G. A., Denenberg, V. H. and Zarrow, M. X. (1968). *Behaviour* **30,** 259–274.
Jacobel, P. and Rodgers, C. H. (1971). *Horm Behav.* **2,** 201–206.
Jarrett, R. (1965). *Endocrinology* **76,** 434–440.
Joffe, J. M., Mulick, J. A. and Rawson, R. A. (1972). *Horm. Behav.* **3,** 87–96.
Kamin, L. J. (1963). *J. comp. Physiol. Psychol.* **56,** 713–718.
Kasper–Pandi, P., Hansing, R. and Usher, D. R. (1970). *Physiol. Behav.* **5,** 361–363.
Kass, E. H., Kendrick, M. I. and Finland, M. (1955). *J. exp. Med.* **102,** 767–774.
Kendall, J. W. (1970). *Horm. Behav.* **1,** 327–336.
Koboyashi, U., Broverman, D. H., Klaiber, E. L. and Vogel, W. (1974). *Psychol. Bull.*
    81, 672–694.
Kostowski, W., Rewerski, W. and Piechocki, T. (1970). *Neuroendocrinology* **6,** 311–318.
Lange, C. G. and James, W. (1922). "The Emotions". Williams and Wilkins,
    Baltimore.
Lawton, I. E. (1972). *Endocrinology* **90,** 575–579.
Leshner, A. I. (1971). *Physiol. Behav.* **6,** 551–558.
Leshner, A. I. (1975). *Physiol. Behav.* **15,** 225–235.
Lesher, A. I. and Candland, D. K. (1972). *Physiol. Behav.* **8,** 441–445.
Leshner, A. I., Walker, W. A., Johnson, A. E., Kelling, J. S., Kreisler, S. K. and
    Svare, B. B. (1973). *Physiol. Behav.* **11,** 705–711.

Leshner, A. I., Moyer, J. A. and Walker, W. A. (1975). *Physiol. Behav.* **15**, 689–694.

Levine, S. (1962). *Science*, N.Y. **135**, 795–796.

Levine, S. (1962). *In:* "Experimental Foundations of Clinical Psychology". (Ed. A. J. Bachrach). Basic Books, New York.

Levine, S. (1968a). *Dev. Psychol.* **1**, 67–70.

Levine, S. (1968b). *In:* "Nebraska Symposium on Motivation". Univ. of Nebraska Press, Lincoln. pp. 85–101.

Levine, S. (1970). *In:* "Pituitary, Adrenal and the Brain". *Progress in Brain Research* **32**, (D. de Wied and J. A. W. M. Weignen eds.) pp. 79–85 Elsevier, Amsterdam.

Levine, S., Haltmeyer, G. C., Karas, G. G. and Denenberg, V. H. (1967). *Physiol. Behav.* **2**, 55–59.

Levitt, L. and Bennett, T. L. (1972). *Psychon. Sci.* **29**, 52–54.

Lidicker, W. Z. (1966). *Ecol. Monogr.* **36**, 27–50.

Liptrap, R. M. (1970). *J. endocr.* **47**, 197–205.

Liptrap, R. M. and Raeside, J. I. (1975). *J. endocr.* **66**, 123–131.

Lloyd, J. A. (1971). *Proc. Soc. exp. Biol. Med.* **137**, 19–22.

Louch, C. D. and Higginbotham, M. (1967). *Gen. comp. Endocr.* **8**, 441–444.

McEwen, B. S., Magnus, C. and Wallach, G. (1971). *In:* "Steroid Hormones and Brain Function". (Sawyer, C. H. and Gorski, R. A, Eds.) pp. 247–258, University of California Press.

Mason, J. W. (1959). *A. Rev. Physiol.* **21**, 353–380.

Matsumoto, K., Takeyasu, K., Mizutani, S., Hamanaka, Y. and Vozumi, T. (1970). *Acta endocr. Copnh.* **65**, 11–17.

Meslow, E. C. and Keith, L. B. (1968). *J. Wildl. Mgmt.* **32**, 812.

Money, J. and Erhardt, A. A. (1968). *In:* "Endocrinology and Human Behaviour" (Ed. R. P. Michael). pp. 32–48. Oxford U.P., London.

Morrison, B. J. and Thatcher, K. (1969). *J. comp. Physiol. Psychol.* **69**, 658–662.

Moyer, K. E. (1958). *J. genet. Psychol.* **92**, 11–16.

Moyer, J. A. and Leshner, A. I. (1976). *Physiol. Behav.* **17**, 297–301.

Murray, K. F. (1965). *Ecology* **46**, 163–171.

Myers, K. (1967). *Nature, Lond.* **213**, 147–150.

Myers, K. (1970). *Proc. Study Inst. Hynam. Num. Popul.* 478–506.

Mykytowycz, R. (1962). *Parasitology* **52**, 375–395.

Nakashima, A., Kochiyama, K., Vozumi, T., Monden, Y., Hamanaka, Y., Kurachi, K., Aono, T., Mizutani, S. and Matsumoto, K. (1975). *Acta endocr. Copnh.* **78**, 258–269.

Nequin, L. G. and Schwartz, N. B. (1971). *Endocrinology* **88**, 325–331.

Ng, L. K. Y., Marsden, H. M., Colburn, R. W. and Thòa, N. B. (1973). *Brain Res.* **59**, 323–330.

Noble, G. A. (1962). *Exp. Parasit.* **12**, 368–371.

Noble, G. A. (1971). *Exp. Parasit.* **29**, 30–32.

Nock, B. L. and Leshner, A. I. (1976). *Physiol. Behav.* **17**, 19–22.

Oliver, L. (1962). *J. Parasit.* **48**, 758–762.

Olton, D. S. and Isaacson, R. L. (1968). *Physiol. Behav.* **3**, 719–724.

Palkovits, M. and Mitro, A. (1968a). *Neuroendocrinology* **3**, 200–210.

Palkovits, M. and Mitro, A. (1968b). *Gen. comp. Endocr.* **10**, 253–262.

Paris, C. A., Resko, J. A. and Gay, R. W. (1971). *Biol. Reprod.* **4**, 23–30.

Pffaf, D. W., Gregory, E. and Silva, M. T. A. (1971). *In:* "Influence of Hormones on the Nervous System". (D. Ford Ed.) Proc. Int. Soc. Psychoneuroendocrinology Brooklyn, 269–281. Karger, Basel.

Phillpott, J. E., Zarrow, M. X. and Denenberg, V. H. (1969a). *Steroids* **14**, 21–31.
Phillpott, J. E., Zarrow, M. X. and Denenberg, V. H. (1969b). *Proc. Soc. exp. Biol. Med.* **131**, 26–29.
Pitelka, F. A. (1957). *Arctic Biol. 18th Biol. Colloq.* 73–88.
Plaut, S. M., Ader, R., Friedman, S. B. and Ritterson, A. L. (1969). *Psychonom. Med.* **31**, 536–552.
Rabe, A. and Haddad, R. K. (1969). *Physiol. Behav.* **4**, 319–323.
Ramirez, V. D. and McCann, S. M. (1963). *Endocrinology* **72**, 452–464.
Ramirez, V. D. and McCann, S. M. (1964). *Endocrinology* **76**, 412–417.
Raps, D., Barthe, P. L. and Desculles, P. A. (1971). *Experientia* **27**, 339–340.
Robson, J. and Sharaf, A. (1952). *J. Physiol. (Lond.)* **116**, 236–243.
Rose, R. M., Gordon, T. P. and Bernstein, I. S. (1972). *Science* **178**, 643–645.
Rosecrans, J. A. (1970). *Arch. int. Pharmacodyn. Ther.* **187**, 349–366.
Russell, P. A. (1971). *Psychol. Bull.* **75**, 192–202.
Sakellaris, P. C. (1972). *Physiol. Behav.* **9**, 495–500.
Sassenrath, E. N. (1970). *Horm. Behav.* **1**, 283–298.
Schachter, S. and Singer, J. E. (1962). *Psychol. Rev.* **69**, 379–399.
Selye, H. (1939). *Endocrinology* **25**, 615–624.
Selye, H. (1950). "The Physiology and Pathology of Exposure to Stress". Acta Inc., Montreal.
Sen, H. G., Joshi, V. N. and Seth, D. (1965). *Trans. R, Soc. trop. Med. Hyg.* **59**, 684–689.
Seyler, L. E. and Reichlin, S. (1973). *Endocrinology* **92**, 295–302.
Sigg, E. B. (1969). *In:* "Aggressive Behaviour". (S. Garattini and E. B. Sigg, Eds.) Excerpta Medica Monograph, Amsterdam.
Sigg, E. B., Day, C. A. and Columbo, C. (1966). *Endocrinology* **78**, 679–684.
Smith, E. R. and Davidson, J. M. (1967). *Am. J. Physiol.* **212**, 1385–1390.
Smith, E. R., Johnson, J., Weick, R. F., Levine, S. and Davidson, J. M. (1971). *Neuroendocrinology* **8**, 94–106.
Sokolov, J. H. (1960). *In:* "The Central Nervous System and Behaviour". (Ed. M. A. B. Brazier), Trans. of the Third Conference of the Josiah Macy Foundation. New York.
Solem, J. H. (1966). *Scand. J. Clin. Lab. Invest.* **18**, Suppl. **93**, 1–36.
Southwick, C. H. and Bland, V. P. (1959). *Am. J. Physiol.* **197**, 111–114.
Stark, E., Fachet, J. and Mihaly, K. (1963). *Can. J. Biochem. Physiol.* **41**, 1771–1777.
Stern, J. M., Erskine, J. S. and Levine, S. (1973). *Horm. Behav.* **4**, 149–162.
Swanson, H. H. and McKeag, A. M. (1969). *Horm. Behav.* **1**, 1–5.
Thiessen, D. D. (1964a). *J. comp. Physiol. Psychol.* **57**, 412–416.
Thiessen, D. D. (1964b). *Tex. Rep. Biol. Med.* **22**, 266–314.
Thompson, R. W. and Lippman, L. G. (1972). *J. comp. Physiol. Psychol.* **80**, 439–448.
Valzelli, L. (1969). *Psychopharma.* **15**, 232–235.
Valzelli, L. (1971). *Psychopharma.* **19**, 91–94.
Varon, H. H., Touchstone, J. C. and Christian, J. J. (1966). *Acta endocr. Copnh.* **51**, 488–496.
Vessey, S. H. (1964). *Proc. Soc. exp. Biol. Med.* **115**, 252–255.
Walch, R. N. and Cummins, R. A. (1976). *Psychol. Bull.* **83**, 482–504.
Ward, I. L. (1972). *Science, N.Y.* **175**, 82–84.
Weiss, J. M., McEwen, B. S., Silva, M. T. A. and Kalkut, M. F. (1970). *Am. J. Physiol.* **218**, 864–868.
Welch, B. L. (1962). Proceedings First National Deer Disease Symposium, Athens,

Ga., Feb. 13–15, Univ. Georgia, Center for Continuing Education Publications.

Welch, B. L. (1964). *In*: "Symposium on Medicine Aspects of Stress in the Military Climate". Walter Reed Army Institute of Research, Washington D.C.

Wells, P. A., Lowe, G., Sheldon, M. H. and Williams, D. I. (1969). *Br. J. Psychol.* **60,** 389–393.

Wied, D. de, (1966). *Proc. Soc. exp. Biol. Med.* **122,** 28–32.

van Wimersma Greidanus, T. B. (1970). *In:* "Pituitary, Adrenal and the Brain". (de Weid, D. and J. A. W. M. Weijnen, Eds.) *Prog. Brain Res.* **32,** 185–191.

Wurtman, R. J. and Axelrod, J. (1966). *J. biol. Chem.* **241,** 2301–2305.

Wynne–Edwards, V. O. (1962). "Animal Dispersion in Relation to Social Behaviour" Oliver and Boyd, Edinburgh and London.

Zarrow, M. X., Campbell, P. S. and Denenberg V. H. (1972). *Proc. Soc. exp. Biol. Med.* **141,** 356–358.

Zarrow, M. X., Phillpott, J. E., Denenberg, V. H. and O'Connor, W. B. (1968). *Nature (Lond.)* **218,** 1264–1265.

7.   The Interrenal Gland in Pisces

Part 1.   Structure

I. Chester Jones and W. Mosley

Department of Zoology, University of Sheffield, Sheffield, England

Part 2.   Physiology

I. W. Henderson and H. O. Garland*

*Department of Zoology, University of Sheffield, Sheffield England and *Department of Physiology, University of Manchester, Manchester, England*

PART 1. STRUCTURE

1. Introduction

Pisces or Fishes is used as a convenient term for non-tetrapod vertebrates and it has no real standing in established systematics. The whole group comprises about half the total number of vertebrates though exact numeration depends on the authority. In the range between 17 000 and 30 000 of living species Cohen (1970) gives a mean of 20 600 and Nelson (1976) an estimate of 18 818: some 50 jawless fish (Agnatha) and for the Gnathostomata, cartilaginous about 515 to 550 and bony fish generally some 18 000 or more. Their sizes are extremely diverse and their habitats enormously varied. All these facts are not reflected in the knowledge of interrenal structure and function, as the paucity of observations on a few species, which will be given below, will clearly demonstrate. A description of the interrelationships of fishes would help, to some degree, in comparative endocrinological classification. This is an area of controversy. Since Hennig (1966) proposed cladism (the possible geneology of taxa to give sister groups of common descent) for classification, two major attitudes have emerged. On the one hand, cladism receives less than enthusiastic support (Halstead, 1978). Against this, it is not accepted that a classification based on hierarchy necessarily reflects relationships (Gardiner et al., 1979). This topic is discussed by Greenwood et al. (1966, 1973), Greenwood (1975), Nelson (1976) and Midgalski and Fichter (1977) among others. For comparative endocrinology, Atz (1973) recommends the cladistic attitude: so taking together all known aspects, for example, the elasmobranchs have less in common with teleosts than these with birds. We find, for the purposes of this chapter, the classification adopted by Nelson (1976) to be convenient, with the proviso that assignments to Families and Orders readily change as authorities grapple with the many problems.

In general terms, the interrenal tissue of fishes cannot be considered apart from their excretory systems (Chester Jones, 1976). Embryological formation inevitably binds both of them to the nephrogenic blastema, the mesodermal blocks being divided up, in varying proportions, into renal, gonadal and adrenocortical organs (see e.g. Witschi, 1956). The excretory system in all vertebrates is built on a fundamentally similar plan (Fraser, 1950). Potentially it may extend from the level of the second post-otic somite to some distance behind the anus. The anterior portion is more or less degenerate with considerable variations in length and complexity of the posterior portion. Confusion has arisen by the use of pronephros, mesonephros and metanephros supposedly laid down all along the trunk and succeeding one another in time. However, it is only possible to distinguish a pronephros when the young has a larval stage and requires an excretory mechanism which can function early in the life history. Thus the term should be used for the larvae of Anamniota and the few adult teleosts and Amphibia where a pronephros may persist into maturity. As to other terminology, opisthonephros is often used (Kerr, 1919). This indicates that in many lower vertebrates there is no separation between mesonephros and metanephros. However there are all gradations in bony fish and to name a separate posterior portion of mesonephros as metanephros brings some confusion. It seems preferable to retain pronephros and mesonephros and to use metanephros to designate the compact kidney of the Amniota which lies some distance anteriorly on the dorsal side of the body cavity (Fraser, 1950). These considerations will be returned to when the varying anatomical relationships of the adrenal cortex and its homologues are considered.

2. Agnatha (Cyclostomata)

Whatever the nomenclature in different classifications (see e.g. Hubbs and Potter, 1971; Greenwood, 1975), living jawless vertebrates divide into two groups, the lampreys and the hagfishes. One order, Petromyzoniformes, has the family Petromyzonidae, lampreys, with nine genera and 31 extant species (Nelson, 1976). These are ecto- or non-parasitic and both types may exist in individuals of the same species or characterize animals closely related. The second order,

Myxiniformes, has the family Myxinidae, hagfishes, with 5 genera and about 32 species. These are scavenger feeders, for the most part eating the insides of dying or dead invertebrates and fish. The cyclostomes demand the attention of comparative endocrinologists because of their individual life styles (Sterba, 1969) and their evolutionary significance. It is a common assumption, though not necessarily completely accepted, that the Agnatha are diphyletic so that lampreys and myxinoids do not really belong to the same class (Greenwood *et al.*, 1973; Greenwood, 1975). Further, it can be argued that fossil Agnatha were already committed to features of organization that debarred them from being considered as ancestors of the jawed vertebrates, the Gnathostomata (Jarvik, 1965). Nevertheless, with a cladistic approach (see Atz, 1973), we can suggest that both these two groups of cyclostomes represent, however blurred the picture, in their differing ways, the transition from protochordate to vertebrate organization, perhaps as far back as the Cambrian (Tailo and Whiting, 1965).

A. PETROMYZONIDAE

The endocrinological literature deals, for the most part, with *Lampetra planeri*, *L. fluviatilis* and *Petromyzon marinus*. Lampreys hatch in fresh water, spend about 4 years as larvae (Ammocoetes), burrowing and blind, to metamorphose to a swimming form with eyes and horny teeth. Sexual maturation occurs either directly or after a period of growth as in river lampreys. In this latter case, the lampreys do not feed after entering fresh water so that there is mobilization of body lipid, liver glycogen and the levels of blood glucose fall (Bentley and Follett, 1965). Secondary sexual characters appear a few weeks before spawning after which death occurs (Larsen, 1973). This type of life cycle would be expected to impinge on interrenal function but this has not been demonstrated. Should the Petromyzonidae have morphological entities homologous with the adrenocortical tissue of tetrapods, it would be anticipated, with recent sensitive assay methods, that corticosteroids would be detected in plasma samples. Here, however, is an area of dubiety. The first work indicated recognizable quantities of plasma cortisol and corticosterone in the land-locked *P. marinus* of the freshwater Lake Superior, in North America (Chester Jones and Phillips, 1960). Thereafter very small quantities were reported to occur (Leloup-Hatey, 1964; Weisbart and Idler, 1970; Idler and Truscott, 1972). Use of developed rigorous methods for the identification of minute amounts of known corticosteroids, rendered their occurrence in doubt. Thus, Buus and Larsen (1975) were certain that serum of adult female river lampreys (*L. fluviatilis*) contained less than 3 ng/100 ml of cortisol, 11-deoxycortisol, corticosterone and less than 5 ng/100 ml cortisone. By and large, there is no definitive evidence that the established corticosteroids are to be found in these cyclostomes (Weisbart *et al.* 1978). On the other hand, such enzymes as 17α- and 21-hydroxylases and 21-desmolase activities may well exist (Weisbart and Youson, 1975). We are left with the impression that gonadal steroidogenesis can be found in extant species of this cyclostome family. Therefore, by extrapolation, it occurred in extinct Ordovician Agnatha some 500 million years ago, yet we are faced with the paradox that absence of corticosteroids is confirmed for the Cephalochordata (Sandor *et al.*, 1976) and in doubt for the lampreys and their relatives.

Morphological evidence is no more definitive than that derived from analysis for corticosteroids. Presumptive interrenal tissue is mostly concentrated, in the larvae and adults of *P. marinus* and *L. fluviatilis* and *planeri*, within the pronephric region on the dorsal side of the pericardium (Giacomini, 1902; Gaskell, 1912; Hardisty, 1972; Figs. 1, 2, 3). Small islets of interrenal cells (presumptive is always implied) are also found along the walls of the cardinal veins (Fig. 3b). After standard histological techniques, these cells follow the pattern seen throughout the vertebrates (Chester Jones, 1957; Holmes and Phillips, 1976; Hanke, 1978; Lofts, 1978; Idelman, 1978). A striking difference is seen in the occurrence of large vacuoles (after processing from wax embedding) or of lipid droplets (in frozen sections) approximating to the size of the nucleus, about 8 μm in section. The cytoplasm, apart from the liposomes, is faintly granular or reticular, the nucleus round with a centrally placed nucleolus (Hardisty, 1972). He noted that in some of the larger groups of interrenal tissue, the cells may be arranged in cords, separated by distinct collagen septae. Sterba (1955) laid the ground work for our approach to the understanding of the cyclostome interrenal. He and his co-workers showed, in the interrenal of *L. planeri*, representatives of organelles characteristic of steroid producing cells (Idelman, 1978). Whilst these organelles show considerable variation, three basic types may be discerned (Seiler *et al.*, 1973). There are dark cells in early stages of development, active cells and inactive or exhausted cells. Such cells, together with the results from the use of mammalian ACTH, metyrapone and cortisol injections, indicate a hormonal activity. Furthermore, the cells contain lipid which is osmophilic, coloured by Sudan stains, and contains cholesterol, phospholipids, unsaturated lipids and acetyl-

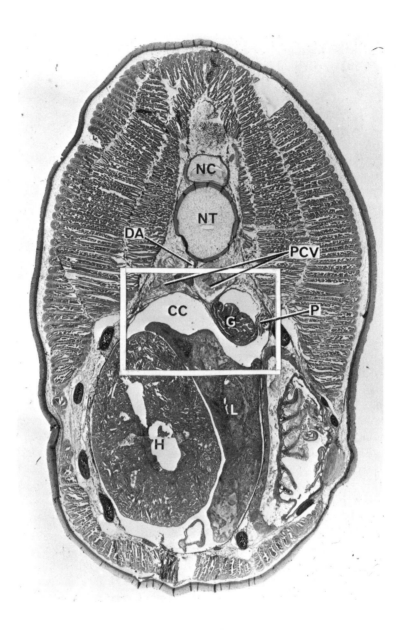

Fig. 1. Transverse section of the lamprey (*Lampetra fluviatilis*) in the cardiac region.
N.C. nerve cord; N.T. notochord; D.A. dorsal aorta; P.C.V. posterior cardinal veins;
G. gut; P. pronephros; L. liver; H. heart; C.C. coelomic cavity. (× 7).

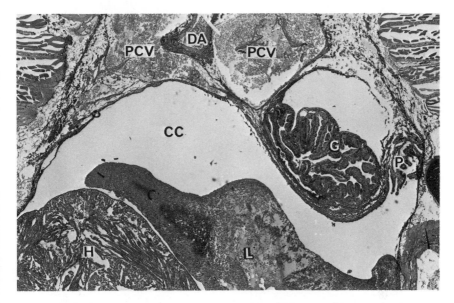

Fig. 2. Box of Fig. 1 showing in more detail the relationship of the pronephros to the gut and other organs. (× 17). Abbreviations as for Fig. 1.

phosphatid (Seiler *et al.*, 1970). So far then, we may be encouraged to regard the presumptive interrenal tissue as the homologue of the adrenal cortex. However one basic identification procedure is the demonstration of Δ^5 3β-hydroxysteroid dehydrogenase and these cells do not react (Seiler *et al.*, 1970; Yousen, 1972; Weisbart, 1975; Weisbart *et al.*, 1978). The presence of this enzyme is basic to the pathways which give steroid hormones and the gonadal cells of cyclostomes give positive reactions (Barnes and Hardisty, 1972; Mosley, 1978). Nevertheless, Ruth and Karl Seiler (personal communication) found this enzyme in interrenal cells of *L. fluviatilis* examined at the upstream migration near Leningrad in 1974. This indicates that the interrenal cells may only be active at certain stages of the life cycle. Furthermore, it should be noted that, although positive 3β-hydroxysteroid dehydrogenase activity can be readily demonstrated in vertebrate steroidogenic tissue, it is not always shown and often unexpectedly absent. Thus Deane and Rubin (1964) noted the absence of positive identification in the corpus luteum of the rabbit though the organ was active in steroidogenesis of the typical kind. It may be that if presumptive interrenal tissue is presented with progesterone from other sources, its corticosteroidogenic pathways

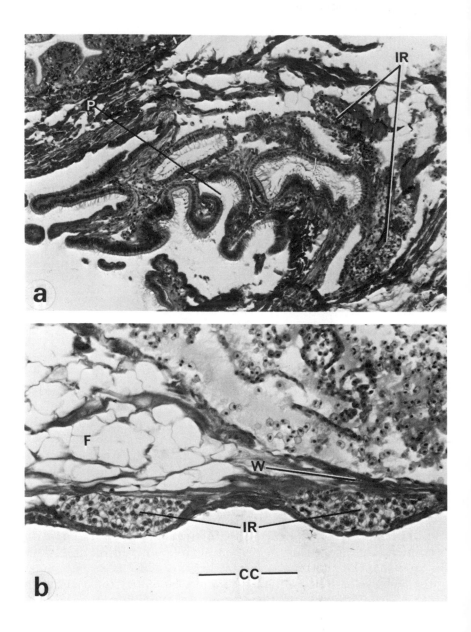

Fig. 3. *Lampetra fluviatilis*. (a) Presumptive interrenal cells shown around the base of the pronephric funnel. P. pronephric funnel; IR. interrenal tissue. (× 117). (b) Example of sites of interrenal cells usually found within the wall of the posterior cardinal vein and existing as islets along the full length of this vein. (× 210). W. wall of posterior cardinal vein; F. fatty tissue; IR. interrenal tissue; C.C. coelomic cavity.

would not give the dehydrogenase reaction. As this is shown by the gonads, they offer themselves as a possible source.

The ultrastructure of cells which secrete steroids has been well documented (Idelman, 1978). Hardisty (1972) discusses the problem in the light of observations on the ammocoetes of *L. planeri*. Each presumptive interrenal cell generally has one, slightly electron dense, liposome up to 8 μm in diameter bounded with a bilaminar membrane, reminiscent of some stages of activity of vertebrate steroidogenic tissue. Their large size, however, is unusual. The endoplasmic reticulum is characteristically agranular with abundant free ribosomes scattered throughout the cytoplasm. A compact Golgi apparatus was occasionally observed, consisting of three or four parallel cisternae with a few associated vesicles. The abundant mitochondria have a variable internal structure from parallel arrays of narrow tubular cristae to irregular tubular to the tubulo-vesicular type.

Youson (1973a) followed up earlier work on *P. marinus* (Youson and McMillan, 1971; Youson, 1972) to find whether or not mammalian ACTH altered the ultrastructure of interrenal cells of both ammocoetes and adults, in ways similar to those of mammalian adrenocortical cells. (Figs. 4a, b; 5a, b). The cells of control animals had irregularly shaped nuclei, with cytoplasm comprising randomly distributed rough and smooth endoplasmic reticulum, abundant free ribosomes, lipid droplets and a few dense granules. The mitochondria contained tubulo-vesicular cristae and the Golgi apparatus had numerous flattened saccules with closely associated small vesicles. After ACTH administration, the nuclei were oval and the number of lipid droplets diminished; there were more mitochondria, many showing "puffs" or dilations, and the cristae were more tubular. There was a striking increase in the amount of smooth endoplasmic reticulum. These changes and other evidence such as the incorporation of injected ^{3}H-cholesterol (Youson, 1975) are in keeping with the unproven role of steroidogenesis by presumptive interrenal tissue.

B. MYXINIDAE

In the second group, family Myxinidae, the serum has a similar ionic composition to that of the environmental seawater (Robertson, 1954; McFarland and Munz, 1958; Bellamy and Chester Jones, 1961). This renders them unique amongst vertebrates and raises questions about the evolution of fishes (Robertson, 1954, 1963; Chester Jones, 1963; Atz, 1973). The original work on 8500 ml of serum from *Myxine*

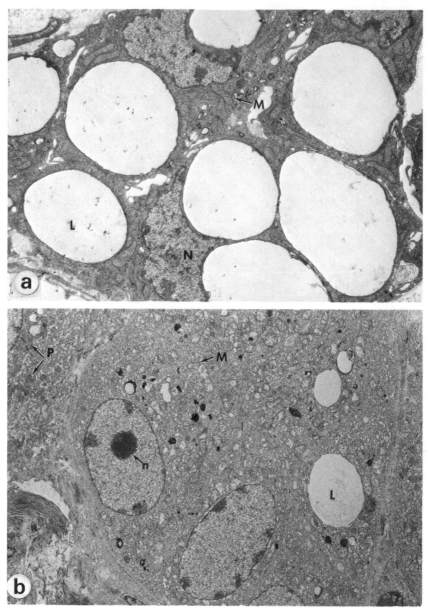

Fig. 4. Effects of mammalian ACTH on presumptive interrenal cells of ammocoete larvae of *Petromyzon marinus*. (a) Control (4 daily injections of isotonic saline). Note: large lipid droplets (L); moderate numbers of mitochondria (M); irregular shaped nuclei (× 6500) (b) ACTH (3 daily injections). Note: more mitochondria; more oval nucleus and prominent nucleolus; fewer, smaller lipid droplets; mitochondrial matrices are not as electron dense as those of proximal tubular epithelium (P) and control mitochondria (× 5500). (From Youson, 1973b).

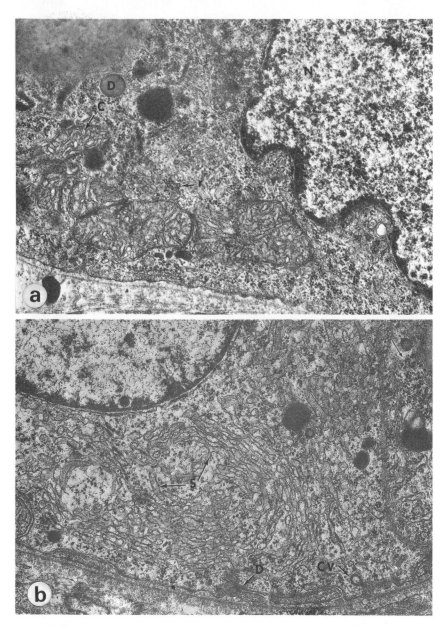

Fig. 5. Effects of mammalian ACTH on presumptive interrenal cells of adult *Petromyzon marinus*. (a) Control (3 daily injections of isotomic saline). Note: irregular nuclear form (N); abundant free cytoplasmic ribosomes (r); few dense granules (D); mitochondria with tubulo-vesicular cristae (C). (b) ACTH (3 daily injections). Note: free surface of plasma membrane close to parietal epithelium of Bowman's capsule (not shown) and not an adjunct to a perivascular space – plasma membrane therefore not folded and desmosomes (D) are on lateral surfaces; abundant tubular cisternae of smooth endoplasmic epithelium (S); glucogen particles (arrow); coated vesicles (CV) (both ×26 500) (From Youson, 1973a).

glutinosa resulted in an identification of cortisol and corticosterone but not aldosterone (Phillips *et al.*, 1962). The levels appeared to be too high (Weisbart and Idler, 1970; Idler *et al.*, 1971). Nevertheless, the presence of cortisol, cortisone, corticosterone and 11-deoxycortisol was confirmed (Idler and Truscott, 1972). The presumptive interrenal cells may lie in the region of the pronephros (Idler and Burton, 1976) but the evidence is even weaker than that for the Petromyzonidae. Our examination of *Myxine glutinosa* (from Göteborg, Sweden, courtesy of Professor Stromberg) revealed traces of corticosterone in the plasma, using standard techniques (Kenyon, 1979), successful for the gnathostomes, from fish to mammals. Search for a histological entity similar to the adrenocortical homologue of vertebrates either in the region of the pronephros or by the posterior cardinal veins and their tributaries gave no certain results (Figs. 6a, b, c).

3. Chondrichthyes

This class of cartilaginous fishes belongs to the Superclass Gnathostomata to which, except for Agnatha, all the Vertebrates, Pisces and Tetrapoda, are assigned.

The Sub-class Elasmobranchii comprise the sharks and rays of about 128 genera and 602 living species. The second Sub-class, Holocephali, have the chimaeras, rabbitfishes, rat fishes of 6 genera and 25 extant species.

The major corticosteroids of sharks and rays are corticosterone and the unusual 1α-hydroxycorticosterone. The latter may be related to Vitamin D and its relatives which have an hydroxyl group at the one position, and the elasmobranch liver is particularly adipose (Yamamura, 1952, 1956; Oguri, 1978). Aldosterone may be present (Bern *et al.*, 1962). Morphologically the important fact is that interrenal tissue is separate from the chromaffin groups of cells, thereby posing both phylogenetic and physiological problems. The interrenal lies in the region of the posterior kidney and displays three general forms (Chester Jones, 1957). Briefly, there is a horse-shoe form found particularly in rays (Fig. 7a), the rod-shaped type lying between the posterior kidney lobes (e.g. sharks and Holocephali; Dittus, 1941; Oguri, 1977; Fig. 7b) and a third type, as in *Torpedo*, mainly concentrated as an elongated oval body on the dorsal part of the left kidney near the mid-line. Indication of the interrenal weight in relation to body weight for some rays and a dogfish species is given in Table I.

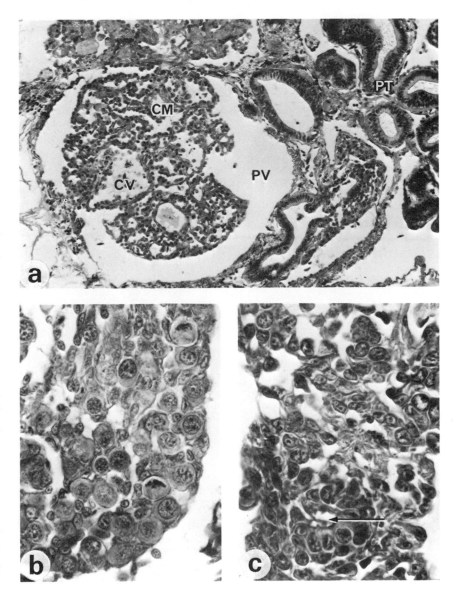

Fig. 6. Areas of *Myxine* in which interrenal tissue might be expected to occur. All 6μm sections, Masson Trichrome stain. (a) Transverse section of the pronephros of *myxine* showing the central mass within the pronephric vein (× 117). PT = pronephric tubules; CM = central mass; PV = pronephric vein; CV = central vacuity. (b) Part of the central mass with haemocytoblasts, including two mitotic cells (× 470). (c) Fibre and spindle cells within the central mass together with cells transforming into tubular cells at the periphery of the central mass (arrow) (× 588).

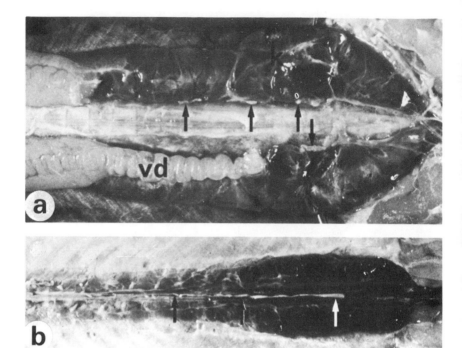

Fig. 7. (a) The interrenal gland (arrows) of *Raia clavata* mature male *in situ* adjacent to the kidneys (K). Dorsal view with alimentary canal and gonads removed, kidney bisected posteriorly; vd. vas deferens; the horse-shoe type. (b) The interrenal gland (arrows) of *Scyliorhinus canicula* mature male, lying between the lobes of the kidney; the rod shaped type. (From Roscoe, 1976.)

Roscoe (1976) found the interrenal cells of *Raia* spp. and *Scyliorhinus canicula* to be homogenous, grouped into cords or lobules usually bounded by a thin connective tissue sheath. He discerned no apparent zonation (Fig. 8a, but see below). Interrenal cells are round or polygonal with a large prominent nucleus containing basophilic chromatin with one or more nucleoli (Fig. 8b). Routine histological examination shows considerable vacuolation, representing lipid droplets demonstrable by histochemical methods (Oguri, 1978). Thus the majority of interrenal cells give positive reactions with Sudan, Nile blue and Oil red colorants and cholesterol is a dominant cytoplasmic component (Roscoe, 1976; Oguri, 1978). The capacity for corticosteroidogenesis is shown by the presence of Δ^5-3β-hydroxysteroid dehydrogenase. At lower powers, electron micrographs indicate two types of cell, dark and light, in immature *R. clavata*. Usually dark

TABLE I

Interrenal/body weight ratios in *Raia* sp. and *Scyliorhinus canicula*.

Raia species *Scyliorhinus*	sex	n	Body weight (B) range in g	Interrenal weight (I) range in mg	I/B ratio mg/100 g
R. clavata[a]					
immature	♂	5	260–2000	2–15	0.82 ± 0.34
	♀	8	175–4150	5–44	3.30 ± 3.60
R. naevus[a]					
mature	♂	8	1675–2375	37–93	2.90 ± 0.42
	♀	10	1325–2350	59–165	5.70 ± 0.86
R. radiata[a]					
immature	♂	13	130–660	8–28	4.80 ± 0.86
	♀	15	120–465	7–36	6.40 ± 0.86
mature	♂	7	560–1050	19–92	6.40 ± 3.60
	♀	8	400–825	27–128	8.10 ± 2.90
R. erinacea[b]					
	♂	12	748 ± 490	11.3 ± 5.32	1.51
non-ovulatory	♀	9	548 ± 214	12.9 ± 5.56	2.36
ovulatory	♀	12	630 ± 97	21.5 ± 3.11	3.41
S. canicula[a]					
mature	♂	13	640–1105	20–101	5.50 ± 1.90
	♀	14	535–950	13–71	4.10 ± 0.90

[a] From Roscoe (1976); [b]from Macchi and Rizzo (1962).

cells are packed with liposomes and mitochondria whilst in light cells, the organelles are more widely spaced (Fig. 9). Smooth endoplasmic reticulum and Golgi material are visible within the electron transparent matrix. The large, usually central, nucleus is bounded by a scolloped membrane. Nearby, the Golgi complex is seen as small piles of lamellae or vesicles. Liposomes are of variable size, with or without a single limiting membrane. The numerous mitochondria are round or oval with cristae of variable appearance (Fig. 10). These are, most commonly, tubulovesicular (cf. types given in Idelman, 1978). The agranular endoplasmic reticulum is often associated with liposomes. Polyribosomes are dispersed throughout the cytoplasm (Fig. 11).

Early work suggested different types of cells and secretory cycles (Vincent, 1897; Fraser, 1929). Fancello (1937) noted changes at maturity in female *S. canicula* and the interrenal may be correlated with viviparity (Ranzi, 1936; Pitotti, 1938; Oguri, 1960) but definitive observations, especially in varying corticosteroid activity, are lacking (see Idler and Truscott, 1972). Moreover amounts of 3β hydroxysteroid dehydrogenase do not appear to alter during the life of *Scyliorhinus* or *Torpedo* species (Chieffi *et al.*, 1963). There are dark and light

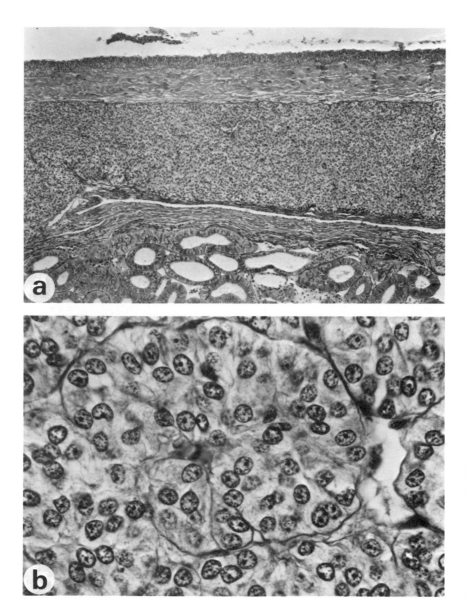

Fig. 8. (a) Parts of longitudinal section of the interrenal gland of a dogfish (*S. canicula*) shown lying between the posterior cardinal vein (above) and the kidney (below). (× 57). (b) High power of part of the interrenal gland of Fig. 8a showing the irregular cords and groups of cells surrounded more or less completely by fine connective tissue. Cells are usually round or polygonal with a prominent nucleus with strongly basophilic chromatin and one or more nucleoli (× 590).

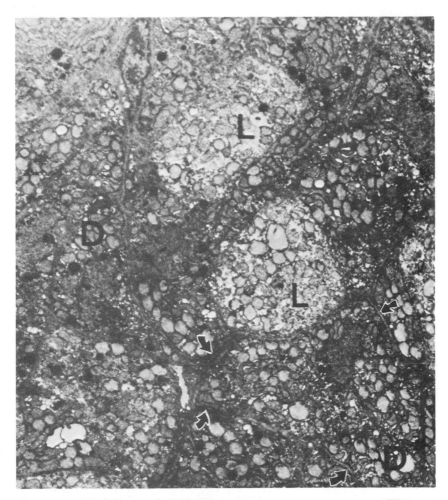

Fig. 9. Low power electron micrograph of interrenal tissue of immature *Raia clavata* showing light (L) and dark (D) cells in close proximity. Small groups of cells are invested in connective tissue (arrows) (× 3400). (From Roscoe, 1976.)

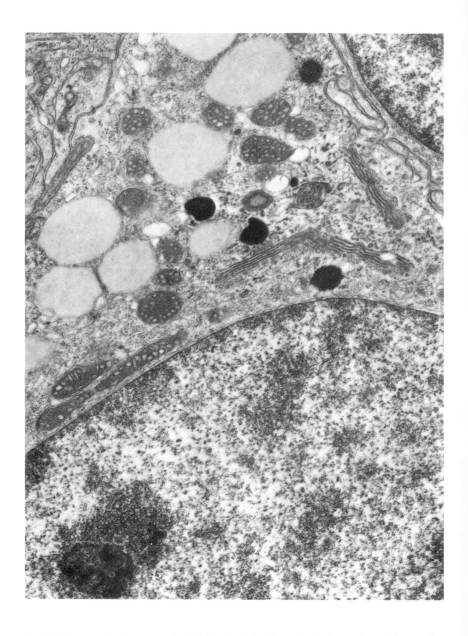

Fig. 10. A part of an interrenal cell of a dogfish (*S. canicula*). A portion of the nucleus shows the scolloped appearance of the membrane. The Golgi complex is seen as small piles of lamellae or vesicles. Liposomes are of variable size with or without a single limiting membrane. There are numerous round or oval mitochondria with cristae of variable appearance (× 12 700). (From Roscoe, 1976.)

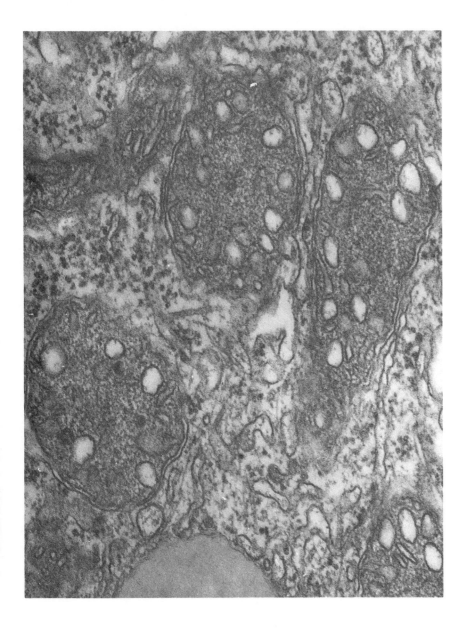

Fig. 11. Detail of Fig. 10 showing the tubulo-vesicular mitochondria. Association of agranular endoplasmic reticulum with a liposome is seen (bottom). Polyribosomes are dispersed throughout the cytoplasm (× 52 000). (From Roscoe, 1976.)

interrenal cells in immature *R. clavata* and *R. naevus* (Roscoe, 1976) though not in mature animals (Mellinger, 1970). Zonation of the interrenal is not apparent in many elasmobranch species though in our experience the peripheral cells, 5 to 10 cells deep, often are the larger and with clearer cytoplasm after routine methods. On the other hand, Taylor *et al.* (1975) presented elaborate pictures of interrenal zonation based on an examination of two female and one male nurse sharks (*Ginglymostoma cirratum*, Bonnaterre) shown in Fig. 12. It would be interesting to have this approach further developed in this and other laboratories.

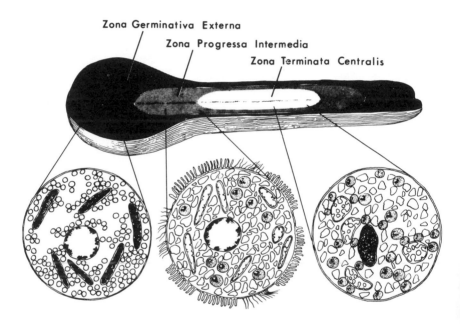

Fig. 12. Proposed zonation of the interrenal gland of the nurse shark (*Ginglymostoma cirratum* Bonnaterre, Orectolobidae). Three zones are shown although not sharply delimited at the interface regions which show transitional morphologies. The external zone has elongated to spherical mitochondria with short lamellar cristae projecting into a dense matrix; numerous spherical vesicles; no smooth endoplasmic reticulum; no lipid droplets; cilia and microvilli not found; cell death not found and all cells display intense basophilia. The intermediate zone (zona progressa intermedia) elongated to spherical mitochondria with dilated cristae; numerous irregularly shaped vesicles; smooth endoplasmic reticulum; lipid droplets; cilia or microvilli present; limited cell death and all cells show intense basophilia. The central zone (zona terminata centralis) elongated to irregularly shaped mitochondria with vesicular cristae; numerous irregularly shaped vesicles; smooth endoplasmic reticulum; lipid droplets; cilia or microvilli not found; extensive cell death and all cells display weak basophilia. (From Taylor *et al.*, 1975.)

A further perplexing problem in elasmobranchs is the relationship of the pituitary gland to its target tissues. In standard nomenclature, four principal lobes are described, rostral lobe (RL), median lobe (ML), ventral lobe (VL) and neurointermediate lobe (NIL) (see Dodd and Kerr, 1963; Holmes and Ball, 1974). It is the general experience that removal of the pituitary in many species produces little or no histological change in the interrenal. Here techniques are difficult in that selective removal of the lobes singly or in combination is demanding. Dodd has mastered the skills and he has shown (personal communication) that in the common species of *Scyliorhinus* total removal leads to some interrenal atrophy in about one year. The changes are not, however, dramatic (Figs. 13a, b). Roscoe (1976) had some success in producing interrenal changes in 114–144 days after removal of the RL + ML of *S. canicula* but data on histochemical and ultrastructural appearances are lacking. It is possible to suppose that the conditions in which experimental elasmobranchs are kept are very important in allowing the induction of endocrine responses; these would include temperature and feeding; their general husbandry is difficult. The warm water stingray (*Dasyatis sabina*), kept at 20–25°C, showed interrenal atrophy 30 days after removal of the RL (De Vlaming and Sage, 1972). The early work of Dittus (1941) certainly gives the impression of a responsive and not a sluggish animal type. However, it is to be noted that standard incubation techniques (see Sandor *et al.*, 1976) gave increased corticosteroid production from *Raia* interrenals only after 12 hours with ACTH (Idler and Truscott, 1972; Idler, 1973). It may well be that, in some species, interrenal function is somewhat independent of the pituitary complex though the activity of the hypothalamus has yet to be investigated.

4. Crossopterygii

This comprises a fossil group and only the coelacanth *Latimeria chalumnae* is extant. The lack of well preserved tissues and fluids render our knowledge limited though the fish is of great evolutionary interest (Lutz and Robertson, 1971; Griffith *et al.*, 1975, 1976; Dingerkus *et al.*, 1978). The site of interrenal tissue is of considerable interest and Lagios and Stasko-Concannon (personal communication) have allowed us to quote the sole observation known to us.

"The presumptive interrenal tissue of the coelacanth *Latimeria chalumnae* has been identified in a glutaraldehyde fixed segment of the mesonephros of an 85 cm immature female caught on 22 March 1972

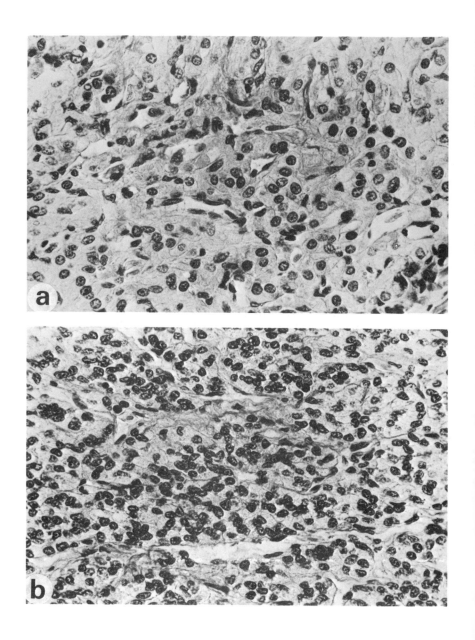

Fig. 13. (a) Part of the interrenal gland of *S. canicula* sham-operated and kept for one year as a control for Fig. 13b (× 400). (b) Part of the interrenal gland of *S. canicula*, totally hypophysectomized for one year. Some interrenal atrophy is seen with dense, pycnotic nuclei (× 400). (J. M. Dodd, personal communication.)

off Iconi, Grande Comore during the course of the International Coelacanth Expedition of that year.

The presumptive interrenal consists of numerous small circumscribed and partially encapsulated bright yellow corpuscles ranging from 0.1 to 2 mm in diameter, distributed within the walls and along the course of the posterior cardinal veins and their major tributaries within the kidney [see Fig. 14]. These corpuscles are easily discerned

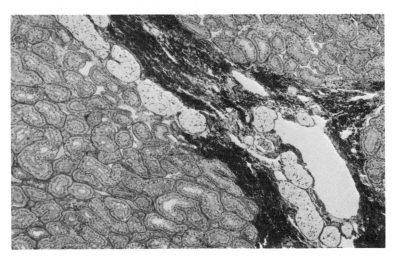

Fig. 14. *Latimeria* mesonephros. Reticulum stain. Small pale presumptive interrenal corpuscles are noted in the wall of an intrarenal vein. Tubules of the mesonephros stain more darkly. (From M. D. Lagios and S. Stasko-Concannon, personal communication.)

with a dissecting microscope at 3× magnification and are commonly associated with myelopoietic tissue. Individual corpuscles are composed of large vacuolated cells with relatively small, angular central nuclear profiles. Nucleoli are focally prominent. The cytoplasm contains non PAS-diastase resistant material or granules but stains intensely with Sudan Black B and Oil red O. Ultrastructural examination of the presumptive interrenal demonstrates a cytoplasm engorged with liposomes with rather small mitochondria which show the characteristic tubulo-vesicular cristae of steroid synthesizing cells. Smooth-surfaced endoplasmic reticulum is similarly poorly developed. Scant desmosomes are noted between adjacent cells and there is a clearly defined plasmalemma noted at the periphery of the corpuscular mass of cells adjacent the fibrous capsule. The development of liposomes is quite striking but comparable to that previously described in the inactive adrenocortical tissue of *Salamandra* (Berchtold, 1969) and in *Acipenser* (Youson and Butler, 1976a)."

The authors go on to point out that such diffuse distribution of small corpuscles of presumptive interrenal associated with the posterior cardinal veins and tributaries is comparable with that of the extant members of the Chondrostei and Holostei (see 5 and 7 below) and of *Polyodon* (Lagios and Stasko-Concannon, unpublished observations). They note that one shark group, such as *Heterodontus francisci*, has similar multiple small corpuscles. However the usual elasmobranch form of aggregated interrenal tissue has been given above (see 4). Lagios and Stasko-Concannon conclude "A rather diffusely distributed interrenal tissue along the course of the cardinal veins would appear to be a primitive (plesiomorph) condition in vertebrates. Accordingly identification of the presumptive interrenal tissue of the coelacanth in a like distribution contributes nothing to systematic analysis of its phylogenetic position."

5. Chondrostei

The family Acipenseridae, the sturgeons, comprise four genera and 23 species but little is known and this concerns principally *Acipenser fulvescens* Raineaque, the Lake Sturgeon, Youson and Butler (1976a) examined 4 animals, 45 to 51 cm long from Lake Huron, Ontario. They found the presumptive interrenal tissue near the posterior cardinal and subcardinal veins consisting of spherical to oblong yellow corpuscles for the most part within a narrow band of tissue connecting the dorsal surfaces of the kidneys (Fig. 15). In one specimen, 81 bodies were seen ranging from slightly posterior to the heart to the most caudad part of the kidney. Given, then, that these corpuscles are adrenocortical homologues, they concluded that these bodies possessed Δ^5-3β-hydroxysteroid dehydrogenase and that most had a length of 200 to 700 μm, and a diameter of 150 to 700 μm (Fig. 16a, b, c, d). Each corpuscle consisted of anastomosing cords of epithelial cells, partially encapsulated at the periphery by fibroblasts. The cells were polyhedral, with pale cytoplasm with many vacuoles, spherical or oval nuclei and prominent nucleoli. Details are given in Fig. 16c, d. Large lipid droplets are common and mitochondria, distributed evenly throughout the cytoplasm, have cristae with tubular or vesicular profiles. The smooth endoplasmic reticulum is in the form of an anastomosing network of tubules or vesicles though the precise appearance differs between cells, even those of the same cord.

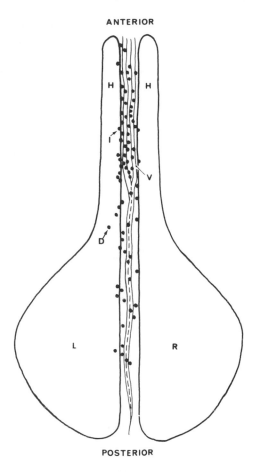

Fig. 15. Diagrammatic representation of the location of yellow corpuscles (I) with respect to the right (R) and left (L) kidneys and accompanying cardinal veins (V) of a sturgeon, *Acipenser fulvescens*. The majority of the corpuscles (I) are located anteriorly near the cardinal veins, where the kidney consists predominantly of haemopoietic tissue (H). These corpuscles are only rarely found deep (D) within the kidney. (From Youson and Butler, 1976a.)

The Golgi apparatus was conspicuous with several sets of from four to six saccules with which numerous vesicles are associated. The free ribosomes are numerous (Fig. 17a, b, c, d).

Whilst the corpuscles can be regarded as presumptive interrenal corticosteroidogenic tissue, only slight, nanogram quantities of these steroids have been identified in the plasma of *A. oxyrhynchus* (Idler *et al.*, 1971; Sangalang *et al.*, 1971). Tsepelovan and Rusakov (1976) had

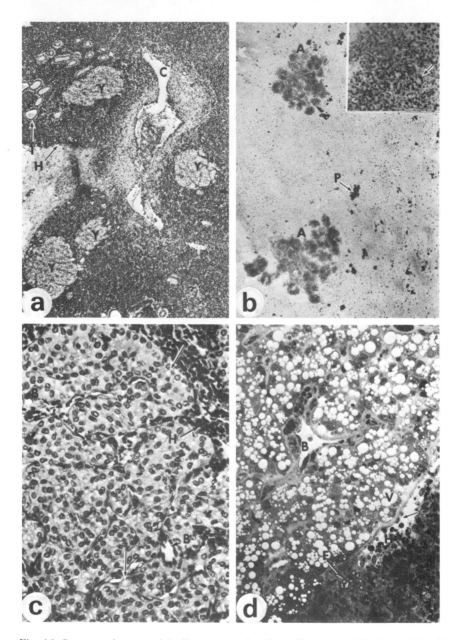

Fig. 16. Sturgeon interrenal (yellow corpuscles; from Youson and Butler, 1976). (a) Transverse section at dorsal junction of kidneys. Note: four yellow corpuscles (Y) surrounded by haemopoietic tissue and nephroi (T); cardinal vein (C). (b) Δ^5-3β-hydroxysteroid dehydrogenase (dehydroepiandrosterone as substrate). Note: diformazan granules over corpuscles (A) but not haemopoietic tissue; pigment granules (P). Unstained ($\times$ 25). Inset: higher magnification. Note: unstained lipid droplets surrounded by diformazan granules (arrow) ($\times$ 208). (c) Section of yellow corpuscles. Note: anastomosing cords of epithelial cells separated by interstitium containing blood vessels (B); large vacuoles (arrows) in epithelium clearly separated from haemopoietic tissue (H). Haemalum, Periodic acid-Schiff, Orange G ($\times$ 208). (d) As (c) showing epithelial cells and blood vessels (B). Note: vacuolated (V) epithelial cells clearly delineated from haemopoietic tissue (H) by flattened cells (arrow); myelinated nerve fibres (F) outside yellow corpuscle. Toluidine Blue ($\times$ 208).

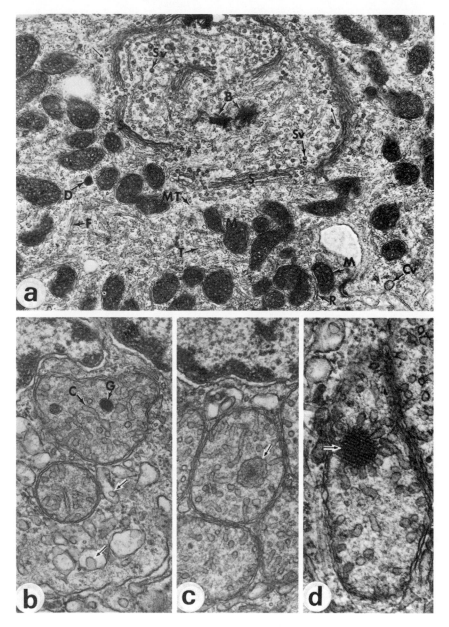

Fig. 17. Sturgeon interrenal. (a) Cell from yellow corpuscle. Note: large Golgi apparatus with several sets of saccules (s) and small coated vesicles (Sv), some of which appear to be continuous with saccules (arrows); numerous mitochondria (M); filaments (F); microtubules (MT); tubular smooth endoplasmic reticulum (T); fragments of rough endoplasmic reticulum (R); dense granules (D); large coated vesicle (Cv) near lateral intercellular space; basal bodies (B) near Golgi (× 16 700). (b) Mitochondria and smooth endoplasmic reticulum. Note: mitochondrial cristae in tubular and vesicular profile (C); matrix granules; vesicles (arrows) in cisternae of smooth endoplasmic reticular cisternae (× 36 700). (c) Mitochondrian with large vesicular body (arrow) (× 34 700). (d) Mitochondrial matrix with tubules in crystalline array (arrow) (× 44 300). (From Youson and Butler, 1976.)

more success with *A. guldenstadti*. Here it was shown that during the anadromous migration before spawning the level of corticosteroids in the blood increased, initiated in the sea phase. Blood levels decreased after spawning. The range of values was about 10 to 30 μg/100 ml plasma, though the methods used might well have given false high values.

6. Dipneusti (Dipnoi)

There are three extant genera of lungfishes—the Australian *Neoceratodus*, the South American *Lepidosiren* and the African *Protopterus*. It is generally agreed that corticosteroids occur in lungfish (Phillips and Chester Jones, 1957). Corticosterone was indicated for *Protopterus* (Janssens *et al.*, 1965), and aldosterone and deoxycorticosterone for *Neoceratodus:* males average aldosterone 1.2 ng/100 ml, deoxycorticosterone 4.4 ng/100 ml plasma; females 5.0 and 31.9 respectively (Blair-West *et al.*, 1977). Pooled blood from 4 female *Lepidosiren* gave aldosterone 0.58, cortisol 0.60, corticosterone 0.16 and 11-deoxycortisol 0.03 all in μg/100 ml plasma (Idler *et al.*, 1972). These findings are not extensive but they do suggest that interrenal tissue is present in dipnoan fish. The difficult problem has been to locate the tissue anatomically. Three attempts have been made: by Gérard (1951) and Janssens *et al.* (1965) for *Protopterus* spp., and by Call and Janssens (1975) for *Neoceratodus*.

In both these genera, the adrenocortical homologue consists of small groups of cells closely associated with the post-cardinal veins and their tributaries where these vessels pass through the kidneys (Fig. 18). In *Protopterus*, these cells were sudanophilic, cholesterol positive and contained Δ^5-3β-hydroxysteroid dehydrogenase, and lay in the region of the kidney that synthetized corticosterone (Fig. 19). Essentially similar findings were reported for *Neoceratodus*. Clearly this identification is not completely definitive. It is sufficiently indicative, however, to emphasize the amphibian characteristics of the presumptive interrenal tissue of lungfish as is shown for the urodele *Pleurodeles waltlii* (Certain, 1961) and for the developing interrenal of *Rana temporaria* tadpole (see Fig. 38b). Janssens *et al.* (1965) and Call and Janssens (1975) discuss these matters in detail.

7. Holostei

Knowledge of the holostean interrenal is confined to *Amia calva*, the bowfin found in freshwaters of eastern North America. The literature

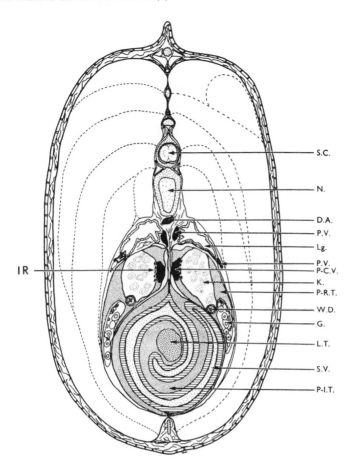

Fig. 18. Diagram of a transverse section through the abdominal region of *Protopterus* showing the relationship of interrenal tissue to the kidneys and the principal organs of the body. I.R. interrenal tissue; D.A. dorsal aorta; G. gonad; K. kidney; Lg. lung; L.T. lymphoid tissue; N. notochord; P-C.V. post-cardinal vein; P-I.T. peri-intestinal tissue; P-R.T. peri-renal tissue; P.V. pulmonary vessels; S.C. spinal cord; S.V. spiral valve of intestine; W.D. Wolffian duct.

is covered by Garrett (1942), de Smet (1962), Youson *et al.* (1976) and Youson and Butler (1976b). Chromaffin cells are described by Youson (1976). The "yellow corpuscles" of Youson *et al.* (1976) are regarded as adrenocortical homologues. They extend along the anterior two-thirds of the mesonephros, along the dorsal aorta and posterior cardinal veins. None was found in the posterior, bulbous, portion of the kidneys. The interrenal tissue, not encapsulated, is clearly deline-

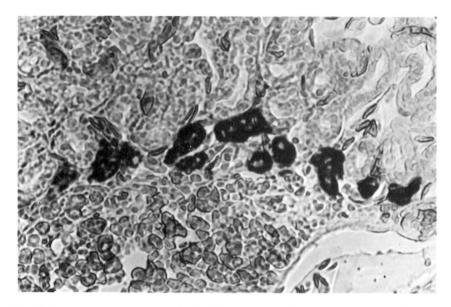

Fig. 19. Tranverse section through the kidney region of *Protopterus* cut on the freezing microtome; formol-calcium fixation. Small groups of sudanophilic cells lie on the border betweeen kidney (top of figure) and perirenal tissue (bottom of figure). Sudan black. (× 140.)

ated from the surrounding haemopoietic and renal tissue, often accompanied by chromaffin tissue (Fig. 20a, b). Each interrenal group comprises cords of pale-staining epithelial cells (e.g. after Mallory–Heidenhain or aldehyde–fuchsin). Cords were separated from one another by connective tissue and tortuous sinusoids which emptied directly into the larger veins. The cellular vacuolated appearance after routine methods probably represent lipid droplets. The interrenal cells gave a positive reaction of Δ^5-3β-hydroxysteroid dehydrogenase. Ultrastructural examination (Fig. 21a, b) showed epithelial cells limited by a plasma membrane, often folded and interdigitated, and with an abundance of small vacuoles, mitochondria and smooth endoplasmic reticulum. The mitochondria were mostly spherical or ovoid, occasionally rod-shaped, with several large matrix granules and numerous tubular cristae. The smooth endoplastic reticulum consisted of anastomosing tubules which were evenly distributed throughout the cytoplasm. The Golgi apparatus, with several sets of flattened saccules, was conspicuous. Nuclei were spherical with finely distributed chromatin and a prominent, eccentrically placed, nucleolus (Youson and Butler, 1976b). The numerous

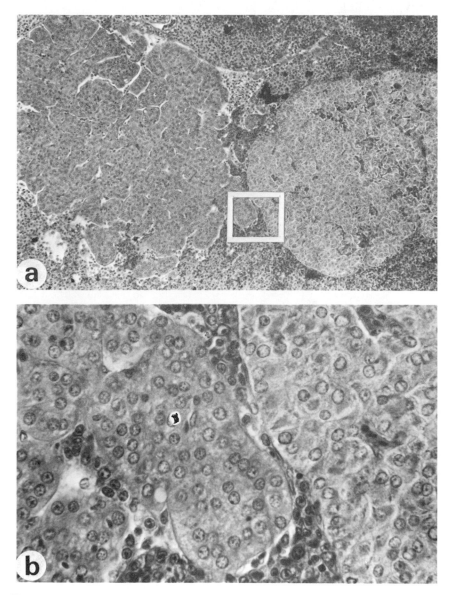

Fig. 20. (a) A group of interrenal cells in the kidney of *Amia calva* (bowfin) on the left of the picture; to the right is a clump of chromaffin tissue (× 110). (b) High power of box in Fig. 20a showing interrenal tissue on the left and chromaffin tissue on the right (× 590).

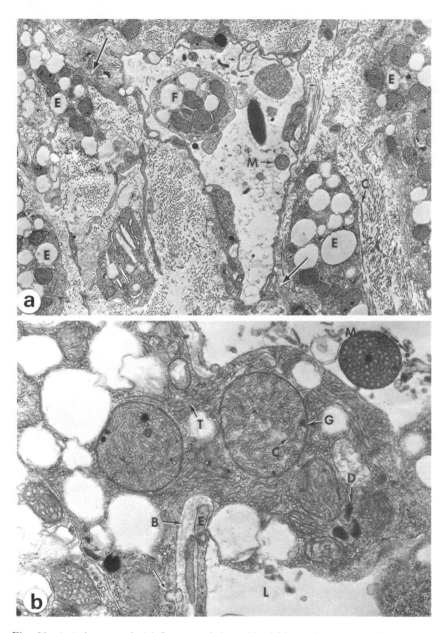

Fig. 21. *Amia* interrenal. (a) Lumen of sinusoid within an interrenal cell containing fragment of cytoplasm (F) with similar content to surrounding epithelial cells (E). Note: fenestrations (small arrows) of endothelial cells of sinusoid in close contact with epithelial cells (large arrows); mitochondrion (M) free within lumen; abundant collagen microfibrils (C) in interstitium (× 7 500). (b) Cytoplasmic projection of interrenal cell extending through fenestra of endothelial cell (E) into lumen (L) of sinusoid. Note: tubular smooth endoplasmic reticulum (T); mitochondria with tubular cristae; several matrix granules (G); a few dense granules (D); projection no longer contained within basal lamina (B); a mitochondrion (M) is free within lumen of sinusoid; coated vesicle (arrow) (× 25 000). (From Youson and Butler, 1976.)

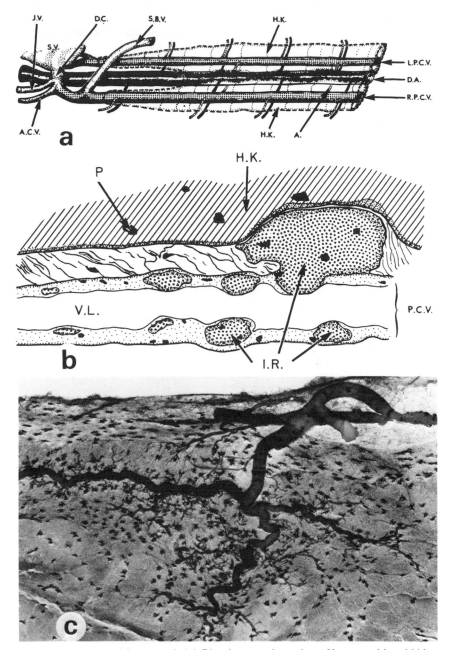

Fig. 22. European eel interrenal. (a) Blood system in region of heart and head kidney which contains interrenal. Drawn from right side; A, intercardinal anastomosis; A.C.V., anterior cardinal vein; D.A., dorsal aorta; H.K., head kidney; J.V., jugular vein; L.P.C.V., left posterior cardinal vein; R.P.C.V., right P.C.V.; S.B.V., swim bladder vein; S.V., sinus venosus; D.C., Ductus Cuvieri. (b) Diagram of longitudinal section of P.C.V. Note: interrenal islets (I.R.) embedded in H.K. and vein wall; pigment cells (P) in H.K. (from Chester Jones *et al.*, 1964). (c) Flattened preparation of P.V.C. showing latex injected branch of D.A. supplying blood to interrenal cells in vein wall. Numerous pigment cells are seen (× 21).

corpuscles of Stannius are not to be confused with interrenal tissue
(see below).

8. Teleostei

These bony fish, an Infraclass of the Class Osteichthyes are the most
abundant and diversified group of all vertebrates. About 18 007
extant species placed in 31 orders, 415 families and 3869 genera
represent about 96% of known living fish (Nelson, 1976). Perch-like
fish, Perciformes, are the largest and most varied of all fish orders with
about 1257 genera and 6880 species (Table II).

Early literature concerning the teleost interrenal, including the
seminal work of Giacomini, 1898 to 1933, is given in Chester Jones
(1957), van Overbeeke (1960) and Nandi (1962). Later contributions
are copious and the following references may be helpful: Olivereau
and Ball (1963); Olivereau (1965, 1966); Ball and Olivereau (1966);
Ball (1969); Ball and Baker (1969); Chester Jones et al. (1969, 1974).
Other sources are to be found in Krauter (1952, 1958); Olivereau and
Fromentin (1954); Oguri (1960a, b, c, d, e,); Yadav et al. (1970);
Chan et al. (1975); Hooli and Nadkarni (1975).

A. HISTOLOGY

In general, the teleost interrenal consists of groups, cords or strands of
cells set around the posterior cardinal veins and their branches, and
vascularized by twigs of the dorsal aorta (Fig. 22a, b, c). More rarely,
the tissue may occur about the anterior cardinal veins as in *Anguilla
anguilla*. The interrenal is usually confined to the anterior part of the
kidney, named the "head kidney", commencing caudad to the last gill
slit. This area is frequently equated to the pronephros but this is not
strictly true for the adult (see Introduction) except perhaps in such
families as the Cyprinodontidae in which the anterior part may be
persistent. In the majority of fish, the head kidney, representing the
cephalad part of the mesonephros, is mostly lymphoid tissue with
pigment and haemopoietic cells. Other tissue elements have, from
time to time, been noted: thyroid follicles (Chavin, 1956; Sathyansen
and Chary, 1962; Baker-Cohen, 1959; Ahuja, 1962; Peter, 1970);
intramesonephric glands (Sharma, 1968); tubule corpuscles (Rasquin, 1969); adreno-melanogenic tissue (Shrivastava and Thakur,
1974). The head kidney is bilateral but sometimes the interrenal may
be larger on, or confined to, one side [e.g. a killifish, *Fundulus
heteroclitus* (Phillips and Mulrow, (1959))].

Nandi (1962) has assigned the morphological location of the interrenal into 4 types, with intermediaries, and her classification is frequently followed (see Banerji, 1973). The main patterns and histological appearance are shown in Figs. 23 to 29:

Type I: interrenal tissue surrounds the postcardinal veins or their largest branches (Figs. 23a, b; 24a, b).

Type II: interrenal tissue surrounds small or medium-sized branches of the veins and therefore is rather widely dispersed throughout the anterior parts of the kidney (Figs. 25a, b; 26a, b).

Type III: interrenal tissue is associated with venous sinuses within the anterior kidney tissue. It often forms strands or cords of cells, sometimes scattered throughout the lymphoid and haemopoietic tissue which it may, occasionally, largely replace (Figs. 27a, b; 28).

Type IV: interrenal tissue forms a solid mass of cells in a localized area (Fig. 29a, b).

Intermediate types occur, as would be expected, in this Infraclass comprising so abundant a number of species. Clearly if each main type be found to be associated with the others, there is the possibility of 12 intermediates. Thus if we take Type I, then type I/II indicates interrenal tissue which is associated with the largest branches of the postcardinal veins but also extending along their smaller ramifications. Type I/II would be regarded as interrenal tissue forming layers or groups of cells around the cephalad parts of the postcardinals with cords extending into surrounding tissue. Type I/IV indicates interrenal tissue associated with the major branches of the postcardinal veins around and near which there is also a solid mass. The remaining 9 intermediate types can readily be constructed. Such a plethora must necessarily throw some dubiety on the usefulness of the classification. On the other hand, it does, in a neat manner, indicate the morphological characteristic of a particular species (Table II). It would be helpful in the overall analysis to assign physiological functions to these differences but this is impossible. Another problem is the relationship of interrenal to chromaffin tissue. Here Nandi (1962) described 5 types of interrelationships. We, however, favour the simplistic approach and only distinguish between those head kidneys which show separate interrenal and chromaffin tissue (S) and those where there is mingling (M) (Table II). This stresses the puzzle of the centralization of chromaffin tissue as the medulla of the adrenal gland of Metatheria and Eutheria. Presenting only the outline of the evolutionary trends as depicted by Greenwood et al. (1966), it is possible to suggest that separate interrenal and chromaffin tissue

TABLE II

Examples of the types of interrenal and chromaffin tissue in the Class Teleostomi, Grade Teleostei, based on Nandi (1962)—see Figs. 22–30. Classification and common names from Greenwood (1975) and Nelson (1976). I/R = interrenal; Chr. = chromaffin tissue; S = interrenal and chromaffin tissue separate and M = mingled; FW = Freshwater; SW = Seawater.

Family or Subfamily	Common names	Tissue types				Predominant Habitat	Approximate number of species
		I/R	Chr.	S.	M.		
COHORT: TAENIOPAEDIA; SUPERORDER: Elopomorpha; ORDER: Elopiformes							
Elopidae	Tenpounders	IV	III	S		SW	5
ORDER: Anguilliformes							
Anguillidae	Eels	I	IV?		M	FW Catadromous	15
Xenocongridae	False Morays	I	IV		M	SW	15
Muraenidae	Moray eels	I	IV		M	SW	100
Congridae	Conger eels	I	IV;II	S	M	SW	100
Opichthyidae	Snake eels	I	IV		M	SW	21
COHORT: CLUPEOCEPHALA; SUPERORDER: Clupeomorpha; ORDER: Clupeiformes							
Clupeidae	Herrings	I/II	V		M	SW	180
Engraulidae	Anchovies	I		M		SW	110
COHORT: ARCHAEOPHYLACES; SUPERORDER: Osteoglossomorpha; ORDER: Osteoglossiformes							
Notopteridae	Featherbacks or Knifefishes	I				FW	4
COHORT: EUTELEOSTEI; SUPERORDER: Protacanthopterygii; ORDER: Salmoniformes							
Salmonidae	Salmonids	II/III; II	I	S		FW Anadromous	68
Osmeridae	Smelts	II	II;I	S		SW Anadromous	6/7
Sternoptychidae	Hatchetfishes	I;I/III	IV		M	SW	27
Esocidae	Pikes	III	I	S		FW	3
Umbridae	Mudminnows	II/III	V		M	FW	5

SUPERORDER: Ostariophysi; ORDER: Cypriniformes							
Characidae	Characins	I	V		M	FW	1000
Cyprinidae	Minnows, carps	II	V		M	FW	1600
Catostomidae	Suckers	I				FW	58
ORDER: Siluriformes							
Ictaluridae	Catfishes		I?	S		FW	35
SUPERORDER: Scopelomorpha; ORDER: Myctophiformes							
Synodontidae	Lizardfishes	III/IV	II	S		SW	34
SUPERORDER: Paracanthopterygii; ORDER: Gadiformes							
Gadidae	Cods	I;II	I	S		SW	55
Ophidiidae	Brotulas and Cusk eels	I	IV		M	SW	155
Carapidae	Carapids	I	IV		M	SW	25
Zoarcidae	Eelpouts	I	IV		M	SW	65
ORDER: Batrachoidiformes							
Batrachoididae	Toadfishes	II	IV		M	SW	55
ORDER: Lophiiformes							
Lophiidae	Goosefishes	IV	V		M	SW	12
Antennariidae	Frogfishes	II/III	IV		M	SW	60
ORDER: Gobiesociformes							
Gobiesocidae	Clingfishes	III	IV		M	SW	100
SUPERORDER: Acanthopterygii; ORDER: Atheriniformes							
Cyprinodontidae	Killifishes	I	IV;V		M	FW	300
Poeciliidae	Livebearers	I	IV		M	FW	136
Atherinidae	Silversides	II/III; II	II	S		SW	156

TABLE II *continued*

Family or subfamily	Common names	I/R	Tissue types Chr.	S.	M.	Predominant Habitat	Approximate number of species
SUPERORDER: Percomorpha; ORDER: Lampridiformes							
ORDER: Beryciformes							
Holocentridae	Squirrelfishes	II	II	S		SW	70
ORDER: Zeiformes							
Zeidae	Dories	I/II	V?		M	SW	50
ORDER: Gasterosteiformes							
Gasterosteidae	Sticklebacks	I	V		M	SW	8
Aulostomidae	Trumpetfishes	II	III	S		SW	4
Solenostomidae	Ghost pipefishes	I/II	V?		M	SW	5
Syngnathidae	Pipefishes and sea horses	I/II	V?		M	SW	150
ORDER: Scorpaeniformes							
Scorpaenidae	Scorpionfishes	II/III; II	IV		M	SW	330
Triglidae	Searobins	II/III	V		M	SW	70
Anoplopomatidae	Sablefishes	II	IV		M	SW	2
Platycephalidae	Flatheads	II	IV		M	SW	Several
Cottidae	Sculpins	I/III	IV		M	SW & FW	300
Agonidae	Poachers	I/II	IV		M	SW	49
ORDER: Dactylopteriformes							
Dactylopteridae	Flying Gurnards	II/III	II	S		SW	4

ORDER: Perciformes

Family	Common name			S	M	Water	No.
Serranidae	Sea basses	I	V		M	SW	370
Centrarchidae	Sunfishes	II/III	V		M	FW	30
Priacanthidae	Bigeyes	II/III	I	S		SW	18
Apogonidae	Cardinalfishes	II/III	V		M	SW	170
Percidae	Perches	I/III	IV		M	FW	126
Carangidae	Jacks and Pompanos	II;IV	III;II	S		SW	200
Lutjanidae	Snappers	III	I;II	S		SW	230
Sparidae	Porgies	II?	IV		M	SW	100
Sciaenidae	Drums (Croakers)	I	V		M	SW	160
Mullidae	Red mullets	II/III	IV		M	SW	55
Kyphosidae	Sea chubs	III	II	S		SW	31
Chaetodontidae	Butterfly fishes	II/III	II;III	S		SW	160
Cichlidae	Cichlids	II/III; II	IV	S		FW	680
Cirrhitidae	Hawkfishes	I/II	IV		M	SW	34
Mugilidae	Mullets (grey)	I/II	IV?III			SW	70
Sphyraenidae	Barracudas	II	II	S		SW	18
Labridae	Wrasses	II;I/ III;II/ III	IV		M	SW	400
Scaridae	Parrotfishes	II;II/ III	IV		M	SW	68
Trichodontidae	Sandfishes	I	II	S		SW	2
Trachinidae	Weaverfishes	I/II	V?		M	SW	4
Uranoscopidae	Stargazers	I/II	V?		M	SW	25
Blenniidae	Combtooth blennies	II?	V		M	SW	276
Stichaeidae	Pricklebacks	I/III	I?	S		SW	57
Acanthuridae	Surgeonfishes	II;IV; II/III III	II;IV	S	M	SW	75
Gobiidae	Gobies	I/III;I	V		M	SW	800

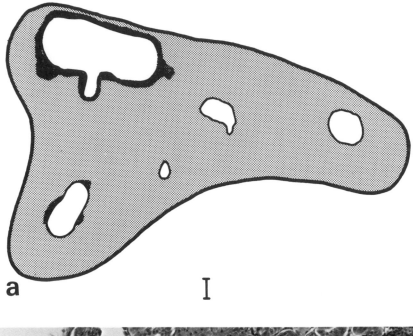

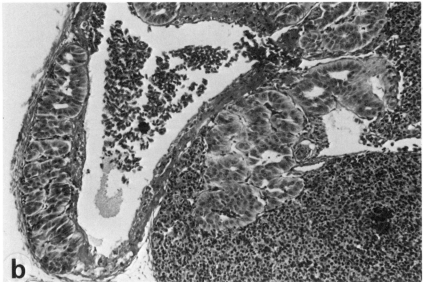

Fig. 23. (a) A diagram of Nandi's Type I in which interrenal tissue surrounds the posterior cardinal veins or their largest branches. (b) A transverse section of the posterior cardinal vein from an eel (*Anguilla anguilla*) adapted to sea water showing interrenal cells in the wall of the surrounding vein as in Type I (× 145).

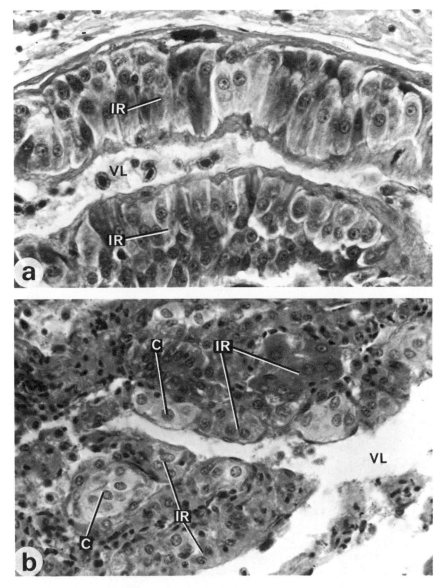

Fig. 24. (a) Interrenal cells within the walls of the posterior cardinal vein of a freshwater yellow eel (× 590). (b) Interrenal cells of *Fundulus heteroclitus* surrounding branch of posterior cardinal vein. V.L. vein lumen; IR. interrenal cells; C. chromaffin tissue (× 590). Note the mingling of interrenal and chromaffin tissue.

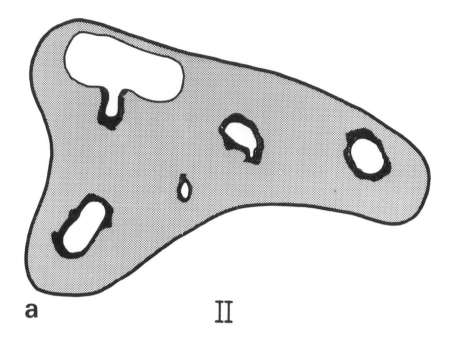

a   II

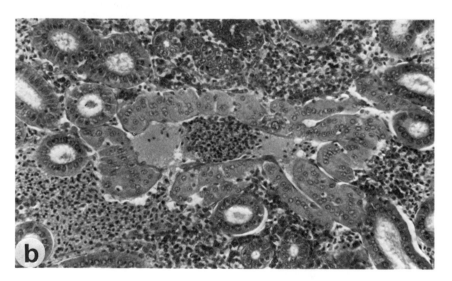

b

Fig. 25. (a) A diagram of Nandi's Type II in which interrenal tissue surrounds small or medium sized branches of the veins. (b) Interrenal cells of the herring (Clupeidae) surrounding a branch of the posterior cardinal vein as in Type II (× 290).

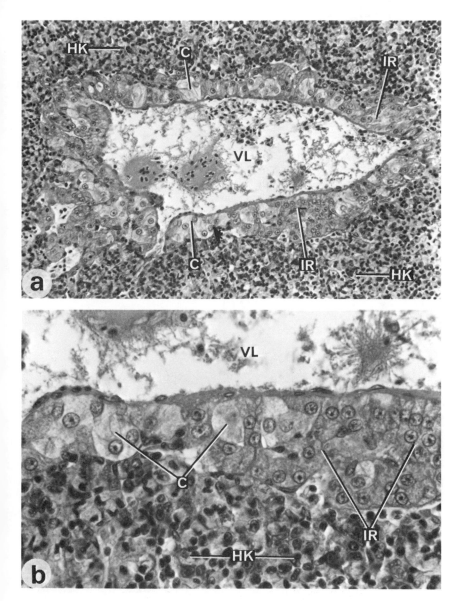

Fig. 26. (a) Interrenal cells of the carp surrounding a branch of the postcardinal vein. Note the intermingling of chromaffin and interrenal cells (× 230). (b) A higher power of Fig. 26a showing the interrenal and chromaffin cells lining a branch of the post cardinal vein. (× 590). VL. vein lumen; IR. interrenal cells; C. chromaffin cells; HK. head kidney.

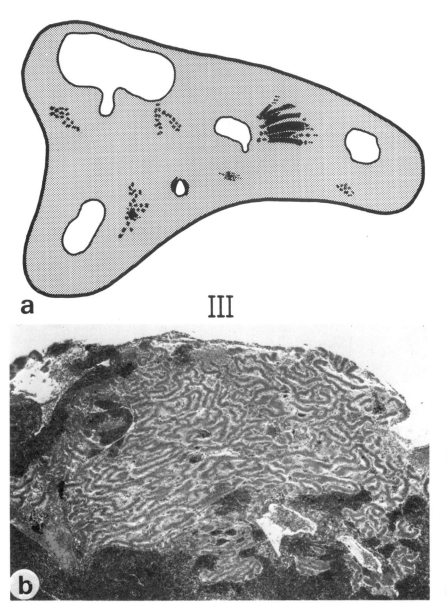

Fig. 27. (a) A diagram of Nandi's Type III in which interrenal tissue is associated with venous sinuses within the anterior kidney tissue. The interrenal often forms strands or cords of cells, sometimes scattered throughout and sometimes appearing to replace large areas of head kidney tissue. Interrenal cells do not surround the veins. (b) Interrenal cells scattered in the head kidney of the plaice (Pleuronectidae). ($\times$ 57).

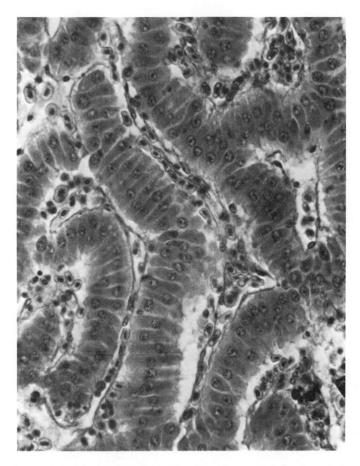

Fig. 28. Interrenal cells of the plaice showing the well defined corded structure
(× 590).

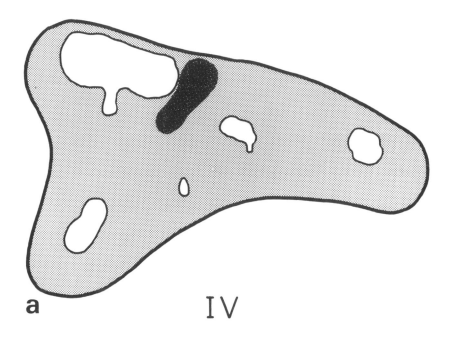

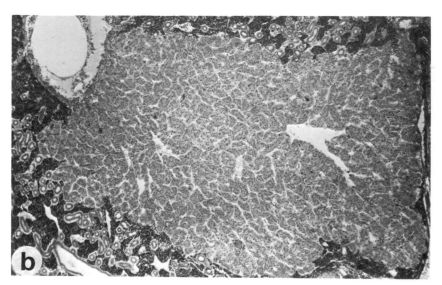

Fig. 29. (a) A diagram of Nandi's Type IV in which the interrenal tissue forms a solid mass of cells in a localized area. (b) The circumscribed group of interrenal cells found in the head kidney of *Lophius piscatorius* (Lophiidae) (× 47).

represent the basal type and the firmly mingled appearance that of the dominant "advanced" groups (e.g. the Perciformes). The Gadiformes of the Paracanthopterygii stand out, amongst the other exceptions. However this order is represented, in the literature, only by *Gadus morhua*, the cod, from about 684 species (Fig. 30).

The interrenal of teleost fish displays most of the histochemical reactions found in the tetrapod adrenal cortex (Holmes and Phillips, 1976; Hanke, 1978; Lofts, 1978; Idelman, 1978). There is now a considerable literature and not all species examined show comparable reactions (Banerji and Ghosh, 1966). The lipid droplets of the cells of the zona fasciculata, seen in most but not all eutherian species (Chester Jones, 1957; Idelman, 1978), are not obvious in the teleost interrenal though sudanophilic reactions have been demonstrated (see Krauter, 1958; Chavin and Kovacevic, 1961; Mahon *et al.*, 1962; Chavin, 1966; Yadav *et al.*, 1970). The presence of cholesterol, ascorbic acid and glucose-6-phosphate dehydrogenase has often been demonstrated (Oguri, 1960a-e; Chieffi and Botte, 1963; Hanke and Chester Jones, 1966; Olivereau, 1966a; Yaron, 1970). The proviso must be mentioned that the results are not necessarily revealing nor the methods always specific (Barka and Anderson, 1965; Yaron, 1970). Localization of Δ^5-3β-steroid dehydrogenase is almost universally made though not too much stress should be placed on the strength of colouring produced by the method nor on its presence or absence (Section 2).

B. VARIATIONS IN HISTOLOGICAL APPEARANCE

It would be anticipated that, as the teleost interrenal possesses vertebrate adrenocortical characteristics and synthesizes cortisol as its major steroid (see below), there would be a fish ACTH whose secretory activities would be controlled by the brain. There is considerable evidence for this hypothesis though of an indirect nature because neither a piscine adrenocorticotrophin nor releasing factors have been isolated, chemically characterized and synthesized as is the case for a few mammals (see Chester Jones *et al.*, 1974; Holmes and Ball, 1970; Fryer and Peter, 1977a, b, c).

1. *Hypophysectomy*

Varying degrees of interrenal atrophy occur after hypophysectomy, depending on species, the length of time after operation and the environment, and these have been fully documented. A few examples

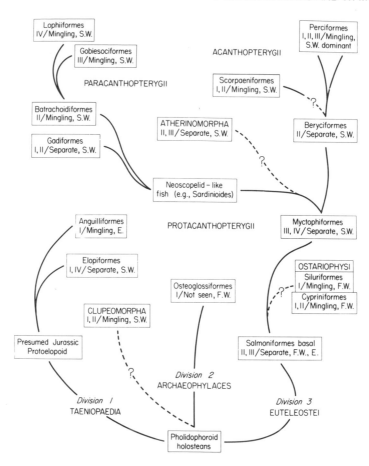

Fig. 30. Diagram showing the conception of the evolutionary relationship of the principal groups of teleostean fishes by Greenwood *et al.* (1966). The relationships are varied in subsequent publications, e.g. Greenwood *et al.*, 1973. There are no clear evolutionary trends but the Perciformes of the Acanthopterygii are a major group and show intermingling of chromaffin and interrenal tissue, whilst for example the Salmonids of the Protacanthopterygii show mainly separate interrenal chromaffin tissue. Mingling: chromaffin cells interspersed with interrenal cells. Separate: chromaffin and interrenal cells separately placed. S.W.: Seawater. F.W.:Freshwater.

are: *Anguilla anguilla* (Fontaine and Hatey, 1953; Olivereau and Fromentin, 1954; Ball and Olivereau, 1966; Fig. 31a, b); *Fundulus heteroclitus* (Pickford, 1953; Pickford *et al.*, 1965); *Carassius auratus* (Chavin and Kovacevic, 1961); *Tilapia mossambica* (Handin *et al.*, 1964; Basu *et al.*, 1965); *Poecilia latipinna* (Ball and Hawkins, 1976); *Salmo gairdneri* (Donaldson and McBride, 1967). Long-term

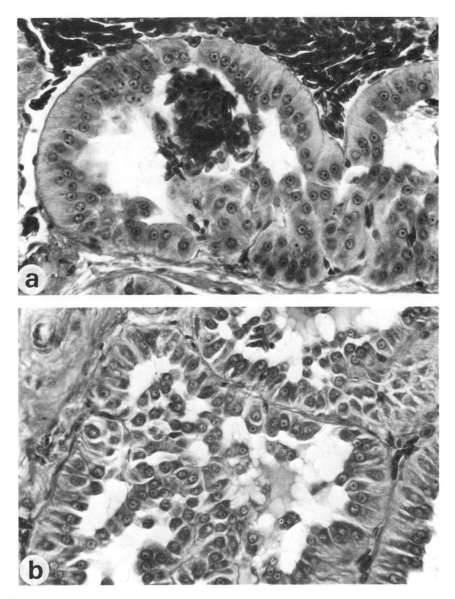

Fig. 31. (a) Part of the interrenal of an untreated (control) freshwater yellow eel (× 470). (b) Part of the interrenal of a freshwater yellow eel hypophysectomized for 1–2 weeks (× 470) illustrating collapse and disintegration (W. Hanke, personal communication).

hypophysectomy of the Indian catfish, *Heteropneustes fossilis*, for one and two years gave about a 37% reduction of the nuclear diameter of the interrenal cells (Sundararaj and Goswami, 1968)

2. *ACTH*

An important contribution gave good evidence for a fish ACTH. Ball *et al.* (1965) demonstrated the hyperplastic actions of ectopic pituitary transplants on the interrenal of the hypophysectomized *Poecilia*. Another work which used heterologous pituitary extracts was by van Overbeeke and Ahsan (1966). Using a freshwater cyprinid, the minnow *Couvesius plumbeaus*, they examined the effects of hypophysectomy 22, 48 and 79 days after operation. By 22 days, the interrenal cells became atrophic with small intercellular spaces and densely stained basophilic cytoplasm. After 48 and 79 days, the interrenal cells were flat, the cytoplasm not homogeneous and the nuclear membranes wrinkled. There was considerable shrinkage of cells and of nuclei but complete pycnosis was no more frequent than in the cells of control tissue. Injections of pituitary extracts into minnows, 30–40 days after hypophysectomy, on alternate days for 12–14 days, restored the interrenal to a normal histological appearance though with some variation in response. This work is particularly significant as the extracts used were made from the pituitaries of mature spawning spring salmon (*Oncorhynchus tshawytscha* and *O. kisutch*). The extracts were partially purified by starch gel electrophoresis and had both adrenocorticotrophic and gonadotrophic properties.

In most experiments, mammalian ACTH is used with effect. This is not surprising as the fish adenohypophysis contains cells with the characteristics of ACTH cells throughout the tetrapods (Holmes and Ball, 1974; Chan *et al.*, 1975). The numerous reports agree that mammalian ACTH stimulates the gland of normal fish and restores the appearance after hypophysectomy (Fig. 32a; Hanke and Chester Jones, 1966; Ball and Hawkins, 1976 and reviews noted above).

3. *Hormones and drugs* (Table III)

The natural hormone exogenously administered generally produces some degree of adrenocortical atrophy in vertebrates. This is true of cortisol for fish though much depends on the doses used and length of time of treatment (Fig. 32b; Hanke and Chester Jones, 1966; Olivereau, 1966b).

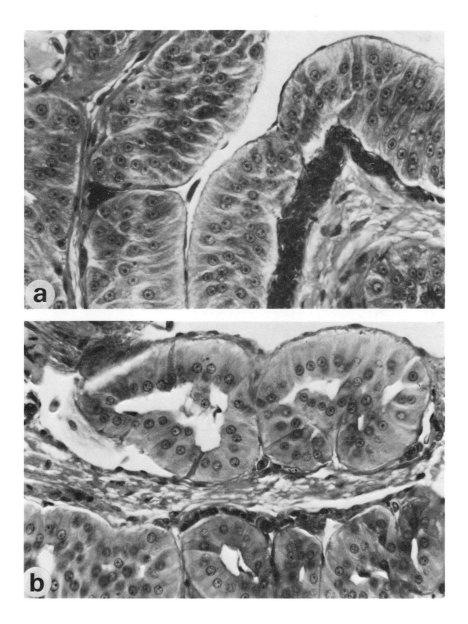

Fig. 32. (a) Part of the interrenal of a freshwater yellow eel given 1 I.U. ACTH daily for 4 days. There is clear evidence of stimulation (× 470). (b) Part of the interrenal of a freshwater yellow eel given 50 μg cortisol daily for 4 days (× 470). There is the appearance of reduction in secretory capacity with the administration of the endogenous type of hormone (W. Hanke, personal communication).

TABLE III

Measurement of interrenal cells and nuclei taken at different stages of the life cycle of the Atlantic Salmon (*Salmo salar*) from Olivereau (1975).

Environment	Month	Stage	Interrenal		total volume mm^3/kg	nuclear diameter μm	nuclear index
			sex	n			
Fresh water	December	Immature Parr	♂	3		6.39 ± 0.04	11.10 ± 0.30
		Immature Parr	♀	8		6.41 ± 0.05	11.58 ± 0.11
	March–April	Ripe parr	♂	7	12.07 ± 0.96	6.72 ± 0.08	8.70 ± 0.18
		Spring parr	♂	4	13.45 ± 0.24	6.34 ± 0.05	10.80 ± 0.40
		Pseudo-smolts	♂	5	21.37 ± 2.15	6.87 ± 0.05	9.60 ± 0.11
		Smolts	♂	6	25.40 ± 3.71	6.91 ± 0.06	8.00 ± 0.28
		Smolts	♀	7	21.87 ± 2.56	7.02 ± 0.12	7.45 ± 0.13
Ocean	September	Labrador	♀	2		7.16	8.08
		Greenland	♂	4		7.56 ± 0.24	5.57 ± 0.08
		Greenland	♀	4		7.37 ± 0.04	6.52 ± 0.26
Fresh water	March–April	Upstream	♂	3		6.57 ± 0.23	9.99 ± 1.14
		Upstream	♀♂	6		6.69 ± 0.17	9.32 ± 0.63
	December	Spawning	♂	4		6.91 ± 0.05	9.46 ± 0.30
		Spawning	♀♂	4		6.78 ± 0.19	9.39 ± 0.80
	March–April	Kelts	♀	3		6.53 ± 0.05	10.66 ± 0.63

Amongst the drugs, the blocker of 11β-hydroxylation—metyrapone (metopirone, SU 4885 CIBA)—is the most used, both experimentally and clinically (Sandor *et al.*, 1976; this volume, Chapter 2). Change in feed back with declination in corticosteroid production, allows increased ACTH secretion to enlarge the interrenal cells. Such results emphasize the dual action of ACTH in vertebrates to act not only as an agent to form and keep the cytomorphology of adrenocortical cells and their vasculature but also to catalyse certain pathways in steroid synthesis.

Metyrapone has been used with teleosts many times (Chester Jones *et al.*, 1974). The work of Ball and Olivereau (1966) gives the essential type of results usually obtained. Male silver eels in freshwater, body weight 50 to 75 g, had interrenal tissue extending for approximately 10 mm in the walls of the anterior and posterior cardinal veins. The nuclei of the interrenal cells of the controls (n = 14) varied in diameter from 4.35 to 4.87 μm (mean 4.61 $\pm$ 0.99 μm) and those injected with saline (n = 8) gave no significant difference (mean 4.67 $\pm$ 0.046 μm). Nuclear count in a standard area (1600 μ^2) gave 18.46 $\pm$ 0.54 and 18.33 $\pm$ 0.87 respectively. The drug was injected over 22 days but by the second to fourth day clear demonstration of its effects were seen. The nuclear diameter was markedly increased (5.20 to 5.97 μm), as was the nuclear count (9.87 to 12.85). These are indications of enlarged cells and cell divisions were frequent. Stimulation was further shown by the tendency of the larger interrenal cells to be arranged in single layers, implying increased vascularization, and the lumina of the lobules become bigger. By the fifth to tenth day maximal reactions were seen with pronounced cellular and nuclear hypertrophy: nuclear count 7.88–13.85, diameter 5.27–6.32 μm. There was continued mitotic activity. Signs of exhaustion were seen after the 12th day and up to day 22. Nuclear diameter diminished (4.88–5.60 μm), with some persistence in cellular hypertrophy (nuclear count 8.58–13.66). They found similar reactions, in principle, when another species. *Poecilia*, was used. As would be expected, the effects of metyrapone were not seen in the hypophysectomized animal.

C. ULTRASTRUCTURE

The most useful descriptions of the teleost interrenal cell are those for the goldfish (*Carassius auratus*) by Wakisaka (1964), Yamamoto and Onozato (1965) and Ogawa (1967). In control fish, interrenal tissue comprises 3 layers of elongated cells with their long axes perpendicular to the walls of the posterior cardinal veins with chromaffin cells occurring conspicuously though randomly. In general the fine structure is similar to that of mammalian adrenocortical cells Idelman, 1978) (Figs. 33, 34).

The nucleus is centrally located and round, with a rather loose outer nuclear membrane. The cytoplasmic ground substance is of low electron opacity and contains numerous mitochondria, a moderate amount of endoplasmic reticulum, a number of ribosomes, a few granules, and, frequently, a large Golgi apparatus and cytoplasmic filaments. The mitochondria measure up to 1.5 μm in diameter, have round or oval profiles and are of the tubulovesicular type. In some cases, they closely surround the nucleus and they appear to be characteristic of steroid-secreting cells. Open form and poorly defined mitochondria are sometimes encountered. Amorphous, electron-dense deposits sometimes occur within mitochondria.

A typical Golgi apparatus in the supra-nuclear region of the cell shows cisternae and vesicles with contents of low electron density. Near the Golgi area, many vesicles of smooth-surface endoplasmic reticulum are observed and free ribosomes are found in this area. A few microbodies occur near or within the Golgi area. The microbody consists of a limiting membrane of moderate electron opacity and a granular or lamellated core. Some homogeneous, rounded granules are also seen within the cytoplasm. The majority of these are situated near or within the Golgi area, especially the small ones (less than 0.15 μm). The electron opacity of these granules is usually high. They have a limiting membrane and measure up to 0.8 μm. The cytoplasm shows neither the lipid droplets nor the cytoplasmic droplets of low electron opacity that have been described in the rat adrenal. Smooth endoplasmic reticulum is well developed throughout the cytoplasm and usually shows a vesicular form. Rough endoplasmic reticulum is absent. Conspicuous cytoplasmic filaments, 40 to 50 Å in diameter, are in contact with the mitochondria and endoplasmic reticulum but they do not reach the mitochondrial matrix.

Hypophysectomy is followed by two major changes (Fig. 35). The tubulovesicular nature of the mitochondria diminishes and a highly electron-opaque cell type appears. The cells themselves and the nuclei are smaller, irregular and more electron dense. Apparently as a result of the loss of cytoplasm, the cell is filled with closely packed mitochondria which are electron opaque. Although the Golgi apparatus is indistinct it does contain a few dense granules. Scattered individual filaments run among the cytoplasmic organelles. It would be reasonable to deduce that cells with all these characteristics are less active than normal.

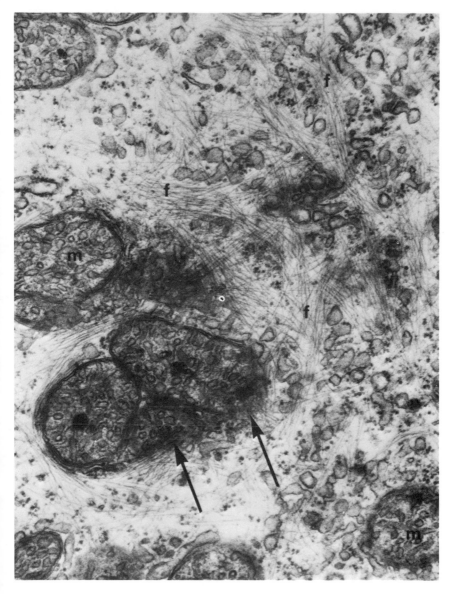

Fig. 33. Interrenal cell of normal goldfish. Numerous individual filaments (f) run randomly through the cytoplasm and are in contact with the mitochondrial outer membrane (arrows). m. mitochondria (× 30 000) (from Ogawa, 1967).

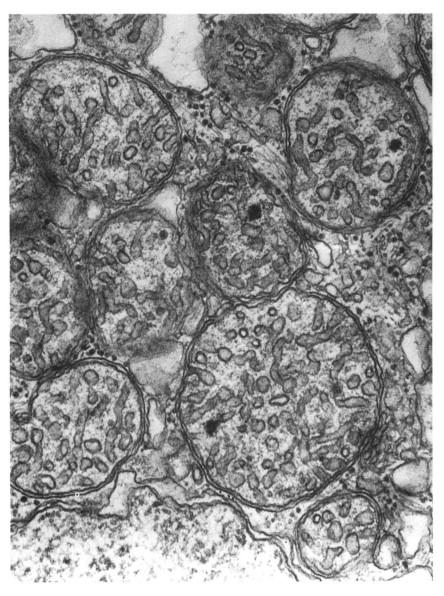

Fig. 34. Mitochondria from the interrenal of the normal goldfish showing their tubulo-vesicular nature (× 78 000).

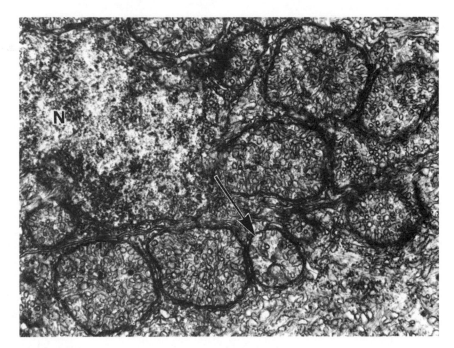

Fig. 35. Interrenal cells from goldfish hypophysectomized from 8 weeks. A decrease of tubulo-vesicular structure is observed in some mitochondria (arrow). Other mitochondria appear almost normal, but with irregular outlines (× 21 600). (Ogawa, 1967).

Goldfish can be adapted to one third seawater. The most striking change in the interrenal cells, 2 to 3 weeks later, is the appearance of circinate tubular structures in the mitochondria, arising from the inner membranes. Granules of high electron density are seen in greater numbers near or within the Golgi area. Numerous individual filaments occur throughout the cytoplasm. Other cytoplasmic structures and the nucleus remain normal (Fig. 36a, b). Circinate mitochondrial structures do not occur in interrenal cells of normal and hypophysectomized goldfish, though they have been seen in the domestic fowl (Fujita *et al.*, 1963), the zona fasciculata of the hamster (De Robertis and Sabatini, 1958) and the zona glomerulosa of the sodium deficient rat (Giacomelli *et al.*, 1965). One interpretation is that they are symptomatic of stimulated cells. The characteristics of teleost interrenal cells suggest a basic zona glomerulosa or intermediate type. Yet, there are difficulties in comprehending differential corticosteroidogenesis by adrenal glands not thrown into definitive

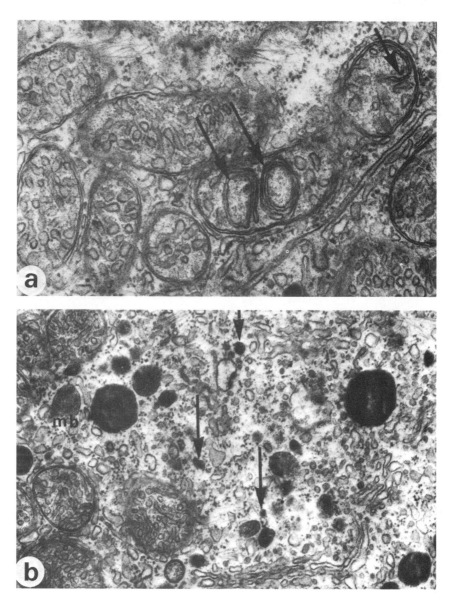

Fig. 36. (a) Mitochondria of the interrenal from goldfish after two weeks in one third sea water showing the appearance of circinate tubular structures (arrows). (× 30 000) (Ogawa, 1967). (b) Part of the interrenal cell from goldfish after two weeks in one third sea water showing Golgi area. Small granules (arrows) are less than 0.2 μm and seen only in this area. These granules may be derived from the Golgi apparatus and developed into the granules of high electron opacity. Microbodies (mb) are also seen (× 17 500) (Ogawa, 1967).

zones, implicated perhaps in some Tetrapods, but not always established (e.g. Pearce *et al.*, 1978).

D. STRESS

Types of stress, however defined, may be anticipated to activate the pituitary–interrenal axis. This is largely the case, based on histological changes and some direct measurement of steroid levels. Interrenal stimulation by handling alone of the carp and Atlantic salmon (Leloup-Hatey, 1958, 1964) and the chinook salmon, *Oncorhynchus tshawytscha* (Hane *et al.*, 1966), has been observed. In experiments designed to produce stressful reactions, forced exercise, cold shock and noxious stimuli are the chief modes used. For example, Fagerlund (1967) forced mature, healthy, unspawned sockeye salmon (*O. nerka*) to swim for 3 h by constant agitation. This procedure was followed by up to tenfold increases in the levels of plasma cortisol compared with controls. Moderate forced swimming for 2–3 days or 12–16 days did not alter cortisol levels. Hill and Fromm (1968) exposed the rainbow trout to the noxious stimuli of hexavalent chromium and to chronic forced exercise. Their fluorescent technique for cortisol determination included some other steroids but nevertheless their results give a pattern of reactions to stress, indeed, throughout vertebrates. Chromium in the tank water produced plasma "cortisol" levels nearly twice those of control fish in one week, returning to normal after 2–3 weeks. The particular method for forced exercise gave some but not a dramatic increase in plasma "cortisol" levels.

Cold shock applied to the freshwater tropical teleost *Colisa fasciatus* (Labyrinth fishes, Belontiidae, Perciformes) was by exposure to waters about 2°C for 171, 267 to 363 minutes. The interrenal tissue (Type I/V see Section G) became hypertrophic (Agrawal and Srivastava, 1978). Another aspect is the impact of social hierarchies. In the light of the discussion in Nowell (this volume, Chapter 6), it is interesting to find that green sunfish *Lepomis cyanellus* are cast into social ranks with the consequences that those "lower down" have significantly greater volumes of interrenal tissue (Erickson, 1967).

E. SEASONAL CHANGES

Apparent stressful reactions may be involved in such natural processes as migration and spawning. In sub-mammalian vertebrates there is a consensus that adrenocortical secretions are intimately concerned with breeding and spawning. Most of the evidence arises from observations on changes in the morphology of the tissue. This is particularly true in teleost fish in which the interrenal undergoes seasonal variations, with enlargement at the reproductive period. The implication is that there is an increased production of corticosteroids but direct evidence is less abundant. Many contributions concern the Salmonids.

In early work the activity of the interrenal tissue of Pacific salmon and of both searun rainbow trout (steelhead) and non-migratory rainbow trout was studied (Hane and Robertson, 1959; Robertson and Wexler, 1959; Robertson *et al.*, 1963). The universal post-spawning mortality of Pacific salmon has been associated with hyperplasia of the interrenal gland and elevated levels of 17-OH corticosteroids (17-OHCS). They also reported that many tissues showed degenerative changes similar to those found in Cushing's syndrome in man, a condition caused by excessive secretion of adrenal steroids. The post-spawning mortality of searun rainbow trout was higher than that of non-migratory rainbow races and was also correlated with elevated interrenal activity during the pre-spawning period. Degenerative changes similar to those that occurred in salmon were produced in immature non-migratory rainbow trout by administration of exogenous cortisol. Idler and Truscott (1963) suggested that elevated steroid blood levels could arise from an impaired clearance mechanism and hence may not precisely indicate interrenal gland activity.

Plasma corticosteroid levels in sockeye salmon were reported to increase about fourfold during the period of the upstream spawning migration (Idler *et al.*, 1959; Schmidt and Idler, 1962). Elevated levels were also found in spawning coho salmon and to a less extent in Atlantic salmon. In captive sockeye salmon after completion of their upstream migration, Fagerlund (1967) found the mean cortisol concentration (μg/100 ml plasma) to be, for males, 2.1 $\pm$ 1.3 and, for females, 2.5 $\pm$ 1.7 (standard deviations and n = 20 in both cases). These low values were not found in fish in other conditions and values for moribund fish ranged from 6.8 to 193 μg/100 ml. In the rainbow trout the control groups for the experiments of Hill and Fromm (1968) showed higher levels of corticosteroids, predominantly cortisol, in December and January and lower levels in July and October.

Olivereau (1975) histologically assessed interrenal activity of the Atlantic salmon at various stages of its life cycle in freshwater and the ocean (Table III). As the immature parr becomes ripe there is an indication of increased interrenal activity. The smolt stage, males and females, is characterized by increased activity together with mitotic divisions (Fontaine and Olivereau, 1957; Olivereau, 1960). Experimentally, transfer of smolt to sea water accentuates the hypertrophy of the interrenal (Olivereau, 1962). However, the big salmon caught off Greenland and Labrador—males 2.6–4.75 kg, females 3.4–10 kg—have a voluminous interrenal with many "clear cells". The measurements indicate interrenal activity associated with many mitotic divisions. Salmon on the upstream migration had two groups of interrenal cells on with a reticulate structure, the other lobular. The overall aspects indicated less activity of the gland as did that of spawning salmon. It should be noted however that the measurements indicate continued activity. The results show that the interrenal–ACTH axis is stimulated in freshwater with paedogenetic male parr. Smoltification is associated with interrenal hypertrophy. In the sea, salmon interrenal shows maximum activity. The gland is still active in the freshwater migration for reproduction, though less so, and the interrenal is heterogenous. The appearance in kelts does not differ significantly.

In another genus of Salmonidae, Honma (1960) studied the change in interrenal tissue during the lifespan of the Ayu, *Plecoglossus altivelis* (Temminck and Schlegel). Most die after spawning but some may over-winter if the degree of fatigue and debility is not too excessive. Interrenal tissue increases in the pre-spawning period (October) and this is quite marked in the breeding season (November). After spawning a marked haemorrhage is often seen in the head kidneys, and interrenal cells collapse with pycnotic nuclei. These fish do not feed during the reproductive period and gonadal maturation and spawning may call on the use of body protein via, amongst other factors, interrenal steroids.

The relationship of interrenal hypertrophy to gonadal maturation is underlined by the work of McBride and Overbeeke (1969) on sexually maturing sockeye salmon. Fully mature fish taken in early October showed considerable interrenal hypertrophy. After completion of spawning, haemorrhages were present and pycnotic nuclei frequent. When fish were fed, interrenal hypertrophy of sexually mature fish was less pronounced. Gonadectomy of sexually mature fish led to a gradual involution of the interrenal tissue. Eight weeks after gonadectomy the sexually mature sockeye tissue was virtually the same as that in sexually immature fish.

In representatives of other orders, histological changes occur in concert with reproductive phases. Yaron (1970) found in *Acanthobrama terraesanctae* that from October to December concomitant with the increase of ovarian weight the interrenal cell nuclear diameters increase and cells are more abundant. During the following winter months there is a gradual decrease in the diameter of interrenal cells and nuclei with minimal values occurring in May. Dixit and Agrawala (1975) studied the seasonal histology of the interrenal of the female *Puntius sophore* (Ham.). In December the interrenal cells start organizing to form almost a uniform lining around the posterior cardinal vein and its major branches. Growth is progressive until the end of May when there are large interrenal clumps around the veins. During April to December the interrenal cells gradually reduce in volume. In October and November the interrenal cells are disorganized. The spawning phase is from April to July or August: there is thus growth in the pre-spawning phase with regression and disorganization after spawning.

A very interesting aspect of teleost endocrinology is hermaphroditism and sex-reversal. Chan (1970) gives details for serramids (seabasses, combers and groupers) and spanids (porgies and sea-breams). Natural sex reversal occurs in, amongst other genera, the maenid *Pagellus acarne* (Reinboth, 1970). One genus to which much attention has been paid is *Monopterus alba*, the freshwater ricefield eel. The small fish are females (body length less than 34 cm), large specimens male (about 45 cm) and intersexual gonads were found only among specimens of medium size (body length 35–45 cm) (Liu, 1944, Chan, 1970). Though this phenomenon is of great interest, in the context of this chapter we are concerned with the involvement of the interrenal. This is not known though cellular reactions after various treatments in female eels 36–56 g in weight and 28–43 cm in length are given in Table IV.

Steroid estimations have been made on *Pleuronectes* sp. Campbell *et al.* (1976) examined the variation in steroids in *P. americanus* found in marine shallow waters with no major migrations. Cortisol concentrations in plasma from females did not fluctuate greatly during the annual cycle but in plasma from males the concentrations found in spent fish (July) were greater than at any other time. The ratio of cortisol to cortisone concentration in both sexes followed a seasonal pattern which was not directly related to the reproductive cycle but high ratios may be associated with the summer feeding activity (April–October) and lower ratios to the period of winter starvation. These authors did not find 11-deoxycorticosterone to be especially active in the oocyte *in vitro* maturation bioassay.

Wingfield and Grimm (1977) examined seasonal changes in plasma steroids in the plaice *Pleuronectes platessa* L. They minimized the effects of stress in capturing, holding and sampling in estimating plasma

TABLE IV

Changes in interrenal tissue of *monopterus* following different treatments
(n = number of animals; means ± standard error)
(from Chan et al., 1975)

		Treatment					
n	Control 10	ACTH, 10 days 10	ACTH, 3 days 6	Cortisol 10	Metyrapone 10	Dexamethasone 10
Interrenal tissue[d]						
Chromaffin cells						
Mean cell diameter	8.57 ± 0.03	8.68 ± 0.05	8.61 ± 0.07	8.62 ± 0.04	9.64 ± 0.08	8.84 ± 0.05[a]
Mean nuclear diameter	4.21 ± 0.02	4.32 ± 0.06	4.28 ± 0.04	4.36 ± 0.05[c]	4.87 ± 0.10	4.66 ± 0.03[a]
Steroidogenic cells						
Mean cell diameter	9.42 ± 0.07	12.90 ± 0.09[a]	11.20 ± 0.05[a]	8.82 ± 0.11[b]	10.07 ± 0.08[a]	7.43 ± 0.09[a]
Mean nuclear diameter	5.12 ± 0.02	6.53 ± 0.09[a]	6.03 ± 0.07	4.82 ± 0.06[a]	6.02 ± 0.05[a]	4.59 ± 0.06[a]

[a] P<0.001 compared with the control; [b] P<0.01 compared with the control; [c] P<0.05 compared with the control; [d] No. of cells measured per fish is 90.

cortisol levels by adapting the fish in tanks for 24–48 h before sampling. In the plaice mature fish do not die after spawning and the peak cortisol levels are accompanied by high titres of oestradiol and testosterone, suggesting a close relationship. However, there is the difficulty that the pronounced cortisol peak in immature fish occurs when testosterone and oestradiol levels remain consistently low. Furthermore, it has been shown in sockeye salmon that sexual maturation *per se* is not necessarily accompanied by elevated plasma cortisol (Fagerlund, 1967). Seasonal elevations in plasma corticosteroids or seasonal interrenal hyperplasia have been found not only in Pacific salmon that migrate, spawn once and die but also in land locked, non-migratory salmonids which spawn several times, including the rainbow trout (Robertson *et al.*, 1961) and the nikko-iwana, *Salvelinus leucomaenis* (Honma and Tamura, 1965). It seems that in the cod *Gadus callarias* the interrenal hypertrophy does not commence until the spawning migration has started (Woodhead and Woodhead, 1965). One possibility is that plasma cortisol levels may rise in any species of fish which through inadequate food intake is required to mobilize stored energy whether for locomotion, gonad maturation, spawning or, in immature plaice, simply for basal maintenance.

The mechanism of action of gonadotrophins coupled with steroids is receiving more and more attention. Colombo *et al.* (1973) examined *Leptocottus*, *Gillichthys* and *Microgadus* species. They considered that ovarian 11-deoxycorticosterone and 11-deoxycortisol may act as local hormones mediating pituitary gonadotrophin-induced maturation and ovulation of oocytes. It is supposed that a pituitary factor would act by stimulating the synthesis of ovulatory corticoids in the ovary but not in the interrenal. Truscott *et al.* (1978) brought together more detailed work.

The interrenal gland is a central focus of an extra-ovarian system for oocyte maturation (Sundararaj and Goswami, 1971, 1977). Sundararaj and his co-workers have done much to advance knowledge of the role of steroids in teleost reproduction, primarily using the Indian catfish, *Heteropneustes fossilis*. This involves the maturation-inducing steroids (MIS). So that there is a gonadotrophin–interrenal–ovarian axis for oocyte maturation in this species. Most of the evidence comes from *in vitro* studies. Those steroids known to be active in *H. fossilis in vitro*, cortisol, 11-deoxycortisol and 11-deoxycorticosterone (Sundararaj and Goswami, 1971; Goswami and Sundararaj, 1974), were measured in the blood plasma of intact and gonadectomized catfish during the spawning and post-spawning seasons after administration of porcine ACTH, ovine LH and partially purified salmon gonadotrophin. They found that the LH preparations used increased cortisol concentrations in the plasma and deduced that the origin was the interrenal gland, rather than the ovary. DOC and cortisol induce ovulation in the hypophysectomized gravid catfish. Paradoxically ACTH which induces enhanced cortisol levels does not bring about oocyte maturation even *in vivo* when given in large doses (Sundararaj and Goswami, 1977). There is clearly much more to know. There may be synergism between cortisol and testosterone in oocyte maturation and one with DOC has been reported (Sundararaj and Goswami, 1971). In the catfish 11-deoxycortisol and DOC are potent maturation-inducing steroids *in vitro* (Goswami and (Sundararaj, 1974). However the plasma values of these steroids were low even after LH or ACTH injections. In *Tilapia*, the concentration of plasma DOC increased from 0.06 to 2.2 μm/100 ml on the initiation of spawning (Katz and Eckstein, 1974), though this is not shown by the catfish.

In the more recent work on fish reproduction and the role of the interrenal, the hypothalamus, so dominant in mammals and other vertebrates, has yet to be brought into the picture. Fryer and Peter (1977a, b, c), using the goldfish, found that the hypothalamic–adenohypophysial axis seems to operate in the control if interrenal activity.

9. Embryology

A. FISH (Fig. 37a, b, c)

The development of adrenocortical tissue and its homologues is intimately bound up with that of the kidney, the gonads and the posterior venous system (see Introduction). These are mesodermal in origin. It is generally considered that interrenal cells come from the columnar epithelial cells which, differentiated from mesoderm, line the coelom. If this differentiation does not arise from the peritoneal epithelia itself, it certainly occurs from the mesenchyme cells in the vicinity of this mesothelium. Thus the cortical anlagen arise from or near the peritoneal epithelium, near the aorta in the region of the angle between the dorsal mesentery and the germinal ridge. The most anterior site of origin of the anlagen is in the region of the posterior part of the pronephros, often in the "Zwischenzone" which lies between the caudal end of the pronephros and the cephalad mesonephros (Figs. 38a, b, elver, tadpole). In essence there is a region, stretching from the second post-otic somite to the anus and

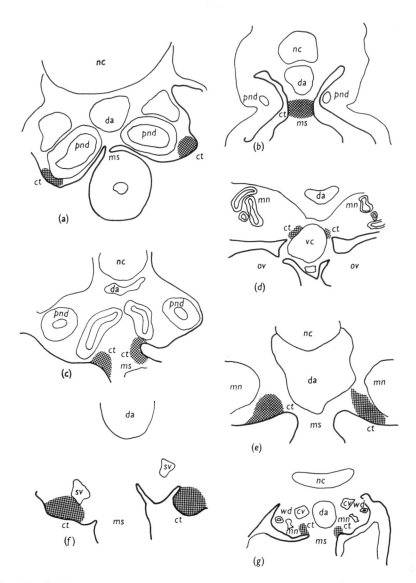

Fig. 37. Early stages in the development of interrenal and cortical tissue in representatives of different vertebrate classes (modified from Chester Jones, 1957). (a) *Petromyzon*; Cyclostomata (b) *Scyllium*; Elasmobranchii (c) *Salmo*; Teleostei (d) *Rana*; Amphibia (e) Turtle; Reptilia (f) Domestic fowl; Aves (g) Mouse; Eutheria.nc, notochord; da, dorsal aorta; pnd, pronephric duct; ms, dorsal mesentery; ct, cortical or interrenal tissue; mn, mesonephros; wd, Wolffian duct; cv, cardinal vein; vc, vena cava; sv, sub-cardinal vein; ov, ovary.

beyond, which is potentially interrenal, gonadal and nephrogenic. This mesodermal potentiality is never fully realized but the cyclostomes approach nearest to this basic evolutionary pattern (Chester Jones, 1957). In the Agnatha (see Giacomini, 1902, Poll, 1906; Fig. 37a), Sterba (1955) found in larval *Lampetra planeri* of 3.5 to 4 mm, pronephric anlagen originating directly behind the pharyngeal gill formation, from numerous segmental stalks. In the 6 mm larva, numerous mitoses occurred, in the angle between the mesenterium and the somatopleure, to give a thickening of the coelomic epithelium, the first signs of interrenal tissue. The diffuse anlagen, six to seven in number, are found at the caudal end of the pronephros corresponding to spinal ganglia six to eight. By the 7.5 mm larva, they comprise many cells of typical amoeboid appearance. In the 8 mm larva the cells can be made out on the ventral side of the blood vessels and, by 9 mm, reaching to the anal region when the mesonephros becomes identifiable. The beginning of metamorphosis is characterized by an increase in the number and size of interrenal cells. By the end of metamorphosis, but before maturity, many islets have changed from about 200 cells to 800. When sexual maturity and spawning occur, the total volume of the interrenal system has increased about 9 times. This may be due to an ACTH-like action (Sterba, 1955) correlated with sexual maturation rather than the process of metamorphosis itself (Hardisty, 1972).

In the dogfish (*Scyliorhinus stellaris*), as a representative of the elasmobranchs, the first indication of cortical tissue is in the 7 mm embryo. Here, a group of cells emerges from the coelomic epithelium at each side at the root of the dorsal mesentery (Fig. 37b). These cells join to form one mass which commences at the level of the posterior region of the pronephros and extends, broken up into many groups though not segmentally, to the cloaca. At 10 mm embryonic length, the cortex appears as a rod of tissue lying below the ventral wall of the aorta and in the root of the mesentery. The longitudinal extent of the interrenal has, however, diminished, now being from the seventh segment behind the end of the pronephros to the edge of the first and second quarters of the cloaca. This covers twenty segments as against the twenty-five originally bearing cortical anlagen. The adult interrenal is achieved by the loss of the anterior interrenal groups and the growth and amalgamation of the posterior ones. Hence in the embryo of 24 mm in length, the interrenal spreads over only the twelve segments lying anterior to the cloaca. At the same time the posterior part of the mesonephros has become predominant.

For the embryology of the interrenal of teleost fish, Giacomini

(1912, 1920, 1922) gives the classical description (see Chester Jones, 1957; Belsare, 1973; Fig. 37c). The sequence of events in *Salmo* species may be regarded as typical. The first sign of cortical tissue is seen in embryos about 4 mm in length, 22 to 23 days after fertilization. The cortical anlagen arise by proliferation of coelomic epithelial cells on each side of the root of the mesentery, ventral to the pronephric glomeruli. In 5 mm embryos, the archinephric duct has reached the cloaca and the cortical anlagen appear as round groups of cells. These proliferate and the cortical tissue spreads irregularly in the region of the pronephros (Fig. 38a). With the degeneration of the pronephros, the interrenal tissue lies in the lymphoid head kidney as in the adult. If we accept the potential of the mesoderm from pronephros to cloaca in proliferating interrenal tissue, the unanswered question still remains as to its sequestration anteriorly in teleosts and posteriorly in elasmobranchs.

B.   OTHER VERTEBRATES (Fig. 37d, e, f, g)

The vertebrate pattern of the embryology of adrenocortical homologues is consistent and that of the eutherian mammals has been described by Idelman (1978). In Amphibia (Dittus, 1936; Branin, 1937; Witschi, 1951, 1953; Witschi *et al.*, 1953; Figs. 37d, 38b) the frog shows interrenal differentiation near the time of closure of the gill sacs, some one or two days before the larvae begin to feed. Now the genital ridges appear, the left and right subcardinal veins have moved together and their fusion has just given origin to the lowest segment of the vena cava. The cortical anlagen arise from the mesodermal bars which lie between the numerous outlets of the mesonephric blood sinuses running into the vena cava (Witschi loc. cit.). In the urodele (*Hemidactylium scutatum*) Branin (1937) found the cortical anlagen in the 8.8 mm embryo as paired or unpaired cell groups in the Zwischenzone, either in contact with or just beneath the mesothelium, on either side of the dorsal mesentery. Further interrenal anlagen arise posteriorly so that while only two to four are seen at the earliest stage, there are thirty or more at metamorphosis. The increase in number of primordia occurs concomitantly with the development of the vena cava and of the mesonephros. A month after hatching, in larvae of 13–15 mm, the interrenal tissue extends over an area bounded by the eighth and fourteenth pairs of spinal ganglia and consists of individual masses of different sizes. At metamorphosis, the interrenal tissue is distributed from the level of the ninth to that of the twentieth spinal ganglia. Accompanying and immediately following

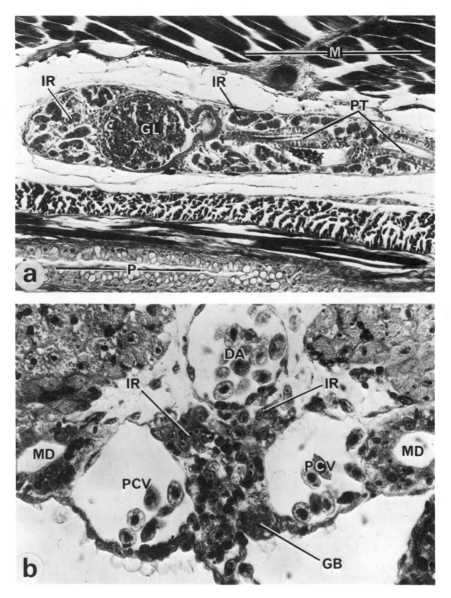

Fig. 38. (a) Horizontal longitudinal section of the pronephric area of an elver, showing the development of interrenal islets. IR. interrenal islets; GL. glomus; PT. pronephric tubules; M. myotome; P. pharynx (× 74). (b) Transverse section of a tadpole of *Rana temporaria* at the external gill stage showing a clump of undifferentiated interrenal cells. DA. dorsal aorta; MD. mesonephric duct; PCV. posterior cardinal veins; GB. gonadal bud; IR. interrenal cells. (× 470).

metamorphosis there is a progressive coalescence and growth of the individual interrenal groups resulting in strands of tissue along the caval system. This leads to the adult form (Hanke, 1978) where the interrenal islets are more or less interconnected in the region of the mesonephros, scattered along the inferior vena cava and its tributaries from the cranial end of the pancreas to the caudal end of this vein.

The reptiles, birds and mammals (Figs. 37e, f, g) are grouped together as the Amniota and now we can use the term adrenal gland. This comes about because of the arrival of the metanephros and, in the adult, the association with the mesonephros ceases, which gives rise to only some reproductive ancillary structures. The separate, encapsulated gland, nevertheless has a similar embryonic origin as in all vertebrates. In reptiles, as represented by the genera *Emys* and *Lacerta* (Poll, 1906; Kuntz, 1912; Bimmer, 1950), cortical anlagen were seen at the earliest stages (10 mm embryos in *Emys*) stretching from the caudal end of the pronephros over the area of eight to nine spinal ganglia (numbers six to fourteen or fifteen). The further development consists of joining up of the cortical anlagen and their concentration so that the primordia extend over only two segments (*Emys*, 28 mm embryo). Further details can be gathered from the development of the chick, the best known representative of the birds. Here the cortical primordia first appear in somite seventeen, at the caudal end of the pronephros in somites five to fifteen or sixteen, and extend over to somite twenty-two. The cephalad region of the mesonephros overlaps the posterior pronephric tubules, originating in somite thirteen or fourteen and reaching to somite thirty. The cortical anlagen appear as a thickening of the peritoneal epithelium, ventral and mesial to the mesonephros, ventral to the abdominal aorta and dorsal to the hindgut which is open at this time (96th hour of incubation). Then the cortical cells form a solid body on each side of the base of the mesentery, just medial to the ventral side of the mesonephros (115th hour of incubation). A dorsal migration occurs so that all cortical cells come to lie slightly dorsal to the ventral level of the aorta, against the mesonephros (120th hour of incubation). A large oval mass forms on each side of the aorta (144th hour of incubation). Thereafter the cortical cords can be seen and the gland continues to enlarge until hatching (Hays, 1914).

Mammals show a similar embryonic history. In these, there is an aggregation of chromaffin tissue which is, to a lesser or greater extent, associated with adrenocortical tissue in the gnathostomes except for the elasmobranchs. Chromaffin tissue has its origin in the neural crest and the cells migrate, insidiously passing between the cells of the growing body to form various structures in addition to their inter-mingling with interrenal and cortical tissue (Chester Jones, 1976; Idelman, 1978).

In the vertebrate classes of the Gnathostomata, in so far as they represent evolutionary patterns, the concentration of chromaffin tissue is matched by that of the adrenocortical homologues. Thus, of the twelve to thirteen somites from the posterior end of the pronephros to the anterior side of the cloaca (spinal ganglia five to seventeen or eighteen) the cortical tissue during some stage of development may occupy 100% in elasmobranchs, 80% in amphibian urodeles, 67% in reptiles, and diminishes in birds and mammals from 33–25% to 8.3% (Chester Jones, 1957).

10.   Corpuscles of Stannius*

These bodies no longer come within a treatise on the adrenal cortex and its homologues. A summary of their impact on interrenal nomenclature and function may be informative and certainly salut-ary. The corpuscles of Stannius (Stannius, 1839a, b; 1846, 1854) are small, irregularly spherical, white or yellowish-pink bodies which lie on or are embedded in the dorsal or ventral surfaces of the posterior region of the kidney in teleosts and holosteans. In most, there is generally a pair of these bodies but in some genera, especially the salmonids and the bowfin, there may be several (Chester Jones, 1957, 1976; Youson et al., 1976). The controversy as to whether or not they represented the interrenal was often bitter and the history is given in Chester Jones (1957). In that year he wrote that he thought it unlikely that . . . "adrenocortical function again be specifically assigned to these engimatic bodies". Interest in the corpuscles was awakened by the finding of Fontaine (1964) and Chester Jones et al. (1965) that their removal was followed by changes in electrolyte values of eel plasma. Nevertheless Diamare (1935) still cast the shadow of the nomenclature that the corpuscles were the "posterior interrenal". Thus, the corpuscles were found to be steroidogenic (Cédard and Fontaine, 1963; Breuer and Ozon, 1964; Idler and Freeman, 1966; Leloup-Hatey, 1970) though denied (Chester Jones et al., 1965). It is now clear that the corpuscles secrete one or two substances, "hypocal-cin" (Pang et al., 1974) being one, possibly a peptide of molecular

* "Stanniectomy" is used to denote their surgical removal but the word means that it is Stannius himself who is cut out. The correct translation is "Stanniosomatiectomy" with "somatiectomy" as the shortened form.

weight in the region of 3000–4000 (Copp and Ma, 1978). Their ultrastructure is consistent with a peptide secreting organ (Meats *et al.*, 1979). Corpuscles, in their action on calcium, probably have an interrelationship with the calcitonin of the ultimobranchial bodies and a secondary axis with interrenal function is not excluded (Chester Jones *et al.*, 1967, 1969; Chester Jones, 1976).

References

Agrawal, U. and Srivastava, A. K. (1978). *Arch. Anat. microsc. morph. exp.* **67**, (1), 1–9.
Ahuja, S. K. (1962). *Curr. Sci. Bangalore* **31**, 466–467.
Atz, J. W. (1973). *Comp. Endocr. and Systematics* **13**, 933–936.
Baker-Cohen, K. F. (1959). *Comp. Endocr.* (A. Gorbman ed.) 283–301.
Ball, J. N. (1969). *In:* "Fish Physiology". Vol. II, Chapter 3. (Eds. W. S. Hoar and D. J. Randall). pp. 207–240. Academic Press, New York and London.
Ball, J. N. and Baker, B. I. (1969). *In:* "Fish Physiology". Vol. II Chapter 1 (Eds. W. S. Hoar and D. J. Randall), pp. 1–110. Academic Press, New York and London.
Ball, J. N. and Hawkins, E. F. (1976). *Gen. comp. Endocr.* **28**, 59–70.
Ball, J. N. and Olivereau, M. (1966). *Gen. comp. Endocr.* **6**, 5–18.
Ball, J. N., Olivereau, M., Slicher, A. M. and Kallman, K. D. (1965). *Phil. Trans. Roy. Soc. Lond.* **249**, 60–99.
Banerji, T. K. (1973). *Anat. Anz.* **133**, 20–32.
Banerji, T. K. and Ghosh, A. (1966). *Curr. Sci.* **35**(23), 596–597.
Barka, T. and Anderson, P. J. (1965). "Histochemistry, Theory, Practice and Bibliography". Harper and Row (Hoeber), New York.
Barnes, K. and Hardisty, M. W. (1972). *J. Endocr.* **53**, 59–69.
Basu, J., Nandi, J. and Bern, H. A. (1965). *J. Exptl. Zool.* **159**, 347–356.
Bellamy, D. and Chester Jones, I. (1961). *Comp. Biochem. Physiol.* **3**, 175–183.
Belsare, D. K. (1973). *La Cellule* **69**, 343–348.
Bentley, P. J. and Follett, B. K. (1965). *Life Sciences* **4**, 2003–2007.
Berchtold, J. P. (1969). *Z. Zellforsch. mikrosk Anat.* **102**, 357–375.
Bern, H. A., DeRoos, C. A. and Biglieri, E. G. (1962). *Gen. comp. Endocr.* **2**, 490–494.
Bimmer, E. (1950). *Anat. Anz.* **97**, 276–311.
Blair-West, J. R., Coghlan, J. P., Denton, D. A., Gibson, A. P., Oddie, C. J., Sawyer, W. H. and Scoggins, B. A. (1977). *J. Endocr.* **74**, 137–142.
Branin, M. L. (1937). *J. Morph.* **60**, 521–561.
Breuer, H. and Ozon, R. (1965) *Arch. Anat. Microsc. Morphol. Exp.* **54**, 17.
Buus, O. and Larsen, L. O. (1975). *Gen. comp. Endocr.* **26**, 96–99.
Call, R. N. and Janssens, P. A. (1975). *Cell Tiss. Res.* **156**, 533–538.
Campbell, C. M., Walsh, J. M. and Idler, D. R. (1976). *Gen. comp. Endocr.* **29**, 14–20.
Cédard, L. & Fontaine, M. (1963). *C. R. Acad. Sci.* **257**, 3095.
Certain, Ph. (1961). *Bull. Biol. Fr. Belg.* **95**, 134–148.
Chan, S. T. H. (1970). *Phil. Trans. Roy. Soc. Lond. Ser. B.* **259**, 59–71.
Chan, S. T. H., O. W–S. and Hui, W. B. (1975). *Gen. comp. Endocr.* **27**, 95–110.
Chavin, W. (1956). *Zoologica* **41**, 101–104.

Chavin, W. (1966). *Gen. comp. Endocr.* **6**, 183–194.
Chavin, W. and Kovacevic, A. (1961). *Gen. comp. Endocr.* **1**, 264–274.
Chester Jones I. (1957). "The Adrenal Cortex". Cambridge University Press, London.
Chester Jones I. (1963). *In:* "The Biology of Myxine" (Eds, Brodal, A. and Fänge, R.) pp. 488–502 Universitetsforlaget, Oslo, Norway.
Chester Jones, I. (1976). *J. Endocr.* **71**, 1P–31P.
Chester Jones, I. and Phillips, J. G. (1960). *Symp. Zool. Soc. Lond.* **1**, 17–32.
Chester Jones I., Henderson, I. W. and Mosley, W. (1964). *J. Endocr.* **30**, 155–156.
Chester Jones, I., Chan, D. K. O., Henderson, I. W., Mosley, W., Sandor, T., Vinson, G. P. and Whitehouse, B. (1965). *J. Endocr.* **33**, 319–320.
Chester Jones, I., Henderson, I. W., Chan, D. K. O. and Rankin, J. C. (1967). Proc. 2nd Int. Cong. Hormonal Steroids Milan 1966. I. C. S. **132** Excerpta Medica Amsterdam 136–145.
Chester Jones, I., Chan, D. K. O., Henderson, I. W. and Ball, J. N. (1969). *In:* "Fish Physiology". Vol. II, Chapter 6. (Eds. W. S. Hoar and D. J. Randall). pp. 323–376. Academic Press, New York and London.
Chester Jones, I., Ball, J. N., Henderson, I. W., Sandor, T. and Baker, B. I. (1974). *In:* "Chemical Zoology" Vol. VIII Chapter 14. (Eds. M. Florkin and B. T. Scheer). pp. 524–593. Academic Press, New York and London.
Chieffi, G. and Botte, V. (1963). *Nature, Lond.* **209**, 793–794.
Chieffi, G., Botte, V. and Visca, T. (1963). *Acta med. romana* **1**, 1–9.
Cohen, D. M. (1970). Proceedings of the California Academy of Sciences. **XXXVIII**, No. 17, 341–346.
Colombo, L., Bern, H. A., Pieprysk, J. and Johnson, D. W. (1973). *Gen. comp. Endocr.* **21** No. 1, 168–178.
Copp, D. H. and Ma, S. W. Y. (1978). *In:* "Comparative Endocrinology" (Eds. P. J. Gaillard and H. H. Boer) pp. 243–253. Elsevier North Holland Biomedical Press, Amsterdam, New York, Oxford.
Deane, H. W. and Rubin, B. L. (1965). *Archs Anat. mikrosk.* **54**, (1), 49–66.
De Robertis, E. and Sabatini, D. (1958). *J. biophys. biochem. Cytol.* **4**, 667–670.
de Smet, D. (1962). *Acta Zool. (Stockholm)* **43**, 201–219.
De Vlaming, V. L. and Sage, M. (1972). *Am. Zool.* **12**, 676, Abstract.
Diamare, V. (1935). *Atti. Accad. Sci. fis. mat. Napoli, Ser. 2* **20**, (6), 1.
Dingerkus, G., Mok, H. K. and Lagios, M. D. (1978). *Nature, Lond.* **276**, 261–262.
Dittus, P. (1936). *Z. wiss. Zool.* **147**, 459–512.
Dittus, P. (1941). *Z. wiss. Zool.* **154**, 40–124.
Dixit, R. K. and Agrawala, N. (1975). *Acta anat.* **93**, 344–350.
Dodd, J. M. and Kerr, T. (1963). *Symp. zool. Soc. Lond.* **9**, 5–27.
Donaldson, E. M. and McBride, J. R. (1967). *Gen. comp. Endocr.* **9**, 93–101.
Erickson, J. G. (1967). *Physiol. Zool.* **40**, 40–48.
Fagerlund, J. H. M. (1967). *Gen. comp. Endocr.* **8**, 197–207.
Fancello, O. (1937). *Pubbl. Staz. zool. Napoli*, **16**, 80.
Fontaine, M. (1964). *C. R. Acad. Sci. (Paris)* **259**, 875–878.
Fontaine, M. and Hatey, J. (1953). *Physiol. Comp. Oecol.* **3**, 37.
Fontaine, M. and Olivereau, M. (1957). *J. Physiol. Paris* **49**, 147–176.
Fraser, A. H. H. (1929). *Quart. J. micr. Sci.* **73**, 121p.
Fraser, E. A. (1950). *Quart. Rev. Biol.* **25**, 159 p.
Fryer, J. N. and Peter, R. E. (1977a, b, c). *Gen. comp. Endocr.* **33**, 196–225.
Fujita, H., Machino, M. and Tokura, T. (1963). *Arch. Histol. (Jap.)* **24**, 77–89.

Gardiner, B. G., Janvier, P., Patterson, C., Furey, P. L., Greenwood, P. H., Miles, R. S. and Jefferies, R. P. S. (1979). *Nature, Lond.* **227,** 175–176.

Garrett, F. S. (1942). *J. Morphol.* **70,** 41–67.

Gaskell, J. F. (1912). *J. Physiol.,* **44,** 59.

Gérard, P. (1951). *Arch. Biol., Paris,* **62,** 371–378.

Giacomelli, F., Wiener, J. and Spiro, D. (1965). *J. Cell Biol.* **26,** 499–519.

Giacomini, E. (1902). *Monitore Zoologico Italiano* **13,** 183–189.

Giacomini, E. (1912). *Mem. R. Accad. Bologner,* (Sezioni delle Scienze Naturali). Serie VI. Vol. **9,** 111.

Giacomini,. E. (1920). *R. C. Accad. Bologner (Classe di Scienze fisiche, Nouva Serie)* **24,** 129.

Giacomini, E. (1922). *Arch. Ital. Anat. Embriol* **18** 548–565.

Goswami, S. V. and Sundararaj, B. I. (1974). *Gen. comp. Endocr.* **23,** 282–285.

Greenwood, P. H. (1975). "A History of Fishes". (3rd Ed.). Ernest Benn Ltd., London.

Greenwood, P. H., Rosen, D. E., Weitzmann, S. H. and Myers, G. S. (1966). *Bull. Am. Mus. of Nat. Hist.* **131,** 341–455.

Greenwood, P. H., Miles, R. S. and Patterson, C. (eds.). (1966) *J. Linn. Soc. (Zool),* **53,** Suppl. 1. Academic Press, New York.

Greenwood, P. H., Miles, R. S. and Patterson, C. (1973). "Interrelationships of Fishes. Zoological Journal of The Linnean Society, Vol. 53, Suppl. 1. (eds. Greenwood, P. H., Miles, R. S. & Patterson, C.). 536 pp.

Griffith, R. W., Mathews, M. B., Umminger, B. L., Grant, B. F., Pang, P. K. T., Thomson, K. S. and Pickford, G. E. (1975). *J. exp. Zool.,* **192,** 165–172.

Griffith, R. W., Umminger, B. L., Grant, F. B., Pang, P. K. T., Goldstein, L. and Pickford, G. E. (1976). *J. exp. Zool.* **196,** 371–380.

Halstead, L. B. (1978). *Nature, Lond.* **276,** 759–760.

Handin, R. I., Nandi, J. and Bern, H. A. (1964). *J. exp. Zool.* **157,** 339–344.

Hane, S. and Robertson, O. H. (1959). *Proc. Nat. Acad. Sci. U.S.A.* **45,** 886–893.

Hane, S., Robertson, O. H., Wexler, B. C. and Krupp, M. A. (1966). *Endocrinology* **78,** 791–800.

Hanke, W. (1978). *In:* "General, Comparative and Clinical Endocrinology of the Adrenal Cortex". (I. Chester Jones and I. W. Henderson, Eds.) pp. 419–497. Academic Press, London and New York.

Hanke, W. and Chester Jones, I. (1966). *Gen. comp. Endocr.* **7,** 166–178.

Hardisty, M. W. (1972). *In:* "The Biology of Lampreys" (Hardisty, M. W. and Potter, I. C. eds.), Vol. 2. Academic Press, London and New York.

Hays, V. J. (1914). *Anat Rec.* **8,** 451.

Hennig, W. (1966). "Phylogenetic Systematics". University of Illinois Press. 263 pp.

Hill, C. W. and Fromm, P. O. (1968). *Gen. comp. Endocr.* **11,** 69–77.

Holmes, R. L. and Ball, J. N. (1974). "The Pituitary Gland, A Comparative Account". Cambridge University Press 397 pp.

Holmes, W. N. and Phillips, J. G. (1976). *In:* "General, Comparative and Clinical Endocrinology of the Adrenal Cortex". (I. Chester Jones and I. W. Henderson, Eds.) Vol. 1. pp. 293–420. Academic Press, London and New York.

Honma, Y. (1960). *Annot. Zool. Jap.* **33,** 234–240.

Honma, Y. and Tamura, E. (1965). *Japan Soc. Sci. Fish.* **31,** 867.

Hooli, M. A. and Nadkarni, V. B. (1975). *Acta anat.* **93,** 367–375.

Hubbs, C. L. and Potter, I. C. (1971). *In:* "The Biology of Lampreys". Vol. 1. (M. W. Hardisty and I. C. Potter, eds.). pp. 1–65. Academic Press, London and New York.

Idelman, I. (1978). *In:* "General, Comparative and Clinical Endocrinology of the Adrenal Cortex". Vol. 2 (I. Chester Jones and I. W. Henderson, Eds.). pp. 1–200. Academic Press, London and New York.

Idler, D. R. (1973). *Am. Zool.* **13,** 881–884.

Idler, D. R. and Burton, M. P. M. (1976). *Comp. Biochem. Physiol.* **53A,** 73–77.

Idler, D. R. and Freeman, H. L. (1966). *J. Fish Res. Bd. Can.* **23,** 1249–1255.

Idler, D. R. and Truscott, B. (1963). *Can. J. Biochem. Physiol.* **41,** 875–878.

Idler, D. R. and Truscott, B. (1972). *In:* "Steroids in Nonmammalian Vertebrates". Chapter 4. (D. R. Idler, ed.), pp. 127–252. Academic Press, New York and London.

Idler, D. R., Ronald, A. P. and Schmidt, P. J. (1959). *Can. J. Biochem. Physiol.* **37,** 1227–1238.

Idler, D. R., Sangalang, G. B. and Weisbart, M. (1971). Proc. 3rd Int. Cong. Hormonal Steroids, Hamburg 1970 I.C.S. 219. (V. H. T. James and L. Martini, eds.) Excerpta Medica Foundation, Amsterdam.

Idler, D. R., Sangalang, G. B. and Truscott, B. (1972). *Gen. comp. Endocr., Suppl.* **3,** 238–244.

Janssens, P. A., Vinson, G. P., Chester Jones, I. and Mosley, W. (1965). *J. Endocr.* **32,** 373–382.

Jarvik, E. (1965). *Annls. Soc. Zool. Belg.* **94,** 11–95.

Katz, Y. and Eckstein, B. (1974). *Endocrinology* **95,** 963–967.

Kenyon, C. J. (1979). Unpublished observations.

Kerr, J. G. (1919). Vertebrata with the exception of Mammalia. Text-book of Embryology. Vol. 2., Macmillan and Co., Ltd., London.

Krauter, D. (1952). *Zool. Anz., Bd.* **148,** Heft 1/2, 23–30.

Krauter, D. (1958). *Roux' Archiv für Entwicklungsmechanik, Bd.* **150,** S. 607–637.

Kuntz, A. (1912). *Am. J. Anat.* **13,** 71–88.

Lagios, M. D. and Stasko-Concannon (1979). Personal communication.

Larsen, L. O. (1973). "Development in Adult, Freshwater River Lampreys and its Hormonal Control. Starvation, Sexual Maturation and Natural Death." Zoophysiological Laboratory A. University of Copenhagen, Denmark.

Leloup-Hatey, J. (1958). *C. R. Acad. Sci., Paris* **246,** 1088–1091.

Leloup-Hatey, J. (1964). *Ann. Inst. Oceanogr.* **42,** 221–338.

Leloup–Hatey, J. (1970). *Gen. comp. Endocr.* **15,** 388–397.

Liu, C. K. (1944). *Sinensia* **15,** 1–8.

Lofts, B. (1978). *In:* "General, Comparative and Clinical Endocrinology of the Adrenal Cortex". (I. Chester Jones and I. W. Henderson, Eds.) Vol. 2. Part I. pp. 292–369. Academic Press, London.

Lutz, P. L. and Robertson, J. D. (1971). *Biol. Bull.* **141,** 553–560.

McBride, J. R. and van Overbeeke, A. P. (1969). *J. Fish. Res. Bd. Can.* **20,** 2975–2985.

McFarland, W. N. and Munz, F. W. (1958). *Biol. Bull.* **114,** 348–356.

Macchi, I. A. and Rizzo, F. (1962). *Proc. Soc. exp. Biol. Med.* **110,** 433–437.

Mahon, E. H., Hoar, W. S. and Tabata, S. (1962). *Can. J. Zool.* **40,** 449–454.

Meats, M., Ingleton, P. M., Chester Jones, I., Garland, H. O. and Kenyon, C. J. (1979). *Gen. comp. Endocr.* **36,** 451–461.

Mellinger, J. C. A. (1970). *Ann. Univ. A.R.E.R.S.* **8,** 27–28.

Midgalski, E. C. and Fichter, G. S. (1977). "The Fresh and Salt Water Fish of the World." Mandarin Publishers Ltd. Hong Kong.

Mosley, W. (1978). Unpublished observations.

Nandi, J. (1962). *University of California Publications in Zoology,* **65,** No. 2, 129–212.

Nelson, J. S. (1976). "Fishes of the World." John Wiley and Sons, New York, London. 416 pp.
Ogawa, M. (1967). Z. Zellforsch. 81, 174–189.
Oguri, M. (1960a). Bull. Jap. Soc. Sci. Fish. 26, 443–447.
Oguri, M. (1960b). Bull Jap. Soc. Sci. Fish. 26, 448–451.
Oguri, M. (1960c). Bull. Jap. Soc. Sci. Fish. 26, 476–480.
Oguri, M. (1960d). Bull. Jap. Soc. Sci. Fish. 26, 481–485.
Oguri, M. (1960e). Bull. Jap. Soc. Sci. Fish. 26, 981–984.
Oguri, M. (1977). Bull. Jap. Soc. Sci. Fish. 43(7), 781–784.
Oguri, M. (1978). Bull. Jap. Soc. Sci. Fish. 44(7), 703–707.
Olivereau, M. (1960). Acta Endocr., Copnh. 33, 142–156.
Olivereau, M. (1962). Gen. comp. Endocr. 2, 565–573.
Olivereau, M. (1965). Gen. comp. Endocr. 5, 109–128.
Olivereau, M. (1966a). 2nd Symp. Internat. Biol. Quant., Helgoland. Helgol. wiss. Meeresunt. 14, 422–438.
Olivereau, M. (1966b). Ann. Endocr. Paris 27 (5), 549–560.
Olivereau, M. (1975). Gen. comp. Endocr. 27, 9–27.
Olivereau, M. and Ball, J. N. (1963). Compt. Rend. 256, 3766–3769.
Olivereau, M. and Fromentin, H. (1954). Annls. Endocr. 15, 805–826.
Pang, P. K. T., Pang, R. K. and Sawyer, W. H. (1974). Endocrinology 94, 548–555.
Pearce, R. B., Cronshaw, J. and Holmes, W. N. (1978). Cell Tiss. Res. 192, 363–379.
Peter, R. E. (1970). Gen. comp. Endocr. 15, 88–94.
Phillips, J. G. and Chester Jones, I. (1957). J. Endocr. 16, iii.
Phillips, J. G. and Mulrow, P. J. (1959). Proc. Soc. exp. Biol. Med. 101, 262–264.
Phillips, J. G., Chester Jones, I., Bellamy, D., Greep, R. O., Day, L. R. and Holmes, W. N. (1962). Endocrinology 71, No. 2, 329–331.
Pickford, G. E. (1953). Bull. Bingham Oceanog. Collection, 14, 46–68.
Pickford, G. E., Robertson, E. E. and Sawyer, M. H. (1965). Gen. comp. Endocr. 5, 160–180.
Pitotti, M. (1938). Pubbl. Staz. zool. Napoli 17, 22.
Poll, H. (1906). In: "Hertwig's Handbuch der vergleichenden und experimentellen Entwicklungslehre der Wirbeltiere," 3, 443–618. Jena: Fischer.
Ranzi, S. (1936). Rend. R. Accad. naz. Lincei cl. S. fis. nat e nat. Ser. 6 24, 528–530.
Rasquin, P. (1969). Copeia, 83–90.
Reinboth, R. (1970). Mem. Soc. Endocr. 18, 515–544.
Robertson, J. D. (1954). Biol. Rev. 32, 156–187.
Robertson, J. D. (1963). In: "Biology of Myxine" (A. Brodal and R. Fänge, eds). pp. 504–515. Universitetsforlaget, Oslo.
Robertson, O. H. and Wexler, B. C. (1959). Endocrinology 65, 225–238.
Robertson, O. H., Krupp, M. A., Favour, C. B., Hane, S. and Thomas, S. F. (1961). Endocrinology 68, 733.
Robertson, O. H., Hane, S., Wexler, B. C. and Rinfert, A. P. (1963). Gen. comp. Endocr. 3, 422–436.
Roscoe, M. J. (1976). Functional Morphology and Physiology of the Pituitary Complex and Interrenal Gland in Elasmobranch, Fishes. Ph.D. Thesis, University College of North Wales, Bangor.
Sandor, T., Fazekas, A. G. and Robinson, B. H. (1976). In: "General, Comparative and Clinical Endocrinology of the Adrenal Cortex". (I. Chester Jones and I. W. Henderson, eds.), Vol. 1, Chapter 2, pp. 25–125. Academic Press, London and New York.

Sangalang, G. B., Weisbart, M. and Idler, D. R. (1971). *J. Endocr.* **50,** 413–421.
Sathyansen, A. G. and Chary, C. S. (1962). *Sci. Cult. Calcutta,* **28**(2), 81–82.
Schmidt, P. J. and Idler, D. R. (1962). *Gen. comp. Endocr.* **2,** 204–214.
Seiler, K., Seiler, R. and Sterba, G. (1970). *Acta Biol. Med. Ger.* **24,** 553–554.
Seiler, K., Seiler, R. and Hoheisel, G. (1973). *Gegenbaurs. morph. Jahrb. Leipzig,* **119,** 6, 823–856.
Sharma, S. (1968). *Nature, Lond.* **217,** (5123), 85.
Shrivastava, R. K. and Thakur, D. P. (1974). *Acta histochem. Bd.* **48,** S. 42–50.
Stannius, H. (1839a). Symbolae ad Anatomiam Piscium Concerning a singular organ inside the abdominal cavity of certain fishes. Literis Adlesianis. Rostock 36–40.
Stannius, H. (1839b). *Arch. Anat. Physiol.* pp. 97–101.
Stannius, H. (1846). Lehrbuch des vergleichenden Anatomie der Wirbelthiere, p. 118. Berlin.
Stannius, H. (1854). Handbuch der Anatomie der Wirbelthiere. Erste Buch Die Fische. pp. 254–267. Berlin. Veit and Comp.
Sterba, G. (1955). *Zool. Anzeig.* **155,** 151–168.
Sterba, G. (1969). *Gen. comp. Endocr. Suppl.* **2.** 500–509.
Sundararaj, B. I. and Goswami, S. V. (1968). *J. exp. Zool.* **168,** 85–104.
Sundararaj, B. I. and Goswami, S. V. (1971). *In:* "Hormonal Steroids". *Proc. 3rd. Int. Congr. Horm. Steroids Excerpta Medica Foundation* (V. H. T. James and L. Martin, Eds.), Excerpta Medica, Amsterdam. *I.C.S.* **219,** 966–975.
Sundararaj, B. I. and Goswami, S. V. (1977). *Gen. comp. Endocr.* **32,** 17–28.
Tailo, L. B. H. and Whiting, H. P. (1965). *Nature, Lond.* **206,** 148–150.
Taylor, J. D., Honn, K. V. and Chavin, W. (1975). *Gen. comp. Endocr.* **27,** 358–370.
Truscott, B., Idler, D. R., Sundararaj, B. I. and Goswami, S. V. (1978). *Gen. comp. Endocr.* **34,** 149–157.
Tsepelovan, P. G. and Rusakov, Yu. I. (1976). *Dokl. Akad. Nauk SSSR* **XII,** 77–80.
van Overbeeke, A. P. (1960). "Histological studies on the interrenal and the phaeochromic tissue in Teleostei." Van Munster's Drukkerijen, N. V., Amsterdam. 102 pp.
van Overbeeke, A. P. and Ahsan, P. P. (1966). *Can. J. Zool.* **44,** 969–979.
Vincent, S. (1897). *Anat. Anz.* **13,** 39.
Wakisaka, N. (1964). *Bull. Yamaguchi med. Sch.* **10,** 113–122.
Weisbart, M. (1975). *Gen. comp. Endocr.* **26,** 368–373.
Weisbart, M. and Idler, D. R. (1970). *J. Endocr.* **46,** 29–43.
Weisbart, M. and Youson, J. H. (1975). *Gen. comp. Endocr.* **27,** 517–526.
Weisbart, M., Youson, J. H. and Weibe, J. P. (1978). *Gen. comp. Endocr.* **34**(1), 26–37.
Wingfield, F. C. and Grimm, A. S. (1977). *Gen. comp. Endocr.* **31,** 1–11.
Witschi, E. (1951). *Recent Prog. Horm. Res.* **6,** 1–27.
Witschi, E. (1953). *J. clin. Endocr.* **13,** 316–329.
Witschi, E. (1956). "Development of Vertebrates." Philadelphia and London: W. B. Saunders Company.
Witschi, E., Bruner, J. A. and Segal, S. J. (1953). *Anat. Rec.* **115,** 381.
Woodhead, A. D. and Woodhead, P. M. J. (1965). *Int. Comm. N.W. Atl. Fish. Spec. Publ. No.* 6, 691–715.
Yadav, B. N., Singh, B. R. and Munshii, J. S. D. (1970). *Mikroskopie* **26,** S. 41–49.
Yamamoto, K. and Onozato, H. (1965). *Annot. Zool. Jap.* **38,** 140–150.
Yamamura, Y. Y. (1952). *Bull. Japan. Soc. Sci. Fish.* **17,** 337–341.
Yamamura, Y. Y. (1956) *Bull. Tohoku Reg. Fish. Res. Lab.* **6,** 1–70.
Yaron, Z. (1970). *Gen. comp. Endocr.* **14,** 542–550.

Youson, J. H. (1972). *Gen. comp. Endocr.* **19,** 56–58.
Youson, J. H. (1973a). *Am. J. Anat.* **138,** No. 2, 235–251.
Youson, J. H. (1973b). *Can. J. Zool.* **51,** 796–799.
Youson, J. H. (1975). *Acta Zool. (Stockh.)* **56,** 219–223.
Youson, J. H. (1976). *Can. J. Zool.* **54,** 843–851.
Youson, J. H. and Butler, D. G. (1976a). *Am. J. Anat.* **145,** 207–224.
Youson, J. H. and Butler, D. G. (1976b). *Acta zool. (Stockh.)* **57,** 217–230.
Youson, J. H. and McMillan, D. B. (1971). *Anat. Rec.* **170,** 401–412.
Youson, J. H., Butler, D. G. and Chan, A. T. C. (1976). *Gen. comp. Endocr.* **29,** 198–211.

PART 2 PHYSIOLOGY

1. Introduction

The heterogeneous nature of the group fishes has been emphasized in the chapter by Sandor *et al.* (1976) and in part I of this chapter, dealing respectively with biosynthesis and distribution of adrenocorticosteroids and the morphology of the piscine adrenocortical homologue. Three major groups of fish will be considered in this part: the Agnatha (Cyclostomata), the Elasmobranchiomorphi and the Teleostomi. The taxonomy of each of these groups is complex and controversial (see pp. 397–406). For the purposes of this discussion, largely because knowledge of interrenal physiology of fish is still in its infancy, statements can only be made about individual genera and species and a phylogenetic trend of corticosteroid function cannot be deciphered; only a few species have been studied, and these have been examined largely because of their ready availability or their ability to withstand laboratory study. (For convenience in this Part "interrenal", "adrenocortical" and equivalent terms are arbitarily interchanged).

Adrenocorticosteroids have been implicated in a variety of homeostatic processes including osmoregulation, carbohydrate and intermediary metabolism and reproduction. Target organs thus include gills, kidney, liver, gastrointestine (gut) and gonads, although the integument, spleen, and the central nervous system are all affected in some species by exogenous corticosteroids.

The predominant secretion from the interrenal is cortisol though other corticosteroids occur (Sandor *et al.*, 1976; Colombo *et al.*, 1977). The capacity of the anterior kidney of a teleost, *Fundulus heteroclitus*, to make aldosterone was indicated, in the first place, by Phillips and Mulrow (1959) and, secondly, was shown in salmon plasma (Phillips *et al.*, 1959; Chester Jones and Phillips, 1960). Nevertheless the occurrence of aldosterone in teleost fish was not readily accepted and the firm statement that this steroid is one of the usual teleost interrenal products still cannot be made. It has been demonstrated, with rigorous techniques, in the herring (Idler and Truscott, 1972) and the whitefish (Whitehouse and Vinson, 1975). We do not know if this compound plays an essential role in the physiology of fish, though it is effective experimentally. On the other hand, Dr. G. Hargreaves (in our laboratories) examined the plasma of 12 goldfish of the same batch. The plasma of 6 had aldosterone, rigorously characterized, but in the rest it was not found. In contrast, Amphibia placed in appropriate environments produce relatively large amounts of aldosterone (see Hanke, 1978) whilst fish in similar circumstances do not. Here cortisol and prolactin (paralactin) appear to fulfil the requirements for osmoregulation, provided, of course, all other systems are normally active.

It can be generally stated that teleost head kidneys give 17α-hydroxylated corticosteroids such as cortisol and in this respect do not differ from tetrapod adrenocortical tissue. However there is apparently limited 18-oxygenation (Arai and Tamaoki, 1967; Truscott and Idler, 1968; Sandor et al., 1970, 1976; Weisbart and Idler, 1971; Colombo et al., 1972; Whitehouse and Vinson, 1975). The values obtained may be found also in Leloup-Hatey (1968, 1974), Hawkins et al. (1970), Liversage et al. (1971) and Sandor et al. (1976).

2.   Agnatha (Cyclostomata)

The jawless vertebrates comprise two disparate taxonomic categories—the Myxinidae and the Petromyzonidae—which are generally considered to be widely separate phylogenetically. There is very limited information on the physiology of interrenal secretions, and fewer than ten studies have been specifically directed towards the functional rôle of these hormones. Indeed there is still disagreement as to the sources and types of steroids (see Part I of this Chapter; Sandor et al., 1976) and their sites of action.

A.   MYXINIDAE

The Myxinidae typically maintain their plasma osmolarity at about that of the environmental sea water (Robertson, 1960; Bellamy and Chester Jones, 1961) and it has been suggested that the intra-cellular environment of *Myxine* more closely resembles that of marine invertebrates than that of typical vertebrates. The kidney may contribute towards ionic regulation (Munz and McFarland, 1964; Morris, 1965) although there may be little or no net water or sodium reabsorption (Munz and McFarland, 1964; Rall and Burger, 1967). Functional morphology of the myxinoid kidney is obscure and possible sites of hormonal action not readily delineated; the pronephros comprises large fused Malpighian corpuscles and a very short tubular portion which may not lead into the urinary excretory duct (Holmgren, 1950; Fangë, 1963). The functional mesonephric portion presents large filtering Malpighian corpuscles arranged segmentally along the archinephric ducts and the latter may be sites for reabsorption of macromolecules from the ultrafiltrate (Ericsson, 1967; Ericsson and Seljelid, 1968). Extrarenal sites of osmoregulation include the liver (Rall and Burger, 1967) and the skin by virtue of its copious mucus production (Munz and McFarland, 1964); the gut does not seem to participate in osmoregulation to a significant extent (Morris, 1960, 1965). Interrenal (or indeed any) hormonal involvement at these sites either as regards osmoregulation or intermediary metabolism is at present doubtful. Aldosterone, DOC and hagfish pituitary extracts influenced body electrolyte content and distribution when injected into *Myxine* held in 60% sea water (Chester Jones *et al.*, 1962), but the effects were relatively undramatic and unquestionably pharmacological, and as pointed out (Chester Jones, 1963) "the degree to which corticosteroids play a role in the normal physiology of hagfish is yet to be determined" (Table I).

In summary the hagfishes present many fascinating problems regarding interrenal physiology; the apparent presence of ACTH-like cells in the pituitary (Fernholm and Olsson, 1969), the lack of definitive demonstration of adrenocortical-type cells (see Part 1), the spasmodic occurrence of corticosteroids (Sandor *et al.*, 1976) as well as the atypical responses to manipulations such as hepatectomy (Inui and Gorbman, 1978) and hypophysectomy (Gorbman and Tsuneki, 1975) must all point to more careful and critical analyses of interrenal physiology of these, according to one view, the progenitorial groups of craniates.

B. PETROMYZONIDAE

The general biology, including the endocrinology of petromyzontian cyclostomes has received greater investigative attention than the Myxinidae. The physiology and biochemistry of lampreys is sufficiently similar to those of actinopterygian fishes for many to consider this group to resemble fairly closely the stem types for gnathostomatous vertebrates.

Perhaps largely because the adrenocortical homologue of lampreys has proven difficult to identify definitively (see Part 1 of this chapter),

TABLE I

Possible involvement of adrenocortical hormones in agnathan osmoregulation (Summarized from Chester Jones et al. (1962), Bentley and Follett (1962, 1963), Chester Jones (1963) and Morris (1972)).

Species	Substance injected	Effects
Myxine glutinosa in 60% sea water (*circa* 600 mOsm/L)		
	Aldosterone	Reduced muscle Na content (serum Na not measured)
	Deoxycorticosterone	Reduced muscle Na content; no effect on serum Na or K
	Cortisone	No effect
	Mammalian ACTH	Reduced Na content; reduced serum K concentration
Lampetra fluviatilis in fresh water (*circa* 15 mOsm/L)		
	Aldosterone	Reduced overall rate of sodium loss; No effect on renal sodium loss; Reduced urine flow; increased total body sodium
	Cortisol	Reduced renal potassium loss; No effect on urine flow or total sodium loss
	SC11927 (an aldosterone antagonist)	Increased total sodium loss
	Mammalian ACTH	Reduced overall sodium loss —actually inducing net positive Na balance over 8 h experiment; no renal effect

CONCLUSIONS:

In Petromyzonidae: renal water but not sodium affected? branchial sodium balance influenced, but whether outflux or influx not known.

In Myxinidae: actions of fluxes of water and electrolytes, either renal or branchial not demonstrated. adrenocorticosteroids influence fluid (osmotic?) distribution within internal fluid compartments by acting on overall cell permeability?

the function of adrenocorticosteroids has been an area of endocrinology that has lagged behind other aspects of hormonal physiology, such as the thyroidal, pancreatic and hypophysial (see Falkmer *et al.*, 1974). Unlike their myxinoid counterparts, lampreys, at various stages of their life cycle, occur in both fresh water and sea water (see Hardisty and Potter, 1971). The well developed osmoregulatory mechanisms maintain plasma hyperosmotic to the environment in fresh water and hypo-osmotic in the sea; the processes are broadly similar to those of euryhaline teleosts under similar circumstances (Wikgren, 1953; Morris, 1960, 1972).

Aldosterone, but not cortisol, decreased overall rates of sodium loss from lampreys in fresh water (Bentley and Follett, 1962) while an aldosterone antagonist increased it (Bentley and Follett, 1963). An extrarenal effect for aldosterone was suggested as the renal responses were slight, although renal water excretory mechanisms may have been affected (Table I). An hypophysial–interrenal axis to maintain osmoregulatory homeostasis has been indicated, in that mammalian ACTH, like aldosterone, decreased overall sodium loss from the fresh water lamprey. Cytophysiological data obtained after metyrapone, aldactone and ACTH injections (Molnár and Szabó, 1968; Larsen and Rothwell, 1972; Youson, 1973), hypophysectomy (Larsen, 1969; Hardisty, 1972a, b) or after variations in environmental salinity (McKeown and Hazlett, 1975) point to, but do not clearly delineate, a role for adrenocorticosteroids in petromyzonidae osmoregulation and intermediary metabolism (Bentley and Follett, 1965).

3. Chondrichthyes (Elasmobranchiomorphi)

This section deals almost exclusively with representatives of the Elasmobranchii, although some, rather limited, reference is made to members of the Holocephali, in particular the ratfish, *Hydrolagus colliei*.

Elasmobranch fishes are predominantly, but not exclusively, marine, and maintain their plasma at, or slightly above, the osmolarity of the environmental sea water as a result of the retention of urea and trimethylamine oxide. Unlike marine teleosts, they do not to any great extent actively imbibe the environmental sea water for osmoregulatory purposes. Water and electrolyte homeostasis is maintained by the conjoint actions of the gills, kidney and rectal gland. The general processes and mechanisms of osmoregulation have been rehearsed at length and it is beyond the terms of reference of this chapter to reconsider them (Smith, 1931, 1936; Burger, 1967; Payan and Maetz, 1970, 1971; Payan et al., 1973). Endocrine, more especially adrenocorticosteroidal, impingement on osmoregulation (or indeed on all aspects of elasmobranch physiology) is poorly understood and few definitive remarks can be made in this area.

Qualitatively, the interrenal steroids differ in certain respects from the usual vertebrate pattern; the spasmodic occurrence of 1-α-hydroxy-corticosterone as a minor or major product, alongside corticosterone, cortisone and DOC, raises many fundamental questions about adrenocortical function (see Sandor et al., 1976). The apparent absence of a renin-angiotensin system (Nishimura et al., 1970), and the relative ineffectiveness of hypophysectomy on interrenal structure (see part 1 of this chapter) are suggestive of perhaps unique adrenocortical controlling mechanisms.

It is equivocal whether or not the adrenocortical homologue is essential for life in elasmobranchs as sham operation in the hands of all investigators has been shown to have effects not unlike those of adrenalectomy itself. Thus Hartman et al. (1944) in the skate, Raia erinacea, were uncertain as to the effects of this operation compared with sham operation, and they drew attention to earlier studies which encountered similar difficulties. Hartman et al. (1944) concluded that "there was no evidence that the interrenal plays any role in electrolyte metabolism. Although the blood sugar changes indicated no interrenal responsibility, the fall in liver glycogen showed that the interrenal is necessary for normal carbohydrate metabolism." Hypernatraemia, hyperkalaemia, hypermagnesaemia and hyperchloraemia alongside reduced plasma glucose and urea concentrations and a lowered haematocrit were suggested but statistical significance was not achieved in any of these when the effects of sham operation were considered. Chester Jones (1957) in a re-examination of the data suggested that perhaps hyperkalaemia, a usual sequela of adrenal insufficiency in mammals at least, was present. Idler and Szeplaki (1968) encountered similar difficulties in Raia radiata in that sham-operated skates gave distinct responses, some of which were similar to adrenalectomy itself (Table II). The most major difference noted by the latter authors was a transient hypercalcaemia between 13 and 23 days after interrenalectomy and there was a suggestion of an impaired regulation in animals placed in diluted sea water; the reduction in plasma urea concentration was greatest in interrenalectomized skate, but the effect was not statistically significant. De Vlaming et al. (1975), investigating the euryhaline elasmobranch, Dasyatis sabina, observed reductions in plasma sodium, calcium, chloride and urea

TABLE II

Plasma composition of *Raia radiata* and *Dasyatis sabina* six days after removal of adrenocortical homologue (from Idler and Szeplaki, 1968, and De Vlaming *et al.*, 1975). Means ± S.E.

	Osmolarity	Sodium	Potassium	Calcium	Magnesium	Chloride	Urea
	mOsm/L			(m Eq/litre)			(mM/L)
Raia radiata (Environmental osmolarity: 920 mOsm/L; n = 6 except where indicated ( )) Pre-operative values of sham operated animals	958 ± 5	269 ± 5	3.7 ± 0.2	7.5 ± 0.6 (5)	2.0 ± 0.4 (5)	294 ± 6	416 (2.5 ± 0.1 g%)
Sham-operated animals	957 ± 6	302 ± 5	3.5 ± 0.1	9.7 ± 0.5	3.1 ± 0.7 (4)	268 ± 5	366 (2.2 ± 0.1 g%)
Pre-operative values of adrenalectomized animals	960 ± 5	268 ± 9	3.5 ± 0.2	7.9 ± 0.3 (3)	1.9 ± 0.2 (4)	295 ± 17	416 (2.5 ± 0.0 g%)
Adrenalectomized	942 ± 14	290 ± 6	3.2 ± 0.3	9.9 ± 0.6	3.2 ± 0.5	255 ± 15	366 (2.2 ± 0.1 g%)
Dasyatis sabina (Environmental osmolarity: 813 mOsm/L) Intact (n = 4)	848 ± 2	257 ± 2	—	4.5 ± 0.2[a]	—	299 ± 5	284 ± 2
Adrenalectomized (n = 4)	815 ± 12	241 ± 1	—	2.8 ± 0.1[a]	—	258 ± 3	266 ± 1

[a] These units, stated to be mEq/L, are probably mM/L.

concentrations after interrenalectomy, and there was an overall reduction in plasma osmolarity in animals held in an environment of 813 mOsm/L; plasma potassium concentrations were not given. These changes reflect a comparison with intact rather than sham-operated animals, and in the light of previously cited studies, caution is necessary before concluding a definitive role for adrenal steroids in osmoregulation of *Dasyatis* (Table II).

Payan and Maetz (1970, 1971) and Payan *et al.* (1973) investigated the effects of hypophysectomy on water and electrolyte metabolism of a number of species of skates, rays and dogfish. There was considerable species variation in response. Very broadly, in some species such as *Scyliorhinus* branchial water permeability and urine flow were diminished by hypophysectomy, while in others such as *Raia* sp. the effects were slight. In those animals in which a lesion could be detected, mammalian ACTH and in some cases prolactin restored normality. In a very limited study, sodium outflux was found to be more than halved by adrenalectomy in *Raia radiata*, while in *R. erinacea* adrenalectomy had no such effect.

Exogenous cortisol or DOC produced no changes in plasma sodium and potassium concentrations of *Hemiscyllium plagiosum*, although the secretory rates of fluid from the rectal gland were diminished (Chan *et al.*, 1967a). In equivalent studies on *Raia ocellata* Holt and Idler (1975) have found reduced rates of secretion and solute excretion by the rectal gland after adrenalectomy, and these changes were returned towards normal by both 1-α-hydroxy-corticosterone and corticosterone (Fig. 1). It is of interest that levels of the enzyme, sodium–potassium-activated adenosine triphosphatase in the rectal gland, are unaffected by sham operation or adrenalectomy (Idler and Kane, 1976); the authors suggest that adrenocortical control of rectal gland function involves a permeability or haemodynamic action rather than an enzyme induction type of effect.

As implied from the flux studies of Payan and Maetz (1970) an hypophysial–adrenocortical axis is involved in regulating fluid and electrolyte balance in elasmobranchs. de Roos and de Roos (1967) demonstrated corticotrophic activity in the pituitary glands of dogfish, skate and ratfish. Hypophysectomy resulted in interrenal atrophy in *Torpedo* (Dittus, 1941) although not in *Scyliorhinus*. In the *Torpedo* studies, mammalian ACTH restored the normal cytological picture of the cortical cells in hypophysectomized forms. Early *in vitro* studies with ACTH and adrenocortical tissue were conflicting. Macchi and Rizzo (1962) described a corticotrophin-stimulation of steroidogenic activity in isolated skate adrenocortical tissue. In the same

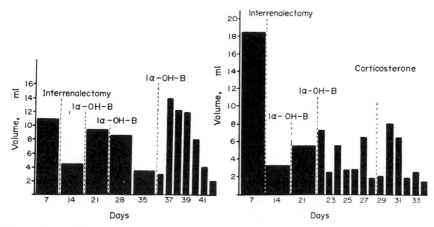

Fig. 1. The effects of interrenalectomy, 1α-hydroxycorticosterone (1α-OH-B) and corticosterone (B) on daily fluid output by the rectal gland of 2 skate, *Raia ocellata*. Rectal gland fluid was collected on the day prior to interrenalectomy, 7 days after operation and again 7 days after injection of steroids. Doses of steroid: 1α-OH-B, 3.5 mg/fish; B, 2 mg/fish. (From Holt and Idler, 1975.)

year, however, Bern *et al.* (1962) could find no definite evidence for ACTH increasing aldosterone, cortisol or corticosterone production by interrenal tissue incubates from dogfish, skate or ratfish. More recently, Klesch and Sage (1973, 1975) showed that addition of whole pituitary homogenates to interrenal incubates of *Dasyatis* significantly increased the production and release of 1-α-hydroxy-corticosterone. Mammalian ACTH similarly stimulated steroidogenesis, and hypophysectomy (as in *Torpedo* (Dittus, 1941)) resulted in interrenal atrophy. (See Roscoe, 1975 for full discussion of these data.) Evidence for involvement of factors other than corticotrophin in the control of the chondrichthyean interrenal is sparse. Renin appears to be absent from the kidneys of those elasmobranchs examined (Capreol and Sutherland, 1968; Nishimura *et al.*, 1970). In the holocephalan ratfish, however, the presence of both juxtaglomerular-like granular epithelioid cells and renin-like activity have been described (Nishimura *et al.*, 1973). Both prolactin and ACTH restored branchial water permeability to normal after hypophysectomy in dogfish, skate and *Torpedo* (Payan and Maetz, 1971), but its role in elasmobranch osmoregulation is uncertain (see Ensor and Ball, 1972; Ensor 1979).

It is perhaps surprising, in view of the key roles that urea and trimethylamine oxide play in osmoregulation of elasmobranchs, that our knowledge of endocrine factors regulating their synthesis is, to say the least, fragmentary. Protein, carbohydrate, fat and intermediary metabolism generally have received scant attention from adrenocortical physiologists. Hartman *et al.* (1944) in their pioneering studies observed a reduced liver glycogen store after interrenalectomy, an effect that was apparent when comparison was made either with intact or sham-operated animals. In similar studies Idler *et al.* (1969) could find no effect on liver glycogen in skates following adrenalectomy. Exogenous administration of various steroids to skate produced few changes in glycogen stores, although brief and transient hyperglycaemia may have occurred (Wright, 1961). In a very extensive study of a gamut of hormones, Patent (1970) was unable to conclude firmly that corticoids had major physiological rôles to play in hepatic metabolism of carbohydrate, although cortisol and corticosterone depleted hepatic lipid stores. Patent (1970) suggests that the abundant hepatic lipid reserves are of greater significance than the glycogen stores which characteristically are relatively sparse and highly variable.

In conclusion there are many and various lines of evidence that tentatively point to interrenal control of fluid, electrolyte and intermediary metabolism in certain Elasmobranchiomorphs. Definitive sites of action for the steroids await careful study, and it is worth emphasizing again that the failure of some species to respond to hypophysectomy or adrenalectomy in the generally accepted vertebrate way may reflect hitherto undescribed modes of metabolism in this, one of the most ancient vertebrate lines.

The biology and metabolism of elasmobranchs have received extensive attention and many of the basic processes are becoming understood. Endocrinological approaches have however remained within the confines of standard laboratory animals (or simply of the rodent!); novel and enlightened hypotheses as to the rôles of hormones are required so that innovative avenues of investigation may emerge.

4. Osteichthyes (Teleostomi)

The bony fish are the largest single category of Vertebrata. They display an immense diversity of forms and function, and taxonomic interrelationships remain controversial; it is hardly surprising therefore that the physiological adaptations are such that only tentative generalizations can be applied to the group as a whole. Rather it seems that families and genera retain quite unique mechanisms superimposed upon an underlying vertebrate pattern of homeostasis.

Most of our current knowledge of the physiology of the adrenal cortex of teleostomes comes from studies on a few representatives of the order Teleostei. The processes whereby these fish maintain body fluid homeostasis in environments hyper-osmotic (marine species) or hypo-osmotic (fresh water species) to their blood plasmas have fascinated students of osmoregulation for more than a century. In many respects non-osmoregulatory processes (growth, reproduction, intermediary metabolism) have received much less attention, and the balance of the present chapter reflects such emphasis.

A. TELEOSTEI

1. *Osmoregulation*

The general mechanisms of renal and extrarenal regulation of body fluid volume and composition have been extensively reviewed (Conte, 1969; Hickman and Trump, 1969; Henderson and Chester Jones, 1972; Johnson, 1973; Maetz, 1976). There is overall agreement as to the mechanisms of changing water permeability, sodium and other electrolyte exchanges with the environment, and the relative import-ance of renal, gastro-intestinal and branchial contributions. The hormonal bases for certain of these responses display intriguing variations and a number of articles have attempted to relate endocrine activities to water and electrolyte metabolism (Chester Jones *et al.* 1959, 1969, 1972; Maetz, 1968; Ball and Olivereau, 1970; Henderson *et al.*, 1970; Mayer, 1970; Bentley, 1971; Butler, 1973; Chester Jones, 1974).

A variety of approaches has clearly indicated rôles for interrenal secretions in osmoregulation, but their exact sites and mechanisms of action and their interactions with other hormones require further elucidation. The earliest attempts to uncover adrenocortical regula-tion of body fluid homeostasis relied on observing the effects of exogenous, frequently heterologous, corticosteroids; the results were somewhat equivocal. Thus Lockley (1957) and Edelman *et al.* (1960) concluded that corticosteroids played no key role in osmoregulatory adaptation, while others (e.g. Holmes and Butler, 1963; Olivereau and Chartier-Baraduc, 1965; Chan *et al.*, 1967b) found that the effects were deleterious and somewhat enigmatic if corticosteroids were to assist body fluid homeostasis. In retrospect these early studies revealed the need for experimental approaches that took cognisance of endogenous levels of steroids as well as possible changes in metabolic clearance and sites of action. Table III gives a selection of effects obtained in intact teleosts after injection of adrenocortical steroids.

TABLE III

Some selected effects of exogenous corticosteroids administered to intact teleost fishes.

Species	Steroid dose (mg/kg body weight)	Effects–conclusions	Reference
Anguilla anguilla	Cortisol: c 10 mg i.p./kg body weight for up to 32 days.	Regressive changes in adrenocortical cells; temporary increase in liver glycogen and then decrease and accumulation of hepatic lipid; stimulation of testis	Olivereau, 1966b
Carassius auratus	Cortisol: 100–200 mg i.p. for 4–6 days.	No effect on liver glycogen; decreased growth; increased potassium and nitrogen excretion and hepatic glutamic-pyruvic-transaminase. Cortisol promotes protein catabolism.	Storer, 1967
Salmo gairdneri	Cortisol; c 150 mg i.p. implant for 1–2 weeks.	No effect on plasma glucose; increased liver glycogen; increased nitrogen excretion in starved fish only.	Hill and From, 1968
Anguilla rostrata	Cortisol: 5 mg i.p. daily for 10 and 21 days.	Hyperglycaemia; increased liver and muscle glycogen. Cortisol gluconeogenic.	Butler, 1968.
Anguilla japonica	Cortisol (c 1 mg) Cortisol acetate (c 5 mg) Corticosterone (c 5 mg) Aldosterone (c 1 mg) DCA (c 10 mg). All i.p.	Cortisol increased *in vitro* intestinal sodium and water transport of fresh water eels. Other steroids without effect.	Hirano and Utida, 1968, 1971.
Anguilla rostrata	Cortisol: 0.5–4 mg. daily for 3–14 days i.m.	Increased branchial and intestinal sodium-potassium-activated adenosine triphosphatase (Na-K-ATPase) with no effect on magnesium activated enzyme in fresh water eels. Aided survival and adjustments of eels transferred to sea water. "Silvered" integument of "yellow" eels.	Epstein *et al.*, 1971.
Carassius auratus	Cortisol: 3 mg daily for 6 days i.p.	No effect on *in vitro* intestinal sodium or fluid transfer of saline adapted fish. Actions in hypophysectomized fish suggest metabolic effects on mucosal cells.	Ellory *et al.*, 1972.
Anguilla japonica	Cortisol: c 5 mg daily for 7 days.	Increased branchial Na-K-ATPase of fresh water eels. No effect on Mg-ATPase.	Kamiya, 1972

Species	Treatment	Results	Reference
Anguilla rostrata	Cortisol: 4 mg daily for 2 or 10–14 days i.m.	Increase branchial Na-K-ATPase after high dose regime. Cortisol aids adaptations to sea water, by affecting above enzyme and by other mechanisms (chloride cell organization)	Forrest et al., 1973a.
Salmo gairdneri	Cortisone, cortisol: 10 and 20 mg for 5 days i.p. respectively.	Cortisone without effect on hepatic glutamic-oxaloacetic transaminase (GOT); higher dose of cortisol increased GOT.	Freeman and Idler, 1973.
Salvelinus fontinalis	Cortisone (i.p. and implant 10 and 20 mg) Cortisol (ditto).	Both implanted steroids increased hepatic GPT. Cortisol increased body weight loss and haematocrit.	Freeman and Idler, 1973.
Carassius auratus	Cortisol, aldosterone: both 1 mg daily for 2 days i.p.	Cortisol reduced serum Na, Cl especially at higher temperatures; hyperglycaemia. Aldosterone without clear effect.	Umminger and Gist, 1973.
Anguilla japonica	Cortisol: 2 or 10 mg daily for 5 days i.m.	Hyperglycaemia; increased liver and muscle glycogen; increased hepatic GOT, fructose-1, 6-diphosphatase and phosphofructokinase. No effects on plasma amino nitrogen or on hepatic GPT. Cortisol gluconeogenic, actions on amino acid metabolism equivocal.	Inui and Yokote, 1975.
Carassius auratus	Cortisol, cortisone, Aldosterone, 11-deoxy-corticosterone: all 1 mg i.p. 4, 24 and 48 h before study.	In hypophysectomized fish, only cortisol restored depressed water and sodium transport across isolated intestine.	Porthé-Nibelle and Lahlou, 1975.
Galichthys mirabilis	Cortisol: 10 mg daily for 3 days i.p.	Increased osmotic permeability, sodium and chloride transport across isolated urinary bladders of sea water fish.	Doneen, 1976.
Anguilla anguilla	Cortisol: 0.5 and 4 mg daily for 3 days i.p.	No effect on branchial Na-K-ATPase in fresh water eels. Slight, considered "equivocal" stimulation of enzyme in sea water eels. Cortisol is only active in absence of the pituitary.	Scheer and Langford, 1976.
Anguilla anguilla	Cortisol: 5 mg daily for 1, 4 and 14 days i.p.	After 4 and 14 days, no effect on haematocrit; decreased white blood cell count and lymphocyte numbers; increased neutrophilic granulocytes. Cortisol influences lymphocytes and neutrophils but not erythrocytes.	Johansson-Sjobeck et al., 1978.

Surgical interrenalectomy, technically difficult in teleostomes, is clearly a pre-requisite if classical endocrine approaches are to be applied. Only two species have thus far been surgically interrenalectomized successfully—the goldfish (Etoh and Egami, 1963) and the eel, *Anguilla* sp. (Chester Jones *et al.*, 1964). Only the latter species has been examined in any detail following interrenalectomy (Chan *et al.*, 1967b; Butler and Langford, 1967; Mayer *et al.*, 1967; Chan and Chester Jones, 1968). In fresh water interrenalectomy is accompanied by a weight increase of about 1% per day, reflecting net water retention; whether this results from impaired renal excretion and/or increased branchial influx is not definitely known (Chan *et al.*, 1969). In addition there is haemodilution with reduced serum concentrations of sodium, magnesium and calcium. Tissue electrolytes also decline (Table IV). Interrenalectomized sea water adapted eels do not survive well, but can be successfully maintained in one third sea water prior to transfer to full strength media (Mayer *et al.*, 1967). Interrenalectomized sea water eels are characterized by gradual weight loss, indicative of net water efflux, decreased intra- and extra-cellular fluid volumes and increased concentrations of sodium, magnesium and calcium in blood. Tissue water content declines, while sodium and magnesium levels increase. Table IV summarizes these effects of adrenalectomy, and full descriptions may be found in Chan *et al.* (1967b, 1969), Chester Jones *et al.*, 1969, 1972, 1974), Butler (1973), and Pang (1973).

Removal of the adrenal cortex, therefore, at least in the eel, profoundly affects the maintenance of water and electrolyte homeostasis. In teleosts in general the gills, the kidney, the urinary bladder and the alimentary tract act in concert to achieve osmotic and ionic balance between the animal and its environment. The interactions of the interrenal secretions with each of the effector systems will be considered separately.

(a) The gills

Interrenalectomy of fresh water adapted eels reduces net extrarenal (branchial) uptake of sodium while reducing the net loss of potassium (Henderson and Chester Jones, 1967; Chan *et al.*, 1969). "Physiological" doses of cortisol, as well as aldosterone (which must be considered heterologous in this species), restore sodium uptakes and potassium losses to, or beyond, control levels. In the goldfish, cortisol injections also accelerate branchial water influx (Lahlou and Giordan, 1970). Very high doses of cortisol and other corticosteroids induce net extrarenal sodium loss in fresh water eels and trout (Holmes, 1959;

TABLE IV

Effects of surgical adrenalectomy on serum and tissue electrolytes and body weight of female eels, *Anguilla anguilla* adapted to fresh water and sea water environments. Means ± S.E. (Chan *et al.*, 1967b).

EXPERIMENTAL	SERUM				(n)	H_2O (%)	MUSCLE			BODY WEIGHT (% change)	(n)
	Na (mmol/L)	K (mmol/L)	Ca (mmol/L)	Mg (mmol/L)			Na (mmol/L)	K (mmol/L)	Mg (mmol/L)		
FRESHWATER Silver Eels											
Controls	150.1 ± 1.2	1.75 ± 0.15	2.29 ± 0.05	2.13 ± 0.22	25	76.43 ± 0.61	43.22 ± 2.11	101.70 ± 3.81	20.95 ± 0.92	+1.67 ± 0.61	6
Adrenalectomized (3 weeks)	122.6 ± .2	1.82 ± 0.22	1.79 ± 0.21	1.88 ± 0.22	6	80.26 ± 0.38	28.25 ± 1.27	76.31 ± 4.76	19.22 ± 0.71	+20.85 ± 3.68	6
SEAWATER Silver Eels											
Controls	183.3 ± 3.0	3.22 ± 0.20	2.37 ± 0.08	3.74 ± 0.30	16	76.32 ± 0.64	46.00 ± 1.38	123.80 ± 5.51	29.51 ± 1.53	-0.80 ± 0.74	7
Adrenalectomized (3 weeks)	222.4 ± 13.3	3.24 ± 0.39	4.23 ± 0.57	10.27 ± 1.88	11	72.14 ± 1.18	74.91 ± 9.90	164.60 ± 12.80	33.47 ± 3.37	-9.55 ± 1.97	8

Chester Jones *et al.* 1959, 1969; Henderson and Chester Jones, 1967; Chan *et al.*, 1969).

In sea water eels, interrenalectomy reduces branchial sodium efflux and small doses of cortisol restore kinetics towards normal (Mayer *et al.*, 1967; Maetz, 1969; Fig. 2). Mayer and Maetz (1967) have discussed the "stress" release of corticosteroids in relation to sodium balance of both fresh water and marine fish, while Mayer (1970) comprehensively summarizes data to show that corticosteroids, in particular cortisol, play a key rôle in regulating total sodium turnover in marine teleosts. Aldosterone, not always an endogenous hormone, influences sodium exchanges across the branchial epithelium of both fresh water and sea water animals, apparently through an action on sodium efflux (Motais, 1967; Fig. 3).

The magnitude of sodium transport (total turnover?) across branchial epithelia has been related to the activity of sodium–potassium-activated adenosine triphosphatase (Na + K-ATPase). Enzyme activities in the gills are greater in sea water-adapted than in fresh water-adapted euryhaline species (*Fundulus heteroclitus:* Epstein *et al.*, 1967; *Anguilla japonica:* Kamiya and Utida, 1968; Kamiya, 1972; *Anguilla rostrata:* Jampol and Epstein, 1970; *Anguilla anguilla:* Motais, 1970; Sargent *et al.*, 1975; Table V). Correlations between enzyme activities and environmental salinities have been made (Utida *et al.*, 1971; Butler and Carmichael, 1972). Utida and Hirano (1973) and Jampol and Epstein (1970), in very extensive surveys of euryhaline and stenohaline teleosts, justifiably relate sodium turnover by the gills to the activities of the Na+K-ATPase, and also show that "residual ATPases" (ATP breakdown in the absence of potassium), succinic dehydrogenase or glutaminase are largely unrelated to sodium status. Hypophysectomy reduces both sodium turnover rates and branchial Na+K-ATPase activities (Maetz *et al.*, 1967a; Pickford *et al.*, 1970a) and prolonged cortisol treatment restores enzyme activities to normal (Pickford *et al.*, 1970b). Similar data have been presented for sea water adapted hypophysectomized eels, and here the cortisol-stimulated enzyme activity has been related to parallel changes in sodium metabolism (Maetz *et al.*, 1967b; Maetz, 1971; Milne *et al.*, 1971). Cortisol also stimulates Na+K-ATPase activity in gill filaments from intact fresh water *Anguilla rostrata*, (Epstein *et al.*, 1971; Doyle and Epstein, 1972), and it was suggested that the chloride cells, a likely source of enzyme activity, were the immediate site of cortisol action.

Two studies have failed to confirm these general observations and resulting conclusions: Kirschner (1969) failed to show branchial

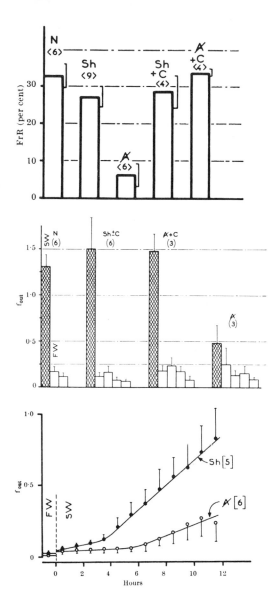

Fig. 2. The effects of adrenalectomy and cortisol injections on sodium metabolism of eels, *Anguilla anguilla* L. Top panel: Renewal rate of exchangeable sodium (F_rR, %) of eels in sea water. Middle panel: Sodium outflux (f_{out}) expressed in mEquiv/h/100 g body weight after transfer of sea water eels to fresh water. Cross hatched columns give sea water eel sodium outflux. Bottom panel: Sodium outflux (expressed as before) of eels transferred from fresh water to sea water as a function of time. Abbreviations: SW–sea water; FW–fresh water; N–normal intact animals; Sh–sham-operated, for adrenalectomy; A –adrenalectomized; C–cortisol injections. Number of animals are given in parentheses. Values are means ± SE. (From Mayer *et al.*, 1967).

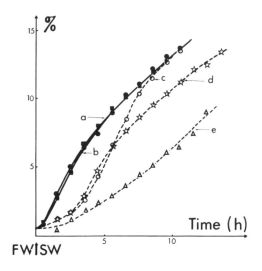

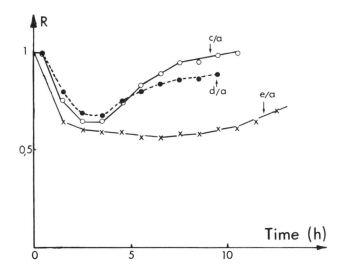

Fig. 3. Effects of aldosterone (2μg/100 g body weight) on sodium fluxes of flounders, *Platichthys flesus*, transferred from fresh water (FW) to sea water (SW). Upper panel: The fraction of exchangeable sodium renewed per hour (%) on ordinate is plotted against time on abscissa for individual fish—a and b are control injected, c, d and e aldosterone injected. Lower panel: The ratio (R) of fluxes in aldosterone injected animals (c, d and e) to a control (fish a) is plotted against time. The greatest effect is seen about 3 h after transfer to sea water (9 h after injection). (from Motais, 1967).

TABLE V

Effects of cortisol and prolactin on branchial adenosine triphosphatase activities (ATPase) of intact, hypophysectomized and sham-hypophysectomized eels. Means ± SE (Kamiya, 1972).

Adaptation	Treatment	Number of fish	ATPase activity (μmoles Pi/mg protein per hr) Na-K-ATPase	Mg-ATPase
Fresh water	None	4	8.2 ± 1.1	9.1 ± 0.9
1 week in	Sham-hypx	4	42.2 ± 2.7	8.2 ± 0.6
sea water	Hypx + saline	5	33.0 ± 2.6	9.1 ± 1.1
	Hypx + cortisol[a] (5 days)	3	48.6 ± 2.0	6.7 ± 0.3
1 week in	Sham-hypx	6	23.1 ± 2.3	17.1 ± 1.9
sea water	Hypx + saline	5	15.5 ± 1.6	17.3 ± 1.4
	Hypx + cortisol[a] (7 days)	6	33.6 ± 4.1	15.9 ± 1.9
1 month in	Sham-hypx	4	37.0 ± 3.0	4.3 ± 0.8
sea water	Hypx	4	34.2 ± 6.4	5.5 ± 0.8
Fresh water	Saline	8	15.3 ± 1.1	17.1 ± 1.0
	Cortisol[b]	10	25.9 ± 1.7	15.1 ± 0.6
1 week in	Hypx + saline	5	49.9 ± 3.6	1.5 ± 0.2
fresh water	Hypx + prolactin	5	36.7 ± 3.7	2.0 ± 0.2
1 week in	Hypx + saline	4	61.2 ± 2.9	3.2 ± 0.4
fresh water	Hypx + prolactin[c]	5	47.1 ± 2.7	2.1 ± 0.3
	Hypx + cortisol[d]	4	81.8 ± 10.8	1.8 ± 0.6
Sea water	Saline	4	59.0 ± 5.9	1.9 ± 0.7
	Prolactin[e]	4	53.2 ± 3.8	1.3 ± 0.6
Sea water	Hypx + saline	6	71.4 ± 8.4	0.7 ± 0.3
	Hypx + prolactin[f]	6	64.3 ± 4.2	1.1 ± 0.3

[a] Sheroson F was injected for 5 or 7 days at a dose of 1 mg per day. [b] Solu-Cortef was injected for 7 days at a dose of 1 mg per day. [c] Bovine prolactin was injected for 7 days at a dose of 0.5 mg per day. [d] Sheroson F was injected for 7 days at a dose of 1 mg per day. [e] Bovine prolactin was injected for 5 days at a dose of 1 mg per day. [f] Bovine prolactin was injected for 5 days at a dose of 0.5 mg per day.

enzymic changes as between fresh water and sea water adapted eels, while Scheer and Langford (1976) concluded that cortisol had slight, if any, effects on the Na+K-ATPase, and that the steroid might only act in the absence of the pituitary.

Apart from these contrary accounts, the general view is thus that cortisol is at least one controlling factor governing branchial modifications—both biochemical and morphological—that allow adaptation of teleosts to increased, hyper-osmotic salinities. The exact nature, particularly as regards coincidence of Na+K-ATPase changes and sodium fluxes is dubious (Forrest *et al.*, 1973a). The molecular bio-chemical and physiological events, involving steroid–receptor interactions are coming under increasing study. Plasma binding of steroids (Freeman and Idler, 1971) and relatively non-specific tissue uptakes (Cameron *et al.*, 1969; Goodman and Butler, 1972) have been indicated. Despite extensive examination of gills and gut of trout and goldfish Porthé-Nibelle and Lahlou (1978) failed to demonstrate the presence of high affinity binding sites for cortisol in cytosolic fractions.

(b) The kidney and urinary bladder
Early studies demonstrated that deoxycorticosterone, cortisol and corticosterone increased renal sodium retention in the face of a sodium load (Holmes, 1959) and significantly reduced the glomerular filtration rates of fresh water trout (Holmes and McBean, 1963). In the fresh water eel, *Anguilla anguilla*, adrenalectomy was followed by a gradual reduction in GFR and urine production rate (Chan *et al.*, 1969; Gaitskell and Chester Jones, 1971). The changes mentioned as well as urinary sodium and potassium excretory rates were returned towards normal by cortisol injections. Indirect experiments examining adrenocortical influence on renal function have employed hypophysectomized animals (Lahlou and Giordan, 1970; Butler, 1966) and broadly confirm that the renal lesions of adrenal insufficiency are rectified by cortisol.

Renal ATPase activity of fresh water fishes is considerably greater than that of marine species (Jampol and Epstein, 1970), with the notable exception of marine aglomerular species. A link between sodium turnover and enzyme activity is once again indicated; in the aglomerular species renal function is of course an entirely secretory process, presumably demanding the ATPase for initial urine preparation. Hypophysectomy of fresh water *Fundulus heteroclitus* reduces renal ATPase, and the defect is restored by exogenous cortisol (Pickford *et al.*, 1970b), an effect linked with increased sodium reabsorption (Epstein *et al.* 1969).

The ability of the urinary bladder of many teleosts to modify the composition of stored urine before excretion has recently come under study (Hirano *et al.*, 1973; Renfro, 1975). The general transport characteristics of water and ions are very broadly similar to those of distal segments of the nephron, which on anatomical grounds might be anticipated, since the bladder is formed by fusion of mesonephric ducts. There is selective passage of water and sodium chloride (Fossat and Lahlun, 1977; Johnson *et al.*, 1972), and the rates are influenced by environmental salinity largely as a result of hormonal changes, in particular prolactin and cortisol. These hormones may act synergistically or antagonistically to change osmotic permeability to water and/or active sodium transport (Utida et al., 1972; Fig. 4). A very broad generalization of this relationship is that prolactin maintains a reduced urinary bladder permeability to water in fresh water fishes, while cortisol increases this flux in marine conditions. Indeed, cortisol increases the permeability of urinary bladders of fresh water species (Doneen and Bern, 1974), and the effect of prolactin (decreased water permeability) requires the presence of cortisol. The sodium transport effects (Fig. 4) demand cortisol, with the changes in Na+K-ATPase

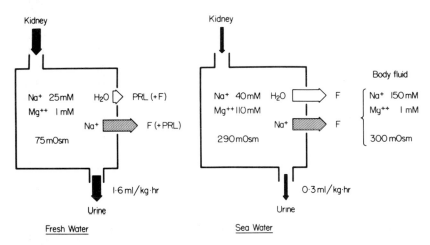

Fig. 4. Effects of cortisol (F) and prolactin (PRL) on sodium and water movements across the urinary bladder of flounder adapted to fresh water and sea water environments. The size of arrows may be taken to indicate relative magnitudes. (From Hirano, 1977).

activity (Utida et al., 1974) relying on a possibly permissive action of prolactin (Doneen, 1976).

(c) Gut

Marine and, contrary to earlier theories, fresh water teleosts (including non-feeding) drink environmental water. The process is essential for survival of marine species, but the physiological significance of the habit in fresh water is unclear (e.g. Maetz and Skadhauge, 1968; Lahlou et al., 1969; Gaitskell and Chester Jones, 1971). The absolute rates of drinking have been correlated with environmental concentration (Dall and Millward, 1969; Shedadeh and Gordon, 1969), and the stimulus may be osmotic rather than salinity *per se* (Sharratt *et al.*, 1964; Motais and Maetz, 1965; Motais, 1967), although recent studies by Hirano and colleagues (Hirano, 1974; Hirano *et al.*, 1978) suggest that the chloride ion is a key environmental dipsogenic factor.

The permeability and transport characteristics of the intestine, and indeed oesophagus (Hirano and Mayer-Gostan, 1976), differ as between fresh water and sea water adapted teleosts (Hirano *et al.*, 1976). As with the urinary bladder, cortisol and prolactin are key endocrine factors influencing the changes (Hirano and Mayer-Gostan, 1978). Compared with fresh water eels, the gut of sea water adapted eels displays a higher water permeability and flux, and greater sodium and chloride transport rates (Utida *et al.*, 1967; Skadhauge, 1969). Fig. 5 summarizes these general changes. The oesophagus in sea water adapted eels effectively dilutes the ingested water, largely as a result of passive solute flow (Kirsch and Laurent, 1975; Hirano and Mayer-Gostan, 1976); in fresh water animals the oesophagus is impermeable to both solute and water. The diluted sea water is thus delivered to the intestine for both water and solute absorption by a process which is presumably less energy demanding than absorption of undiluted sea water.

Various enzymes, located on both mucosal and serosal surfaces of the teleostean gut, show changes in activity concomitant with altered rates of water and solute metabolism. Transfer of euryhaline teleosts from fresh water to sea water is associated with increased intestinal $Na+K$-ATPase activity, although residual ATPase does not change (Oide, 1967; Jampol and Epstein, 1970). Other enzymes, including alkaline phosphatase and bicarbonate activated ATPase, also increase on sea water adaptation, and these increments have been correlated with stimulated solute and water absorption (Fig. 6; Utida *et al.*, 1968; Oide, 1973; Morisawa and Hirano, 1978).

Adrenalectomy of fresh water adapted eels reduces the rate of water absorption from the intestine, and cortisol restores the water permeability (Gaitskell and Chester Jones, 1970; Table VI). The enhanced water absorption seen when eels are transferred from fresh water to sea water does not occur after hypophysectomy and a lowered cortisol concentration has been held responsible for such a failure (Hirano and Utida, 1968, 1971). Similar results have been obtained in goldfish in saline environments after hypophysectomy (Ellory *et al.*, 1972; Porthé-Nibelle and Lahlou, 1975) although saline adaptation in this species tends to reduce intestinal sodium transport. The actions of cortisol on intestinal solute and water metabolism are

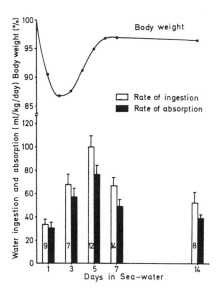

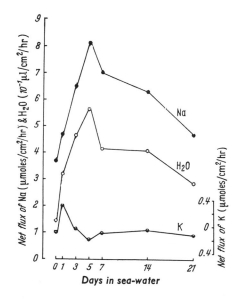

Fig. 5. Changes in body weight and gut function in eels *Anguilla japonica* transferred from fresh water to sea water environments. Upper panel shows body weight changes (as % of pre-transfer values), and the increases in rates of water imbibition and absorption at various times after transfer. Lower panel shows *in vitro* mucosal to serosal fluxes of water, sodium and potassium in intestines removed from eels at various times after transfer. (From Oide and Utida, 1967a, b).

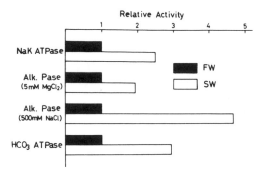

Fig. 6. The relative activities of sodium–potassium activated adenosine triphosphatase (NaK ATPase), alkaline phosphatases (Alk. Pase) and bicarbonate activated
adenosine triphosphatase (HCO_3 ATPase) in the intestinal mucosae of eels, *Anguilla
japonica*, adapted to fresh water (FW) or sea water (SW) environments. (From Hirano
et al., 1976).

summarized in Figs. 7 and 8: there is a net increase in sodium,
chloride and water absorption; a serosa-negative potential difference
develops as a result of an active chloride pump; and permeability to
both sodium and chloride increases as does osmotic water permeability (Ando, 1974, 1975; Ando *et al.*, 1975). Although beyond the terms
of reference of this chapter it should be noted that, as in the urinary
bladder and gills, prolactin interacts in a delicate fashion to, in many
respects, antagonize these cortisol effects (Utida *et al.*, 1972; Hirano
and Bern, 1972; see also Table VI).

2. *Protein and intermediary metabolism*

The life histories of teleost fishes, which frequently involve extensive
migratory cycles, prolonged starvation, spawning sequences leading

TABLE VI

Rates of water transport by intestinal segments of the fresh water eel. Rate = Increase in weight of sac
contents (mg)/[Dry tissue weight (g) × 10] in 90-min experimental period. A, anterior intestine; P_I first
(cephalad) segment of posterior intestine; P_2 second (Caudad) segment of posterior intestine; n, number of
animals; n_T, number of intestinal segments. Results are means ± SE. (Gaitskell and Chester Jones, 1970.)

Group	n	A	P_1	P_2	n_T	Total
Normal	6	63.12 ± 5.14	68.41 ± 12.59	49.80 ± 10.85	18	60.44 ± 5.68
Adrenalectomized	6	47.90 ± 5.47	34.69 ± 5.98	5.68 ± 1.96	18	29.42 ± 5.01
Sham-adrenalectomized	5	66.54 ± 10.53	76.03 ± 15.86	67.71 ± 18.34	15	70.09 ± 8.24
Adrenalectomized + cortisol	5	89.02 ± 21.17	67.95 ± 19.09	49.12 ± 10.78	15	68.69 ± 10.36
Saline injected	5	55.95 ± 8.03	72.00 ± 7.74	43.29 ± 4.07	15	57.08 ± 4.83
Cortisol injected	5	93.09 ± 17.30	131.23 ± 17.28	75.91 ± 17.48	15	100.07 ± 11.14

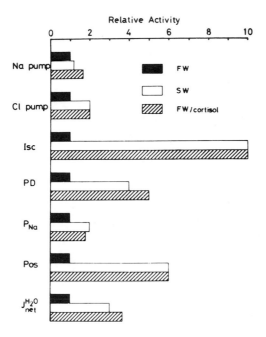

Fig. 7. The relative activities of sodium (Na) and chloride (Cl) pumps (net ion movement under short circuit conditions), short circuit current (I_{sc}), potential difference (PD), Na (P_{Na}, serosa-mucosa) water permeabilities (P_{os}) and net water movement ($J^{H_2O}_{net}$) in isolated intestines of fresh water (FW) and sea water (SW) adapted eels and of FW eels treated with cortisol. (From Hirano *et al.*, 1976).

to moribundity and extremes of environmental stress (e.g. food supply, temperature cycles etc.), impose great demands on the mechanisms regulating the processes of growth and energy supply. Interrenalectomy, hypophysectomy, pharmacological blockade of adrenocorticosteroidogenesis have all been found to influence intermediary metabolism of teleosts in one way or another, and the manner conforms in some, but not all, respects with the usual vertebrate pattern. Curious and in many ways enigmatic patterns have however emerged to regulate some processes; hepatic and tissue metabolism of amino acids, hepatic enzyme induction and glycogenolysis are of especial note in this regard (Hatey, 1951a, b; Chan and Cohen, 1964; Kenyon, 1967).

In mammals, adrenalectomy or hypophysectomy produce hypoglycaemia and depletion of hepatic glycogen stores. Surgical interrenalectomy or inhibition of corticosteroidogenesis of eels give qualita-

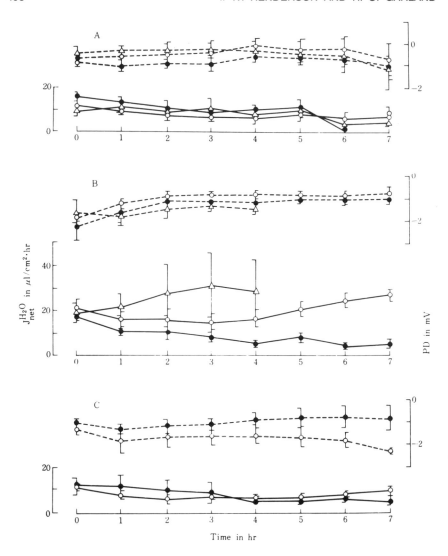

Fig. 8. *In vitro* actions of cortisol on potential differences (PD) and net water flux ($J_{net}^{H_2O}$) across posterior intestines of fresh water (Top panel, A) and sea water (Middle panel, B) adapted eels and hypophysectomized fresh water adapted eels (Bottom panel, C). Means ± SE. ● indicates no hormone added, ○ 1 μg cortisol/ml and △ 1 mg cortisol/ml. (From Ando, 1974.)

tively similar although perhaps less severe effects (Butler, 1968; Butler
et al., 1969). Hypophysectomy in a number of species has given
inconsistent effects on blood glucose concentrations. Some authors
(e.g. Falkmer and Matty, 1966—*Cottus scorpio:* Butler, 1968—*Anguilla
rostrata;* Mazeaud, 1969—*Cyprinus carpio*) failed to observe changes in
blood glucose up to two weeks after hypophysectomy. Other studies,
however, have demonstrated a profound hypoglycaemia following
hypophysectomy (e.g. Umminger, 1970, 1971; Srivastava and Pickford, 1972—*Fundulus heteroclitus*; Chidambaram *et al.*, 1973—*Ictalurus
melas;* Chan and Woo, 1978b—*Anguilla japonica*).

Hypophysectomy of eels results in a progressive reduction in liver
glycogen (Hatey, 1951a; Butler, 1968; Chan and Woo, 1978a), while
the hepatic glycogen stores are unaffected in *Tilapia mossambica*
(Swallow and Fleming, 1965). In contrast, *Poecilia latipinna*
and *Poecilia formosa* (Ball *et al.*, 1965, 1966), *Fundulus heteroclitus* (Ball
et al., 1965; Umminger, 1972) and *Carassius auratus* (Walker and
Johansen, 1975) show an unchanged liver glycogen after hypophysectomy.

Chan and Woo (1978b) have drawn up a detailed metabolic
balance sheet for hypophysectomized fresh water eels (Table VII).
This extensive analysis examined the chemical composition of liver
and muscle, rate of body weight loss, oxygen consumption and carbon
dioxide production rates. On the assumption that the major lesion
resulting from hypophysectomy is the failure to produce adequate
amounts of cortisol, a broad conclusion is that this hormone is
responsible either wholly, or in part, for maintenance of gluconeogenesis and regulation of the balance between protein and lipid
metabolism towards carbohydrate utilization. Fishes in general have
much higher levels of tissue lipids than other vertebrates and it would
seem that circulating levels come under the control of pituitary
factors, either primarily or secondarily, via adrenocortical steroids
and catecholamines (Mazeaud, 1971; Srivastava and Pickford, 1972;
Walker and Johansen, 1975).

In tetrapod vertebrates, adrenocorticoids promote gluconeogenesis,
to give a general depletion of non-carbohydrate reserves, a loss in
body weight, hyperglycaemia and elevated hepatic glycogen reserves.
Cortisol and/or mammalian ACTH produce a loss in body weight of
goldfish (Storer, 1967), *Poecilia* (Ball and Ensor, 1969) and trout
(Freeman and Idler, 1973). Moreover in various teleost species
hyperglycaemia and/or elevated liver glycogen follow exogenous
corticosteroids or ACTH (Nace, 1956; Robertson *et al.*, 1963; Oguri
and Nace, 1966; Butler, 1968; Hill and Fromm, 1968; see also Table

The metabolic balance sheet of unfed eels taken 2 weeks after
(Chan and

Intact eels		

Rate of muscle catabolism
Estimated from:

Na$^+$ balance:	Muscle Na$^+$ content	$=$ 29.8 μmol/g (wet wt.)
	Na$^+$ net flux: Gill	$=$ +40.0 μmol/kg/day
	Na$^+$ net flux: Kidney	$=$ $-$90.0 μmol/kg/day

Net $=$ $-$50.0 μmol/kg/day
$=$ $-$1.678 g of muscle kg/day

K$^+$ balance:	Muscle K$^+$ content	$=$ 83.5 μmol/g (wet wt.)
	K$^+$ loss: Gill	$=$ $-$136.0 μmol/kg/day
	K$^+$loss: Kidney	$=$ $-$4.2 μmol kg/day

Net $=$ $-$140.2 μmol/kg/day
$=$ $-$1.679 of muscle kg/day

Ca^{2+} balance:	Muscle Ca^{2+} content	$=$ 7.48 μmol/g (wet wt.)
	Ca^{2+} loss: Kidney	$=$ $-$12.3 μmol/kg/day
		$=$ $-$1.644 g of muscle kg/day

N balance:	Muscle protein content	$=$ 0.153 g/g (wet wt.)
	Muscle N content	$=$ 23.6 mg/g (wet wt.)
	NH$_3$ loss: Gill	$=$ $-$36.06 mg of N kg/day
	NH$_3$ loss: Kidney	$=$ $-$0.08 mg of N kg/day
	Urea loss: Kidney	$=$ $-$2.31 mg of N kg/day
	α-Aminoacid: Kidney	$=$ $-$0.18 mg of N kg/day

Net $=$ $-$39.63 of N kg/day
$=$ $-$1.679 g of muscle kg/day

Calculated respiratory exchange
Estimated from muscle catabolic rate $=$ $-$1.675 g/kg/day
(muscle N $=$ 23.6 mg/g; fat $=$ 183 mg/g; glycogen $=$ 0.37 mg/g)

	(mg/kg/day)	(mg/kg/hr)	QO$_2$	QCO$_2$	(ml/kg/hr)
N	39.6	1.65	9.78	7.85	
Fat	307	12.8	25.46	18.24	
Glycogen	0.62	0.026	0.02	0.02	
			35.26	26.11	R.Q. $=$ 0.74

Observed respiratory exchange
QO^2 $=$35.96 $\pm$ 2.01 ml/kg/hr; R.G. $=$0.74

hypophysectomy and maintained in fresh water at 20°C
Woo, 1978b).

Hypophysectomized eels

Rate of muscle catabolism
  Estimated from nitrogen balance

Muscle N content	= 18.1 mg/g (wet wt.)
Loss via gills	= −28.81 mg of N/kg/day
Loss via kidney	= −2.01 mg of N/kg/day

Net	= −30.82 mg of N/kg/day
	= −1.70 of muscle/kg/day
Muscle fat content	= 107 mg/g (wet wt.)
Fat catabolism	= −1.81 mg/kg/day
Muscle glycogen	= 0.186 mg/g (wet wt.)
Glycogen catabolism	= −0.32 mg/kg/day

Glycogen catabolism from other sources
  From decline in muscle glycogen concentration

	Muscle weight	= 144 g/kg (fat-free dry wt.)
	Decline in muscle glycogen concentration in 2 weeks	
		= −0.28% (fat-free dry wt.)
	Total glycogen	= −403.2 mg/kg over 2 weeks
		= −28.8 mg/kg/day
From liver:	Total glycogen	= −343 mg/kg over 2 weeks
		= −24.5 mg/kg/day
Total glycogen catabolism from all sources		= −53.62 mg/kg/day

Calculated respiratory exchange

	(mg/kg/day)	(mg/kg/hr)	QO_2	QCO_2	(ml/kg/hr)
N	30.82	1.28	7.61	6.10	
Fat	181	7.54	14.99	10.74	
Glycogen	53.62	2.23	1.85	1.85	
Net			24.45	18.69	R.Q. = 0.764

Observed respiratory exchange
  QO_2 = 23.80 ± 0.97 ml/kg/hr; R.Q. = 0.77

III). In addition both ACTH and cortisol oppose the changes induced by hypophysectomy as described above in *Poecilia* (Ball *et al.*, 1966; Ball and Hawkins, 1976).

Corticosteroids are involved in the increased blood glucose and changed blood parameters indicative of "stress" in a number of teleosts. A variety of stressful factors (simple mechanical disturbance, cold shock, descaling) induce changes that include hyperglycaemia, changes in circulating red and white cells, elevated plasma corticosteroids and catecholamines (Chavin, 1964; Black and Tredwell, 1967; Nakano and Tomlinson, 1967; Narasimhan and Sundararaj, 1971a; Pickford *et al.*, 1971; Liversage *et al.*, 1971). Narasimhan and Sundararaj (1971b) have also described a circadian rhythm in blood carbohydrates which may be related to a stress release of corticosteroids in *Notopterus notopterus* and *Colisa fasciata* (see also Section 4, A, 3).

The action of corticosteroids on hepatic metabolism as such has come under increasing scrutiny. Cortisol increases liver glutamic-pyruvic transaminase activity in goldfish and trout (Storer, 1967; Freeman and Idler, 1973), although similar treatment in *Roccus chrysops* and *Promoxis nigromaculatus* failed to alter hepatic tyrosine transaminase activity (Chan and Cohen, 1964). ACTH and/or cortisol increased the rate of nitrogen elimination in *Carassius auratus*, *Gobius cephalarges*, *G. melanostomum* and *Trachurus mediterraneus* (Storer, 1967; Pora and Precup, 1971). An extensive study by Inui and Yokote (1975) demonstrated that cortisol influences a number of hepatic enzymes of eels (Fig. 9) while Chan and Woo (1978a, b) and Chan *et al.* (1978) have implicated cortisol as a key hormone in a series of events, including basal respiratory metabolism, cardio-ventilatory adaptations and nitrogen metabolism associated with adaptation to sea water (see also Table VII).

Finally, the changed interrenal activity observed during the spawning migrations of anadromous teleosts (Section 4, A, 3) can be correlated with parallel alterations in protein and carbohydrate metabolism. Thus in Sockeye salmon, *Oncorhynchus*, up to 60% of the body protein is catabolized during the non-feeding spawning migration (Duncan and Tarr, 1958; Idler and Clemens, 1959). This is associated with a six-fold increase in circulating corticosteroid (Idler *et al.*, 1959), and an elevation of liver glycogen, so that levels at spawning grounds were around twice those at the mouth of the river (Chang and Idler, 1960). In such a way, fasting fish are provided with the energy necessary for normal metabolism and the development of the maturing gonads.

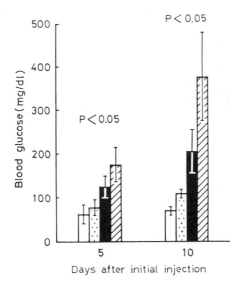

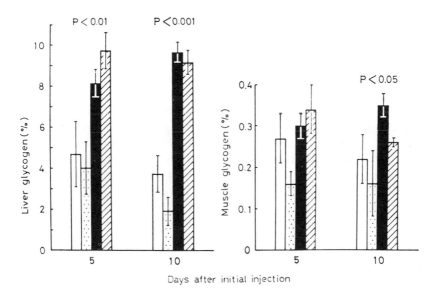

Fig. 9. Effects of cortisol on plasma glucose (upper panel) and on hepatic and muscle glycogen contents (Lower panel) of fresh water adapted eels, *Anguilla japonica.* □ Intact controls; ▨ saline injected controls; ■ 2 mg cortisol/kg body weight/day; ▨ 10 mg cortisol/kg body weight/day. Periods after injections are given. (From Inui and Yokote, 1975).

3. *Interactions between the interrenal, other endocrines and the environment*

Environmental salinity and temperature, together with a variety of miscellaneous "stress" factors that induce shock reactions, bring about structural and functional changes in the teleostean interrenal. In addition the secretory patterns of the gland are subject to natural circadian and circannual rhythms in common with other endocrine glands.

(a) Salinity

Studies relating interrenal histology with variations in environmental salinity in teleosts have been conflicting. Olivereau (1966a) described a transitory interrenal stimulation following the transfer of eels to distilled water. In animals adapted for 4 weeks or more, however, adrenal cells showed reduced nuclear diameters and smaller nucleoli (Hanke and Chester Jones, 1966; Hanke *et al.*, 1969). The latter authors also described a transitorily reduced adrenal cell nuclear size in eels adapted to sea water. Olivereau (in Olivereau and Ball, 1970), however, reported a transitory stimulation of the ACTH–interrenal system in similarly transferred animals. Long term exposure to sea water may stimulate or fail to alter the histology of eel interrenal cells (Hanke and Chester Jones, 1966; Hanke *et al.*, 1969). In the marine teleost *Gobius paganellus* adrenocortical cells appeared stimulated after transfer to fresh water (Hanke *et al.*, 1969). Similarly, in natural populations of mullet, *Mugil cephalus*, fresh water forms showed greater interrenal activity than their marine counterparts (Johnson, 1972).

There is very little difference between the plasma concentrations of cortisol in eels adapted for long periods to fresh water or sea water (Hirano, 1969; Ball *et al.*, 1971; Henderson *et al.*, 1974; Leloup-Hatey, 1974, 1976; see Table VIII). A transitory increase in plasma cortisol levels occurs when eels are transferred to sea water (Hirano, 1967, 1969; Ball *et al.*, 1971; Utida and Hirano, 1973, Forrest *et al.*, 1973a). Adaptation of goldfish to 0.09% sodium chloride solution produces a similar effect (Singley and Chavin, 1971). Eels transferred from sea water to fresh water or from fresh water to distilled water do not show such a rise (Hirano, 1969; Ball *et al.*, 1971). The metabolic clearance and derived production rates for cortisol in the eel are significantly elevated in sea water animals (Henderson *et al.*, 1974; Leloup-Hatey, 1974; Table VIII). Recent studies have employed a more direct method of estimating cortisol secretion rates in the eel by quantitatively collecting cardinal venous blood and measuring both the flow rate

TABLE VIII

Cortisol dynamics in fresh water and sea water adapted eels. (Henderson *et al.*, 1974, 1975)

	Fresh water eels	(n)	Sea water eels	(n)
Plasma cortisol concentration (μg/100 ml plasma)	4.1 ± 0.8	18	4.8 ± 0.4	19
Cortisol metabolic clearance rate (ml/h/Kg body weight)	20.6 ± 1.3	18	28.4 ± 2.4	19
Cortisol production rate (μg/h/Kg body weight)	0.72 ± 0.08	18	1.29 ± 0.21	19
Cortisol secretory rate (μg/h/Kg body weight)	0.6 ± 0.1	9	1.6 ± 0.4	6

and cortisol concentrations (Henderson *et al.*, 1975; Table VIII). Cortisol secretory rates for sea water adapted eels were 2–3 times higher than those of fresh water animals. Moreover, in adaptation experiments, eels transferred from fresh water to sea water showed a gradual increase in cortisol secretory rate over 24 h; transfer from sea water to fresh water was accompanied by a gradual decrease in steroid production. It would appear, therefore, certainly from the more direct studies of cortisol production and secretion rates, that this steroid hormone is fundamentally involved in the osmoregulatory adaptation to varying environmental salinity.

(b) Stress and shock
A variety of shock-inducing factors, ranging from simple handling stresses to the administration of noxious chemicals, are known to affect the teleost interrenal. Increased plasma corticoid levels have been found in carp and salmon following handling, confinement in aquaria, and even after transfer from one tank to another (Leloup-Hatey, 1958; Hane *et al.*, 1966; Fagerlund, 1967). General excercise and forced swimming in salmonids, goldfish and eels similarly induce significant, if transitory, increases in plasma cortisol concentrations (Fagerlund, 1967; Hill and Fromm, 1968; Forster, 1970; Fryer, 1975). Phenoxyethanol-anaesthesia increased plasma cortisol levels in Sock-

eye salmon, *Oncorhynchus nerka* (Fagerlund, 1967), whereas electric-immobilization, cold-narcosis, and MS222-anaesthesia failed to alter steroid concentrations significantly in goldfish, *Carassius auratus* (Singley and Chavin, 1975). In salmonids, both disease (bacterial and fungal infection) and exposure to low levels of noxious chemicals (environmental hexavalent chromium) significantly enhanced circulating cortisol levels (Fagerlund, 1967; Hill and Fromm, 1968).

(c)   Circadian and circannual rhythms

Evidence is now accumulating to suggest the existence of daily and seasonal rhythmical activity in teleost interrenal function. Such rhythms are known to be present in reptiles, birds and mammals (see Peterson, 1957; McCarthy *et al.*, 1960; Jørgensen, 1976). Boehlke *et al.* (1966) first reported a diurnal rhythm for plasma cortisone and corticosterone (but not cortisol) in the catfish *Ictalurus punctatus*, with a peak for *total* glucocorticoids at 14.00 h. Srivastava and Meier (1972) found a similar unimodal circadian rhythm for cortisol in the gulf killifish, *Fundulus grandis*, the peak occurring eight hours after the onset of the light period. In the same species, Garcia and Meier (1973) have further described a bimodal rhythm in plasma cortisol, peculiar to male fish, with the additional peak occurring one hour after the onset of the light period. Interestingly, peak adrenal activity in diurnal mammals is around 08.00–09.00 (Harwood and Mason, 1956; Peterson, 1957). More recently, Singley and Chavin (1975) have described a cortisol circadian rhythm for goldfish, *Carassius auratus*, comprising two periods of high cortisol levels and two low periods (see Fig. 10), while Ikeda *et al.* (1976) found corticosterone to show one nadir during hours of darkness in *Seriola quinqueradiata*; however, the normal rhythm for fish fed a frozen fish diet virtually disappeared when fish were fed a vitamin supplemented diet.

Forrest *et al.* (1973b) could find no evidence for a diurnal variation in plasma cortisol in fresh water or sea water eels *Anguilla rostrata*. Leloup-Hatey (1972) and Leloup-Hatey and Hardy (1976), however, have reported a circannual variation in cortisol production in *A. anguilla*. Similar annual variations have been described in natural populations of *Pleuronectes platessa* (Wingfield and Grimm, 1977) and *Mugil cephalus* (Johnson, 1972), where interrenal cells appear histologically more active in winter than summer. This correlates well with the studies of Hill and Fromm (1968), who found a seasonal variation in cortisol levels in the rainbow trout, *Salmo gairdneri*, with elevated steroid titres during the winter months.

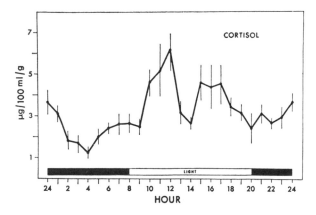

Fig. 10. Serum cortisol concentrations (μg/100 ml/g body weight) in normal fasted goldfish, *Carassius auratus* L. over a 24 hour period. Two peaks are apparent. The shaded portions at the bottom of the figure indicate hours of darkness and the open section, hours of light. Each point is Mean ± SE of nine fish. (Singley and Chavin, 1975).

(d) Interaction with other glands

(i)   *The pituitary.* Injections of corticosteroids into a variety of teleost species characteristically produce regressive changes in inter-renal histology, suggestive of a negative feedback system with the pituitary (Basu *et al.*, 1965; Hanke and Chester Jones, 1966; Olivereau, 1966b; Hanke *et al.*, 1967; Subhedar and Rao, 1974). The dependence of the teleost interrenal upon pituitary function has been discussed by Jørgensen (1976). Briefly, therefore, hypophysectomy typically results in interrenal cell atrophy (see, for example, Chester Jones *et al.*, 1969) with reduced plasma cortisol levels (see Hirano, 1969; Fenwick and Forster, 1972) and blood production rates (Henderson *et al.*, 1974; Table IX). The metabolic clearance rates for cortisol are however elevated after hypo-physectomy, and perhaps enigmatically are restored to normal values after infusions of mammalian ACTH (Fig. 11). Administration of mammalian ACTH causes hypertrophy of adrenocortical cells (Hanke and Chester Jones, 1966; Hanke *et al.*, 1967), prevents interrenal regression in hypophysectomized forms (see Basu *et al.*, 1965) and increases plasma cortisol titres (Hawkins and Ball, 1970). Adrenalectomy results in nuclear enlargement and degranulation of ACTH cells (Hanke *et al.*, 1967). It would appear, therefore, that there exists in teleosts a negative feedback system between the interrenal and the pituitary, and that both the structure and function

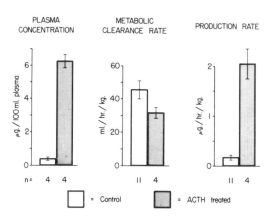

Fig. 11. Effects of mammalian ACTH infusions on cortisol dynamics in hypophysectomized sea water adapted eels, *Anguilla anguilla* L. Means ± SE. n gives numbers of animals. (From Henderson *et al.*, 1974).

of the former gland are controlled by pituitary corticotrophin, which itself is regulated by hypothalamic neuroendocrine loops including CRF (see Ball and Hawkins, 1976).

(ii) *The gonads.* A progressive interrenal hyperplasia occurs in both Pacific and Sockeye salmon (*Oncorhynchus tshawytscha* and *O. nerka*) during sexual maturation (Robertson and Wexler, 1959, 1962; McBride and Van Overbeeke, 1969). At the same time, corticosteroid concentrations, metabolic clearances and secretory rates also increase (Idler *et al.*, 1959; Robertson *et al.*, 1961; Donaldson and Fagerlund, 1970; Fagerlund and Donaldson, 1970). DOCA levels were similarly elevated during the initiation of spawning in *Tilapia aurea* (Katz and Eckstein, 1974). In salmonids gonadectomized before reaching maturity, such enhanced steroid titres were reversed (Donaldson and Fagerlund, 1970; Fagerlund and Donaldson, 1970), and interrenal hypertrophy prevented (McBride and van Overbeeke, 1969). Adrenocortical stimulation by exogenous ACTH did not significantly change during maturation in salmon whose endogenous corticotrophin had been suppressed by dexamethasone (Fagerlund, 1970). Moreover, sexual maturation was not accompanied by any apparent changes in ACTH cells (McBride and van Overbeeke, 1969), suggesting a direct pathway between the gonads and the interrenal rather than one via the pituitary. Maturation and subsequent ovulation of oocytes can be induced in both normal and hypophysectomized catfish *Heteropneustes fossilis* by injections of DOCA (Sundararaj aand Goswami, 1966;

TABLE IX

Changes in cortisol dynamics and plasma electrolyte concentrations in fresh water and sea water adapted eels after hypophysectomy (Henderson et al., 1974).

Experimental group	Plasma cortisol concentration (μg/100 ml plasma)	Cortisol metabolic clearance rate (ml/h/kg body weight)	Cortisol production rate (μg/h/kg body weight)	Plasma electrolyte concentrations (mmol/l)			
				Na	K	Ca	Mg
SEA WATER EELS							
Intact control (n = 19)	4.8 ± 0.4	28.4 ± 2.5	1.29 ± 0.21	178 ± 3	3.5 ± 0.2	2.3 ± 0.05	3.7 ± 0.2
Hypophysectomized (n = 11)	0.43 ± 0.1^a	45.57 ± 5.3^a	0.18 ± 0.04^a	189 ± 3^a	2.6 ± 0.06^a	3.2 ± 0.2^a	4.1 ± 0.6
FRESH WATER EELS							
Intact control (n = 18)	4.1 ± 0.8	20.6 ± 1.3	0.72 ± 0.08	143 ± 1	1.8 ± 0.15	2.3 ± 0.06	2.4 ± 0.2
Hypophysectomized n = 10	0.51 ± 0.08^a	31.5 ± 3.3^a	0.17 ± 0.04^a	113 ± 2^a	2.8 ± 0.1^a	1.75 ± 0.2^a	2.4 ± 0.3

Values are means ± standard error; [a] Values statistically different from appropriate control groups.

Goswami and Sundararaj, 1971), again indicative of a direct action on the gonads.

(e) The renin–angiotensin system
In mammals, adrenocortical function is influenced by the renin–angiotensin system. Although such a system is thought to have appeared during the early evolution of bony fishes (Nishimura et al., 1973; Nishimura and Ogawa, 1973), its precise rôle in the control of the teleostean adrenocortical homologue remains unclear. By analogy with mammals, sodium-depletion would be expected to be accompanied by an increase in plasma renin levels. Investigations attempting to relate plasma renin activity in teleosts to changing environmental salinity, however, have been somewhat conflicting. The early studies of Kaley et al. (1963) and Kaley and Donshik (1965) indicated that the renal renin activity (RRA) of marine teleosts was considerably higher than in fresh water forms. Subsequent histological studies of juxtaglomerular (JG) cells in euryhaline fishes similarly implied a greater physiological rôle for the teleost renin in marine environments (Lagios, 1968; Wendelaar Bonga, 1973). In contrast, in *Anguilla japonica* and *Tilapia mossambica* renal renin activity was found to decrease significantly after transfer from fresh water to sea water (Sokabe et al., 1966, 1968). Similarly, the six fresh water teleosts examined by Mizogami et al. (1968) showed higher RRA than nine marine species. Capelli et al. (1970) further demostrated that migratory herrings showed a 315 times increased RRA when caught in fresh water tributaries than in full strength sea water. Other studies have indicated renin levels to be fairly similar in teleosts regardless of habitat (Malvin and Vander, 1967; Nolly and Fasciolo, 1972). Other studies of Nishimura (1971), Nishimura et al. (1976), and Henderson et al. (1976), however, described a *decreased* plasma renin activity (PRA) in eels *Anguilla rostrata* and *A. anguilla* transferred from sea water to hypo-osmotic media. Conversely, an *increase* in PRA occurred in eels transferred from fresh water to sea water (Sokabe et al., 1973; Henderson et al., 1976; Fig. 12). Such changes may parallel those transitory ones described for cortisol in similarly manipulated animals (Hirano, 1969; Ball et al., 1971; Henderson et al., 1974; 1975; Leloup-Hatey, 1974). Moreover, a transitory increase in size and number of JG cells of *Tilapia mossambica* has been found following transfer to sea water (Krishnamurthy and Bern, 1973). Unfortunately, however, when PRA and plasma cortisol levels were measured together in eels adapted to different salinities, no clear correlations between the two parameters could be found (Nishimura et al., 1976).

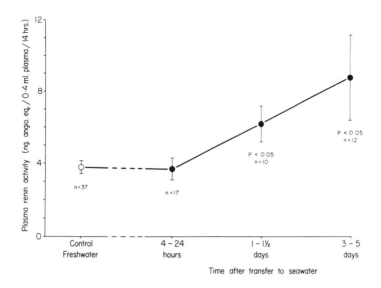

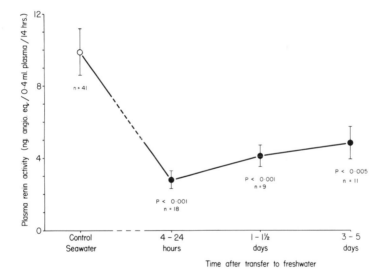

Fig. 12. Changes in plasma renin activity of eels transferred from fresh water to sea water (upper panel) or from sea water to fresh water (lower panel). Eels were transferred directly from one environment to the other and animals were killed at the times given on the abscissae. Points are means ± SEM; the number of eels is indicated in parentheses at each point. Statistical significance from grouped data analysis is given for each point. Plasma renin activity is expressed in ng equivalents angiotensin II/0.4 ml eel plasma/14 h incubation as bioassayed in the nephrectomized, pentobarbitone anaesthetized, pentolinium-blocked rat. (From Henderson *et al.*, 1976.)

Moreover, there was no relationship between PRA and either plasma sodium levels or plasma volume, as has been found in mammals (Figs. 13 and 14). Obviously further studies are necessary before a physiolo-

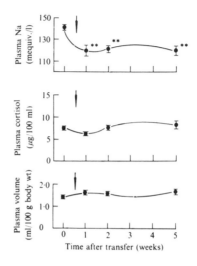

Fig. 13. Plasma sodium concentrations, plasma cortisol concentrations, and plasma volumes in sea water eels and eels after adaptation to fresh water. Vertical lines indicate ± SEM. Changes between sea water control group and groups after the fresh water transfer (arrows) were analysed statistically by Student's t-test (** $P<0.01$); plasma Na and cortisol concentrations were determined in 16–20 eels, and plasma volume in seven eels in each group. (From Nishimura *et al.*, 1976).

gical renin–angiotensin–adrenocortical axis is established but evidence is accumulating that the kidney factors affect interrenal function (Fig. 15 and 16; see also Taylor and Davis, 1971). Indeed, from the existing evidence available, Nishimura *et al.* (1976) have concluded that renin and angiotensin may serve a completely different function in these animals.

B. HOLOSTEI

The functions of adrenocortical steroids in holostean fishes are unknown. Indeed only one study (Hanson *et al.*, 1976), on possible pituitary influences on water and electrolyte metabolism, impinges, and then only peripherally, on adrenocortical involvement. Hypophysectomized young, growing bowfin lost large amounts of sodium and the effect was reversed by prolactin. The latter hormone, rather than ACTH, was considered to be the active hormone with

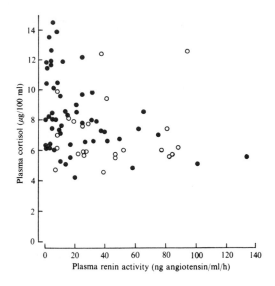

Fig. 14. Relation of plasma renin activity to plasma cortisol concentrations in eels. Open circles indicate sea water eels, and closed circles eels adapted to fresh water for various periods. Sea water eels: r = 0.07; fresh water-adapted eels; r = 0.40. (From Nishimura *et al.*, 1976.)

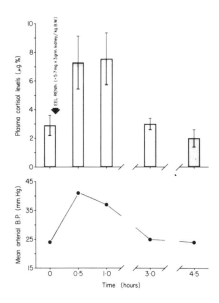

Fig. 15. Effects of eel renin on arterial plasma cortisol concentrations and arterial blood pressure in six intact fresh water-adapted eels. After a control sample and reading had been taken, the renin was injected i.v. and measurements were made for the subsequent 4.5 h. Values are means ± SEM. (From Henderson *et al.*, 1976).

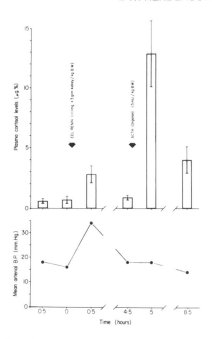

Fig. 16. Effects of eel renin and mammalian ACTH on arterial plasma cortisol concentrations and arterial blood pressure in eight hypophysectomized fresh water-adapted eels. Control readings were taken 30 min and immediately before the i.v. injection of eel renin, and further values were obtained 0.5 h and 4.5 h after the initial injection. ACTH (Organon, 5 mu./kg body weight) was then injected and further readings taken. Values are means ± SEM. (From Henderson *et al.*, 1976).

respect to water and electrolyte imbalance following removal of the pituitary.

C. CHONDROSTEI

There have been no studies on adrenocortical steroid actions on the physiology of this group of fishes.

D. DIPNEUSTI (DIPNOI)

The lungfishes which conceptually span the tetrapod emergence from aquatic piscine ancestors have not been extensively studied in terms of the functional rôles of the adrenocortical steroids (Sandor *et al.*, 1976). Of particular interest is the fact that *Lepidosiren* displays circulating concentrations of aldosterone akin to those seen in amphibians (Idler *et al.*, 1972). The other extant lungfishes (*Protopterus* sp. and *Neocerato-*

dus forsteri) possess adrenocortical characteristics similar to amphibians but the blood values for corticosteroids are lower than their South American relation (Janssens *et al.*, 1965; Blair-West *et al.*, 1977). Experimental manipulations of these various crossopterygian species with regard to corticosteroid function are few and firm conclusions cannot be made.

5. Conclusions

Contemporary understanding of the physiology of adrenocorticosteroids in fishes centres almost exclusively on a few species of teleosts. Even in this group, only one hormone, cortisol, has received intensive study, and other corticosteroids, undoubtedly present, have thus far escaped thorough study. In the case of cortisol, its role in water and electrolyte regulation has been emphasized, and it is only relatively recently that a functional relationship with growth, reproduction, behaviour and protein–carbohydrate metabolism has been experimentally demonstrated. Of the other fish groups, information is sparse to say the least. Agnathous vertebrates require much more careful and critical study, and corticosteroid secretory patterns alongside rigorous identification of the steroids present must have a high priority. Such also applies to present knowledge of Elasmobranchii, and here perhaps newer notions should be applied, taking into account the rather idiosyncratic mode of osmoregulation involving urea and trimethylamine oxide.

It is timely to consider the application of the many new and sophisticated techniques currently available for the study of hormone secretory patterns, their transport, metabolism, tissue uptake and mode of action. Corticosteroid–hormone receptor proteins have so far eluded identification in fishes and the application of some of these techniques will undoubtedly lead to considerable advances in the understanding of how corticosteroids, a key group of perhaps ubiquitous chemicals of the animal kingdom, influence the day to day and generation to generation physiology of the largest, if most heterogeneous, group of vertebrates.

References

Ando, M. (1974). *Endocr. Japon.* **21**, 539–546.
Ando, M. (1975). *Comp. Biochem. Physiol.* **52A**, 229–233.

Ando, M., Utida, S., and Nagahama, J. (1975). *Comp. Biochem. Physiol.* **51A,** 27–32.

Arai, R. and Tamaoki, B. I. (1967). *J. Endocr.* **39,** 453–454.

Ball, J. N. and Ensor, D. M. (1969). *Colloq. Int. Centre Nat. Rech. Sci., Paris* **177,** 216–224.

Ball, J. N. and Hawkins, E. F. (1976). *Gen. comp. Endocr.* **28,** 59–70.

Ball, J. N. and Olivereau, M. (1970). *Mem. Soc. Endocr.* **18,** 57–85.

Ball, J. N., Olivereau, M., Slicher, A. M. and Kallman, K. D. (1965). *Phil. Trans. Roy. Soc. Lond.* **B 249,** 69–99.

Ball, J. N., Giddings, M. R. and Hancock, M. P. (1966). *Am. Zoologist,* **6,** 595.

Ball, J. N., Chester Jones, I., Forster, M. E., Hargreaves, G., Hawkins, E. F., Milne, K. P. (1971). *J. Endocr.* **50,** 75–96.

Basu, J., Nandi, J. and Bern, H. A. (1965). *J. Exp. Zool.* **159,** 347–356.

Bellamy D. and Chester Jones, I. (1961). *Comp. Biochem. Physiol.* **3,** 125–135.

Bentley, P. J. (1971) "Endocrines and Osmoregulation". Springer-Verlag, Berlin, Heidelberg and New York.

Bentley, P. J. and Follet, B. K. (1962). *Gen. comp. Endocr.* **2,** 329–335.

Bentley, P. J. and Follet, B. K. (1963). *J. Physiol. Lond.* **1969,** 902–918.

Bentley, P. J. and Follett, B. K. (1965). *J. Endocr.* **31,** 127–137.

Bern, H. A., de Roos, C. C. and Biglieri, E. G. (1962). *Gen. comp. Endocr.* **2,** 490–494.

Black, E. C. and Tredwell, S. J. (1967). *J. Fish Res. Bd. Can.* **24,** 939–953.

Blair-West, J. R., Coghlan, J. P., Denton, D. A., Gibson, A. P., Oddie, C. J., Sawyer, W. H. and Scoggins, B. A. (1977). *J. Endocr.* **74,** 137–142.

Boehlke, K. W., Church, R. L., Tiemeier, O. W. and Eleftheriou, B. E. (1966). *Gen. comp. Endocr.* **7,** 18–21.

Burger, J. W. (1967). In "Sharks, Skates and Rays" (P. W. Gilbert, R. F. Mathewson, and D. P. Rall, eds.) pp 177–185. Johns Hopkins Press, Baltimore.

Butler, D. G. (1966). *Comp. Biochem. Physiol.* **18,** 773–782.

Butler, D. G. (1968). *Gen. comp. Endocr.* **10,** 85–91.

Butler, D. G. (1973). *Am. Zool.* **13,** 839–880.

Butler, D. G. and Carmichael, F. J. (1972). *Gen. comp. Endocr.* **19,** 421–427.

Butler, D. G. and Langford, R. W. (1967). *Comp. Biochem. Physiol.* **22,** 309–312.

Butler, D. G., Clarke, W. C., Donaldson, E. M. and Langford, R. W. (1969). *Gen. comp. Endocr.* **12,** 502–514.

Cameron, I. L., Tolman, E. L. and Harrington, G. W. (1969). *Texas Rep. Biol. Med.* **27,** 367–380.

Capelli, J. P., Wesson, L. G. and Aponte, G. E. (1970). *Am. J. Physiol.* **218,** 1171–1178.

Capreol, S. V. and Sutherland, L. E. (1968). *Can. J. Zool.* **46,** 249–256.

Chan, D. K. O. and Chester Jones, I. (1968). *J. Endocr.* **42,** 109–117.

Chan, D. K. O. and Woo, N. Y. S. (1978a). *Gen. comp. Endocr.* **35,** 160–168.

Chan, D. K. O. and Woo, N. Y. S. (1978b). *Gen. comp. Endocr.* **35,** 169–178.

Chan, D. K. O., Phillips, J. G. and Chester Jones, I. (1967a). *Comp. Biochem. Physiol.* **23,** 185–199.

Chan, D. K. O., Chester Jones, I., Henderson, I. W. and Rankin, J. C. (1967b). *J. Endocr.* **37,** 297–317.

Chan, D. K. O., Rankin, J. C. and Chester Jones, I. (1969). *Gen. comp. Endocr.* Suppl. **2,** 342–353.

Chan, D. K. O., Ho, S. M. and So, S. T. C. (1978). *In* "Comparative Endocrinology"

(P. J. Gaillard and H. H. Boer, eds.). pp. 227–230. Elsevier/North Holland Biomedical Press, Amsterdam, New York and Oxford.

Chan, S. K. and Cohen, P. P. (1964). *Archs Biochem. Biophys.* **104,** 335–337.

Chang, V. M. and Idler, D. R. (1960). *Can. J. Biochem. Physiol.* **38,** 553–558.

Chavin, W. (1964) Proc. 7th Conf. Great Lakes Res., University of Michigan. Great Lakes Res. Div. Publication **11,** 54–67.

Chester Jones, I. (1957). "The Adrenal Cortex", Cambridge University Press, London.

Chester Jones, I. (1963). *In* "The Biology of Myxine" (A. Brodal and R. Fangë, eds.), pp. 488–452. Univeristetsforlaget, Oslo.

Chester Jones, I. (1974). *J. Endocr.* **71,** 3P–30P.

Chester Jones, I. and Phillips, J. G. (1960). *Symp. Zool. Soc. Lond.* **1,** 17–32.

Chester Jones, I., Phillips, J. G. and Holmes, W. N. (1959). *In* "Comparative Endocrinology" (A. Gorbman, ed.), pp. 582–612. Wiley, New York.

Chester Jones, I., Phillips, J. G. and Bellamy, D. (1962). *Gen. comp. Endocr.* Suppl. **1,** 36–47.

Chester Jones, I., Henderson, I. W. and Mosley, W. (1964). *J. Endocr.* **30,** 155–156.

Chester Jones, I., Chan, D. K. O., Henderson, I. W. and Ball, J. N. (1969). *In* "Fish Physiology" (W. S. Hoar and D. J. Randall, eds.), Vol. 1, pp. 322–356. Academic Press, London and New York.

Chester Jones, I., Bellamy, D., Chan, D. K. O., Follett, B. K., Henderson, I. W., Phillips, J. G. and Snart, R. S. (1972). *In* "Steroids in Non-mammalian Vertebrates" (D. R. Idler, ed.), pp. 414–480. Academic Press, London and New York.

Chester Jones, I., Ball, J. N., Henderson, I. W., Baker, B. I. and Sandon, T. (1974). *In* "Chemical Zoology" (M. Florkin and B. T. Scheer, eds.), Vol. 8, pp. 523–593. Academic Press, London and New York.

Chidambaram, S., Meyer, R. K. and Hasler, A. D. (1973). *J. exp. Zool.* **184,** 75–80.

Colombo, L., Bern, H. A., Pieprzyk, J. and Johnson, D. W. (1972). *Endocrinology* **91,** 450–462.

Colombo, L. Colombo Belvedere, P. and Cisotto, T. (1977). *Comp. Biochem. Physiol.* **57B,** 89–93.

Conte, F. P. (1969). *In* "Fish Physiology" (W. S. Hoar, and D. J. Randall, eds.) Vol. 1 pp. 241–292. Academic Press, New York and London.

Dall, W. and Millward, N. E. (1969). *Comp. Biochem. Physiol.* **30,** 247–260.

de Roos, R. and de Roos, C. C. (1967). *Gen. comp. Endocr.* **9,** 267–276.

De Vlaming, V. L., Sage, M. and Beitz, B. (1975). *Comp. Biochem. Physiol.* **52A,** 505–513.

Dittus, P. (1941). *Z. wiss. Zool.* **154,** 40–124.

Donaldson, E. M. and Fagerlund, U. H. M. (1970). *J. Fish. Res. Bd. Can.* **27,** 2287–2296.

Doneen, B. A. (1976). *Gen. comp. Endocr.* **28,** 33–41.

Doneen, B. and Bern, H. A. (1974). *J. exp. Zool.* **187,** 173–179.

Doyle, W. L. and Epstein, F. H. (1972). *Cytobiologie,* **6,** 58–73.

Duncan, D. W. and Tarr, H. L. A. (1958). *Can. J. Biochem. Physiol.* **36,** 799–803.

Edelman, I. S., Young, H. L. and Harris, J. B. (1960). *Am. J. Physiol.* **199,** 666–670.

Ellory, J. C., Lahlou, B. and Smith, M. W. (1972). *J. Physiol., Lond.* **222,** 497–509.

Ensor, D. M. (1979). "The Comparative Endocrinology of Prolactin". Chapman and Hall, London.

Ensor, D. M. and Ball, J. N. (1972). *Fed. Proc. Fedn Am. Socs exp. Biol.* **31,** 1615–1623.

Epstein, F. H., Katz, A. I. and Pickford, G. E. (1967). *Science* **156,** 1245–1247.

Epstein, F. H., Manitius, A. J., Weinstein, E., Katz, A. I. and Pickford, G. E. (1969). *Yale J. Biol. Med.* **41,** 388–393.
Epstein, F. H., Cynamon, M. and McKay, W. (1971). *Gen. comp. Endocr.* **16,** 323–328.
Ericsson, J. L. E. (1967). *Z. Zellforsch. mikrosk. Anat.* **83,** 219–230.
Ericsson, J. L. E. and Seljelid, R. (1968). *Z. Zellforsch. mikrosk. Anat.* **90,** 263–272.
Etoh, H. and Egami, N. (1963). *Proc. Jap. Acad.,* **39,** 503–506.
Fagerlund, U. H. M. (1967). *Gen. comp. Endocr.* **8,** 197–207.
Fagerlund, U. H. M. (1970). *J. Fish. Res. Bd. Can.* **27,** 1169–1172.
Fagerlund, U. H. M. and Donaldson, E. M. (1970). *J. Fish. Res. Bd. Can.* **27,** 2323–2331.
Falkmer, S. and Matty, A. J. (1966). *Acta Soc. Med., Uppsala* **71,** 156–172.
Falkmer, S., Thomas, N. W. and Boquist, L. (1974). *In* "Chemical Zoology" (M. Florkin and B. T. Scheer, eds.) Vol. 8, pp. 195–260. Academic Press, New York and London.
Fangë, R. (1963). *In* "The Biology of Myxine" (A. Brodal and R. Frangë, eds.) pp. 516–529. Universitetsforlaget, Oslo.
Fenwick, J. C. and Forster, M. E. (1972). *Gen. comp. Endocr.* **19,** 184–191.
Fernholm, B. and Olsson, R. (1969). *Gen. comp. Endocr.* **13,** 336–356.
Forrest, J. N., Cohen, D. A., Schon, D. A. and Epstein, F. H. (1973a). *Am J. Physiol.* **224,** 709–713.
Forrest, J. N., MacKay, W. C., Gallagher, B. and Epstein, F. H. (1973b). *Am. J. Physiol.* **224,** 714–717.
Forster, M. E. (1970). "Hormonal effects on the physiology of the eel, *Anguilla anguilla* L." Ph.D. Thesis, University of Sheffield, England.
Fossat, B. and Lahlou, B. (1977). *Am. J. Physiol.,* **233,** F525–F531.
Freeman, H. C. and Idler, D. R. (1971). *Steroids* **17,** 233–350.
Freeman, H. C. and Idler, D. R. (1973). *Gen. comp. Endocr.* **20,** 69–75.
Fryer, J. N. (1975). *Can. J. Zool.* **53,** 1012–1220.
Gaitskell, R. E. and Chester Jones, I. (1970). *Gen. comp. Endocr.* **15,** 491–493.
Gaitskell, R. E. and Chester Jones, I. (1971). *Gen. comp. Endocr.* **16,** 478–483.
Garcia, L. E. and Meier, A. H. (1973). *Biol. Bull.* **144,** 471–479.
Goodman, J. H. and Butler, D. G. (1972). *Comp. Biochem. Physiol.* **42A,** 277–296.
Gorbman, A. and Tsuneki, K. (1975). *Gen. comp. Endocr.* **26,** 420–422.
Goswami, S. V. and Sundararaj, B. I. (1971). *J. exp. Zool.* **178,** 457–466.
Hane, S., Robertson, O. H., Wexler, B. C. and Krupp, M. A. (1966). *Endocrinology* **78,** 791–800.
Hanke, W. (1978). *In* "General, Comparative and Clinical Endocrinology of the Adrenal Cortex" (I. Chester Jones and I. W. Henderson, eds.) Vol. 2, pp. 417–495. Academic Press, London, New York and San Francisco.
Hanke, W. and Chester Jones, I. (1966). *Gen. comp. Endocr.* **7,** 166–178.
Hanke, W., Bergerhoff, K. and Chan. D. K. O. (1967). *Gen. comp. Endocr.* **9,** 64–75.
Hanke, W., Bergerhoff, K. and Chan. D. K. O. (1969). *Gen. comp. Endocr.* Suppl. **2,** 331–341.
Hanson, R. C., Duff, D., Brehe, J. and Fleming, W. R. (1976). *Physiol. Zool.* **49,** 376–385.
Hardisty, M. W. (1972a). *In* "The Biology of Lampreys" (M. W. Hardisty and I. C. Potter eds.). Vol. 2, pp. 171–192. Academic Press, London and New York.
Hardisty, M. W. (1972b). *Gen. comp. Endocr.* **18,** 501–514.
Hardisty, M. W. and Potter, I. C. (1971) *In* "The Biology of Lampreys" (M. W.

Hardisty and I. C. Potter, eds.). Volume 1, pp. 85–126 and 127–206. Academic Press, London and New York.

Hartman, F. A., Lewis, L. A., Brownell, K. A., Angerer, C. A. and Sheldon, F. (1944). *Physiol. Zool.* **17,** 228–238.

Harwood, C. T. and Mason, J. W. (1956). *Am. J. Physiol.* **186,** 445–452.

Hatey, J. (1951a). *C.r. Seanc. Soc. Biol.* **145,** 172–175.

Hatey, J. (1951b). *C.r. Seanc. Soc. Biol.* **145,** 315–318.

Hawkins, E. F. and Ball, J. N. (1970). *J. Endocr.* **48,** xxvii–xxviii.

Hawkins, E. F., Hargreaves, G. and Ball, J. N. (1970). *J. Endocr.* **48,** lxxiv–lxxv.

Henderson, I. W. and Chester Jones, I. (1967). *J. Endocr.* **37,** 319–325.

Henderson, I. W. and Chester Jones, I. (1972). *Annls Inst. Michel Pacha.* **5,** 69–235.

Henderson, I. W., Chan, D. K. O., Sandor, T. and Chester Jones, I. (1970). *Mem. Soc. Endocr.* **18,** 31–55.

Henderson, I. W., Sa'di, M. N. and Hargreaves, G. (1974). *J. Steroid Biochem.* **5,** 701–707.

Henderson, I. W., Jackson, B. A. and Hargreaves, G. (1975). *Trans. Biochem. Soc.* **3,** 1168–1171.

Henderson, I. W., Jotisankasa, V., Mosley, W. and Oguri, M. (1976). *J. Endocr.* **70,** 81–95.

Hickman, C. P. and Trump, B. F. (1969). *In* "Fish Physiology" (W. S. Hoar and D. J. Randall, eds.). Vol. 1, pp. 91–240. Academic Press, New York and London.

Hill, C. W. and Fromm, P. O. (1968). *Gen. comp. endocr.* **11,** 69–77.

Hirano, T. (1967). *Proc. Jap. Acad. Sci.* **43,** 793–796.

Hirano, T. (1969). *Endocr. japon.* **16,** 557–560.

Hirano, T. (1974). *J. exp. Biol.* **61,** 737–747.

Hirano, T. (1977). *Gunma Symp. Endocr.* **14,** 45–59.

Hirano, T. and Bern, H. A. (1972). *Endocrinol. japon.* **19,** 41–45.

Hirano, T. and Mayer-Gostan, N. (1976). *Proc. natn. Acad. Sci. U.S.A.* **73,** 1348–1350.

Hirano, T. and Mayer-Gostan, N. (1978). *In* "Comparative Endocrinology" (P. J. Gaillard and H. H. Boer, eds.). pp. 209–212. Elsevier/North Holland Biomedical Press, Amsterdam, New York and Oxford.

Hirano, T. and Utida, S. (1968). *Gen. comp. Endocr.* **11,** 373–380.

Hirano, T. and Utida, S. (1971). *Endocrinol. japon.* **18,** 47–52.

Hirano, T., Johnson, D. W., Bern, H. A. and Utida, S. (1973). *Comp. biochem. Physiol.* **45A,** 529–540.

Hirano, T., Morisawa, M., Ando, M. and Utida, S. (1976). *In* "Intestinal Ion Transport" (J. W. L. Robinson, ed.). pp. 301–317. MTP Press, Lancaster.

Hirano, T., Takei, Y. and Kobayashi, H. (1978). *In* "Osmotic and Volume Regulation" (C. Barker Jørgensen and E. Skadhauge, eds). pp. 123–128. XI Alfred Benzon Symposium, Munksgaard, Copenhagen.

Holmgren, N. (1950). *Acta Zool, Stockh.* **31,** 233–348.

Holt, W. F. and Idler, D. R. (1975). *Comp. Biochem. Physiol.* **50C,** 111–119.

Holmes, W. N. (1959). *Acta Endocr. Copnh.* **31,** 587–602.

Holmes, W. N. and Butler, D. G. (1963). *J. Endocr.* **25,** 457–464.

Holmes, W. N. and McBean, R. L. (1963). *J. exp. Biol.* **40,** 335–341.

Idler, D. R. and Clemens, W. A. (1959). "International Pacific Salmon Fisheries Commission Progress Report". New Westminster, B. C., Canada. 80 pp.

Idler, D. R. and Kane, K. M. (1976). *Gen. comp. Endocr.* **28,** 100–102.

Idler, D. R. and Szeplaki, B. J. (1968). *J. Fish. Res. Bd. Can.* **25,** 2549–2560.

Idler, D. R. and Truscott, B. (1972). *In* "Steroids in Non-mammalian Vertebrates" (D. R. Idler, ed.) pp. 127–252. Academic Press, London and New York.

Idler, D. R., Ronald, A. P. and Schmidt, P. J. (1959). *Can. J. Biochem. Physiol.* **37,** 1227–1238.

Idler, D. R., O'Halloran, M. J. and Horne, D. A. (1969). *Gen. comp. Endocr.* **13,** 303–312.

Idler, D. R., Sangalang, G. B. and Truscott, B. (1972). *Gen. comp. Endocr.* Suppl. **3,** 238–244.

Ikeda, Y., Ozaki, H. and Sawada, S. (1976). *Jap. J. Ichthyol.* **23,** 93–99.

Inui, Y. and Gorbman, A. (1978). *Comp. Biochem. Physiol.* **60A,** 181–183.

Inui, Y. and Yokote, M. (1975). *Bull. jap. Soc. Sci. Fish.* **41,** 973–981.

Jampol, L. M. and Epstein, F. H. (1970). *Am. J. Physiol.* **218,** 607–611.

Janssens, P. J., Vinson, G. P., Chester Jones, I. and Mosley, W. (1965). *J. Endocr.* **32,** 373–382.

Johansson-Sjobeck, M–J., Dave, G., Larsson, A., Lewander, K. and Lidman, U. (1978). *Comp. Biochem. Physiol.* **60A,** 165–168.

Johnson, D. W. (1972). *Gen. comp. Endocr.* **19,** 7–25.

Johnson, D. W. (1973). *Am. Zool.* **13,** 799–818.

Johnson, D. W., Hirano, T., Bern, H. A. and Conte, F. P. (1972). *Gen. comp. Endocr.* **19,** 115–128.

Jørgensen, C. B. (1976). *In* "General, Comparative and Clinical Endocrinology of the Adrenal Cortex" (I. Chester Jones and I. W. Henderson, eds). Vol. 1, pp. 143–206. Academic Press,, London, New York and San Francisco.

Kaley, G. and Donshik, P. C. (1965). *Biol. Bull.* **129,** 411.

Kaley, G., Robinson, A. and Lubben, B. (1963). *Biol. Bull.* **125,** 381.

Kamiya, M. (1972). *Endocr. Japon.* **19,** 489–494.

Kamiya, M. and Utida, S. (1968). *Comp. Biochem. Physiol.* **26,** 675–686.

Katz, Y. and Eckstein, B. (1974). *Endocrinology* **95,** 963–967.

Kenyon, A. J. (1967). *Comp. Biochem. Physiol.* **22,** 169–176.

Kirsch, R. and Laurent, P. (1975). *C.r. hebd. Seanc. Acad. Sci., Paris* **280,** 2013–2015.

Kirschner, L. B. (1969). *Comp. Biochem. Physiol.* **29,** 596–604.

Klesch, W. and Sage, M. (1973). *Comp. Biochem. Physiol.* **45A,** 961–967.

Klesch, W. and Sage, M. (1975). *Comp. Biochem. Physiol.* **52A,** 145–146.

Krishnamurthy, V. G. and Bern, H. A. (1973). *Acta Zool.* **54,** 9–14.

Lagios, M. D. (1968). *Gen. comp. Endocr.* **11,** 248–250.

Lahlou, B. and Giordan, A. (1970). *Gen. comp. Endocr.* **14,** 491–509.

Lahlou, B., Henderson, I. W. and Sawyer, W. H. (1969). *Comp. Biochem. Physiol.* **28,** 1427–1434.

Larsen, L. O. (1969). *Gen. comp. Endocr. Suppl.* **2,** 522–527.

Larsen, L. and Rothwell, B. (1972). *In* "The Biology of Lampreys" (M. W. Hardisty and I. C. Potter, eds). Vol. 2, pp. 1–67. Academic Press, London and New York.

Leloup-Hatey, J. (1958). *C.r. hebd. Seanc. Acad. Sci., Paris* **246,** 1088–1091.

Leloup-Hatey, J. (1968). *Comp. Biochem. Physiol.* **26,** 997–1013.

Leloup.-Hatey, J. (1972). *Gen. comp. Endocr.* **18,** 603.

Leloup-Hatey, J. (1974). *Gen. comp. Endocr.* **24,** 28–37.

Leloup-Hatey, J. (1976). *Can. J. Physiol. Pharmacol.* **54,** 262–276.

Leloup-Hatey, J. and Hardy, A. (1976). *J. Physiol., Paris,* **72,** 103A.

Liversage, R. A., Price, B. W., Clarke, W. and Butler, D. G. (1971). *J. exp. Zool.* **178,** 23–28.

Lockley, A. S. (1957). *Copeia* 241–242.
McBride, J. R. and van Overbeeke, A. P. (1969). *J. Fish. Res. Bd. Can.* **26**, 2975–2985.
McCarthy, J. L., Corley, R. C. and Zarrow, M. K. (1960). *Proc. Soc. exp. Biol. Med.* **104**, 787–789.
McKeown, B. A. and Hazlett, C. A. (1975). *Comp. Biochem. Physiol.* **50A**, 379–381.
Macchi, I. A. and Rizzo, F. (1962). *Proc. Soc. exp. Biol. Med.* **110**, 433–436.
Maetz, J. (1968) *In* "Perspectives in Endocrinology" (E. J. W. Barrington and C. Jørgensen, eds.) pp. 47–162. Academic Press, London and New York.
Maetz, J. (1969). *Gen. comp. Endocr., Suppl.* **2**, 229–316.
Maetz, J. (1971). *Phil. Trans. Roy Soc., Lond.* B **262**, 209–249.
Maetz, J. (1976). *In* "Lung Liquids" (R. Porter and M. O'Connor, eds.). CIBA Fdn. Symp. **38**, 133–155. Elsevier/Excerpta Medica,/North Holland, Amsterdam, Oxford and New York.
Maetz, J. and Skadhauge, E. (1968). *Nature, Lond.* **217**, 371–372.
Maetz, J., Mayer, N. and Chartier-Baraduc, M. M. (1967a). *Gen. comp. Endocr.* **8**, 177–188.
Maetz. J., Sawyer, W. H., Pickford, G. E. and Mayer, N. (1967b). *Gen. comp. Endocr.* **8**, 163–176.
Malvin, R. L. and Vander, A. J. (1967). *Am. J. Physiol.* **213**, 1582–1584.
Mayer, N. (1970). *Bull, Infs tech. Commt. Energ. atom.* **146**, 45–73.
Mayer, N. and Maetz, J. (1967). *C.r. hebd. Seanc. Acad. Sci., Paris* **264**, 1632–1635.
Mayer, N., Maetz, J., Chan, D. K. O., Forster, M. E. and Chester Jones, I. (1967). *Nature, Lond.* **214**, 1118–1120.
Mazeaud, F. (1969). *C.r. Seanc. Soc. Biol.* **163**, 24–28.
Mazeaud, F. (1971). *C.r. Seanc. Soc. Biol.* **165**, 539–544.
Milne, K. P., Ball, J. N. and Chester Jones, I. (1971). *J. Endocr.* **49**, 177–178.
Mizogami, S., Oguri, M., Sokabe, H. and Nishimura, H. (1968). *Am. J. Physiol.* **215**, 991–994.
Molnár, B. and Szabó, Zs. (1968). *Acta biol. hung.* **19**, 373–379.
Morisawa, M. and Hirano, T. (1978). *Comp. Biochem. Physiol.* **59C**, 111–115.
Morris, R. (1960). *Symp. zool. Soc. Lond.* **1**, 1–16.
Morris, R. (1965). *J. exp. Biol.* **42**, 359–371.
Morris, R. (1972). *In* "The Biology of Lampreys" (M. W. Hardisty and I. C. Potter, eds.). Vol. 2, pp. 193–239. Academic Press, London and New York.
Motais, R. (1967). *Annls. Inst. Oceanogr., Paris* **45**, 1–83.
Motais, R. (1970). *Comp. Biochem. Physiol.* **34**, 497–501.
Motais, R. and Maetz, J. (1965). *C. r. Hebd. Seanc. Acad.`Sci., Paris* **261**, 532–535.
Munz, F. W. and McFarland, W. N. (1964). *Comp. Biochem. Physiol.* **13**, 381–400.
Nace, P. F. (1956). *Anat. Rec.* **124**, 340.
Nakano, T. and Tomlinson, N. (1967). *J. Fish. Res. Bd. Can.* **24**, 1701–1715.
Narasimhan, P. V. and Sundararaj, B. I. (1971a). *J. Fish. Biol.* **3**, 441–451.
Narasimhan, P. V. and Sundararaj, B. I. (1971b). *Comp. Biochem. Physiol.* **39B**, 89–99.
Nishimura, H. (1971). *J. Med. Soc. Toho, Japan* **18**, 946–951.
Nishimura, H. and Ogawa, M. (1973). *Am. Zool.* **13**, 823–838.
Nishimura, H., Oguri, M., Ogawa, M., Sokabe, H. and Imai, M. (1970). *Am. J. Physiol.* **218**, 911–915.
Nishimura, H., Ogawa, M. and Sawyer, W. H. (1973). *Am. J. Physiol.* **224**, 950–956.
Nishimura, H., Sawyer, W. H. and Nigrelli, R. F. (1976). *J. Endocr.* **70**, 47–59.
Nolly, H. L. and Fasciolo, J. C. (1972). *Comp. Biochem. Physiol.* **41A**, 249–254.

Oguri, M. and Nace, P. F. (1966). *Chesapeake Science* **7**, 198–202.
Oide, M. (1967). *Annotnes zool. japon.* **40**, 130–135.
Oide, M. (1973). *Comp. Biochem. Physiol.* **46A**, 639–645.
Oide, M. and Utida, S. (1967a). *Marine Biol.* **1**, 102–106.
Oide, M. and Utida, S. (1967b). *Marine Biol.* **1**, 172–177.
Olivereau, M. (1966a). *Annls. Endocr., Paris* **27**, 665–678.
Olivereau, M. (1966b). *Annls. Endocr. Paris* **27**, 549–560.
Olivereau, M. and Ball, J. N. (1970). *Mem. Soc. Endocr.* **18**, 57–85.
Olivereau, M. and Chartier-Baraduc, M–M. (1965). *C.r. Seanc. Soc. Biol.* **159**, 1498–1502.
Pang, P. K. T. (1973). *Am. Zool.* **13**, 775–792.
Patent, G. J. (1970). *Gen. comp. Endocr.* **14**, 215–242.
Payan, P. and Maetz, J. (1970). *Bull. Infs tech. Commt. Energ. atom.* **146**, 77–96.
Payan, P. and Maetz, J. (1971). *Gen. comp. Endocr.* **16**, 535–554.
Payan, P., Goldstein, L. and Forster, R. P. (1973). *Am. J. Physiol.* **224**, 367–372.
Peterson, R. E. (1957). *J. clin. Endocr. Metab.* **17**, 1150–1157.
Phillips, J. G. and Mulrow, P. J. (1959). *Proc. Soc. exp. Biol. Med.* **101**, 262–264.
Phillips, J. G., Holmes, W. N. and Bondy, P. K. (1959). *Endocrinology*, **65**, 811–818.
Pickford, G. E., Griffith, R. W., Torretti, J., Hendler, E. and Epstein, F. H. (1970a). *Nature, Lond.* **228**, 378–379.
Pickford, G. E., Pang, P. K. T., Weinstein, E., Torretti, J., Hendler, E. and Epstein, F. H. (1970b). *Gen. comp. Endocr.* **14**, 524–534.
Pickford, G. E., Srivastava, A. K., Slicher, A. M. and Pang, P. K. T. (1971). *J. exp. Zool.* **117**, 109–118.
Pora, E. A. and Precup, O. (1971). *Marine Biol.* **11**, 77–81.
Porthé-Nibelle, J. and Lahlou, B. (1975). *Comp. Biochem. Physiol.* **50A**, 801–806.
Porthé-Nibelle, J. and Lahlou, B. (1978). *J. Endocr.* **78**, 407–416.
Rall, D. P. and Burger, J. W. (1967). *Am. J. Physiol.* **212**, 354–356.
Renfro, J. L. (1975). *Amer. J. Physiol.* **228**, 52–61.
Robertson, J. D. (1960). *J. exp. Biol.* **37**, 879–888.
Robertson, O. H. and Wexler, B. C. (1959). *Endocrinology* **65**, 225–238.
Robertson, O. H. and Wexler, B. C. (1962). *Gen. comp. Endocr.* **2**, 458–472.
Robertson, O. H., Krupp, M. A., Thomas, S. F., Favour, C. B., Hane, S. and Wexler, B. C. (1961). *Gen. comp. Endocr.* **1**, 473–484.
Robertson, O. H., Hane, S., Wexler, B. C. and Rinfret, A. P. (1963). *Gen. comp. Endocr.* **3**, 422–436.
Roscoe, M. J. (1975). Functional morphology and physiology of the pituitary complex and interrenal gland in elasmobranch fishes. PhD Thesis, University College of North Wales, Bangor 115 pp.
Sandor, T., Chan, S. W. C., Phillips, J. G., Ensor, D. M., Henderson, I. W. and Chester Jones, I. (1970). *Can. J. Biochem.* **48**, 553–558.
Sandor, T. Fazekas, A. G. and Robinson, B. H. (1976). *In* "General, Comparative and Clinical Endocrinology of the Adrenal Cortex" (I. Chester Jones and I. W. Henderson, eds.). pp. 25–142, Vol. 1. Academic Press, London, New York, San Francisco.
Sargent, J. R., Thomson, A. J. and Bornancin, M. (1975). *Comp. Biochem. Physiol.* **51B**, 75–79.
Scheer, B. T. and Langford, R. W. (1976). *Gen. comp. Endocr.* **30**, 313–326.
Sharratt, B. M., Bellamy, D. and Chester Jones, I. (1964). *Comp. Biochem. Physiol.* **11**, 19–30.

Shedadeh, Z. H. and Gordon, M. S. (1969). *Comp. Biochem. Physiol.* **30**, 397–418.
Singley, J. A. and Chavin, W. (1971). *Am. Zool.* **11**, 653.
Singley, J. A. and Chavin, W. (1975). *Comp. Biochem. Physiol.* **50A**, 77–82.
Skadhauge, E. (1969). *J. Physiol. Lond.* **204**, 135–158.
Smith, H. W. (1931). *Am. J. Physiol.* **98**, 279–295.
Smith, H. W. (1936). *Q. Rev. Biol.* **7**, 1–26.
Sokabe, H., Mizogami, S., Murase, T. and Sakai, F. (1966). *Nature, Lond.* **212**, 952–953.
Sokabe, H., Mizogami, S. and Sato, A. (1968). *Jap. J. Pharmacol.* **3**, 332–343.
Sokabe, H., Oide, H., Ogawa, M. and Utida, S. (1973). *Gen. comp. Endocr.* **21**, 1061–1667.
Srivastava, A. K. and Meier, A. H. (1972). *Science* **177**, 185–187.
Srivastava, A. K. and Pickford, G. E. (1972). *Gen. comp. Endocr.* **19**, 290–303.
Storer, J. H. (1967). *Comp. Biochem. Physiol.* **20**, 939–948.
Subhedar, N. and Rao, P. D. P. (1974). *Gen. comp. Endocr.* **23**, 403–414.
Sundararaj, B. I. and Goswami, S. V. (1966). *J. exp. Zool.* **163**, 49–54.
Swallow, R. L. and Fleming, W. R. (1965). *Am. Zool.* **5**, 688.
Taylor, A. A. and Davis, J. O. (1971). *Am. J. Physiol.* **221**, 652–657.
Truscott, B. and Idler, D. R. (1968). *J. Fish Res. Bd. Can.* **25**, 431–435.
Umminger, B. L. (1970). *Am. Zool.* **10**, 299.
Umminger, B. L. (1971). *Experientia* **27**, 701–702.
Umminger, B. L. (1972). *J. exp. Zool.* **181**, 217–222.
Umminger, B. L. and Gist, D. H. (1973). *Comp. Biochem. Physiol.* **44A**, 967–977.
Utida, S. and Hirano, T. (1973). *In* "Responses of Fish to Environmental Changes" (W. Chavin, ed.) pp. 240–269. Charles Thomas, Springfield, Illinois.
Utida, S., Oide, M., Saishu, S. and Kamiya, M. (1967). *C.r. Seanc. Soc. Biol.* **161**, 1201–1204.
Utida, S., Oide, M. and Oide, H. (1968). *Comp. Biochem. Physiol.* **27**, 239–250.
Utida, S., Kamiya, M. and Shirai, N. (1971). *Comp. Biochem. Physiol.* **38A**, 443–447.
Utida, S., Hirano, T., Oide, H., Ando, M., Johnson, D. W. and Bern, H. A. (1972). *Gen. comp. Endocr.* Suppl. **3**, 317–327.
Utida, S., Kamiya, M., Johnson, D. W. and Bern, H. A. (1974). *J. Endocr.* **62**, 11–14.
Walker, R. M. and Johansen, P. H. (1975) *Experientia* **31**, 1252–1253.
Weisbart, M. and Idler, D. R. (1971). *J. Endocr.* **17**, 416–423.
Wendelaar Bonga, S. E. (1973). *Z. Zellforsch. mikrosk. Anat.* **137**, 563–588.
Whitehouse, B. J. and Vinson, G. P. (1975) *Gen. comp. Endocr.* **27**, 305–313.
Wikgren, B. J. (1953). *Acta zool. fenn.* **71**, 1–102.
Wingfield, J. C. and Grimm, A. S. (1977). *Comp. Biochem. Physiol.* **31**, 1–11.
Wright, P. A. (1961). *Biol. Bull.* **121**, 414.
Youson, J. H. (1973). *Am. J. Anat.* **138**, 235–252.

8. The Adrenal Cortex and its Homologues in Vertebrates; Evolutionary considerations

R. J. Balment,* I. W. Henderson and I. Chester Jones

*Department of Zoology, University of Sheffield, Sheffield S10 2TN, England and *Department of Zoology, University of Manchester M13 9PL, England*

1. Introduction

The living processes are regulated by a variety of agents and some of these are termed hormones. Classically, hormones emanate from traditional ductless glands and are delivered directly into blood vessels (Chester Jones and Henderson, 1976a). Modern research has revealed a host of bioregulators which widen the concept of those blood borne chemical messengers and bring in neuroendocrinological and other activators. Furthermore, many compounds which act in related ways, nevertheless do not lend themselves to precise definition. For example there is a spectrum of "tissue" hormones. The universality of materials such as amino and nucleic acids is of course notable, although the evolutionary stages at which some chemical compounds could be regarded as hormonal must, of necessity, be deduced from extant species. Here insurmountable difficulties arise because animals adapt and specialize to particular environments, thereby overshadowing the precise delineation of progressive steps in lines of evolutionary change. That small part of the problem represented by adrenocortical/interrenal tissue does give some broad trends which may help in understanding wider issues. This immediately brings another unit of life, the steroids, under examination.

The naturally occurring steroids comprise compounds of diverse structures with the carbon nucleus, cyclopentanoperhydrophenanthrene, as the common feature (Kime and Norymberski, 1976). Steroids are found in plants and in both invertebrate and vertebrate animals. Their role is, for the most part, unknown. Indeed Sandor and his co-workers advanced the premise that "steroids are very ancient bioregulators which evolved prior to the appearance of eukaryotes or were even synthesized abiotically" (Sandor *et al.*, 1975). A fundamental problem is thus immediately presented. Whatever the origin of life (Wald, 1973; Folsome, 1979), its units are presumed to have been present prior to forming organisms as such.

With the assembly of these units, sequestration of steroid producing, or perhaps merely concentrating, units ensued to give the characteristic secretions of vertebrate adrenocortical and gonadal tissues. Here we have morphological evolution coupled with changes in the structure of a primordial chemical compound.

Such changes may be regarded as evolutionary ones in the adoption of a hormonal role. If this is so, this modifies the dictum (Medawar, 1953) that evolution of endocrine glands is not one of chemical formulae, but reactivities, reaction patterns and tissue competence (see Chester Jones, 1976). In the jawed vertebrates, the Gnathostomata, biosynthetic pathways are similar and give the predominant adrenal steroids cortisol and corticosterone and, from the gonads, testosterone, 4-androstene-3, 17-dione, progesterone, oestrone and oestradiol-17β. Eighteen-oxygenated corticosteroids of which aldosterone is physiologically the most potent, can be readily detected in the Tetrapoda. In the fish groups, Teleostei and Selachii, its presence is variable, infrequently identified and often with some dubiety (Chapter 7). Whilst 18 hydroxylation of steroids is not an uncommon occurrence, it may be thought that aldosterone, integrated into the control of sodium and potassium and perhaps secondarily water homeostasis, was associated with the conquest of the terrestrial habitat by proto-amphibians.

A further problem is not that the cells of living species are incapable of many types of steroidal biosynthesis, but that their function is often not clear in plants, in invertebrates and in the Agnatha. Steroids and steroid metabolizing systems are widespread. The phylogeny of terpenoids, sterols, steroids, their associated enzymes, their products and intermediates has been exhaustively discussed (see for example, Sanderman, 1962a, b; Bergmann, 1962; Grant, 1962; Florkin, 1974; Swain, 1974; de Ley and Kersters, 1975). The suggested evolutionary processes clearly demonstrate a close interplay between plants and animals as well as between the living systems and the environment regarding the supply and transformation of steroids. For example it has been supposed that dytiscid beetles produce vertebrate-type adrenocortical and gonadal steroids, possibly from terpenes synthesized by plants, which act as defence mechanisms (Schildknecht, 1971). Capacities demonstrated in insect male and female gonads for steroidogenesis as in molluscs and echinoderms testify that similar biosynthetic pathways are well-nigh universal (Lehoux et al., 1967, 1968, 1970; Dubé et al., 1968; Lehoux and Sandor, 1969, 1970; Dubé and Lemonde, 1970). Ecdysones are formed from cholesterol both in arthropods and plants. Particularly well known are the moulting hormones of insects; α-ecdysone from pupae of *Bombyx mori* was the first to be characterized as 2β, 3β, 14α, 22R, 25-pentahydroxy-5β-cholest-7-en-6-one (see Highnam, 1975; Highnam and Hill, 1977). This compound is produced by the thoracic glands and the very active form, β-ecdysone, may occur by hydroxylation in peripheral tissues, especially the oenocytes of the haemolymph (*op. cit*). For steroids to have a high moulting activity, the following structural features are indispensable: (1) cis fusion of rings A and B of the steroid skeleton; (2) β-hydroxylic function at position 3; (3) keto group at position 6 in conjugation with Δ^7-doubled bond; (4) steroid side chain with an appropriately (r)-orientation hydroxylic function at position 22. This example from the dominant invertebrate group emphasizes that the evolution of the adoption of a hormonal role for steroids in vertebrates demands a precise spatial configuration (Kime and Norymberski, 1976). In summary, then, of the ubiquitous presence of steroids in living matter, it is possible to group them as: (i) steroids which contribute a property to the life of the individual; (ii) steroids to which a hormonal or pheromonal role can be assigned; (iii) steroids which occur in copious quantities with no known role; and (iv) prokaryotes which, when presented with the appropriate steroid substrate, give microbiological conversions to steroid hormones of the vertebrate types. Steroids, as far as is known, require the presence of receptor complexes if a biological effect is to occur. Such receptors, of somewhat variable specificity, probably display an evolution of equal complexity, but their existence in all vertebrate groups, let alone the animal kingdom as a whole, has yet to be verified.

2. Adrenocortical homologue throughout the vertebrates

A. MORPHOLOGY

The Gnathostomata present a recognizable morphological evolution of the adrenal gland that is bound up with the kidney, the gonads and the neural crest (Chester Jones, 1976) with a general trend towards aggregation of isolated adrenocortical islets (Fig. 1). The eventual fate of the adrenocortical homologue is closely related to the fate of mesonephric blastema (Witschi, 1956), a feature exemplified during embryology and apparent in the adult.

Amphibia have a persistent mesonephros in the adult so that it is in

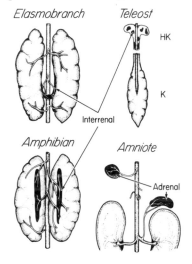

Fig. 1. The formation of kidney, gonadal and interrenal/adrenocortical tissue from one block of embryonic mesoderm inevitably leads to the morphology found in the adult vertebrate. In elasmobranchs the interrenal lies posteriorly on or near the mesonephros whose anterior portion is given over to some reproductive tracts. In the largest order of bony fish, the Teleostei, interrenal tissue remains anteriorly in the mesonephros which in this region is, for the most part, non-renal. In Amphibia generally, the interrenal tissue lies firmly on the ventral surface of the mesonephros though with islets here and more anteriorly in some orders. The amniote adrenal—reptiles, birds, mammals—is a separate encapsulated organ so formed because of the "new" kidney, the metanephros, with the evolutionary erstwhile functional mesonephros now mostly vestigial donating only a few persistent structures. HK, head kidney; K, kidney.

close association with interrenal tissue which is found chiefly on the ventral surface, fed by branches of the dorsal aorta and draining through venules of the renal veins (Hanke, 1978). The other Classes, Mammalia, Aves and Reptilia (the tetrapod amniotes), have discrete encapsulated adrenal glands vascularized from the major blood vessels coursing through the abdominal cavity (Holmes and Phillips, 1976; Idelman, 1978; Lofts, 1978). This is an inevitable concomitant of the appearance of the metanephros leaving the adrenocortical tissue separated from the mesonephros which has only vestigial remnants in the adults. Pisces, as a general term for fish, present many problems. The major groups of the jawed fish, teleosts and elasmobranchs have great differences in the morphological appearance of the interrenal tissues. The common factor is their association with the cardinal vein and, in different ways, with the pronephric/mesonephric segments (Chapter 7, this volume). This is also the case with Amiiformes and Acipenseriformes, while on the other hand the Dipneusti may represent, in the living form, an off-shoot not unlike the proto-Amphibia. Here the interrenal is of the amphibian type, associated with kidney tissue, similar to that found in the tadpole stage of this Class (Chapter 7, this volume). In the cladistic approach it might be possible to say that a line of evolution in the conquest of the land habitat was by neotenic forms of proto-Amphibia.

Two of the major features of the gnathostomatous vertebrate are first the microanatomy of adrenocortical tissues and secondly its association with chromaffin tissue (Fig. 2). It has been advanced that the basic adrenocortical unit is a cord of cells and indeed cords are fundamental modes of cellular aggregation in kidneys and gonads (Chester Jones, 1957, 1976). This leads to the phenomenon of zonation of the adrenal cortex seen clearly in the eutherian mammal. Peripherally lies the zona glomerulosa persistent without the secretions of the anterior lobe of the pituitary yet responsive to them, and secreting aldosterone and 18-hydroxycorticosterone. The more centripetal zona fasciculata, in parallel line of cords, produces cortisol/corticosterone. The innermost region, the zona reticularis, may be inactive or may be a corticosteroid reservoir or a site of some sex steroid secretion (Idelman, 1978). The X zone and fetal zones are special enigmatic features (Idelman, 1978). Zonation is not readily explicable and any consideration is inevitably bound up with the search for secretory products. Hartman and Brownell (1949) give the early history and conceptual advances and reference should be made to this book for the bibliography. Adrenalectomized laboratory eutherians were used, for the most part, for assay purposes. By 1927

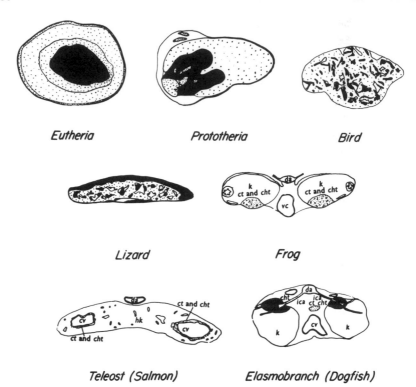

Fig. 2. Diagrammatic representation of the relationships between chromaffin and adrenocortical tissues in various vertebrate types. In the eutherian mammal the chromaffin tissue is centralized. One example of the Prototheria, *Echidna*, shows eccentric aggregation of the chromaffin tissue. In birds, reptiles, amphibians and teleost fish chromaffin and cortical tissue intermingle, the Elasmobranchii differ in that cortical and chromaffin tissue are completely separated. Stippled, cortical tissue; black, chromaffin tissue; k, kidney; hk, head kidney; cv, cardinal vein; vc, vena cava; cht, chromaffin tissue; da, dorsal aorta; ica, intercostal artery; ct, cortical tissue. (Modified from Chester Jones, 1957.)

adrenal extracts were shown to increase the survival time of such animals. By 1939, and indeed somewhat earlier, adrenal cortical extracts were clearly divided into two factors: (1) the sodium factor, a term now used rather than the "salt and water hormone" of C. N. H. Long and F. D. W. Lukens; (2) the gluconeogenic factor. The development of the technique to hypophysectomize the rat by P. E. Smith showed the persistence of the zona glomerulosa later considered to secrete a "sodium factor" (see Deane and Greep, 1946). Thus the hypophysectomized rat is in sodium balance in contrast to the condition of the adrenalectomized animal. Failure to characterize the sodium factor was later overcome by the development of further techniques. It was characterized as aldosterone and its chemical formula determined (see Tait and Tait, 1979). Though the secretion of aldosterone seemed to demand the presence histologically of a zona glomerulosa, this was principally due to the restriction of the search to the eutherian laboratory animal. Some type of zonation is apparent in many vertebrates, from elasmobranchs (Chapter 7, this volume) to Amphibia (Varma, 1977) and birds (Holmes *et al.*, 1978; Pearce *et al.*, 1978, 1979; Holmes, personal communication). The imprecise nature of zonation in non-mammalian vertebrates may be resolved by the theory that the basic unit of adrenocortical cells is a cord and that the arrangement of such cords reflects different mechanical associations (Fig. 3). It is interesting in this respect that Varma (1977) has given ultrastructural evidence for aldosterone- and corticosterone-secreting cells in the adrenocortical tissue of the American bullfrog, *Rana catesbeiana*. It is thus possible that the interrenal/adrenocortical cord of tetrapods has cells some of which always possess 18-hydroxylases; in the Pisces such cells are only occasionally present.

Once encapsulated outside and confined then to a constricted area, conspicuous zonation is inevitable. Obvious zonation is seen when there is a strong centralized tissue of completely different embryological origin, the medulla. Chromaffin tissue has its origin in the neural crest, cells occurring above the neural tube in the embryo. Thence they migrate, passing through the growing cells· of the body to condense in several sites to give a variety of structures (Weston, 1970; Chester Jones, 1976). The question is why should chromaffin and cortical tissue come together in differing degrees and yet, paradoxically, be completely separate in the elasmobranchs. There has been much speculation about the possible functional significance of this association and separation but no helpful facts. Secretion of adrenalin and nor-adrenalin is not confined to the adrenal medulla and its vertebrate homologues. Separation of the two tissues may be the basic vertebrate pattern (Chapter 7, this volume). One speculation is that the adrenal cortex (and its homologues) depends for its morphology and differential activities on the vascular pattern . Vaso-constriction and vaso-dilation are then essential factors in the different secretions and the centralization in the eutherian mammals with clear zonation may represent the apotheosis of this association. The selachian interrenal remains separated from the chromaffin tissue and it is just possible that its notorious slowness of response to experimental, if not natural, changes may, at least in part, be determined by this.

The extension of an examination of evolutionary changes in

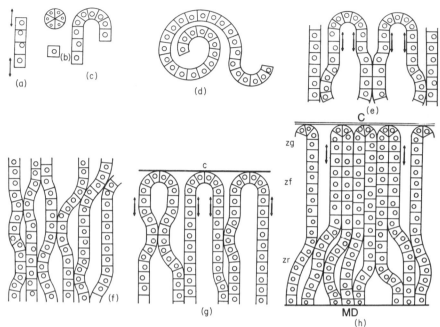

Fig. 3. The organization of the basic cord of adrenocortical cells throughout the vertebrates. (a) the basic unit: a cord of cells capable of growth in each direction; (b) transverse sections of (a) showing that the cord can be either a single line of cells or grouped to give a radial pattern; (c) simple looping of the basic cord; (d) more complex looping (as in teleosts); (e) illustrating the capacity to form loops, all parts of which are not equally capable of growth; it is easier for growth to take place in the direction away from the loop; (f) organization of cords of cells in longitudinal runs (as in amphibians and reptiles); (g) looping of cords peripherally against a well-formed outer connective tissue capsule (C) (as in birds) when the consideration given in (e) applies; (h) looping of cords peripherally against a well-formed capsule (C) (when (e) again applies) and also the termination of the cords against a central medulla (MD) (with necessary alteration of growth pattern)—as in Metatheria and Eutheria (zg, zona glomerulosa; zf, zona fasciculata; zr, zona reticularis). (From Chester Jones, 1957.)

morphology to the Agnatha brings unexplained problems. The Pet-romyzonidae are more tractable for inclusion in a continuing evolutionary scheme and they can be regarded, certainly on the hypothesis of cladism (Chapter 7 this volume), as separate from the second Order, the Myxinoidea. In the lampreys (Petromyzonidae), both from the embryology and the histological appearance, interrenal cells occur, particularly in the region of the pronephros. This would be in keeping with the interdependent relationship of the interrenal–kidney–gonad (see below). However we are faced with the absence of the characteristic enzyme Δ^5, 3β-ol steroid dehydrogenase though indeed it was once demonstrated in lampreys on an upstream migration, when it might be expected (Seiler, K. and Seiler, R. personal communication). The quantities of typical vertebrate corticosteroids in plasma are slight or not demonstrable. It is conceivable that the extant lampreys represent that stage in evolution when steroids occur but are not harnessed in a positive way to the essential physiology of the animal. Morphological expression appears before functional although extrapolation to the Placodermi is too conjectural. This is emphasized by the Myxinoidea, perhaps with randomly situated interrenal cells in the walls of the cardinal veins (Chester Jones, 1963) or in the pronephric region (Idler and Burton, 1976). Our recent histological investigations do not reveal interrenal cells though it would be more satisfactory to show them to be present. This would accord with the identification of some plasma corticosteroids. Perhaps the appearance of the vertebrate from some invertebrate ancestor did not demand, at the outset, regulation by hormones now identified in extant gnathostome species. Extant lower chordates (Hemi-, Uro- and Cephalo-chordate), insofar as they are representative of fossil animals, certainly do not appear to have an active and complete complement of vertebrate hormones.

Embryology provides the background for the interpretation of the various morphological manifestations of adrenocortical tissue. In vertebrate development, blocks of tissue form along the central axis to make mesoderm with a basic segmentation partially or fully express- ed. From the bilateral fingers of tissue are carved the kidney, gonads and adrenal cortex (Kerr, 1919; Goodrich, 1958; Witschi, 1956; Chester Jones, 1957). There is competition between the proportion of the block on each side of the developing animal which is to be taken by the three structures. Furthermore there is competition for the use of ducts for renal excretion and shedding of gametes. The adrenal cortex, as an organ of internal secretion using only vascular outlets, awaits passively the outcome of the battle which will delineate its morphological appearance over evolutionary time. The results of this interplay of renal and gonadal tissue and their ducts have varied in the vertebrate orders, and the differences in the general anatomy of the adrenal gland and its homologues reflect the consequences of this interplay.

The ancestral craniate may be envisaged as a worm-shaped creature provided with a complete set of segmented tubules through- out the whole trunk region, opening into a longitudinal duct leading to a posterior cloaca. Goodrich (1958) considers that the Craniata were originally provided with a continuous series of segmental excretory tubules opening by peritoneal funnels into the coelom and that by the backwards growth and coalescence of their outer ends they formed a longitudinal duct leading to the cloaca. Such a primitive uniform archinephros no longer exists in any living form, for specialization of the series led to differentiation into pronephric and mesonephric (opisthonephric) regions. The tendency for the pronephric tubules to develop early and to be replaced by later, more posterior, tubules led to further specialization of the pronephros with early completion of the duct with which the later developing mesonephric tubules fused. That is, instead of the duct being formed by the fusion of a succession of segmental rudiments, it tended more and more to be produced by the anterior tubules which grew freely back to the cloaca. Meanwhile, the increasing importance of the mesonephros induced the tubule-forming tissue (nephrotome) to produce numbers of secondary tubules and the original segmentation was lost. The exit from the coelom, and the coelomic funnel, though originally developed to transmit genital products, would also allow the passage of superfluous coelomic fluids, incidentally helping to remove excretory matter. The nephridial tube thus came to transmit (i) reproductive cells; and (ii) excretory products. In vertebrates there are several familiar examples of part of the kidney system associated with transmission of genital cells becoming separated off from those which retain a renal function. The two main authors from which this account is taken differ in that Goodrich considered that evolution brought about a more and more intimate connexion between originally independent genital funnels and nephridial tubes. On the other hand, Kerr favoured the view that the funnel opened into the nephridial tube at the time of its first appearance, the progress of subsequent evolution having been in the direction of separating genital funnel and nephridial tube—not uniting them.

To summarize, there is an archinephros with its archinephric duct and the development takes place from the anterior backwards to give regions designated, for convenient nomenclature, pronephros, mesonephros and metanephros (Kerr, 1919). The pronephros may be functional in larval vertebrates but the metotic anterior and posterior segments to not develop tubules, and in vertebrates generally the pronephros degenerates. The mesonephros (opisthonephros) is the functional kidney in the adult Anamniota. In the Amniota (reptiles, birds and mammals), whilst mesonephros and metanephros may be discerned, the former is functional only in early development. Here the metanephros and its separate duct comprise the adult kidney. This definitive kidney is devoted to the elimination of waste products and there are separate genital ducts but in lower Gnathostomata the kidney and its ducts serve also for the passage to the exterior of gonadal products. The gross anatomy of the adrenal gland is thus greatly influenced by the ontogeny of the adult kidney (Witschi, 1956).

B.   ADRENOCORTICAL SECRETIONS AND THEIR BIOLOGICAL ACTIVITIES

The chemical class of steroids comprises numerous compounds in plants and animals though there are many with unassigned physiological functions. Thus more than 50 different steroids have been isolated from and attributed to the adrenal cortex. Yet only a small number appear in the venous blood carrying adrenocortical secretions and of them, few have relevant biological activity. This means that the majority of steroids found by processing large numbers of the available adrenal glands (cattle, sheep, pig) are intermediate in the biosynthetic pathways or catabolic products (Sandor et al., 1976; Kime, 1978). Functional adrenal steroids influence carbohydrate, lipid and protein metabolism and have anti-inflammatory effects; these are cortisol and corticosterone to which are added, though less dominant, cortisone, 11-dehydrocortisone, 11-deoxycortisol and 21-deoxycortisol. A second functional category of adrenocortical steroid regulates electrolyte relationships, especially those of sodium and potassium, and here aldosterone and deoxycorticosterone are especially important. The distinction between these two steroidal activities is one of convenience but frequently not of fact, and once we leave the eutherian mammals their relative importance varies from group to group. Representatives of the Tetrapoda so far examined have aldosterone as do lungfishes. However its production in gnathostomatous fish is in doubt. Birds, reptiles and elasmobranchs also have 11-hydroxycorticosterone in significant amounts. This latter group, the cartilaginous fish of modern times, diverged from the proto-bony fish some 300 million years ago and may produce 1α-hydroxycorticosterone as a major corticosteroid along with corticosterone. The former has activity in "mineralocorticoid" assays using the laboratory rat and the toad bladder (Sandor et al., 1976) but its relevance to elasmobranch physiology remains to be fully described (see Chapter 7, this volume).

The two extant orders of cyclostomes, the Agnatha, are taken together but the Myxinidae (hagfishes) and Petromyxonidae (lampreys) are not closely related (Chapter 7). In both cases corticosteroids have not been shown to be normally involved in their physiology nor, indeed, have they been shown to occur without reservations and to be synthesized by identified interrenal tissue. It is intriguing to consider that the vertebrate body evolved, both jawed and jawless, by prototypes which did not rely on standard hormone production. Yet the major endocrine organs are represented in living forms. The hagfish and lamprey have recognizable pituitary bodies, but trophic functions are equivocal and an hypothalamic control has not been clearly demonstrated. It may be suggested that the cyclostomes are secondarily degenerated from placoderms with a full complement of endocrine activities. There is no real answer to this. Yet the lampreys, with osmoregulation to give plasma electrolyte values in keeping with those of the bony teleosts, suggest an original basic capacity which other lines modified in evolution by the accretion of hormonal function.

A theme to which comparative endocrinology constantly returns is one that considers the evolution of hormonal patterns in relation to tissue response. The latter may, indeed, not be a central question in contemplating the appearance of complex endocrine inter-relations. It is seen that mammalian ACTH acts in cyclostomes to raise titres of corticosteroids (Idler *et al.*, 1971), but we have discussed earlier that the presence of these steroids as normally circulating is in doubt. Tissue reactivity does not have to evolve, but hormone production of familiar, widespread, chemical types does. The evolution of hormones has been refinement and perhaps multiplication of factors that control their secretion. Thus the mammalian adrenal cortex comes under the influence of many factors including ACTH, the renin-angiotensin, potassium and serotonin. At what stage in evolution these influences gained predominance cannot be readily judged and there remains a great deal to be done before even tentative generalizations can be made. The conceptual problem is that the adrenal cortex produces materials that influence certain key physiological events at specific target organs, although it is itself a specific target organ for other hormones with feedback between the homeostatic events and adreno-cortical function.

3.   Adrenocortical homologue and its interrelationship with sense organs required for physiological adaptation

Adrenocortical/interrenal secretions are essential for life in most if not all vertebrates, although the underlying reasons for morbidity in their absence are not always clear. Physiological adaptations associated with, for example, the maintenance of body fluid homeostasis in the presence of either excess or insufficient environmental sodium are complex, involving and affecting many organs. Here we view corticosteroids as central to the adaptive processes of the CNS, the kidney and the liver, systems which reciprocally interact with adrenocortical structure and function. The previous volumes (Chester Jones and Henderson, 1976b, 1978) cover the literature and further details are to be found in Blaschko *et al.* (1975). Our discussion must, of necessity, be speculative.

A.   ADRENOCORTICAL HOMOLOGUES AND THE CENTRAL NERVOUS SYSTEM

Nerves are predominant in endocrinological activities both because of the accepted direct excitatory and inhibitory roles and because of their neurosecretory function in producing readily defined hormones. These are important controlling factors within the framework of internal secretion.

Pre-eminent are the connexions, principally examined in a few species of Eutheria, between the hypophysis and hypothalamus which in turn receive signals from the brain to respond, *inter alia,* to such environmental changes as season, temperature etc. The hypothalamic control of adrenocorticotrophic hormone (ACTH) release from the pituitary is central to adrenocortical regulation and is well established for most vertebrates, although the manner in which control and adjustment of steroidogenesis is achieved is more controversial. Afferent neural and humoral stimuli are integrated to allow the adrenal cortex to adapt to changes in both internal and external environments.

Ball *et al.* (1980) have considered aspects of the hypothalamoadenohypophysial systems in lower vertebrates. In ancestral chordates, even before the pituitary developed, the ventral floor of the primitive "brain" may have received neurosecretory terminals which discharged systemically. Neurosecretory cells are widely distributed in invertebrates. Examining the chordates, we find, for example, in a tunicate (*Ciona*) that there are cells in the cerebral ganglion that exhibit activity related to the ovarian cycle and which appear to control daily rhythms of the sub-neural gland during the breeding season, in phase with tides (Holmes and Ball, 1974; Georges, 1977). Speculation has assigned the sub-neural gland as, in part, the homologue of the adenohypophysis. In another group, Cephalochordata, *Amphioxus* has two pairs of peptidergic neurosecretory nuclei lying in the "brain bladder" producing axons which terminate synaptically on the ventral basement lamella (Obermüller–Wilen, 1979). Thus it is perhaps an archaic chordate feature to develop neurosecretory neurons in the brain which discharge on its ventral surface to control extracerebral structures and functions, that is by neurohormones. Extant cyclostomes demonstrate an intermediate condition reminiscent of the *Amphioxus* "brain bladder" on the one hand and the vascular median eminence on the other. Whether primitive or degenerate, this condition may illustrate an early stage in the association of the adenohypophysis and the median eminence. The brain–pituitary interrelationships seen in the gnathostomes and more precisely depicted in the Eutheria, seem, as it were, sketched out in the Agnatha. Yet here the hagfish hypothalamus does not appear to control whatever meagre functions reside in the adenohypophysis. Nor in lampreys (Petromyzonidae) are the more obvious actions of the adenohypophysis conspicuously controlled by the hypothalamus.

The Gnathostomata, the jawed vertebrates, form a structural and physiological group however much the details may differ. It is inevitable that, within the Classes, the terms "primitive" and "specialized" are used, often erroneously but obviating tortuous conditional descriptions. The median eminence is an ancient and conservative structure (Holmes and Ball, 1974; Lagios, 1975). Differences as given by Ball *et al.* (1980) and Ball (1980) include the actual innervation of the adenohypophysis, the median eminence relationship and neurohumoral transport to the pars distalis by capillaries. These in the well documented eutherian mammals comprise portal "veins" and are vastly enlarged vessels typified by fenestrated endothelial cells (Bergland and Page, 1979). Teleost fish show the most specialized type of brain–pituitary control mechanisms in that there is apparent incorporation of the median eminence and portal system as the rostral neurohypophysis, adjacent to the pars distalis. Nevertheless in more primitive teleosts, such as salmonids, control of pars distalis function is typically exerted by nerve endings on the capillary plexus in the rostral neurohypophysis, carrying the controlling hypophysiotrophic materials by short vessels.

We can see the pattern emerging in a consideration of all living vertebrates, but reliance must, at present, be put on the hypothalamic–hypophysial system of eutherian mammals. These, in the light of the sparse information about non-mammalian vertebrates, must provide the model to demonstrate the functional complexities of the central nervous system–pituitary–adrenal cortex axes. The end products, the corticosteroids, are not secreted at constant rates. Wide fluctuations in the titres of plasma steroids occur, many following the inherent, basic circadian rhythms. This is inversely related to photo-periods in all groups of vertebrates, from fish to mammals, which have been examined. Moreover, seasonal variations in adrenocortical/interrenal activity may be superimposed upon the daily cycles. Circadian rhythms in eutherian mammals oscillate in close temporal relation with centrally controlled sleep/wake or locomotor patterns. It is usual for the levels of circulating steroids to rise during the resting hours to reach maximum concentrations at the onset of the phases of daily activity. Such corticosteroidal cycles are paralleled, though preceded by cyclic rise and fall in corticotrophic releasing factor (CRF) and ACTH secretion (Iscart et al., 1977). However, variation in ACTH secretion is not necessarily the sole factor in producing the cycle of adrenal activity for there is also a diurnal rhythm in the response or reactivity of the adrenal gland to its trophic hormone. Also the splanchnic nerve, which supplies the adrenal, may play a permissive role in the regulation of the synthesis and/or release of 17 hydroxycorticosteroids (Henry et al., 1976). Thus central stimuli may alter adrenocortical stimulation by either autonomic or endocrine pathways.

The hypothalamus dominates the link between neural and endoc-rine systems. It is salutary that, despite its early general characteriza-tion, CRF has not been chemically identified in contrast to other hypothalamic releasing/inhibitory factors (Brodish and Lyman-grover, 1977). It is likely that mammalian researches demonstrating the relationship of αCRF to αMSH and βCRF to vasopressin present a pattern common to the Vertebrata, however much of the details may be modified (Schally et al., 1960; Yates and Maran, 1974). Significance may be apportioned to vasopressin, vasotocin and MSH as peptides associated with ACTH secretion concomitantly with the hydro-osmotic actions of corticosteroids. This may be a common pattern in vertebrates, discerned though not positively delineated (Yates et al., 1971a, b; Buckingham and Leach, 1979).

The pituitary–adrenal system may, however, be activated not only by hypothalamic CRF but also by extrahypothalamic corticotrophin releasing factors (CRFs), one of which Brodish (1973) has termed tissue-CRF. Tissue-CRF may be distinguished from hypothalamic CRF on the basis of its physical-chemical properties, its extreme potency, its prolonged action and its presence even after the entire hypothalamus has been removed. Witorsch and Brodish (1972) have demonstrated that extracts of liver, kidney and thymus can induce ACTH secretion. It has been postulated that intense and prolonged stress, which results in tissue damage, may evoke the release of such tissue-CRFs to supplement the rapid hypothalamic mechanism and thereby produce a prolonged and massive output of adrenocortical steroids. The demonstration that corticosterone administration may suppress the elaboration of tissue-CRF indicates that a negative feedback system could exist at the tissue level. Although this concept is somewhat speculative, it is clear that CRFs are produced at sites other than the CNS. Certainly, there is a modulating effect of the negative feedback of circulating titres of corticosteroids upon ACTH and CRF production (Yates and Maran, 1974). A common occurrence in vertebrates may be the short feed back loop, that is a circumscribed entity between pituitary and hypothalamus (Motta *et al.*, 1969).

Neural inputs, both stimulatory and inhibitory, to the hypothalamo–hypophysial–adrenocortical axis converge on CRF-neurones, such that rapid responses to afferent stimuli produce a circadian adrenocortical function.

Photoperiodic cues are perhaps the most obvious influences on ACTH release, and these are quite separate from those evoking stress release of the hormone. For example, anterior hypothalamic lesions abolish circadian 17-hydroxycorticosteroidal rhythmicity without altering responsiveness to stress (Slusher, 1964). The retinohypothalamic projection, common to all vertebrates, is essential for the light–dark cycling of corticosteroids in those species examined (Raisman and Brown–Grant, 1977; Moore, 1978).

Pathways towards the CRF centres are complex, involving spinal cord, brain stem, with oligosynaptic and multisynaptic connexions, and probably cerebral cortex, basal ganglia, rhinencephalon and thalamus. Such pathways have been demonstrated in a few species, but on morphological grounds are likely to be common to all vertebrates. Indeed it is of interest that many stimuli which increase adrenocortical activity also elicit changes in diverse areas of physiological function (vasopressin, growth hormone, blood pressure, melanophores, respiration, prolactin). Common neural pathways for endocrine and visceral function are thus likely (Sayers, 1961).

The complexity of the pathways themselves is further magnified by consideration of the suggested neurotransmitters involved. Vermes and Smelik (this volume, Chapter 1) and Bohus and de Wied (Chapter 5) have exhaustively examined the possible adrenergic, serotoninergic, aminergic, dopaminergic and cholinergic mechanisms involved in the mammalian CRF–ACTH system. It is pertinent to this chapter simply to state that avian ACTH release requires both serotoninergic and cholinergic pathways while in the teleost fish ACTH release is governed by monaminergic innervation of the pituitary. A common thread of vertebrate systems is at present difficult to discern and the modalities within each group may indeed reveal singular species and group idiosyncrasy.

The integrity and function of the CNS and peripheral nervous system is itself governed by corticosteroids, possibly as a result of their actions on myelogenesis and electrolyte distributions (Timiras *et al.*, 1968).

In addition to any effects of corticosteroids on centrally controlled behaviour, they appear to regulate the level at which sensory input is perceived and the manner in which this information is integrated by the central nervous system (Henkin, 1975). The ability to detect sensory signals of taste, olfaction and audition significantly increase in adrenal insufficiency. The 17-hydroxycorticosteroids inhibit both central and peripheral nervous systems impairing both detection and integration. Such changes again follow a circadian rhythm.

ACTH release rarely takes place in isolation and many other factors are released or inhibited at least under experimental circumstances. Among these vasopressin has been mentioned, but growth hormone, MSH and prolactin are also involved. The physiological relevance of these observations is not clear but a close physiological interplay is perhaps apparent. Thus there is a close seasonal rhythm of adrenocortical activity with seasonal breeding patterns of many vertebrates, while oestrogens and progesterone enhance corticosteroid synthesis, possibly by actions on the renin-angiotensin system. Indeed, ACTH synthesis and pituitary sensitivity to CRF may be affected by oestrogens (Coyne and Kitay, 1969). In sexually maturing salmon increased rates of cortisol secretion occur alongside elevated androgen (Fagerlund and Donaldson, 1969) and oestrogen (Donaldson and Fagerlund, 1969) secretion. In contrast an antagonism between testicular and adrenal secretions has been described in the duck (Assenmacher and Boissin, 1970). It is probable that there are shared neural inputs to the hypothalamus stimulating secretion of gonodotrophic and corticotrophic releasing hormones. In addition, corticosteroids influence phenylethanolamine transferase, tyrosine dehydroxylase and dopamine hydroxylase in the peripheral nervous system and the eutherian adrenal medulla (e.g. Mueller *et al.*, 1970; Thoenen and Otten, 1978). Such actions follow diurnal rhythms in association with a dose dependent influence of corticosteroids on enzyme synthesis. The higher levels of corticosteroids in the adrenal medullary circulation of eutherian mammals (2 orders of magnitude greater than in the periphery) may have differential effects at this site compared with actions in peripheral sympathetic ganglia.

This difference between the adrenal medulla and peripheral sympathetic ganglia is apparent in early post natal life. During the first two weeks after birth the rat pituitary–adrenal axis is not fully developed, the steroid production is generally low and does not follow a diurnal rhythm. During this period adrenal medullary catecholamine synthesizing enzymes are inducible by short-term cold stress. In contrast, the sympathetic ganglia do not respond at all, unless the animals are preinjected with glucocorticoids.

Whether such an interaction between corticosteroids and catecholamine synthesis is present in non-mammalian vertebrates remains to be investigated. The necessity for the close proximity of the adrenal cortex to the medulla for such an effective modulation of catecholamine production in mammals, could be examined by consideration of interrenal and chromaffin cell function in other vertebrate groups, where such a close morphological association is not apparent.

Prolactin is probably one of the most important pituitary factors that influences adrenocortical function. Prolactin has many actions at target tissues shared with corticosteroids (Table I) particularly those involving water and electrolyte balance (Nicoll, 1974). Prolactin may also influence steroidogenic activity of the adrenal cortex more directly in many vertebrate groups. Conversely adrenal steroids may alter pituitary prolactin secretion in mammals, as indicated by the elevation of plasma prolactin levels following adrenalectomy in the rat (Ben–David et al., 1971). A trophic action of prolactin on steroidogenic tissue may be a general phenomenon among vertebrates in view of prolactin's gonadotrophic activity. On such organs the basic action of prolactin could be to maintain precursor pools and inhibit steroid catabolizing enzymes.

A recent and as yet not fully explored finding that has great relevance to CNS–adrenocortical function is the description of an independent renin–angiotensin system within the brain of the rat and dog. Isorenin, the enzyme substrate angiotensinogen, angiotensin I, angiotensin I converting enzyme, angiotensin II, angiotensin II receptors and the angiotensinases are present in brain tissue (Ganten et al., 1978). Angiotensin II-like immunoreactivity is evident in both the adenohypophysis and hypothalamus. A specific, direct, action of angiotensin on the hypothalamo–pituitary–adrenal axis waits demonstration, but fluctuations in blood or CSF levels of angiotensin are associated with changes in CRF secretion (Gann, 1969). Indirect effects of the isorenin–angiotensin system may arise through its

TABLE I

Actions of Prolactin involving synergism with corticosteroid hormones or on organs also influenced by corticosteroids. (Modified from Nicoll, 1974)

Cyclostomes
    Electrolyte metabolism in fish

Teleosts
    Na retention by gills
    Na retention by kidney
    Salt and water movement in the gut
    Dispersal of yellow pigment in xanthophores
    Maintenance of the brood pouch in the seahorse

Amphibians
    Na transport across the anuran bladder
    Skin and electrolyte changes associated with water drive

Reptiles
    Restoration of plasma Na levels in hypophysectomised lizard

Birds
    Stimulation of nasal (orbital) gland secretion

Mammals
    Milk secretion
    Sebaceous and preputial gland secretion
    Renal Na reabsorption

influence upon neurohypophysial peptide secretion and its interference with central control of thirst and salt appetite which may in turn alter the tone of the hypothalamo–pituitary–adrenal axis. The presence of such an isorenin–angiotensin system in non-mammals has not been established but such mechanisms if universally present will be of great significance to our understanding of renal–adrenocortical endocrine evolution (see below).

Although the brain isorenin system has probably no direct effect on adrenocortical steroidogenesis, its extension in the renal renin–angiotensin system (RAS) is apparently central to the regulation of mammalian aldosterone secretion. Control of aldosterone follows a similar pattern to that of 17-hydroxycorticosteroids in most non-mammalian groups examined so far, that is a negative feedback control based upon the CRF–ACTH system. The Mammalia differ in that aldosterone synthesis shows considerable independence of pituitary control and that the RAS is the major influence (see Davis, 1975).

Although aldosterone titres in fish may be slight and spasmodic, it may be the major component of steroid secretion of some tetrapod vertebrates. In amphibians, reptiles and birds aldosterone has a full complement of glucocorticoid activity, and because of the large amounts secreted it may have a physiological role in carbohydrate metabolism in these groups. In mammals, although aldosterone still retains its glucocorticoid activity because of the low levels produced, it has no physiological role in carbohydrate metabolism. Indeed, it is perhaps then not surprising to find that the mode of aldosterone secretion has changed and shows some independence of the pituitary. The basic elements of the renal RAS are present in all other jawed vertebrates except elasmobranchs, but its physiological influence upon adrenal steroidogenesis remains equivocal in non-mammalian vertebrates.

Not all the observed dynamics of mammalian aldosterone secretion can, however, be correlated with changes in the RAS (Table II; Coghlan *et al.*, 1979). Angiotensin II and the heptapeptide angiotensin III may become permissive, rather than stimulatory agents, when

TABLE II

Examples of the breakdown of the angiotensin–aldosterone nexus. (Modified from Coghlan *et al.*, 1979)

1. Hypoxia

2. Ascent to high altitude

3. Small postural change

4. Congestive cardiac failure

5. Essential hypertension

6. Treatment with β-blockers

7. Inappropriate secretion of antidiuretic hormone

8. Familial dysautonomia

9. Idiopathic hyperaldosteronism

10. Starvation and refeeding

11. Fasting

sodium challenge is severe. The change in adrenal sensitivity induced by a falling sodium level is more effective in increasing aldosterone secretion than angiotensin under these conditions. Factors as yet unrecognized, with direct central nervous system connections, can influence aldosterone secretion. For example, the reduction in aldosterone secretion, which follows drinking sodium bicarbonate solution in sodium-depleted sheep, occurs before any significant amount of sodium has been absorbed (Denton *et al.*, 1977). Raising the sodium level in the cerebro-spinal fluid artificially similarly produces a reduction in aldosterone secretion in sodium-depleted sheep. This again suggests that the central nervous system can influence aldosterone secretion without alterations in the accepted stimuli (Abraham *et al.*, 1976). It is now clear that even in mammals ACTH secretion is essential for maximal aldosterone secretion, and normally ACTH may have an important supportive role for aldosterone synthesis and release.

The phylogeny of these differences in the mode of control of aldosterone is unclear. The reliance of the so-called mineralocorticoid release upon a renal hormone, renin, may indeed by an early acquisition, and central nervous control of ACTH release may be an attribute of these species that began to delicately control less obvious features of their metabolism such as liver glycogen.

B.   INTERRELATIONSHIPS BETWEEN THE ADRENOCORTICAL HOMOLOGUE AND THE LIVER

The liver is the largest visceral organ (with the possible exception of the mature gonad of some species) of the vertebrate body, and is almost certainly homologous in all groups. Variations in cellular arrangement, in vascular supply and in ultrastructure exist (Elias and Bengelsdorf, 1952; Mugnaini and Harboe, 1967). However, the generalized function of secretion into the circulation, storage of metabolic supplies, excretion of waste products into the bile are universal and more than likely these were the original attributes of an organ derived from an endodermal diverticulum of the foregut and mesoderm. Its particular relationships with the nervous system, the gut and its vasculature and the circulation in general have, in recent years, focused attention on hepatic sensory systems (see Sawchenko and Friedman, 1979).

The major homeostatic role of the liver cannot be questioned, but it is useful to examine some of the functions with respect to the adrenal cortex and the evolution of corticosteroidal actions in the vertebrates.

Data on hepatic function in non-mammalian vertebrates are scanty, and in this discussion speculative extrapolation is necessary to allow generalizations on evolutionary specialization and adaptive radiation. Table III lists most of the pertinent functions of the liver with respect

TABLE III

Functions of the vertebrate liver which may primarily or secondarily affect adrenocortical physiology.

Sensory functions: sodium receptor, baroreceptor, chemoreceptor

Angiotensinogen production

Hormone degradations: primary—steroid breakdown to influence half life and metabolic clearance; conjugation of steroids
secondary—degradation of trophic hormones (e.g. ACTH, angiotensin)
Excretion: steroid metabolites etc.

Plasma composition: osmolarity—urea, sodium, potassium, protein, lipids, cholesterol

Production of hormone binding protein

Cholecalciferol metabolism

---

to adrenocortical function; not all these functions have been demonstrated in all the major groups of vertebrates, but circumstantial evidence is available for most of the major items.

The liver of the hagfish appears to carry out certain functions that the renal tubule performs in other vertebrates (Fänge and Krog, 1962; Rall and Burger, 1967), in particular the excretion of organic acids and the excretion of magnesium. In this animal, it may be that the liver performs the primary function of maintaining plasma composition, while the kidney merely regulates volume. At this stage it may be that the liver, ideally sited to monitor dietary intake and blood composition, is the primary site of action for corticosteroids (given that they are physiologically active in this group!). Moreover hepatic metabolism of cholesterol and its derivatives as various salts and conjugates for eventual excretion in bile is an early characteristic (Haslewood, 1962). Corticosteroid metabolism and excretion by this route occurs but the factors that control the enzymes are poorly understood. The liver regulates the biological activity of corticosteroids in a variety of ways including influences on (i) rates of destruction and excretion in bile (conjugation, sulphation etc.); (ii) "pool" size, by virtue of the production of corticosteroid-binding protein; (iii) rates of destruction or inactiveness of trophic hormones (e.g. ACTH, angiotensin II) which secondarily affects adrenocortical output.

Another suggested homeostatic role of the liver that impinges on adrenocortical physiology of the vertebrates is brought out by consideration of elasmobranch physiology. Hepatic and possibly other sites of synthesis and turnover of urea and trimethyl-amine oxide as "osmotic ballast" are affected by osmotic stress (e.g. Watts and Watts, 1966; Goldstein et al., 1968). It is attractive to suppose that the liver, as a well established target organ of corticosteroids, regulates plasma composition by removal or release of osmotically active substances to influence plasma osmolarity; the liver thus becomes part of the vertebrate osmoregulatory system. Release of specific protein in response to osmotic stress in some teleosts (Yamashita, 1970a, b), the changing urea concentrations as a function of environmental salinity in elasmobranchs and amphibians (Janssens, 1964; Goldstein and Forster, 1971) and the differential renal responses to infusions into hepatic portal compared with femoral vein in laboratory mammals (see Sawchenko and Friedman, 1979) are suggestive of such an interrelationship.

There are many other aspects of hepatic physiology that logically could impose themselves, either primarily or secondarily, upon the adrenal cortex; these include the presence of chemo-, baro- and iono-receptor sytems with neuroendocrine outputs which ultimately act on corticosteroid function. One most illustrative example is that of angiotensinogen (renin substrate) production. The liver is the primary source of angiotensinogen which as a result of renin and angiotensin I converting enzyme activities produces angiotensin II to stimulate aldosterone production. In addition to synthesizing angiotensinogen, the liver also metabolizes renin and angiotensin I, and of course is responsible in part for aldosterone degradation. The secondary effects that the liver must have on sodium metabolism are obvious. How each component of the loop of activity between sodium retention and angiotensin production interrelates is largely unknown. The complexity is magnified when it is recognized that angiotensin production and/or release is itself influenced by corticosteroids, by oestrogens and possibly by antiotensin II (Reid and Ganong, 1974).

The speculative and known interactions between adrenocortical and hepatic function, just outlined briefly, have stemmed largely from experimental data obtained in mammals, and the degrees to which other vertebrates may utilize the liver for such regulation are largely unknown. The constant presence of the liver in the vertebrates suggests that many of its functions were established at an early stage. Investigations in the areas mentioned will clearly provide insight into the evolutionary significance of adrenocortical function and its control.

C. INTERRELATIONSHIPS BETWEEN THE ADRENOCORTICAL HOMOLOGUE AND THE KIDNEY

Section 2 of this chapter emphasized the propinquity between renal and adrenocortical tissue in vertebrate phylogeny and ontogeny. Teleologically therefore it is hardly surprising that there are intimate physiological associations between the two organs. The kidney, as a key osmoregulatory organ throughout the vertebrates, has been held as a central pre-adaptation that allowed the spectacular radiation of vertebrates as a whole. Recent years have however provided evidence that the kidney is not only an homeostatic excretory device but that it is also an endocrine organ with hormones that regulate such diverse aspects as blood pressure, sodium excretion, bone metabolism, water intake and erythrocyte production. In many respects the kidney forms a link between the adrenocortical secretions and the rest of the endocrine system, an association that perhaps stems from the condition of topographic closeness seen in contemporary anamniotes (see Fig. 1).

In this section, the kidney is treated in a similar way to the liver. Table IV lists a number of ways in which the kidney directly or indirectly affects the adrenal cortex and vice versa. Once again not all the items listed have been definitively shown for vertebrates as a whole and indeed there are some notable discrepancies in the vertebrate series. For example a renin–angiotensin system has not been demonstrated in elasmobranch and cyclostome fishes, and unequivocal actions of corticosteroids on renal tubular function in anamniotes await demonstration.

The endocrine function of the kidney has perhaps received most attention with respect to the renin–angiotensin system and vitamin D metabolism; it is within these two areas that statements can be made that apply to the vertebrate series as a whole. As noted, the renin–angiotensin system is reportedly absent from elasmobranch and cyclostome fishes—both histological and pharmacological approaches have failed to reveal renin-like granules or angiotensin-like pressor (assayed in rats) materials. Negative observations of this sort must be set against observations on effects of mammalian angiotensin in

TABLE IV

Functions of the vertebrate kidney which may directly or indirectly affect adrenocortical physiology

Renin–angiotensin system
Blood pressure
Sodium excretion
Erythropoeitin
Prostaglandins
Kinins
1, 25-$(OH)_2$-cholecalciferol
Steroid excretion, conjugation etc.

elasmobranchs (see Sokabe *et al.*, 1978). It could of course be that Agnatha lack a genuine renin–angiotensin system, and that the peculiar mode of osmoregulation adopted by elasmobranchs in early gnathostome evolution pre-empted an adaptive benefit for the endocrine control exerted in other vertebrates.

In mammals and birds renin is released in response to sodium depletion. This contrasts with representative teleosts and amphibians in which *elevated* environmental salinities may not change or indeed may increase plasma renin activities. Phylogenetically therefore there seems to be a dichotomy between homeothermic and poikilothermic vertebrates; in the former the renin release results in a stimulation of aldosterone production and thence renal sodium retention. In amphibians there is a peculiar anatomical arrangement in which the renal tubule, the adrenocortical cells and renin-producing cells closely appose one another. There is thus the possibility that very local control of aldosterone exists and peripheral changes in renin may not be discernible; thus the renal tubular fluid composition may be assayed by the renal tubular epithelium (macula densa?) which signals renin release to stimulate aldosterone production, almost at its site of action. This whole loop could well exist within the volume of a few hundred cubic microns.

Teleost fishes present further problems in that sodium loading provokes an increase in plasma renin activities and an increase in adrenocortical activity. This contrasts markedly with tetrapod responses. It may be that the elevated renin activities in marine environments are related not so much to renal sodium excretion, but to the relative polydipsia and reduced rates of glomerular filtrations (Hirano *et al.*, 1978; Henderson and Brown, 1979).

The dipsogenic activity of angiotensin, at least in mammals and birds and even perhaps in teleost fishes, reflects an interplay between the kidney, the adrenal cortical steroids and neurohypophysial antidiuretic peptides. Angiotensin and antidiuretic hormone in mammals are in negative feed back with one another and this can satisfactorily explain general aspects of water balance   (isorenin in brain, see Section 3, A). In the mammalian kidney urinary dilution in the ascending limb of the loop of Henle and distal tubular convolutions is dependent upon adrenal steroids, while subsequent water reabsorption (the extent of which depends upon the corticomedullary gradient established by the diluting process) is governed by an antidiuretic hormone. The amount of water available at these various sites is of course governed by glomerular filtration rate. Angiotensin may thus be viewed phylogenetically as a renal product that changes glomerular filtration by vasoconstriction, and influences electrolyte and water balance by acting on both adrenal steroid output and neurohypophysial hormone release. Whether these functions were acquired simultaneously is not known, but an attractive hypothesis is that the sites of renin production were related to primarily intrarenal roles, controlling the renal vasculature and interrenal steroid production (local hormone?). With the gradual separation of renal and adrenocortical tissue, angiotensin took on a more systemic role to influence not only antidiuretic hormone release but also to regulate systemic blood pressure. It is of interest that other renal hormones such as prostaglandins and kinins also influence systemic blood pressure as well as local intrarenal blood flow patterns. In the un-zoned kidneys of anamniotes these too may have been local hormones. With the evolution of metanephric kidneys, their local actions become more specialized, literally shunting blood from one area to another and by so doing produced massive changes in blood volume and hence pressure. Systemic actions were thus essential.

The study of evolutionary processes involving steroid hormones has come to include vitamin D metabolism, a key system regulating calcium metabolism (Norman and Henry, 1979). This is perhaps an ideal example with which to end these speculations on the interplay between the adrenocortical homologue and other organs. In the currently accepted scheme, vitamin D is 25-hydroxylated in the liver, then hydroxylated at position 1 in the kidney and it is this 1, 25-hydroxylated material that affects calcium homeostasis (De-Luca, 1973). The renal enzyme like those in the adrenal cortex is a classical mixed function hydroxylase: it is also a cytochrome P-450-containing enzyme with adrenoxin, and incorporates molecular oxygen. In an extensive survey, Henry and Norman (1975) found renal 1-hydroxylase in representatives of all vertebrate groups studied: unfortunately the survey did not include either cyclostome or elasmobranch representatives. In the latter the adrenocortical homologue contains such an enzyme to produce 1-α-OH corticosterone.

If the adrenocortical and renal enzymes are identical, it is not clear why their substrates differ, and indeed it is not known how specific their actions are (e.g. will the renal enzyme hydroxylate corticosterone?). The vitamin D enzyme is clearly intimately involved in calcium homeostasis at least of tetrapods, while the function of 1-α-OH-corticosterone is not fully elucidated. The endocrine control of skeletal development and calcium homeostasis in the vertebrates is complex and beyond the terms of reference of this discussion, but a broad comparison of elasmobranch systems with other gnathostomes is pertinent in the context of steroid 1-hydroxylation. Tetrapods employ parathyroid hormone, calcitonin and vitamin D metabolites as principal regulators. Bony fish lack parathyroid glands, possess calcitonin, contain large hepatic stores of vitamin D, which perhaps significantly are greater in fish living in hyperosmotic environments. Teleost fishes in addition utilize secretion(s) from the corpuscles of Stannius for calcium homeostasis. Specific vitamin D transport proteins have been demonstrated in the plasma of representatives of the major vertebrate groups (Hay and Watson, 1976); unfortunately in this study elasmobranch plasma was not examined. A final significant aspect is that elasmobranchs are hyper-osmoregulators and have a cartilaginous skeleton, while other gnathostomes are hypo-osmoregulators and have a bony skeleton. One may speculate that in the protovertebrate the adrenocortical homologue–renal enzymic make-up was shared, and with the separation to reach the interrenal gland of elasmobranch, the 1-hydroxylase remained within the interrenal. In other cases the development saw the kidney retaining the system. Until more is known about the physiology of the cartilaginous skeleton of the elasmobranch, the role of vitamin D metabolites in fish generally, and indeed the regulation of their plasma calcium, and the functional significance of this curious enzyme shared by the kidney on the one hand and the adrenocortical on the other must remain speculative.

The kidney and adrenocortical homologue thus maintain intimate endocrine relationships in adult vertebrates, which mirror the ontogenic and phylogenetic association. The original, perhaps intra-renal/adrenocortical mechanisms have now become systemic. Such broadening of renal–adrenocortical interactions may also be seen when the excretory function of the kidney is considered alongside the evolution of influenced hormones.

The influence of corticosteroids on ionic and osmoregulation is the dominant theme of the maintenance of the vertebrate body. In essence the corticosteroids facilitate ionic movement across cell membranes, particularly sodium, potassium and chloride. The regulation of body fluids occurs because of the actions of many hormones with whose functions the corticosteroids must interplay. The body composition of water and electrolytes in modern adult vertebrates is very alike, though elasmobranchs and myxinoids mar this pattern. Mechanisms are directed to maintain a constant internal environment, deviations in one direction bring opposite reactions, the principle of negative feedback for homeostasis (Bernard, 1855, 1859; Cannon, 1932; Greene, 1957; Langley, 1965). Maintenance of body fluid composition comprises three areas which involve individual solute concentration, the volume and osmotic concentration of fluids. Of paramount importance is the management of sodium and chloride coupled with potassium. Thence follows osmotic pressure of the blood and volume regulation.

In eutherian mammals, various receptors have been identified which are sensitive to changes in fluid volume (left atrial, carotid sinus baroreceptors), sodium content (adrenocortical cells; renin–angiotensin system) and osmotic concentration (hypothalamo–neurohypophysial system). In addition, there is the possibility of influences from dietary intake via receptors in the gut and liver. Receptors of these kinds are regarded as evoking release of secretions from the adrenal cortex (and its homologues), prolactin from the adenohypophysis and vasotocin or vasopressin from neurohypophysis as principle controllers of osmoregulation.

The conquest of a terrestrial habitat demanded an increasing precision of the control of intake and output to maintain constant body composition. Fish in water hyper-or hypo-osmotic to their blood plasma use basic mechanisms for homeostasis, yet on land, there is emphasis on the need for water economy and the greater capacity of the kidney as a controlling organ. Thus the kidney assumes an increasingly important role in osmoregulation as we follow presumptive lines of evolution from fish to mammals. In these lines are seen a variety of organs involved in osmoregulation—gills, bladders, rectal gland, skin and salt glands—but with the emergence of the mammals, particularly the Eutheria, the kidney is the central organ for ionic, osmotic and volume regulation of body fluids. The elaboration of kidney function is the predominant linking thread in evolution.

The basic renal unit, the nephron or renal tubule, has a glomerular component for filtering blood and tubular components for both reabsorption from, and secretion into, the tubular fluid—the urine (Fig. 4). The simplest form of nephron (Goodrich, 1958) consists of a

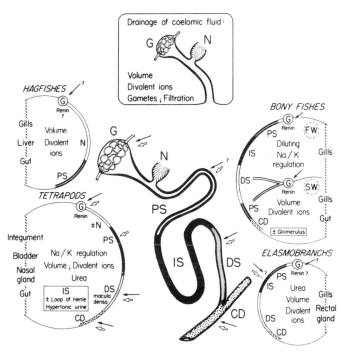

Fig. 4. Scheme summarizing the phylogenetic relationships between the functional activities of the vertebrate renal tubule and hormones. The original coelomoduct is considered to have given rise to the nephron and various segments evolved to carry out some of the functions mentioned (see Goodrich 1958). The nephrostome remains on the nephron of some types. Hagfishes have the simplest nephron, while elasmorbranch fishes have developed very complex segments. Bony fishes in SW and FW differ with respect to the degree of glomerular emphasis and the presence or absence of a distal, diluting segment. Extrarenal organs accessory to the kidney are also given. Renin is not universally present, and of especial note is its apparent absence from agnathans and elasmobranchs. General sites of peptide, protein and steroid hormonal actions are shown. ← : action of polypeptide, peptide hormones; ⇐ : action of steroid hormone. G: glomerulus; N: nephrostome; PS: proximal segment; DS: distal segment; CD: collecting duct; FW: fresh water; SW: sea water; IS: intermediate segment.

coelomoduct with an inserted glomerular tuft. Such a structure is similar to that found as the nephron of larval hagfish. In adult myxinoids, the vertebrate type of nephron is seen with glomerular tuft and simple tubule leading to the cloaca. Later vertebrate groups have segments additional to this basic tubular structure. These components, joined together to give the kidney, are so ideally suited to the excretion of large amounts of unwanted water that inevitably the origin of vertebrates was assigned to a fresh water environment (Marshall and Smith, 1930; Smith, 1932). This theory has given rise to much controversy (see Robertson, 1957). A great amount of dependence is placed on fossil records and deductions from the physiology of extant species. Two considerations bear upon this in significant ways. First, the Agnatha are clearly a diphylectic group (Stensiö, 1968). Secondly, there is no evidence from comparative anatomy or from the fossil record that gnathostomes were derived from any agnathan line (e.g. Greenwood, 1975).

Avoiding the designation of the hagfish as a secondary and degenerate form, it can be proposed that it is representative of an interphase animal of the invertebrate/vertebrate epoch. The contrasting blood composition of the two extant Agnathan groups may reflect differences in the sea water composition at the geological time of their appearance (Conway, 1942, 1943; Hutchinson, 1944). The myxinoid line may have embraced fluids of a higher tonicity and as a result their osmotic exchange is slow and limited. As in invertebrates there are osmoregulators and osmoconformers. This approach suggests a marine ancestry for the vertebrates and that the vertebrate nephron developed for reasons other than those of volume regulation. Moreover the "kidney" of decapod crustaceans functions in much the same way as the vertebrate kidney, that is, as a filtration–reabsorption–secretion system (Ramsey, 1968). In this case, it must have arisen from a marine ancestory. In some freshwater crustaceans an ion absorption segment has been added. The picture presented is that of a crustacean kidney developed primitively as an organ of ionic balance and secondarily to assume the function of volume regulation in those animals which entered estuaries and fresh water niches.

Similar considerations might suggest, contrary to the embedded Homer Smith concept, that the glomerulus of the vertebrate kidney appeared in a sea water environment as a device for the control of ionic balance. This later proved to be a useful preadaptation to fresh water habitats (Robertson, 1957). In the conquest of the new hypotonic environment, the function of volume regulation was added to the kidney together with an extra segment, the proximal tubule for the major reabsorption of ions. The evolution of the kidney involves the addition of extra distal segments thereby modifying the basic renal unit. Thus the volume of urine can be regulated by controlling glomerular filtration, while various other functions, divalent ion secretion, urinary dilution etc. became established and retained in one form or another by all vertebrates.

The marine origin of vertebrates is a matter of debate but the early movement to fresh water is agreed, and gives the line of gnathostome evolution, with the characteristic osmotic profile. Fish in sea water retain urea, as in elasmobranchs, or reduce filtration as in teleosts. The need to reduce the volume of fluid lost in urine when in sea water allows the appearance of the aglomerular kidney so that urine is formed by secretion alone. The movement to land is associated with the addition of further renal distal segments. Selective specific reabsorption and secretion of sodium, potassium and water is thus possible. The importance of these segments increases from Amphibia to reptiles and birds. With the emergence of the mammals, particularly the dominant Eutheria, a further distinctive segment, the loop of Henle, is inserted in the nephron. This is capable of generating concentration gradients within the renal interstitium and augmenting water reabsorption from the urine via the collecting duct. Loops of Henle are also found in some birds where they are interspersed with the simpler reptilian nephron types.

The gradual emergence of the kidney, above the various extra-renal devices of lower vertebrates, as the major organ of osmoregulation is reflected in the actions of the neurohypophysial and adrenocortical hormones. In mammalian species they are involved in the renal management of water and electrolyte, acting, in different circumstances, in a complementary, reciprocal or converse manner. In other vertebrates either or both these sets of hormones act upon extra-renal osmoregulatory organs with kidney functions showing less and varying degrees of control depending on the various groups of non-mammalian vertebrates and the differing osmotic conditions of the habitats.

The neurohypophysis of jawed vertebrates contains two octapeptides. They epitomize, as do steroids, the use of basic chemical compounds, universally found in living matter, for changing into recognizable hormones. The interplay of the vertebrate neurohypophysial and adrenocortical hormones (Table V) is in particular seen in their complicated interactions on the metanephric kidney of eutherian mammals. The antidiuretic effect of vasopressin is specifically on the water content of body fluids and is coincident with the

TABLE V

Summary of influences of neurohypophysial and adrenocortical steroid hormonal actions on vertebrate soidum metabolism. —, not investigated; ?, equivocal actions.

Group	Neurohypophysial peptide	Adrenal steroid
EXTRARENAL SODIUM HANDLING		
Mammalia	?	Decreased loss
Aves	?	Loss by the nasal gland
Reptilia	—	Loss of "salt" gland
		Uptake across bladder
Amphibia	Uptake across skin and bladder	Uptake across skin and bladder
Dipnoi	Uptake across gills	—
Teleostei FW	Uptake across gills	Uptake across gills
SW	Extrusion across gills	Extrusion across gills
Elasmobranchi	?	?
Cyclostomata	?	?
RENAL SODIUM HANDLING		
Mammalia	(Fractional excretion)	Fractional excretion
Aves	Fractional excretion	Fractional excretion
	Filtered load	
Reptilia	Fractional excretion	Fractional excretion
	Filtered load	
Amphibia	Fractional excretion	?
	Filtered load	
Dipnoi	Fractional excretion	—
	Filtered load	?
Teleostei	Filtered load	?
Elasmobranchi	?	—
Cyclostomata	Fractional excretion	?

emergence of precise tubular reabsorptive capacity in the kidney. The corticosteroids have a profound influence on the total osmotic concentration of body fluids. In particular, especially as aldosterone, they act upon the movements of sodium, potassium and chloride. Thereby the total osmotic concentration of the body fluids and hence their total water content is controlled, so underlying the conjugation of actions of vasopressin and aldosterone. A similar intertwining of influences may be claimed for all Amniota though with less force. In so far as extant species represent evolutionary history both aldosterone and corticosterone have renal electrolyte effects, but so also do the neurohypophysial hormones in addition to their control of water. The other major group of the tetrapods, the Amphibia, have a mesonephros which does not show dominant reactions to the two sets of hormones under discussion. Certainly glomerular filtration rates may change under their influence but it is the extra-renal sites which attract the major focus—especially the skin and the bladder. Thus the pattern in the amphibian tetrapod is one showing less significant reactions to steroid and peptide hormones by the mesonephros, but dominant ones by the skin and the bladder. This is a question of degree for it is clear that the kidney is an essential organ; there is a sharing responsibility for the homeostasis of the organism between various structures. The lung-fishes which may have some physiological similarities with proto-amphibians show no net effect on water uptake after vasopressin, vasotocin or oxytocin (Sawyer, 1972). Aldosterone is present but no clear role has been assigned. Not enough is known about the Dipnoi nor, indeed, the elasmobranchs to fit them, with any certainty, into this type of discussion. We have to rely on the bony fish, the Teleostei, to gain insight as to how the interplay of the neurohypophysial and steroid hormones impinged on the need for body fluid homeostasis. In these fish glomerular filtration and urine flow rates can be controlled by neurohypophysial hormones without tubular components. This does not rule out the contribution of tubular responsiveness to AVT and isotocin in respect of osmotically important ions. It can be said, then, that regulation of renal excretion by changes in glomerular filtration rates and by specific tubular electrolyte handling is an early "primitive" feature of gnathostomes. The electrolyte and water contents of teleosts are greatly influenced by neurohypophysial pep-tides acting on both renal and extra-renal tissues. It may be supposed that these actions are short term ones giving quick responses to environmental changes whilst the corticosteroids, having parallel influence, provide a longer term control. Ionic control is primarily displayed by the gills and both uptake and excretion of sodium, depending on the tonicity of the external fluid, are affected by neurohypophysial and adrenocortical hormones.

References

Abraham, S. F., Blair–West, H. R., Coghlan, J. P., Denton, D. A., Munro, D. R. and Scoggins, B. A. (1976). *Acta endocr.* **81,** 120–132.
Assenmacher, I. and Boissin, J. (1970). *Gen. comp. Endocr.* Supplement **3,** 489–498.
Ball, J. N. (1980). Hypothalamic control of the pars distalis in lower vertebrates. *Gen. comp. Endocr.* (In Press)

Ball, J. N., Batten, T. F. C. and Young, G. (1980). Evolution of hypothalamo-adenohypophysial systems in lower vertebrates. *In:* Proceedings of an International Symposium on Hormones and Evolution ed. S. Ishii). Tokyo 1979. Japanese Acad. Sci. and Springer–Verlag, Berlin. (In Press)

Ben–David, M., Danon, M., Benveniste, R., Weeler, C. P. and Sulman, F. G. (1971). *J. Endocr.* **50,** 599–606.

Bergland, R. M. and Page, R. B. (1979). *Science, N.Y.* **204,** 18–24.

Bergmann, W: (1962). *In:* "Comparative Biochemistry", Vol. 3 (Eds. M. Florkin and H. S. Mason). pp. 103–162. Academic Press, New York and London.

Bernard, C. (1855). Leçons de Physiologie expérimentale au Collège de France. Paris.

Bernard, C. (1859). Leçons sur les propriétés physiologiques et les altérations pathologiques des liquides de l'organisme, **2,** 441.

Blaschko, H., Sayers, G. and Smith, A. D. (Eds.) (1975). "Handbook of Physiology", Section 7, Endocrinology, Vol. VI, Adrenal Gland. American Physiological Society. Washington D.C.

Brodish, A. (1973). *In:* "Brain–Pituitary–Adrenal Interrelationships". (Eds. A. Brodish and F. S. Redgate) p. 128–163. Karger Publishing Company, Basel.

Brodish, A. and Lymangrover, J. R. (1977). *In:* "International Review of Physiology. Endocrine Physiology II". Vol. 16. (Ed. McCann, S. M.) pp. 93–149. University Park Press, Baltimore.

Buckingham, J. C. and Leach, J. H. (1979). *J. Physiol. Lond.,* 117p.

Cannon, W. B. (1932). "The Wisdom of the Body". W. B. Norton and Co., New York.

Chester Jones, I. (1976). *J. Endocr.,* **71,** 1P–31P.

Chester Jones, I. and Henderson I. W. (1976a). *In:* "General, Comparative and Clinical Endocrinology of the Adrenal Cortex". pp. vii–viii, Vol. 1 Academic Press, London, New York and San Francisco.

Chester Jones, I. and Henderson, I. W. (Eds.) (1976b). "General, Comparative and Clinical Endocrinology of the Adrenal Cortex". Vol. 1. Academic Press, London, New York and San Francisco.

Chester Jones, I. and Henderson (1976). *In:* "General, Comparative and Clinical Endocrinology of the Adrenal Cortex". pp. vii–viii, Vol. 1 Academic Press, London, New York, San Francisco.

Chester Jones, I. and Henderson, I. W. (Eds.) (1978). "General, Comparative and Clinical Endocrinology of the Adrenal Cortex". Vol. 2. Academic Press, London, New York and San Francisco.

Coghlan, J. P., Blair–West, J. R., Denton, D. A., Fei, D. T., Fernley, R. T., Hardy, K. J., McDougall, J. G., Puy, R., Robinson, P. M., Scoggins, B. A. and Wright, R. D. (1979). *J. Endocr.* **81,** 55–67.

Conway, E. J. (1942). *Proc. Roy. Irish Acad.* **48,** Section B, 119–159.

Conway, E. J. (1943). *Proc. Roy. Irish Acad.* **48,** Section B, 161–212.

Coyne, M. D. and Kitay, J. I. (1969). *Endocrinology* **85,** 1095–1102.

Davis, J. O. (1975). *In:* "Handbook of Physiology", Section 7 Endocrinology, Vol. VI. (Eds. H. Blaschko, G. Sayer, and A. D. Smith) pp. 77–106. American Physiological Society, Washington, D.C.

Deane, H. W. and Greep, R. O. (1946). *Am. J. Anat.* **79,** 117–137.

DeLuca, H. F. (1973). *Kidney Int.* **4,** 80–88.

Denton, D. A., Blair–West, J. R., Coghlan, J. P., Scoggins, B. A. and Wright, R. D. (1977). *Acta endocr. Copnh.* **84,** 119–132.

Donaldson, E. M. and Fagerlund, U. H. M. (1969). *J. Fish. Res. Bd. Can.* **26,** 1789–1799.

Dubé, J. and Lemonde, A. (1970). *Gen. comp. Endocr.* **15,** 158–164.
Dubé, J. and Velleneuve, J. L. and Lemonde, A. (1968). *Arch. Int. Physiol. Biochem.* **76,** 64–70.
Elias, H. and Bengelsdorf, J. (1952). *Acta anat. (Basel)* **14,** 297–337.
Fagerlund, U. H. M. and Donaldson, E. M. (1969). *Gen. comp. Endocr.* **12,** 438–448.
Fänge, R. and Krog, J. (1962). *Nature, Lond.* **199,** 713.
Florkin, M. (1974). *In:* Comprehensive Biochemistry", (Eds. M. Florkin and E. H. Stolz) Vol. 29, Part A pp. 1–124. Elsevier Scientific Publishing Co. Amsterdam, London and New York.
Folsome, C. E. (1979). "The Origin of Life". W. H. Freeman and Co., San Francisco.
Gann, D. S. (1969). *Ann. N.Y. Acad. Sci.* **156,** 740–755.
Ganten, D., Fusce, K., Phillips, M. I., Mann, J. F. E. and Ganteu, U. (1978). *In:* "Frontiers in Neuroendocrinology", Vol. 5 (Eds. W. F. Ganong and L. Martini) pp. 61–99 Raven Press, New York.
Georges, D. (1977). *Gen. comp. Endocr.* **32,** 454–473.
Goldstein, L. and Forster, R. P. (1971). *Am. J. Physiol.* **220,** 742–746.
Goldstein, L., Oppelt, W. W. and Maren, T. (1968). *Am. J. Physiol.* **215,** 1493–1497.
Goodrich, W. S. (1958). "Studies on the Structure and Development of Vertebrates". Vol. 2, Chapter 13. Dover Publications Inc., New York: Constable and Co Ltd., London.
Grant, J. K. (1962). *In:* Comparative Biochemistry" (Eds. M. Florkin and H. S. Mason), Vol. 3, pp. 163–204. Academic Press, New York and London.
Greene, H. C. G. (1957). An introduction to the study of experimental medicine. Translation of Claude Bernard's work of 1855, 1857. Dover Publications Inc: New York.
Greenwood, P. H. (1975). "A History of Fishes". Ernest Benn Ltd., London. 3rd Edition.
Hanke, W. (1978). *In:* General, Comparative and Clinical Endocrinology of the Adrenal Cortex". (Eds. I. Chester Jones and I. W. Henderson). Vol. 2, pp. 419–497. Academic Press, London, New York and San Francisco.
Hartman, F. A. and Brownell, K. A. (1949) "The Adrenal Gland". Henry Kimpton. London.
Haslewood, G. A. D. (1962). *In:* "Comparative Biochemistry". (Eds. M. Florkin and H. S. Mason), Vol. 3, Part A, pp. 205–229. Academic Press, New York and London.
Hay, A. W. M. and Watson, G. (1976). *Comp. Biochem. Physiol.* **53B,** 375–380.
Henderson, I. W. and Brown, J. A. (1979). *In:* "Epithelial Transport in the Lower Vertebrates", (Ed. B. Lahlou). Cambridge University Press, Cambridge (In Press).
Henkin, R. I. (1975). *In:* "Handbook of Physiology". Section 7, Endocrinology, Vol VI (Eds. H. Blaschko, G. Sayers and A. D. Smith) pp. 209–230. American Physiological Society, Washington D.C.
Henry, H. and Norman, A. W. (1975). *Comp. Biochem. Physiol.* **50B,** 431–434.
Henry, J. P., Kross, M. E., Stephens, P. M. and Watson, F. M. C. (1976). *In:* "Catecholamines and Stress", (Eds. E. Usclin, R. Kyatonansky and I. J. Kopin) pp. 457–468. Pergamon Press, Oxford.
Highnam, K. C. (1975). *Biochem. Soc. Trans.* **3,** 1160–1164.
Highnam, K. C. and Hill, L. (1977). "The Comparative Endocrinology of the Invertebrates". 2nd edition. Edward Arnold, London.
Hirano, T., Takei, Y. and Kobayashi, H. (1978). *In:* "Osmotic and Volume

Regulations". (C. Barker Jørgensen and E. Skadhuage, eds.) pp. 123–128. XI Alfred Benzon Symposium. Munksgaard, Copenhagen.

Holmes, R. L. and Ball, J. N. (1974). "The Pituitary Gland". Cambridge University Press.

Holmes, W. N. and Phillips, J. G. (1976). In: "General, Comparative and Clinical Endocrinology of the Adrenal Cortex" (Eds. I. Chester Jones and I. W. Henderson), Vol. 1, pp. 293–420. Academic Press, London, New York and San Francisco.

Holmes, W. N., Cronshaw, J. and Gorsline, J. (1978). Environmental Research 17, 177–190.

Hutchinson, G. E. (1944). Am. J. Sci. 242, 272–280.

Idelman, I. (1978). In: "General, Comparative and Clinical Endocrinology of the Adrenal Cortex". (Ed. I. Chester Jones and I. W. Henderson), Vol. 2, pp. 1–200. Academic Press, London, New York and San Francisco.

Idler, D. R. and Burton, M. P. M. (1976). Comp. Biochem. Physiol. 53A, 73–77.

Idler, D. R., Sangalang, G. B. and Weisbart, M. (1971). In: "Hormonal Steroids", (Eds. V. H. T. James and L. Martini), pp. 983–989. Excerpta Medica I.C.S. 219. Amsterdam.

Iscart, G., Szafarczyk, A. Belugua, J. and Assenmacher, I. (1977). J. Endocr. 72, 113–120.

Janssens, P. A. (1964). Comp. Biochem. Physiol. 13, 217–224.

Kerr, J. G. (1919). "Textbook of Embryology", Vol. 2, Vertebrata with the exception of Mammalia. McMillan, London.

Kime, D. E. (1978). In: "General, Comparative and Clinical Endocrinology of the Adrenal Cortex" (Eds. I. Chester Jones and I. W. Henderson), Vol. 2, pp. 265–290. Academic Press, London, New York and San Francisco.

Kime, D. E. and Norymberski, J. K. (1976). In: "General, Comparative and Clinical Endocrinology of the Adrenal Cortex". Vol. 1, Chapter 1 (Eds. I. Chester Jones and I. W. Henderson). Academic Press, London, New York, San Francisco.

Lagios, M. D. (1975). Gen. comp. Endocr. 25, 126–146.

Langley, L. L. (1965). "Homeostasis". Reinhold Book Corporation. New York, Amsterdam, London.

Lehoux, J. G. and Sandor, T. (1969). Endocrinology, 84, 652–657.

Lehoux, J. G. and Sandor, T. (1970). Steroids 16, 141–171.

Lehoux, J. G., Sandor, T., Lūsis, O. and Lanthier, A. (1967). Proc. Can. Fed. Biol. Soc. 10, 62.

Lehoux, J. G., Sandor, T., Lanthier, H. and Lūsis, O. (1968). Gen. comp. Endocr. 11, 481–488.

Lehoux, J. G., Chapdelaine, A. and Sandor, T. (1970). Can. J. Biochem. 48, 407–411.

Ley, J. de, and Kersters, K. (1975). In: "Comprehensive Biochemistry" (Eds. M. Florkin and E. H. Stoz). Vol. 29 Part B pp. 1–78. Elsevier Scientific Publishing Co. Amsterdam, London and New York.

Lofts, B. (1978). In: "General, Comparative and Clinical Endocrinology of the Adrenal Cortex", (Eds. I. Chester Jones and I. W. Henderson). Vol. 2, pp. 292–369. Academic Press, London, New York and San Francisco.

Marshall, E. K. and Smith, H. W. (1930). Biol. Bull. mar. biol. Lab. Woods Hole, 59, 135–153.

Medawar, P. B. (1953). Symp. soc. Exp. Biol. VII, 320–338.

Moore, R. Y. (1978). In: "Frontiers in Neuroendocrinology", Vol. 5 (Eds. W. F. Ganong and L. Martini) pp. 185–206. Raven Press, New York.

Motta, M., Fraschini, F. and Martini, L. (1969). In: "Frontiers in Neuroendocrinology", (Eds. W. F. Ganong and L. Martini) pp. 211–253. Oxford University Press: New York.

Mueller, R. A., Thoenen, H. and Axelrod, J. (1970). *Endocrinology* **86,** 751–755.

Mugnaini, E. and Harboe, S. B. (1967). *Z. Zellforsch mikrosk Anat.* **78,** 341–369.

Nicoll, C. S. (1974). *In:* "Handbook of Physiology", Section 7, Vol. IV, part 2 (Eds. E. Knobil and W. H. Sawyer) pp. 253–292. American Physiological Society, Washington D.C.

Norman, A. W. and Henry, J. L. (1979). *Trends in Biochemical Sciences,* **4,** 14–18.

Obermüller-Wilén, H. (1976). *Acta Zool. Stockh.* **60,** 187–196.

Pearce, R. B., Cronshaw, G. and Holmes, W. N. (1978). *Cell Tiss. Res.* **192,** 363–379.

Pearce, R. B., Cronshaw, J. and Holmes, W. N. (1979). *Cell Tiss. Res.* **196,** 429–447.

Raisman, G. and Brown–Grant, K. (1977). *Proc. Roy Soc. Lond. (Biol.)* series B. **198,** 297–314.

Rall, D. P. and Burger, J. W. (1967). *Am. J. Physiol.* **212,** 354–356.

Ramsey, J. A. (1968). "A Physiological Approach to the Lower Animals". 2nd Edition. Cambridge University Press, London.

Reid, I. A. and Ganong, W. F. (1974). *In:* "Endocrine Physiology". (Ed. S. M. McCann), pp. 205–237. Physiology Series One, Vol. 6. MTP International Review of Science. Butterworths, London.

Robertson, J. D. (1957). *Biol. Rev.* **32,** 156–187.

Sanderman, W. (1962a). *In:* "Comparative Biochemistry" (Eds. M. Florkin and H. S. Mason), Vol. 3, pp. 503 590. Academic Press, New York and London.

Sanderman, W. (1962b). *In:* "Comparative Biochemistry", (Eds. M. Florkin and H. S. Mason) Vol. 3, pp. 591–630. Academic Press, New York and London.

Sandor, T., Sonea, S. and Mehdi, A. Z. (1975). *Am Zool.* **15** (Suppl 1) 227–253.

Sandor, T., Fazekas, A. G. and Robinson, B. H. (1976). *In:* "General, Comparative and Clinical Endocrinology of the Adrenal Cortex" (I. Chester Jones and I. W. Henderson, eds.) Vol. 1, pp 25–142. Academic Press, London, New York and San Francisco.

Sawchenko, P. E. and Friedman, M. I. (1979). *Am. J. Physiol.* **236,** R5–R20.

Sawyer, W. H. (1972). *Fed. Proc.* **31,** 1609–1614.

Sayers, G. (1961). *Physiologist* **4,** 56–61.

Schally, A. V., Anderson, R. N., Lipscomb, H. S., Long, J. M. and Guillemin, R. (1960). *Nature, London.* **188,** 1192–1193.

Schildknecht, H. (1971). *Schwimmkäfergifte Bild. Wiss* 333.

Slusher, M. A. (1964). *Am. J. Physiol.* **206,** 1161–1164.

Smith, H. W. (1932). *Q. Rev. Biol.* **7,** 1–26.

Sokabe, H., Pang, P. K. T. and Gorbman, A. (1978). *Jap. Heart J.* **5,** 783–790.

Stensiö, E. (1968). *In:* "Current Problems of Lower Vertebrate Phylogeny", 13–71. Proc. 4 Nobel Symp., June, 1967. (Tor Ørvig, ed), Almquist and Wiksell, Stockholm.

Swain, T. (1974). *In:* "Comprehensive Biochemistry", (Eds. M. Florkin and E. H. Stoltz). Vol. 29, Part A, 125–298. Elsevier Scientific Publishing Co. Amsterdam, London and New York.

Tait, S. A. and Tait, J. F. (1979). *J. Endocr.* **82,** 3–31.

Thoenen, H. and Otten, U. (1978). *In:* "Frontiers in Neuroendocrinology". Vol. 5., ed. W. F. Ganong and L. Martini. pp. 163–184. Raven Press, New York.

Timiras, P. S., Vernadakis, A. V. and Sherwood, N. M. (1968). *In:* "Biology of Gestation". (Ed. N. S. Assali) Vol. 2, pp. 261–319. Academic Press: New York.

Varma, M. M. (1977). *Gen. comp. Endocr.* **33,** 61–75.

Wald, G. (1973). *Sci. Am.* 9–81.

Watts, D. C. and Watts, R. L. (1966). *Comp. Biochem. Physiol.* **17,** 785–798.

Weston, J. A. (1970). *In:* "Advances in Morphogenesis". Vol. 8, Chap. 2. (Eds. M. Abercrombie, J. Brachet and T. J. King). pp. 4–108. Academic Press, New York and London.

Witorsch, R. J. and Brodish, A. (1972). *Endocrinology* **90,** 1160–1167.

Witschi, E. (1956). "Development of Vertebrates". pp. 148–150. W. B. Saunders, Philadelphia.

Yamashita, H. (1970a). *Bull. Jap. Soc. scient. Fish.* **36,** 439–449.

Yamashita, H. (1970b). *Bull. Jap. Soc. scient. Fish.* **36,** 450–454.

Yates, F. E. and Maran, J. W. (1974). *In:* "Handbook of Physiology". Section 7, Vol. IV, Part 2, pp. 367–404. (Eds. E. Knobil and W. H. Sawyer). American Physiological Society, Washington D.C.

Yates, F. E., Russell, S. M. and Maran, J. W. (1971a). *A. Rev. Physiol.* **33,** 393–444.

Yates, F. E., Russell, S. M., Dallman, M. F., Hedge, G. A., McCann, S. M. and Dhariwal, A. P. S. (1971b). *Endocrinology* **88,** 3–15.

Author Index

Numbers in italics are those pages on which references are listed

A

Aakvaag, A., 204, *249, 262*
Abdul-Karim, R.-W., 248, *249*
Abe, K., 22, 24, 29, *44, 48*, 61, 85, *111, 114*, 213, *255*, 324, *333*
Abel, M., 247, *259*
Aber, C. P., 105, *111*
Abraham, A. D., 127, 158, 159, *165, 180*
Abraham, G. E., 203, 204, *249, 256*
Abraham, S. F., 545, *557*
Acevedo, H. F., 201, *249*
Acs, G., 119, 125, 155, 156, 161, 162, *166, 171*
Adams, J. B., 201, *249*
Addison, T., 58, 61, *111*
Ader, R., 34, 43, *44*, 297, *333, 338*, 366, 377, 378, 381, 382, *387, 392*
Adey, W. R., 313, *336*
Afonja, A.-O., 111, *115*
Agarwal, M. K., 139, 156, *179*
Agate, F. J., 215, *249*
Agnello, E. J., 301, *337*
Agrawal, U., 453, *466*
Agrawala, N., 456, *467*
Ahjua, S. K., 428, *466*
Ahmad, N., 190, 196, 198, *249, 256*
Ahsan, 444, *471*
Aida, M., 107, *114*
Aillon, J., 25, *44*
Akert, K., 313, *345*
Albe-Fessard, D., 304, *336*
Alberga, A., 154, *166*
Alberts, B., 153, *180*
Alberts, B. M., 147, 154, *180*

Albrechtsen, R., 206, *250*
Albright, F., 88, *111*
Albritton, W. L., 160, *172*
Alescio, T., 131, *166*
Alexander, W. D., 285, *344*
Alexinsky, T., 304, *333*
Algeri, S., 310, *334*
Allard, C., 123, *180*
Allen, C., 19, *48*
Allen, C. F., 32, 34, 42, 43, *44, 47*
Allen, J. P., 32, 42, 43, *44*, 84, 103, *112*, 307, *333*
Allen, W., 34, 35, *49*
Alleyn, G. A. O., 121, 124, *166, 170*
Allfrey, V. G., 159, *166, 178*
Alpert, M., 215, *249, 258*
Altman, K., 156, *166*
Alvarez, J. A., 200, 207, *259*
Amaral, L., 137, *166, 180*
Amer, B., 359, *389*
Amer, I., 359, *389*
Ammedick, U., 126, 160, *179*
Amorosa, L., 108, *114*
Amos, H., 130, 131, *171, 177*
Amsel, B., 90, *114*
Anand, B. K., 9, *44*
Anand, T. C., 230, *249*
Andersen, R. N., 11, *53*, 322, *338*
Anderson, A. B. M., 188, 245, *249, 251*
Anderson, C. O., 379, *387*
Anderson, D. C., 271, *333*
Anderson, E., 9, 30, *44, 49*
Anderson, J. M., 196, *259*
Anderson, J. N., 149, 150, *166*
Anderson, M., 160, *170*

Suyemitsu, T., 137, *179*
Suzuki, T., 22, *54*, 146, *172*
Svare, B. B., 269, 291, 298, *341*, 370, *390*
Swaab, D. F., 322, 323, *345*
Swain, T., 527, *561*
Swale, J., 194, *251*
Swallow, R. L., 499, *523*
Swanson, H. H., 296, *345*, 381, *392*
Sweat, M. L., 189, 190, 196, 197, 201, *251*
Sweeney, E. W., 136, 162, *178*
Swell, L., 225, *252*
Swinyard, C. A., 208, *262*
Swygert, N. H., 216, 217, 218, *257*
Sylvester, S., 224, *262*
Symchowicz, S., 137, 139, *176*
Symington, T., 60, 81, 83, *115*, *116*, 187, 189, 190, 191, *259*
Szabó, Zs, 477, *521*
Szafarczyk, A., 539, *560*
Sze, P. Y., 27, *51*, 326, *336*, *342*
Szentágothai, J., 11, *47*
Szeplaki, B. J., 478, 479, *519*

T

Tabata, S., 441, *469*
Taeusch, H. W., 134, *166*
Taeusch, H. W., Jr, 134, *168*
Tailo, L. B. H., 398, *471*
Tait, A. D., 74, *113*
Tait, J. F., 189, 191, 204, *250*, *255*, 531, *561*
Tait, S. A., 531, *561*
Tajima, H., 196, *249*
Takabe, K., 12, 13, 35, *54*
Takebe, K., 17, *52*
Takei, Y., 494, *519*, 550, *559*
Takeyasu, K., 359, *391*
Tal, E., 247, *263*
Taleisnik, S., 12, *46*
Tam, T., 271, *333*
Tamaoki, B., 196, *249*
Tamaoki, B. I., 474, *516*
Tamásy, V., 269, *345*
Tamura, E., 458, *468*
Tan, C. H., 158, *179*
Tanake, F., 37, *51*
Tani, F., 30, *50*
Tanner, J. M., 228, *263*
Tanney, H., 225, *259*

Tárnok, F., 287, 289, 312, *340*
Tarnowski, W., 122, 124, 161, *173*
Tarr, H. L. A., 502, *517*
Tashjian, A. H., Jr, 160, *166*
Tauber, O. E., 363, *388*
Taylor, A. A., 512, *523*
Taylor, A. N., 32, *50*, 312, *340*
Taylor, J. D., 414, *471*
Taylor, K. M., 28, *54*
Taylor, W., 224, *260*
Telegdy, G., 26, 27, 30, 31, 35, 36, 38, 41, *46*, *54*, 240, *263*, 298, 315, 325, 330, *336*, *340*
Telford, R., 327, *333*
Teller, W. M., *254*
Teplan, I., 322, *340*
Terasawa, E., 31, 33, *48*
Terayama, H., 137, *179*
Terenius, L., 308, 323, *342*, *345*
Terqui, M., 231, *254*
Teysseyre, J., 289, *345*
Thakur, D. P., 428, *471*
Thatcher, J., 297, *338*
Thatcher, K., 380, *391*
Thatcher, W. W., 248, *263*
Theissen, D. D., 380, *392*
Thi, L., 145, *175*
Thienhaus, R., 126, 160, *179*
Thierry, A. M., 25, *54*
Thijssen, J. H. H., 78, *116*
Thoa, N. B., 374, *391*
Thody, A. J., 324, 327, *333*, *344*
Thoenen, H., 541, *560*, *561*
Thoman, E. B., 248, *263*
Thomas, A., 152, *172*
Thomas, B. S., 187, 194, *263*
Thomas, J. P., *263*
Thomas, N. W., 477, *518*
Thomas, S. F., 229, *260*, 454, 458, *470*, 508, *522*
Thomas Smith, W., 106, *112*
Thompson, B., 136, 137, *170*
Thompson, E. B., 119, 126, 128, 138, 145, 149, 150, 156, 160, 162, 163, 164, *174*, *179*, *180*
Thompson, J. Q., 93, *112*
Thompson, J. W., 327, *333*
Thompson, K. W., 83, *116*
Thompson, R., 304, *345*
Thompson, R. W., 381, *392*

Subject Index

A

Acetylcholine = ACh, 21–23, 36
  acetylcholinesterase, 22
  acetyltransferase, 22
  carbachol, 22–23
  CRF blockage, atropine, 23
    hexamethonium, 23
  hypothalamus, in, 22–23
Actinomycin D, 122–124, 127–128,
  130–131, 155–159, 161–163
Addison's Disease, 266
  Addisonian crisis, 76
  adrenocorticol insufficiency, 58–79
    primary, 59–64
    secondary, 60, 64
  aetiology, 59–61
  anaemia, 63–64
    treatment, 73–75
  anti-bodies, 59
  hypoglycaemia, 62
  pathogenesis, 59–61
  pregnancy in, 75–76
  psychological changes, 266
  surgery in, 76
Adenohypophysectomy
  and behaviour, 271–272
Adipose tissue, 128–129
Adrenalectomy
  and agonistic behaviour, 269
    avoidance behaviour, 267–269
    exploratory behaviour, 269
    replacement therapy, 269
    sleep, 316–317
  aminoglutethimide, 107

behavioural effects, 267–270
bilateral, 107
gluconeogenesis, 122
metyrapone, 107
salt balance and maze learning, 268
sub-total, Cushing's syndrome, 107
Adrenal gland
  ACTH control, 235
  adrenalectomy, 248
  anencephaly and hCG, 343
  androgens, 243
  demedullation and running activity,
    300
  fetal, transient cortex, X zone,
    235–247
    hypothalamus, 246
  glucocorticoids, 248
  hydrocephaly, 243
  11β-hydroxylase, 218
  hydroxysteroid dehydrogenases, 3α,
    3β and 17β, 218
  imprint with gonadal steroids, 218
  in anencephaly, 237, 243, 244
  in lactation, 248
  in maternal oestrogen excretion, 243
  in parturition, 244–248
  oestrogens, 245, 248–249
    effects, 218
  oxytocin, 246
  placental progesterone, 244
  progesterone, 245, 249
  prolactin, 248–249
  prostaglandin, $F_2\alpha$, 246
  5α-reductase, 218
  sheep, 244–247